Thermodynamics and Fluid Mechanics of Turbomachinery
Volume I

NATO ASI Series

Advanced Science Institutes Series

A Series presenting the results of activities sponsored by the NATO Science Committee, which aims at the dissemination of advanced scientific and technological knowledge, with a view to strengthening links between scientific communities.

The Series is published by an international board of publishers in conjunction with the NATO Scientific Affairs Division

A	**Life Sciences**	Plenum Publishing Corporation
B	**Physics**	London and New York
C	**Mathematical and Physical Sciences**	D. Reidel Publishing Company Dordrecht and Boston
D	**Behavioural and Social Sciences**	Martinus Nijhoff Publishers Dordrecht/Boston/Lancaster
E	**Applied Sciences**	
F	**Computer and Systems Sciences**	Springer-Verlag Berlin/Heidelberg/New York
G	**Ecological Sciences**	

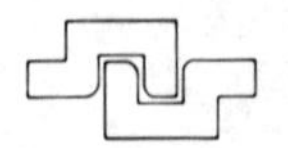

Series E: Applied Sciences – No. 97A

Thermodynamics and Fluid Mechanics of Turbomachinery

Volume I

Edited by

A.Ş. Üçer

Mechanical Engineering Department
Middle East Technical University
(ODTU)
Ankara, Turkey

P. Stow

Theoretical Science Group
Rolls-Royce Limited
Derby, UK

Ch. Hirsch

Free University of Brussels
Brussels, Belgium

1985 **Martinus Nijhoff Publishers**
Dordrecht / Boston / Lancaster
Published in cooperation with NATO Scientific Affairs Division

Proceedings of the NATO Advanced Study Institute on Thermodynamics and Fluid Mechanics of Turbomachinery, Izmir, Turkey, September 17-28, 1984

Library of Congress Cataloging in Publication Data

Main entry under title:

Thermodynamics and fluid mechanics of turbomachinery.

(NATO ASI series. Series E, Applied sciences ; no. 97A-)
"Proceedings of the NATO Advanced Study Institute on "Thermodynamics and Fluid Mechanics of Turbomachinery," Izmir, Turkey, September 17-29, 1984"--V. 1, p. iv.
Bibliography: p.
Includes index.
1. Turbomachines--Congresses. 2. Thermodynamics--Congresses. 3. Fluid mechanics--Congresses. I. Üçer, A. Ş. II. Stow, P. III. Hirsch, Ch. IV. NATO Advanced Study Institute on "Thermodynamics and Fluid Mechanics of Turbomachinery" (1984 : Izmir, Turkey) V. Series: NATO ASI series. Series E, Applied sciences ; no. 97A, etc.
TJ267.T57 1985 621.406 85-18778
ISBN 90-247-3223-9 (set)
ISBN 90-247-3221-2 (v. 1)
ISBN 90-247-3222-0 (v. 2)

ISBN 90-247-3221-2 (this volume)
ISBN 90-247-2689-1 (series)
ISBN 90-247-3223-9 (set)

Distributors for the United States and Canada: Kluwer Boston, Inc., 190 Old Derby Street, Hingham, MA 02043, USA

Distributors for the UK and Ireland: Kluwer Academic Publishers, MTP Press Ltd, Falcon House, Queen Square, Lancaster LA1 1RN, UK

Distributors for all other countries: Kluwer Academic Publishers Group, Distribution Center, P.O. Box 322, 3300 AH Dordrecht, The Netherlands

Printed in The Netherlands

PREFACE

During the last decade, rapid advances have been made in the area of flow analysis in the components of gas turbine engines. Improving the design methods of turbomachine blade rows and understanding of the flow phenomena through them, has become one of the major research topics for aerodynamists. This increase of research efforts is due to the need of reducing the weight and fuel consumption of turbojet engines for the same thrust levels. One way of achieving this is to design more efficient components working at high local velocities. Design efforts can lead to desired results only if the details of flow through the blade rows are understood.

It is also known that for aircraft propulsion systems development, time and cost can be reduced significantly if the performance can be predicted with confidence and enough precision. This generally needs sophisticated two or three dimensional computer codes that can give enough information for design and performance prediction. In the recent years, designers also started to use these sophisticated codes more and more with confidence, in connection with computer aided design and manufacturing techniques. On the other hand, the modelling and solution of flow and the measurement techniques inside the blade rows are difficult, and present challenging problems of fluid dynamics.

It was with these facts in mind that the organizers of the NATO Advanced Study Institute, held in İzmir (Çeşme), Turkey from 17 to 28 September 1984, invited lecturers to review the current state of knowledge in the advanced design and performance prediction of turbomachines. The lecturers were: H.A. Akay, Purdue University at Indianapolis, Indiana, U.S.A.; R.V.D. Braembussche, von Karman Institute, Belgium; T. Cebeci, Douglas Aircraft Company

Long Beach, California, U.S.A.; J. Chauvin, Institute de Mécanique des Fluides, Marseille, France; J. D. Denton, Whittle Lab., Cambridge University, Cambridge, U.K.; E.M. Greitzer, Massachusetts Institute of Technology, Cambridge, Mass., U.S.A.; Ch. Hirsch, Vrije Universiteit Brussel, Brussels, Belgium; E. Macchi, Politecnico di Milano, Milano, Italy; G. Meauze, ONERA, Chatillon, France; R. Schodl, DFVLR-Institut für Antriebstechnik, Köln, W. Germany; G. K. Serovy, Iowa State University, Ames, Iowa, U.S.A.; C. Sieverding, von Karman Institute, Belgium; P. Stow, Rolls-Royce Limited, Derby, U.K.; A.Ş. Üçer, Middle East Technical University, Ankara, Turkey; H. Weyer, Deutsch-Niederlaendischer Windkanal, the Netherlands. This book is compiled from the lectures of the above scientists and their co-workers, and it is aimed to review the experimental and theoretical developments in the field of aerothermodynamics and design aspects of turbomachinery. Naturally the coverage is not complete. However, lecturers have been very conscientious in adopting their work to the aims set down by the organizers. As an outcome of this, the product has exceeded, both in breadth and coherence, the expectations of the initiators of the Institute. The topics are arranged so that the material compiled will serve as an advanced reference book to the turbomachine designers and research workers.

The material is grouped into seven parts. The first part gives the fundamentals of compressible flow in turbomachines. In the second part, solution methods for inviscid and viscous representations are included. The third part presents existing test facilities and the measurement techniques for turbomachinery research and development. In the fourth part, problems associated with secondary flows in radial and axial flow machines and end wall boundary layer modelling and solution methods are discussed. The fifth part covers performance prediction methods and recent design techniques for advanced turbomachinery. In the sixth part unsteady effects and flow instabilities are discussed. In the final part, the round table discussion which covers suggestions for future research in the turbomachinery aerothermodynamics is presented.

The editors would like to thank the NATO Scientific Affairs Division, the Middle East Technical University and the Scientific and Technical Research Council of Turkey for making the Advanced Study Institute possible, and to Martinus Nijhoff Publishers for publishing the edited proceedings. We wish to extend a word of appreciation to DFVLR, General Dynamics Corporation, MTU, Nouvo Pignone, Rolls Royce Limited, U.S. Army Research, Development and Standardization Group, Dokuz Eylül University, and to Turkish firms, GAMA, LAMAŞ, Teknik Komandit Şirketi, TURBOSAN and Yazar Pompa for their support. We are also grateful to the authors for their contributions and their cooperativeness when revisions were requested. Our special gratitute goes to Professor G.K. Serovy, a member of the organizing committee, because of his invaluable contributions at all stages of this endeavour. We also extend our gratitute to the members of the programme committee, session chairmen, co-chairmen, and scientific secretaries in programming, organizing, and executing the technical sessions. The participants of the Institute are acknowledged for their attendance and for their questions and comments which made the sessions lively and convinced everybody that the undertaking had been worthwhile.

Finally we want to express our special appreciation to Drs. H. Aksel, N. Alemdaroğlu, C. Eralp, E. Paykoç, and to Messrs. H. Arıtürk, O. Boratav, M. Çetin, Ü. Enginsoy, O. Oğuz, E. Hatip for their valuable contributions and assistance and to Mrs. G. Beyaz for her efficient secretarial work.

A.Ş. ÜÇER
P. STOW
Ch. HIRSCH

CONTENTS

FUNDAMENTALS

DESCRIPTION OF VARIOUS FLOW MODELS FROM NAVIER-STOKES TO POTENTIAL FLOW MODELS

Ch. Hirsch, H. Deconinck

Vrije Universiteit Brussel, Department of Fluid Mechanics, Belgium

INTRODUCTION

The requirements towards a reduction of the complexity of the complete set of equations governing a three dimensional fluid flow, call for an analysis of the approximations which could be introduced, in order to simplify the mathematical formulation.

The nature of these approximations depends on various factors, the most important ones being the level of accuracy of the numerical flow simulation and the available computational ressources. Depending on these conditions,one can perform an analysis of the flow properties of the considered configuration and investigate the nature and order of magnitude of the various forces dominating the fluid behaviour. In function of this analysis one could neglect certain forces or force components and define in this way a large variety of approximations which we will consider as a <u>dynamical</u> level of approximations.

Similarly, an analysis of the different approximations with regard to the spatial representation of the flow can be performed, ranging from full three-dimensional to quasi three-dimensional, two-dimendional down to one-dimensional approximations. These different possibilities define the <u>spatial</u> level of approximation.

The resulting mathematical model for the numerical flow simulation will be obtained by an appropriate combination of these two levels of approximation.

It is considered that the system of Navier-Stokes equations, supplemented by empirical laws for the dependence of viscosity and thermal conductivity with other flow variables and by a constitutive law defining the nature of the fluid, completely describes all flow phenomena. For laminar flows, no additional information is required and one can consider that any experiment in a laminar flow regime can be accurately duplicated by computations. However, most of the flow situations occuring in practice are turbulent and semi-empirical information about the turbulence structure and properties must be introduced.

In the future with increasing computer power in speed and memory one could be able to simulate the large turbulent eddies, or even the small scale turbulent motion from the time-dependent Navier-Stokes equations, reducing considerably the need for additional information to the basic equations. An estimate of the computer requirements connected to this level of approximation has been given by Chapman [1] , Kutler [2], showing that it is presently outside the reach of our computational capabilities.

Hence, we will consider as the highest level of approximation the Reynolds averaged Navier-Stokes equations supplemented by some models for the Reynolds stresses. These models can range from simple eddy viscosity or mixing length models to transport equations for the turbulent kinetic energy and dissipation rates, the so-called k-ε model, or to still more complicated models computing directly the Reynolds stresses, see e.g. Kline [3], for a recent review of many turbulence models.

Considering the various stages within the dynamical level of approximation, a first reduction in complexity can be introduced for flows with small amount of separation or back-flow and with a predominant mainstream direction at high Reynolds numbers. This allows to neglect the viscous and turbulent diffusion in the mainstream direction and hence to reduce the number of shear stress terms to be computed, considering that they have a negligible action on the flow behaviour. This is the "Thin Shear Layer Approximation".

Close to this level, several methods have been developed, the so-called "Pressure Correction" method or "Parabolic Approximation" of the steady state Navier-Stokes equations, in which the elliptic character of the flow is put forward through the pressure field, while all other variables are considered as transported or as having a parabolic behaviour. These methods solve an elliptic equation for the pressure correction defined such as to satisfy continuity, assuming thereby that the pressure forces are dominant together with the inertia forces, while the viscous and turbulent forces are simplified and reduced to the transverse diffusion of momentum excluding thereby any separation.

The next level to be considered is the "Boundary Layer"

approximation. As is well known, this analysis of the effect of the viscosity on a flow by L. PRANDTL is a most spectacular example of the impact a careful investigation of the order of magnitude of force components can have on the description of a flow system.

For flows with no separation and thin viscous layers, that is at high Reynolds numbers, a decoupling of the viscous and inviscid parts of the flow can be introduced, whereby the pressure field is decoupled from the viscous effects, showing that the influence of the viscous and turbulent shear stresses is confined to small regions close to the walls and that outside these layers the flow behaves as an inviscid flow. This analysis, which was perhaps the greatest breakthrough in fluid mechanics since the discovery of the Navier-Stokes equations, showed that many of the flow properties can be described by the inviscid approximation, e.g. the pressure distributions, and that a simplified boundary layer approximation allows for the determination of the viscous effects. The calculation of the inviscid and the boundary layer components of the flow can be performed interactively, taking into account the influence of the boundary layers on the inviscid flow.

Recently a series of approaches in this direction have been developed, the viscid-inviscid interaction methods whereby attempts are made to calculate or to model separated regions in an approximate way, while keeping the advantages with regard to the computational effort of the boundary layer approximations, see LE BALLEUR [4] for a recent review on these approximations.

When this influence or interaction is neglected, one enters the field of the "Inviscid Approximations" which allows generally a good approximation of the pressure field and hence of lift.

An intermediate level between the partially or fully viscous description and the inviscid approximation is the "Distributed loss model", used in internal flow situations, in particular in the representation of multistage turbomachinery flows. Due to the large number of flow passages between the blades and their rotation with respect to the next blade row, one can consider that a downstream blade row sees, in first approximation, an averaged flow in the sense that the influence of the boundary layers and the wakes of the previous blades mix out. Their overall effect on the next blade can therefore be assimilated to a distributed force. As will be seen more in detail in the following, this approximation level amounts to take into account only the entropy creating part of the shear stresses, and neglect their effect in the energy equation.

With regard to the space level, next to the obvious choice between two-dimensional and three-dimensional representations in function of the geometrical configuration of a given flow system, the particular treatment for internal flows, known as the quasi-three dimensional representation, forms an intermediate step between the two and three dimensional representation, since it approaches a three-dimensional flow field by an iterative superposition of two-dimensional flow descriptions.

Table 1 summarizes the various levels of approximation defined in the present approach and the mathematical models corresponding to these various approximations will be presented and discussed in details in the following sections.

The basic flow equations, valid in all generality for any Newtonian compressible fluid, are summarized in table 2 and 3. Table 2 corresponds to an absolute coordinate system, while table 3 considers the equations written in a steadily rotating relative frame of reference, frequently used in turbomachinery applications. The equations of table 2 and 3 are the starting point for the various approximation models discussed in the remainder of this section.

1. THE NAVIER-STOKES EQUATIONS

The most general description of a fluid flow is obtained from the full system of Navier-Stokes equations. Referring to table 2, one can group the conservation forms for the three basic flow quantities $\rho, \rho\vec{v}$, ρE into a compact form

$$\partial_t \begin{vmatrix} \rho \\ \rho\vec{v} \\ \rho E \end{vmatrix} + \vec{\nabla} \begin{vmatrix} \rho\vec{v} \\ \rho\vec{v} \times \vec{v} + p\bar{\bar{I}} - \bar{\bar{\tau}} \\ \rho\vec{v}H - \bar{\bar{\tau}}\vec{v} - k\vec{\nabla}T \end{vmatrix} = \begin{vmatrix} 0 \\ \rho\vec{f}_e \\ W_f + q_H \end{vmatrix} \qquad (1.1)$$

This equation defines the column vector U of the conservative variables, a generalized flux vector $\vec{F}$ and the source vector $\vec{Q}$, such that

$$\frac{\partial U}{\partial t} + \vec{\nabla}\vec{F} = \vec{Q} \qquad (1.2)$$

The source terms express the effect of the external forces $\vec{f}_e$, the heat sources q_H and the work performed by the external forces $W_f = \rho\vec{f}_e\vec{v}$.

The system of equations (1.1) can be written in conservation form in Cartesian coordinates. In practical configurations however, the geometrical complexity of the boundaries of the flow regions such as airfoils, intakes, turbomachinery passages and others, calls for meshes adapted to the curved boundaries of the flow domain. This leads to curvilinear meshes, generally body fitted in the sense defined by THOMPSON [5], which even when numerically generated, can be considered as forming a family of curvilinear coordinates lines ξ, η, ζ.

The system of Navier-Stokes equations (1.1) has still to be supplemented by the constitutive laws, defining the shear stress tensor in function of the other flow variables, as well as by the thermodynamic gas laws and the laws of dependence of the fluid properties k and μ. For instance, for Newtonian fluids considered here, the shear stress tensor is defined by

$$\tau_{ij} = \mu \left[(\partial_i v_j + \partial_j v_i) - \frac{2}{3} (\vec{\nabla}\vec{v})\, \delta_{ij} \right] \qquad (1.3)$$

The system of Navier-Stokes equations is valid in particular for the laminar flow of a viscous, Newtonian fluid. In reality, the flow will remain laminar up to a certain critical value of the Reynolds number $V.L/\nu$, where V and L are representative values of velocity and length scales for the considered flow system Above this critical value, the flow becomes turbulent and is characterised by the appearance of fluctuations of all the variables, velocity, pressure, density, temperature, etc. around mean values. These fluctuations are of statistical nature and hence cannot be described in a deterministic way. However they can be computed numerically in direct simulations of turbulence including the "large eddy simulation" approach whereby only the small scale turbulent fluctuations are modeled and the larger scale fluctuations being computed directly. See for instance [6] and [7] for a recent review on the subject.

Presently both these approaches are still far from being applicable for practical calculations in industrial environments, due to the tremendous requirements they put on the computational resources. However there is no doubt that these methods will become more and more important in the future, since they require the lowest possible amount of external information in addition to the basic equations of Navier-Stokes.

Therefore there is a need to resort to a lower level of approximation, whereby the equations are averaged out, in time, over the turbulent fluctuations. This leads to the so-called "<u>Reynolds averaged Navier-Stokes equations</u>" which require, in addition, empirical or at least semi-empirical informations on the turbulence structure and its influence on the averaged flow.

2. THE REYNOLDS AVERAGED NAVIER-STOKES EQUATIONS

The turbulent averaging process is introduced in order to obtain the laws of motion for the "mean" or "time-averaged" turbulent quantities. Hence, for any quantity A, the separation

$$A = \bar{A} + A'$$

is introduced with

$$\bar{A}(\vec{r},t) = \frac{1}{T}\int_{t-T/2}^{t+T/2} A(\vec{r},\tau)\, d\tau \qquad (2.1)$$

where T is to be chosen large enough compared to the time scale of the turbulence but still small compared to the time scales of all other unsteady phenomena. For compressible flows, a density weighted average can be introduced, through

$$\tilde{A} = \frac{\overline{\rho A}}{\bar{\rho}} \quad \text{with} \quad A = \tilde{A} + A'' \quad \text{and} \quad \overline{\rho A''} = 0 \qquad (2.2)$$

Performing the averaging process on the continuity equation one obtains

$$\partial_t \bar{\rho} + \vec{\nabla}\overline{\rho \vec{v}} = 0 \qquad (2.3)$$

and for the momentum equations, in absence of body forces, from eq. (T2.5)

$$\partial_t \overline{\rho \vec{v}} + \vec{\nabla}\left(\overline{\rho \vec{v}} \times \tilde{\vec{v}} + \bar{p} - \bar{\bar{\tau}}^V - \bar{\bar{\tau}}^R \right) = 0 \qquad (2.4)$$

where the <u>Reynolds stresses</u> τ^R, defined by $\tau^R_{ij} = -\overline{\rho v''_i v''_j}$ are added to the <u>averaged</u> viscous shear stresses τ^V.

The relations between the Reynolds stresses and the mean flow quantities are unknown and therefore the application of the Reynolds averaged equations to the computation of turbulent flows requires the introduction of some semi-empirical modelisation of these unknown relations.

In a similar way, the averaged energy conservation equation can be obtained under different forms according to the definition taken for the averaged total energy.

If one defines the <u>mean turbulent total energy</u> by the straight-forward relation

$$\bar{\rho}\tilde{E} = \overline{\rho E} = \overline{\rho \left(e + \frac{\vec{v}^2}{2} \right)} \quad \text{and} \quad \bar{\rho}\tilde{H} = \overline{\rho H} = \overline{\rho \left(h + \frac{\vec{v}^2}{2} \right)} \tag{2.5}$$

one obtains

$$\begin{aligned} \tilde{E} &= \tilde{e} + \tilde{k} + k \quad \text{and} \quad E'' = e'' + \vec{v}''.\tilde{\vec{v}} + k'' - k \\ \tilde{H} &= \tilde{h} + \tilde{k} + k \quad \text{and} \quad H'' = h'' + \vec{v}''.\tilde{\vec{v}} + k'' - k \end{aligned} \tag{2.6}$$

where $\tilde{k}$ is the kinetic energy of the mean flow and k the turbulent kinetic energy defined as the average of the kinetic energy k'' of the turbulent fluctuations.

$$\rho\tilde{k} = \frac{\rho\, \tilde{\vec{v}}^2}{2} \quad \text{and} \quad \bar{\rho}k = \frac{\overline{\rho \vec{v}''^2}}{2} = \overline{\rho k''} \tag{2.7}$$

A conservative form of the turbulent energy equation is obtained by averaging the energy equation (T2.6), leading to

$$\partial_t \bar{\rho}\tilde{E} + \vec{\nabla}(\bar{\rho}\tilde{H}\tilde{\vec{v}}) = \vec{\nabla}(-\overline{\vec{F}_D} + \overline{\vec{v}.\bar{\bar{\tau}}} - \overline{\rho H''\vec{v}''}) \tag{2.8}$$

where the heat diffusive flux $\vec{F}_D = -k\vec{\nabla}T$ is introduced. This equation can be written in another form by an explicit calculation of the turbulent flux term using eq. (2.6), leading to the following form of the energy equation

$$\partial_t \bar{\rho}\tilde{E} + \vec{\nabla}(\bar{\rho}\tilde{H}\tilde{\vec{v}}) = \vec{\nabla}(-\overline{\vec{F}_D} + \tilde{\vec{v}}.\bar{\bar{\tau}}^T - \overline{\rho h''\vec{v}''} - \overline{\rho k''\vec{v}''} + \overline{\tau.\vec{v}''}) \tag{2.9}$$

where the total averaged shear stress tensor $\bar{\bar{\tau}}^T$ is introduced : $\bar{\bar{\tau}}^T = \bar{\bar{\tau}}^V + \bar{\bar{\tau}}^R$.

Hence, the system of equations in table 2 remains unchanged provided the shear stress tensor is replaced by the total, viscous plus turbulent stress tensor $\bar{\bar{\tau}}^T$ in the momentum and energy equations and if in equation (2.9) the diffusive flux term is replaced by a turbulent heat diffusion flux

$$\vec{F}_D^T = -\overline{k\vec{\nabla}T} + \overline{\rho\vec{v}''\left[h'' + k'' - \left(\frac{\tau}{\rho}\right)'' \right]} \tag{2.10}$$

It should be noted that the definitions (2.6) of the averaged total energy and enthalpy contain a contribution from the turbulent kinetic energy k. If one would define a <u>total energy of the</u>

averaged flow $\hat{E}$ resp. $\hat{H}$

$$\hat{E} = \tilde{e} + \tilde{k} = \tilde{E} - k \quad \text{and} \quad \hat{H} = \tilde{h} + \tilde{k} = \tilde{H} - k \tag{2.11}$$

a different expression of the energy conservation equation would be obtained. This alternate form is obtained after substracting the equation for the turbulent kinetic energy k from equation (2.9), Cebeci and Smith [8] :

$$\partial_t \bar{\rho}\hat{E} + \vec{\nabla}(\bar{\rho}\hat{H}\vec{\tilde{v}}) = \vec{\nabla}(-\overline{\vec{F}_D} + \vec{\tilde{v}}.\overset{=}{\tau}^T - \overline{\rho h''\vec{v}''}) + \overline{\vec{v}''\vec{\nabla}p} + \bar{\rho}\varepsilon - P \tag{2.12}$$

where $\bar{\rho}\varepsilon = \overline{(\tau\vec{\nabla})\vec{v}''}$ is the viscous dissipation of turbulent fluctuations and P the production of turbulent energy by the Reynolds stresses $P = (\overset{=}{\tau}^R\vec{\nabla})\vec{\tilde{v}}$.

It is common practice to introduce certain simplifications in the energy equations (2.9) and (2.12), Cebeci & Smith [8]. If one neglects the higher order fluctuation terms in H" and $\tilde{H}$, one can write

$$\tilde{H} \approx \tilde{h} + \tilde{k} = \hat{H} \quad \text{and} \quad H'' \approx h'' + \vec{v}''.\vec{\tilde{v}} \tag{2.13}$$

Moreover, within the same order of approximation one can neglect the higher order term $\overline{\overset{=}{\tau}''.\vec{v}''}$, describing the diffusion of turbulent energy by the turbulent viscous forces, in equation(2.9). This leads to the simplified energy equation for the averaged turbulent flow, from eq. (2.9)

$$\partial_t \bar{\rho}\hat{E} + \vec{\nabla}(\bar{\rho}\hat{H}\vec{\tilde{v}}) = \vec{\nabla}(-\overline{\vec{F}_D} + \vec{\tilde{v}}\overset{=}{\tau}^T - \overline{\rho h''\vec{v}''}) \tag{2.14}$$

Comparing with equation (2.12) one sees that these assumptions imply that the three last terms of equation (2.12) are negligible. Indeed, it is known from experimental observations, that the pressure fluctuations are negligible for adiabatic flows below Mach numbers of the order of five. On the other hand, it is also observed in turbulent boundary layers that for the most part, the dissipation practically balances the production of turbulent energy. Hence, this set of assumptions seem to form a consistent unit, at least for attached boundary layers and at Mach numbers below five. But for more complex flow situations and (or) at higher Mach numbers the more complete form of the energy equation (2.12) might have to be used.

Another implication of these simplifications is that the

turbulent heat-flux term F_D^T of equation (2.10) reduces to

$$\vec{F}_D^T = -\overline{k\vec{\nabla}T} + \overline{\rho h''\vec{v}''} \tag{2.15}$$

Turbulence Models

Many different models ranging from simple algebraic to second-order closure models involving the simultaneous solution of a large number of partial differential equations for the Reynolds stresses have been developed.

Many approaches are based on the so-called first order closure models in which the Reynolds stresses are expressed through an eddy viscosity model, following BOUSSINESQ's (1877) original assumption,

$$\tau_{ij}^R = -\overline{\rho v_i'' v_j''} = \mu_T \left(\partial_i \tilde{v}_j + \partial_j \tilde{v}_i - \frac{2}{3} (\vec{\nabla}\tilde{\vec{v}}) \quad \delta_{ij} \right) - \frac{2}{3} \bar{\rho} k \delta_{ij} \tag{2.16}$$

where μ_T is a turbulent eddy viscosity coefficient. Similarly, the turbulent heat flux vector (2.10) will be modeled, with the assumption (2.15), by

$$\vec{F}_D^T = -k_T \vec{\nabla}\bar{T} \qquad \text{or} \qquad \vec{F}_D^T = -\gamma \frac{\mu_T}{Pr} \vec{\nabla}\tilde{e} \tag{2.17}$$

defining a turbulent thermal conductivity coefficient k_T or the corresponding turbulent diffusivity coefficient κ_T through a turbulent Prandtl number

$$Pr_T = \mu_T \frac{c_p}{k_T} = \frac{\nu_T}{\kappa_T} = \frac{\mu_T}{\bar{\rho}\kappa_T} \tag{2.18}$$

where the kinematic turbulent viscosity ν_T has been introduced.

Introducing these expressions into the turbulent averaged Navier-Stokes equations, leads to a system which is formally identical to the Navier-Stokes equations for a laminar flow where the laminar viscosity ν is multiplied by $(1 + \nu_T/\nu)$ and the thermal diffusivity coefficient κ is multiplied by $(1 + \kappa_T/\kappa)$. In addition the turbulent kinetic energy contribution to the normal stresses, the last term in equation (2.16) will generally be absorbed by and included as part of the mean pressure $\bar{p}$. This is the easiest approach to the turbulent averaged Navier-Stokes equations and the various models in this group are distinguished from one another by the way these two coefficients k_T and μ_T are estimated.

We refer the reader to the literature for a description of various turbulence models [3], [8].

3. THE THIN SHEAR LAYER APPROXIMATION (TNS)

At high Reynolds numbers, the shear layers along solid walls are very thin compared to the body length and if the extension of the viscous regions is not becoming too large as would be the case in the regions of separation, in zones of shock-boundary layer interactions or in trailing edge regions, then the dominating influence of the shear stresses will be obtained from the gradients normal to the surface.

If one considers an arbitrary curvilinear system of coordinates with ξ^1 and ξ^2 along the surface and $\xi^3 = n$ directed towards the normal, then the Thin Shear Layer approximation of the Navier-Stokes equations consist in neglecting all ξ^1 and ξ^2 derivatives occuring in the turbulent and viscous shear stress terms, Pulliam & Steger [9], Steger [10].

This approximation is also justified by the fact that generally, at high Reynolds numbers (typically $Re > 10^4$) the mesh is made dense only in the direction normal to the surface and therefore the accuracy of the computation of the neglected terms is of a lower order of magnitude than the normal derivatives.

The general form of the conservation equations keeps the same form of equation (1.1) and (1.2) in the thin shear layer approximation

$$\partial_t U + \vec{\nabla}\vec{F} = Q \tag{3.1}$$

but the flux $\vec{F}$ in equation (1.1) is simplified such that, in the contributions to the viscous terms all derivatives with respect to ξ^1 and ξ^2 are neglected. That is, only the normal derivatives are maintained.

This leaves as the remaining contribution of the shear stress term

$$\vec{\nabla}\bar{\bar{\tau}} \approx \partial_n \cdot \tau^{in} \tag{3.2}$$

in which the derivatives with respect to ξ^1 and ξ^2 in the calculation of $\bar{\bar{\tau}}$ are considered as neglectable, and n is the normal.

This approximation is actually close to a boundary layer approximation, since viscous terms which are neglected in the boundary layer approximation are also neglected here. However, the momentum equation in the directions normal to the wall is retained instead of the constant pressure rule over the boundary layer thickness along a normal to the wall. Therefore the transition from viscous dominated regions to the inviscid region outside the wall layer is integraly part of the calculation, and one has here somehow a "higher order" boundary layer approximation. The classical boundary layer approximation is obtained when the momentum equation in the direction ξ^3 is replaced by the condition

$$\partial_n p = 0 \tag{3.3}$$

The thin shear layer approximation amounts to neglect the viscous diffusion in the direction parallel to the surface and keep only the contributions from the diffusion in the direction normal to the body surface.

In Cartesian coordinates, with x,y coordinates in the plane and z normal to the surface x,y the shear stresses are approximated by

$$\tau_{ij} \approx \mu\,(\delta_{3i}\ \partial_z v_j + \delta_{3j}\ \partial_z v_i) - \frac{2}{3}\ \delta_{ij}\ (\partial_z v_z) \tag{3.4}$$

where i=3 corresponds to z.

4. PARABOLIZED NAVIER-STOKES APPROXIMATION

Starting from the steady state Navier-Stokes equations, a parabolic formulation can be obtained if a coordinate system can be defined aligned with the flow direction, such that one of the velocity components, say in the ξ^1 direction is sufficiently larger than the transverse components in the ξ^2 and ξ^3 direction.

It is further assumed that :

1. the main stream velocity component suffers no reversal.
2. diffusive transport processes in the streamwise ξ^1 direction are of higher order and can be neglected.
3. the pressure field in the transversed plane is governed by an elliptic behaviour.This means that the convective effects in the transverse plane are negligible.

Considering the momentum equation (2.4) in the ξ^1 direction, the transverse velocity derivatives are discarded, together with the diffusion terms in the ξ^1 direction, eliminating in this way the elliptic boundary value character in the ξ^1 direction.

The momentum equations in the transverse direction are combined to a Poisson equation for the pressure in the transverse plane by taking the divergence. The continuity equation serves as a constraint to the solution of this equation.

The system is closed by the energy equation and by a k-ε closure model for the turbulence.

See f.i. Patankar [11] for a detailed description.

5. THE BOUNDARY LAYER APPROXIMATION

It was the great achievement of L. PRANDTL to recognize that, at high Reynolds numbers the viscous regions remain of limited extension δ (of the order of $\delta/L \approx \sqrt{\nu/UL}$, for a body length L) along the surfaces of the solid bodies immersed in or limiting the flow. Hence, in all cases where these viscous regions remain close to the body surfaces, that is in absence of separation, the calculation of the pressure field may be decoupled from the calculation of the viscous velocity field.

In general curvilinear coordinates, with ξ^3 taken along the normal to the surface, the boundary layer equations consist of the ξ^1 and ξ^2 momentum equations in identical form to the one used in the thin shear layer approximation where the pressure field is replaced by the inviscid pressure field obtained from an inviscid calculation.

The inviscid region is limited by the edge of the boundary layer, which is initially unknown since the interactive computational proces has to start by the calculation of the pressure field. In the classical boundary layer approximation the limits of the inviscid region are taken on the surface, which is justified for small boundary layer thickness. This leads to a complete decoupling between the pressure field and the velocity field since the pressure, in the remaining ξ^1 and ξ^2 momentum equations, is equal to the values of the inviscid pressure field at the wall and is known at the moment these equations are to be solved.

When the influence of the boundary layers on the inviscid flow field is considered as non-negligible, this interaction can be taken into account in an iterative way, by recalculating the inviscid pressure field with the limits of the inviscid region at the edge of the boundary layer obtained at the previous interaction. This procedure is applied for thick boundary layers up to small separated regions and is known as the viscid-inviscid interaction approximation, see f.i. Le Balleur [4] for a recent review on the subject.

The system of equation obtained in this way has only the velocities as unknowns and this represents a significant simplification of the full Navier-Stokes equations. Therefore the boundary layer equations are much easier to solve, being basically close to standard parabolic second order partial differential equation and many excellent numerical methods have been developed, e.g. Cebeci & Bradshaw [12]. The difficulties in these methods are essentially connected to the uncertainties of the turbulence models introduced in order to close the system of equations.

A still greater simplification is often introduced if one is satisfied with the knowledge of some global boundary layer parameters such as the integral thicknesses. The boundary layer equations are integrated over ξ^3 from the wall to the edge of the viscous wall layer and a system of first order differential equations is obtained in ξ^1, ξ^2 and time. Of course the knowledge of the boundary layer velocity field is completely lost in this approach.

6. THE DISTRIBUTED LOSS MODEL

The distributed loss model is an approximation used since many years in the field of internal flows in turbomachinery. This model is defined by the assumption that the effect of the shear stresses on the equations of motion is equivalent to a distributed friction force.

Due to the complexity of the internal flows in a multistage turbomachine, its three-dimensional, unsteady, viscous nature, the full three-dimensional description of the flow through the various stages of the machine, including effects such as presence of end walls, tip clearance flows, relative motions between rotating and non-rotating blade rows, is a considerable computing task which cannot be achieved presently within reasonable computing times and costs. Since such a calculation would however be the only way of obtaining, with an adequate turbulence model, a reliable estimation of the loss mechanisms and loss distributions, empirical informations with regard to these losses will have to be introduced in order to replace this computation.

In the practice of turbomachinery flow calculations, these losses are defined on an overall or averaged basis as the stagnation pressure drop between the inlet and the outlet of a blade row without taking into account the details of the physical mechanism or the exact location of the producing regions.

The resulting approximation is then basically of inviscid nature but not isentropic since the entropy variation along the path of a fluid particle will be connected to the energy losses along this path.

Obviously, a certain number of three-dimensional flow details will be lost in such an approximation, in particular all flow aspects which can be attributed to or are strongly influenced by viscous effects. However, in the turbomachinery environment considered here, it is assumed that the main effects, eventually considered on an averaged basis as in the quasi-three-dimensional approximation, are correctly described at least in the main body of the flow.

Mass, momentum and energy conservation in a relative frame of reference rotating with a uniform angular velocity $\vec{\omega}$, take the following form, see Table 3., with the assumption that <u>all quantities</u> have been <u>averaged</u> for the <u>turbulent fluctuations</u> according to section 2, and discarding heat sources and external forces.

$$\partial_t \rho + \vec{\nabla} \rho \vec{w} = 0 \tag{6.1}$$

$$\partial_t \vec{w} - \vec{w} \times \vec{\zeta} = T \vec{\nabla} s - \vec{\nabla} I + \frac{1}{\rho} \vec{\nabla} \bar{\bar{\tau}} \tag{6.2}$$

$$\rho \frac{dI}{dt} = \frac{\partial p}{\partial t} + \vec{\nabla} (k_T \vec{\nabla} T) + \vec{\nabla} (\bar{\bar{\tau}} \cdot \vec{w}) \tag{6.3}$$

with $\vec{\zeta}$ the absolute vorticity and and I the rothalpy. The additional (but not independent) relation for the entropy variation, eq. (T3.11) is also of interest

$$\rho T \frac{ds}{dt} = \vec{\nabla} (k_T \vec{\nabla} T) + \varepsilon \tag{6.4}$$

where the dissipation term ε is defined by

$$\varepsilon = \frac{1}{2 \mu_T} (\bar{\bar{\tau}} \times \bar{\bar{\tau}}) = (\bar{\bar{\tau}} \cdot \vec{\nabla}) \vec{w} \tag{6.5}$$

The basic assumption of the distributed loss model lies in the consideration of the shear stress term in the momentum equation as a <u>distributed friction</u> <u>force</u> $\vec{F}_f$ responsible for the overall entropy increase in the flow

$$\rho\vec{F}_f = \vec{\nabla}\bar{\bar{\tau}} \tag{6.6}$$

In addition, it is considered that the thermal conduction does not affect the flow behaviour significantly and that the heat transfer between fluid and blade walls need not to be taken account. This reduces the entropy equation (6.4) to

$$\rho T \frac{ds}{dt} = \varepsilon = (\bar{\bar{\tau}} \cdot \vec{\nabla})\,\vec{w} = \vec{\nabla}\,(\bar{\bar{\tau}} \cdot \vec{w}) - \vec{w}\,(\vec{\nabla} \cdot \bar{\bar{\tau}}) \tag{6.7}$$

By <u>definition of the distributed loss model</u>, one has

$$T \frac{ds}{dt} = -\rho\vec{w} \cdot \vec{F}_f \tag{6.8}$$

expressing that the work of the distributed friction forces $\vec{F}_f$ can only be transformed into non-reversible heat and hence serves only to increase the entropy. Comparing these last two relations, one finds that the work done by the turbulent stresses is neglected, that is $\vec{\nabla}(\bar{\bar{\tau}}\vec{w})=0$.

Obviously, this cannot be valid in boundary layers and in any region with strong shear stress gradients and therefore the present model is not able to describe correctly the flow in the wall regions. The friction force $\vec{F}_f$ will then be considered as an <u>external force</u> defined by equation (6.8) and not by equation (6.6). In addition $\vec{F}_f$ is assumed to be directed in the direction opposite to the local velocity vector $\vec{w}$. Hence, for positive values of F_f

$$\vec{F} = -F_f \cdot \vec{1}_w = -F_f \frac{\vec{w}}{w} \tag{6.9}$$

and equation (6.8) becomes

$$T \frac{ds}{dt} = \rho\, w\, F_f \tag{6.10}$$

The assumption (6.9) for the direction of the friction force $\vec{F}_f$ is in agreement with the shear layer approximation and boundary layer theory.

An important consequence of the assumptions made in the distributed loss model concerns the energy equation (6.3), which takes a strongly simplified form, and the general equations defining the present model are

$$\partial_t \rho + \vec{\nabla}\rho\vec{w} = 0$$

$$\partial_t \vec{w} - \vec{w} \times \vec{\zeta} = T\vec{\nabla}s - \vec{\nabla}I - F_f \cdot \vec{1}_w$$

$$\rho \frac{dI}{dt} = \frac{\partial p}{\partial t} \qquad (6.11)$$

$$T \frac{ds}{dt} = \rho w F_f$$

Since the details of the loss mechanism, that is of the shear stresses, are not considered, these equations are to be taken as describing an inviscid model, however with an entropy producing term. The boundary conditions for the velocity field are therefore the inviscid conditions of vanishing normal velocity components at the walls with a non-vanishing tangential velocity along these boundaries.

Steady state formulation

This model is generally further simplified by the assumption of steady relative flows. In a turbomachinery environment, this will be justified as long as the inlet flow conditions considered as uniform in the direction of blade rotation, which implies that the presence and effects of the wakes shed from the upstream blades is neglected. Recent detailed data, Dring et al [13] on rotor stator interaction show that the unsteady pressure generate by upstream wakes can be locally very significant although the time averaged flow is in good agreement with steady state calculations. Therefore one can consider the steady state model for the relative flow as a valid approximation for the time averaged flow at constant angular velocity. This leads to the following simplified model,

$$\vec{\nabla}\rho\vec{w} = 0$$

$$- \vec{w} \times \vec{\zeta} = T\vec{\nabla}s - \vec{\nabla}I + \vec{F}_f$$

$$\vec{w} \cdot \vec{\nabla}I = 0 \qquad (6.12)$$

$$T\vec{w} \cdot \vec{\nabla}s = - \rho\vec{w}\vec{F}_f = \rho w F_f$$

In particular, the energy equation reduces to the constancy of the rothalpy I along a flow-path. Note that the system (6.11) or (6.12) contains 6 equations for 5 unknowns $\rho, \vec{w}, I$. Therefore, when use is made of the entropy equation, one of the momentum equations can be dropped out of the system.

The variation of the relative stagnation pressure is determined by the knowledge of an overall pressure loss coefficient, Π defined for instance by

$$\Pi = \frac{\Delta p_0}{q} \tag{6.13}$$

where q is a reference dynamic head and Δp_0 the stagnation pressure drop between two points on a flow path, such as the inlet and the outlet of an arbitrary channel of a blade row in a rotating machine.

The introduction of the friction force in order to be consistent with the entropy production was first discussed by Horlock [14] and further by Bosman & Marsh [15].

The model described above is used extensively in the field of turbomachinery flow calculations with the introduction of empirical data for the loss coefficients, without the computation of the shear stresses. However, a more sophisticated version can be developed in which the shear stresses are computed explicitly with an adequate turbulence model and where the losses are determined within the calculation.

Obviously, if the friction force is not considered one obtains a purely inviscid model describing a steady three-dimensional, rotational, but isentropic flow where an inlet non-uniform distribution of stagnation pressure may exist, but where no new losses are generated within the channel.

7. THE INVISCID FLOW MODEL - EULER EQUATIONS

The most general flow situations for a non-viscous fluid are described by the set of Euler equations, obtained from the Navier-Stokes equations (1.1) by neglecting all shear stresses.

This approximation introduces a drastic change in the mathematical formulation with respect to all the previous models since the system of partial differential equations describing the inviscid flow model reduces from second order to first order. This is of paramount importance since it will determine the numerical and physical approach to the computation of these flows. Also, the number of allowable boundary conditions is modified by passing from the second order viscous equations to the first order inviscid system.

The time-dependent Euler equations, in conservation form and in an absolute frame of reference,

$$\frac{\partial U}{\partial t} + \vec{\nabla}\vec{F} = Q \qquad (7.1)$$

form a system of first order partial differential equations hyperbolic in time , where the flux vector $\vec{F}$ has the Cartesian components f, g, h given by equation (1.1) without the shear stress terms

$$f = \begin{vmatrix} \rho u \\ \rho u^2 + p \\ \rho uv \\ \rho uw \\ \rho Hu - k \frac{\partial T}{\partial x} \end{vmatrix} \quad g = \begin{vmatrix} \rho v \\ \rho uv \\ \rho v^2 + p \\ \rho vw \\ \rho Hv - k \frac{\partial T}{\partial y} \end{vmatrix} \quad h = \begin{vmatrix} \rho w \\ \rho uw \\ \rho vw \\ \rho w^2 + p \\ \rho Hw - k \frac{\partial T}{\partial z} \end{vmatrix} \qquad (7.2)$$

and the source term Q is also given by equation (1.1)

As is known, this set of equations allows also discontinuous solutions in certain cases, namely contact discontinuities or shock waves, occuring in supersonic flows. Of course these discontinuous solutions can only be obtained from the integral form of the conservation equations, since the gradients of the fluxes are not defined at the shock wave surfaces. In local form, one obtains the Rankine-Hugoniot shock relations, f.i. for steady state flows and without heat conduction

$$|\rho\vec{v}|\ \vec{n} = 0$$

$$\rho\vec{v}.\vec{n}\ |\vec{v}| + |p|\ \vec{n} = 0$$

$$|H| = 0$$

where $\vec{n}$ is the normal to the discontinuity surface and $|\ |$ denotes the jump over the discontinuity.

An alternative set of equations for the inviscid flow model is obtained from the equations derived in the previous section by setting the friction force $\vec{F}_f$ to zero. One obtains, in an absolute frame of reference, the following model, neglecting heat conduction

$$\partial_t \rho + \vec{\nabla}\rho\vec{v} = 0$$

$$\partial_t \vec{v} - \vec{v} \times \vec{\zeta} = T\vec{\nabla}s - \vec{\nabla}H \qquad (7.3)$$

$$\frac{dH}{dt} = \partial_t H + (\vec{v}\vec{\nabla})\, H = \frac{1}{\rho}\, \partial_t p$$

together with the entropy equation $\frac{ds}{dt} = 0$, obtained by combining eq. (7.3b) and (7.3c), using $dh = Tds + dp/\rho$.

This set of equations becomes particularly interesting in a steady state flow model where the following simplification is obtained in an absolute resp. relative frame of reference, rotating with a constant angular velocity

$$\begin{aligned}
&\vec{\nabla}\rho\vec{v} = 0 && \vec{\nabla}\rho\vec{w} = 0 \\
&-\vec{v} \times \vec{\zeta} = T\vec{\nabla}s - \vec{\nabla}H && -\vec{w} \times \vec{\zeta} = T\vec{\nabla}s - \vec{\nabla}I \\
&\vec{v} \cdot \vec{\nabla}H = 0 && \vec{w}\vec{\nabla}I = 0 \\
&\vec{v} \cdot \vec{\nabla}s = 0 && \vec{w}\vec{\nabla}s = 0
\end{aligned} \tag{7.4}$$

The last two equations express the constancy of stagnation enthalpy resp. rothalpy and of entropy along a flow-path. However, since the entropy equation is not independent of the other equations, across a discontinuity such as shock waves H (or I) remains constant along a flow path, but the entropy will not since $s = s\ (p_o, H)$ and the discontinuous variation of the stagnation pressure implies a discontinuous variation of the entropy. However, in the regions upstream and downstream of shocks the entropy will remain constant along a flow path according to equation (7.4d). The momentum equation, (7.4b) shows an important property with regard to the origin of vorticity in the steady flow. Since the variations of stagnation enthalpy and entropy can occur only in the directions transverse to the flow directions, these variations are the sources of the vorticity creation in the flow. Therefore a three-dimensional flow will generally be rotational, unless the initial and boundary conditions have uniform total enthalpy and entropy.

Further, a non uniform discontinuity, such as a curved shock, will generate a non uniform entropy field in the direction normal to the velocity and the flow becomes irrotational in the entire region downstream of the discontinuity.

For a flow with constant total enthalpy and entropy, the steady inviscid flow equations reduce to

$$\vec{v} \times \vec{\zeta} = 0 = \vec{v} \times (\vec{\nabla} \times \vec{v}) = 0 \tag{7.5}$$

The most general flow field with non-zero vorticity satisfying this equation is called a Beltrami flow. A Beltrami flow has therefore only streamwise vorticity and the vorticity is aligned with the velocity in every point. The existence of Beltrami flows shows that, even in isoenergetic and isentropic flows, vorticity can exist and be maintained although only in very special circumstances. From the known properties of perfect fluids with regard to vorticity, the Kelvin-Helmholtz theorems (stating that the vorticity flux through a streamtube is constant) it follows that in order to obtain a Beltrami flow, the inlet flow must possess only streamwise vorticity.

However, the most general steady velocity field, satisfying constant energy and entropy will be of zero vorticity leading to the family of potential flows.

8. STEADY INVISCID ROTATIONAL FLOWS - CLEBSCH REPRESENTATION

Before discussing the potential flow models, a general representation of the steady, inviscid but rotational flows can be given which allows in many cases a most economical description in terms of the number of flow variables [16], [17], [18], [19].

Since the vorticity is generated, with the exclusion of Beltrami flows, by the gradients of entropy and total energy, equation (7.4b) one can write for the general absolute velocity field $\vec{v}$, a form of the Clebsch representation

$$\vec{v} = \vec{\nabla}\phi + \psi_1 \vec{\nabla}s + \psi_2 \vec{\nabla}I \tag{8.1}$$

where ϕ is a potential function and where ψ_1 and ψ_2 are two additional functions describing the rotational part of the flow, since

$$\vec{\zeta} = \vec{\nabla} \times \vec{v} = \vec{\nabla}\psi_1 \times \vec{\nabla}s + \vec{\nabla}\psi_2 \times \vec{\nabla}I \tag{8.2}$$

Introducing this relation into the momentum equation (7.4b) and taking into account the energy and entropy equations, one obtains

$$(\vec{w}\vec{\nabla})\, \psi_1 \,.\, \vec{\nabla}s + (\vec{w}\vec{\nabla})\, \psi_2 \,.\, \vec{\nabla}I = - \vec{\nabla}I + T\vec{\nabla}s \tag{8.3}$$

For arbitrary and independent entropy and rothalpy gradients, one obtains two equations for ψ_1 and ψ_2

$$(\vec{w} \,.\, \vec{\nabla})\, \psi_1 = + T \qquad \text{and} \qquad (\vec{w} \,.\, \vec{\nabla})\, \psi_2 = - 1 \tag{8.4}$$

These equations are purely convective and require only the initial knowledge of ψ_1 and ψ_2 in the inlet surface of the flow region. The equation for the third function ϕ is obtained from the continuity equation

$$\vec{\nabla}\rho\vec{\nabla}\phi = \vec{u}\vec{\nabla}\rho - \vec{\nabla}\,(\rho\psi_1\vec{\nabla}s + \rho\psi_2\vec{\nabla}I) \tag{8.5}$$

Generally, the specific mass can be expressed in function of two thermodynamic variables, e.g.

$$\rho = \rho\ (s\ ,\ h) \quad \text{or} \quad \rho = \rho\ (s\ ,\ I)$$

and the three equations above (8.4) and (8.5) have to be closed by the equations of energy and entropy conservation

$$\vec{w}\vec{\nabla}s = 0 \quad \text{and} \quad \vec{w}\vec{\nabla}I = 0 \tag{8.6}$$

In this representation, one has still 5 unknown functions to solve comprising the three velocity components, S and I. A simplified representation can be obtained if a unique relation between S and I exists in the inlet field, that is

$$s = s\ (I) \quad \text{or} \quad I = I\ (s) \tag{8.7}$$

then equation (8.3) reduces to

$$\vec{\nabla}I - T\vec{\nabla}s = (\frac{dI}{dS} - T)\ \vec{\nabla}s = (1 - T\,\frac{ds}{dI})\ \vec{\nabla}I \tag{8.8}$$

since the relations (8.7) are valid in the whole flow field due to equations (8.6).

Consequently, one out of two "rotational functions" ψ_1 and ψ_2 is not necessary and the three-dimensional rotational flow field can be described by the representation

$$\vec{v} = \vec{\nabla}\phi + \psi_1\vec{\nabla}s \tag{8.9}$$

or equivalently by

$$\vec{v} = \vec{\nabla}\phi + \psi_2\vec{\nabla}I \tag{8.10}$$

In the first representation, the function ψ_1 satisfies the equation

$$(\vec{w}\ .\ \vec{\nabla})\ \psi_1 = -\frac{dI}{ds} + T \tag{8.11}$$

and in the second case, one has

$$(\vec{w}\ .\ \vec{\nabla})\ \psi_2 = -1 + T\,\frac{ds}{dI} \tag{8.12}$$

The simplified representation has been used by Lacor & Hirsch [16], [20] to describe three-dimensional non-viscous rotational flows. It is easily seen that the assumptions (8.7) for the simplified representation (8.9) or (8.10) is equivalent to the conditions

$$\vec{\zeta} \cdot \vec{\nabla}s = \vec{\zeta} \cdot \vec{\nabla}I = 0 \tag{8.13}$$

which states that the entropy and total energy are constant along vorticity lines. This condition had been shown shown by Cazal [17] to be necessary and sufficient for the representation (8.9) or (8.10) to be valid.

This is a very economical representation of a rotational flow since the three-dimensional velocity field is described only be two scalar functions, and the complete flow description requires only three equations for ϕ, ψ_1 and s, since I = I (s) is known from the inlet conditions.

It is shown by Lacor & Hirsch [20] that these conditions are satisfied if either H or s are constant in the inlet flow field or, more generally, if the inlet velocity field is uniform in at least one direction transverse to the inlet velocity direction.

However, this restriction on the inlet flow, necessary to obtain a simpler representation for the velocity field is not needed for non-rotating flows or for incompressible flows, Hawthorne [18]. Indeed, in these two cases the number of independent unknowns can be reduced, through the introduction of scaled quantities (indicated by a subscript s), see for instance Lacor and Hirsch [21].

<u>Non rotating stationary flow</u>

One obtains the following representation for the scaled velocity,

$$\vec{v}_s = \vec{\nabla}\phi + \psi\, \vec{\nabla}p_o \tag{8.14}$$

where the p_o is the stagnation pressure. The two scalar functions ϕ and ψ satisfy the following equations

$$\vec{\nabla}\rho_s \vec{\nabla}\phi = - \vec{\nabla}\,(\rho_s \psi \vec{\nabla}p_o) \tag{8.15}$$

$$\vec{v}_s \vec{\nabla}\psi = - \rho_{o_A}\,(\frac{p_o}{p_{o_A}}) \tag{8.16}$$

together with

$$\vec{v}_s \vec{\nabla}p_o = 0 \quad \text{and} \quad \vec{v}_s \vec{\nabla}H = 0 \tag{8.17}$$

where the subscript A indicates reference values. This is a

completely general formulation for steady, non rotating inviscid rotational flows allowing a description with the aid of two scalar functions for the three scaled velocity components.

Incompressible Fluid

For incompressible flows, the scaling is actually not necessary since the momentum equation becomes, with the continuity equation for incompressible flows

$$\vec{w} \times \vec{\zeta} = \frac{1}{\rho} \vec{\nabla} p^* \quad \text{with} \quad \vec{\nabla}\vec{w} = 0 \tag{8.18}$$

It follows that

$$\vec{w}\vec{\nabla} p^* = 0 \tag{8.19}$$

where p* is often called the rotary stagnation pressure.

$$p^* = p + \frac{1}{2} \omega (\vec{w}^2 - \omega^2 r^2) \tag{8.20}$$

Hence, one obtains the representation

$$\vec{v} = \vec{\nabla}\phi + \psi\vec{\nabla} p^* \tag{8.21}$$

and the equations

$$\Delta\phi = - \vec{\nabla} (\psi\vec{\nabla} p^*) \tag{8.22}$$

$$\vec{w}\vec{\nabla}\psi = - \frac{1}{\rho} \tag{8.23}$$

which, together with equation (8.19) for constancy of the rotary stagnation pressure along a streamline, form the complete set of equations for the three-dimensional, steady, iniscid incompressible flow in a steadily rotating system. This flow system is therefore completely determined by the knowledge of the three scalar functions ϕ, ψ and p*.

9. THE POTENTIAL FLOW MODEL

The most impressive simplification of the mathematical model for a flow system is obtained for the approximation of a non-viscous, irrotational flow. From

$$\vec{\zeta} = \vec{\nabla} \times \vec{v} = 0 \tag{9.1}$$

the three-dimensional velocity field can be described by a single scalar function ϕ, the potential function defined by

$$\vec{v} = \vec{\nabla}\phi \tag{9.2}$$

reducing the knowledge of the three velocity components to the determination of a single potential function ϕ.

If the initial and boundary conditions are compatible with uniform entropy on the inflow boundary, then for continuous flows ds/dt=0, eq. (7.4) implies that entropy is constant over the whole flow field. Hence, for the isentropic flows, the momentum equation (7.3b) becomes

$$\partial_t \vec{\nabla}\phi + \vec{\nabla}H = 0 \qquad \text{or} \qquad H + \partial_t \phi = H_0 \tag{9.3}$$

the constant H_0 having the same value along all the streamlines.

This equation shows that the energy equation is no more independent from the momentum equation, and the flow will be completely determined by the initial and boundary conditions on one hand and by the knowledge of the single function ϕ on the other hand.

The equation for the potential function is obtained from the continuity equation, taking into account the isentropic conditions to express the density in function of velocity and hence in function of the gradient of the potential function. On obtains

$$\vec{\nabla}\rho\vec{\nabla}\phi + \partial_t \rho = 0 \tag{9.4}$$

with for a perfect gas, where A is an arbitrary reference state

$$\frac{\rho}{\rho_A} = \left(\frac{h}{h_A}\right)^{1/(\gamma-1)} = \left[\left(H_0 - \frac{\vec{v}^2}{2} - \partial_t \phi\right) / h_A\right]^{1/(\gamma-1)} \tag{9.6}$$

A further simplification is obtained for <u>steady</u> potential flows since the potential equation reduces to, with $H = H_0 = ct$, equation (9.3)

$$\vec{\nabla}\rho\vec{\nabla}\phi = 0 \tag{9.6}$$

with the density given by equation (9.5) where h_A can be chosen equal to H_0 , hence

$$\frac{\rho}{\rho_0} = \left(1 - \frac{(\vec{\nabla}\phi)^2}{2\ H}\right)^{1/\gamma-1} \tag{9.7}$$

and where ρ_0 is the stagnation density, constant throughout the whole flow field.

Both for steady and unsteady flows, the boundary condition along a solid boundary is the condition of vanishing relative velocity between flow and solid boundary in the direction n normal to the solid wall

$$v_n = \partial_n \phi = \vec{u} \cdot \bar{1}_n \tag{9.8}$$

where $\vec{u}$ is the velocity of the solid boundary with respect to the system of reference considered.

For a steadily rotating turbomachinery blade row, for instance, $\vec{u}$ will be the wheel speed (in the tangential direction) if ϕ is the potential of the absolute flow. For oscillating airfoils, $\vec{u}$ will be equal to the velocity of the blade wall oscillations.

Irrotational flow with circulation - Kutta Joukowski condition.

To achieve a non-zero lift on a body, it is necessary to impose a circulation Γ around any closed curve surrounding the body, corresponding to a vorticity production inside the closed curve, which is physically generated in the boundary layer of the body. It follows that Γ cannot be determined from irrotational theory and is an externally given value for potential flow.

However, for aerodynamiclly shaped bodies such as airfoil profiles or blades, a fairly good approximation of the circulation and hence the lift, may be obtained by the Kutta condition, assuming that no boundary layer separation occurs in the physical flow. The Kutta-Joukowski condition states that the value of the circulation which approximates best the real flow is obtained if the stagnation point at the downstream end of the body is located at the trailing edge.

The non-zero circulation requires the introduction of an artificial boundary or cut, emanating from the body to the farfield boundary, and over which a jump in the potential function is allowed. One has the following jump relations across the cut

$$|\phi| = \Gamma \quad \text{and} \quad |\partial_n \phi| = 0 \tag{9.9}$$

where n is the direction normal to the cut.

Limitations of the potential flow model for transonic flows.

For continuous flow, the potential model, given by the conditions of isentropicity and isenergeticity coupled to irrotationality forms a set of conditions fully compatible with the system of Euler equations. Therefore continuous inviscid flow, with an initial distribution satisfying the conditions $s=s_0$, $H=H_0$ and which is vortex-free will be exactly described by the potential model.

However, in presence of discontinuities such as shock waves, this is no longer valid, since the Rankine-Hugoniot relations lead to an entropy increase over a shock. If the shock intensity is uniform, then the entropy will remain uniform downstream of the shock, but at an other value than the initial constant value, although, according to eq. (7.4b) the flow remains irrotational. In addition, if the shock intensity is not constant, which is most likely to occur in practice, e.g. for curved shocks, then eq. (7.4b) shows that the flow is not irrotational anymore and hence the mere existence of a potential function downstream of the discontinuitiy cannot be justified rigorously.

Therefore, the potential model in presence of shock discontinuities cannot be made fully compatible with the system of Euler equations, since it contains no mechanisms to generate entropy variations over discontinuities [22], [23], [24].

The shock discontinuities allowed by the potential model are governed by isentropic shockrelations, where n is the normal and l the tangential direction to the discontinuity

$$|\phi|=0, \qquad |\vec{\nabla}\phi.\vec{1}_l|=0 \qquad |\rho\vec{\nabla}\phi.\vec{1}_n|=0 \tag{9.10}$$

The latter equation is the Rankine-Hugoniot relation for mass conservation however, with the density obtained from the isentropic relation (9.7). Figure 1 illustrates the difference between the Machnumbers downstream of a normal shock for the Rankine-Hugoniot relations compared with the isentropic relations (9.10).

The "physical" non-uniqueness of transonic isentropic potential models

The isentropic restriction of potential models introduces also another limitation connected to the fact that in transonic internal flows, an infinite number of equally valid solutions can be obtained for the same isentropic physical outlet variables such as isentropic backpressure or outlet Mach number. These solutions differ by the position of the isentropic shock, as is shown schematically for a one-dimensional nozzle in figure 2, [23], [24].

Similar non-uniqueness problems have been observed for external flows around airfoils, e.g. by Steinhoff and Jameson [25] and Salas et al. [26]. In these computations, multiple solutions where obtained with different shock positions for the same physical boundary conditions at the farfield.

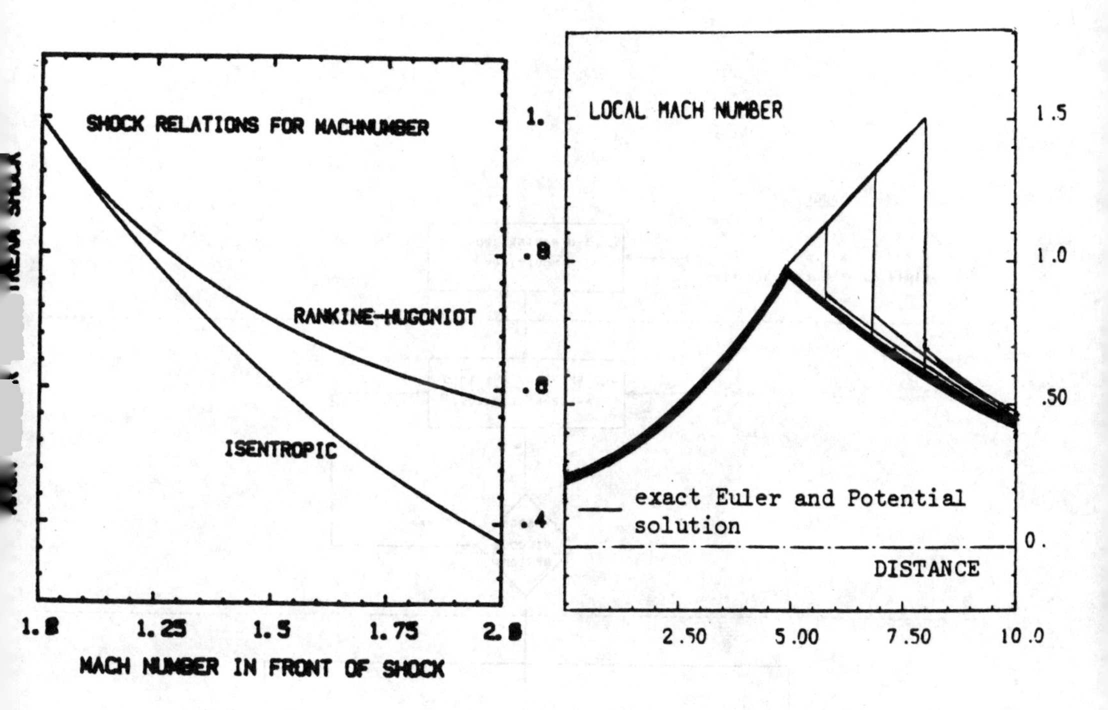

Figure 1 : Shock relations Figure 2 : Isentropic shock in Laval nozzle

The small disturbance approximation of the potential equation

In steady or unsteady transonic flows around wings and airfoils with thickness to chord ratios of a few percent, one can generally consider that the flow is predominantly directed along the chordwise direction, taken as the x-direction.

In this case the velocities in the transverse direction can be neglected and equation (9.4) reduces to the so-called small disturbance potential equation

$$(1 - M^2)\,\phi_{xx} + \phi_{yy} + \phi_{zz} = \frac{1}{a^2}\,(\phi_{tt} + 2\,\phi_x\,\phi_{xt}) \qquad (9.11)$$

Historically, the steady state, two-dimensional form of this equation was used by Murman & Cole (1971) to obtain the first numerical solution for a transonic flow around an airfoil with shocks.

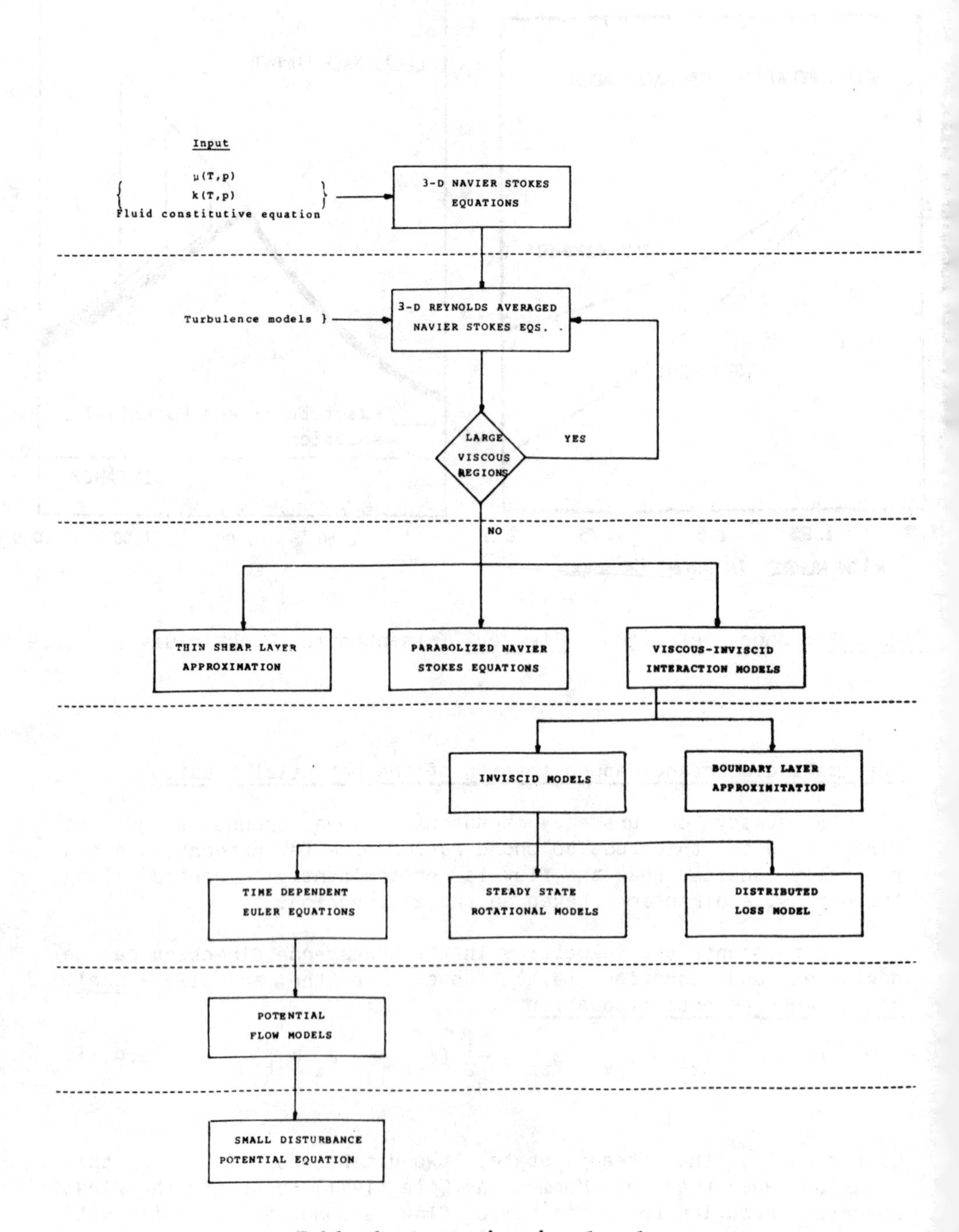

Table 1. Approximation levels

EQUATION	INTEGRAL FORM	DIFFERENTIAL FORM	
		CONSERVATION FORM	NON-CONSERVATION FORM
Conservation of mass	$\frac{\partial}{\partial t}\int_V \rho\, dV + \oint_S \rho\vec{v}\, d\vec{S} = 0$ (T2.1)	$\frac{\partial\rho}{\partial t} + \vec{\nabla}\rho\vec{v} = 0$ (T2.4)	$\frac{d\rho}{dt} + \rho\vec{\nabla}\vec{v} = 0$ (T2.7)
Conservation of momentum	$\frac{\partial}{\partial t}\int_V \rho\vec{v}\, dV + \oint_S (\rho\vec{v} \otimes \vec{v} + p\bar{\bar{I}} - \bar{\bar{\tau}})\, d\vec{S} = \int_V \rho\vec{f}_e\, dV$ (T2.2) $\tau_{ij} = \mu\ [\ (\partial_j v_i + \partial_i v_j) - \frac{2}{3}(\vec{\nabla}\vec{v})\ \delta_{ij}\ ]$	$\frac{\partial}{\partial t}\rho\vec{v} + \vec{\nabla}(\rho\vec{v} \otimes \vec{v} + p\bar{\bar{I}} - \bar{\bar{\tau}}) = \rho\vec{f}_e$ (T2.5)	$\rho\frac{d\vec{v}}{dt} = -\vec{\nabla}p + \vec{\nabla}\bar{\bar{\tau}} + \rho\vec{f}_e$ (T2.8) $\vec{\nabla}\bar{\bar{\tau}} = \mu\ [\ \Delta\vec{v} + \frac{1}{3}\vec{\nabla}(\vec{\nabla}\vec{v})\ ]$ CROCCO'S FORM $\frac{\partial\vec{v}}{\partial t} - \vec{v} \times \vec{\zeta} = T\vec{\nabla}s - \vec{\nabla}H + \frac{1}{\rho}\vec{\nabla}\bar{\bar{\tau}} + \vec{f}_e$ (T2.9)
Conservation of energy	$\frac{\partial}{\partial t}\int_V \rho E\, dV + \oint_S (\rho\vec{v}H - k\vec{\nabla}T + \bar{\bar{\tau}}\vec{v})\, d\vec{S}$ $= \int_V (\rho\vec{f}_e\vec{v} + q_H)\, dV$ (T2.3)	$\frac{\partial\rho E}{\partial t} + \vec{\nabla}\,(\rho\vec{v}H - k\,\vec{\nabla}T - \bar{\bar{\tau}}\vec{v}) = W_f + q_H$ (T2.6)	$\rho\frac{de}{dt} = -p\vec{\nabla}\vec{v} + (\bar{\bar{\tau}}.\vec{\nabla})\vec{v} + \vec{\nabla}(k\vec{\nabla}T) + q_H$ (T2.10) or $\rho\frac{dH}{dt} = \frac{\partial p}{\partial t} + \vec{\nabla}(k\vec{\nabla}T + \bar{\bar{\tau}}.v) + W_f + q_H$ (T2.11) or $\rho\frac{dH}{dt} = \frac{\partial p}{\partial t} + \rho T\frac{ds}{dt} + \vec{v}(\vec{\nabla}.\bar{\bar{\tau}}) + W_f$ (T2.12) ENTROPY EQUATION $\rho T\frac{ds}{dt} = \varepsilon + \vec{\nabla}(k\vec{\nabla}T) + q_H$ (T2.13)
Definitions	$H = h + \frac{\vec{v}^2}{2}$ stagnation enthalpy $E = e + \frac{\vec{v}^2}{2}$ total energy	$W_f = \rho\vec{f}_e\vec{v}$ work of external forces $\vec{f}_e$ q_H heat sources	$\vec{\zeta} = \vec{\nabla} \times \vec{v}$ vorticity $\varepsilon = \frac{1}{2\mu}(\bar{\bar{\tau}} \otimes \bar{\bar{\tau}}) = (\bar{\bar{\tau}}\vec{\nabla})\vec{v}$ viscous dissipation

Table 2 : The system of flow equations in an absolute frame of reference

EQUATION	INTEGRAL FORM	DIFFERENTIAL FORM	
		CONSERVATIVE FORM	NON-CONSERVATIVE FORM
Conservation of mass	$\frac{\partial}{\partial t}\int_V \rho\, dV + \oint_S \rho\vec{w}\, d\vec{S} = 0$ (T3.1)	$\frac{\partial\rho}{\partial t} + \vec{\nabla}\rho\vec{w} = 0$ (T3.4)	$\frac{d\rho}{dt} + \rho\vec{\nabla}\vec{w} = 0$ (T3.7)
Conservation of momentum	$\frac{\partial}{\partial t}\int_V \rho\vec{w}\, dV + \oint_S (\rho\vec{w} \otimes \vec{w} + p\bar{\bar{I}} - \bar{\bar{\tau}})\, d\vec{S}$ $= \int_V \rho[\vec{f}_e - 2\,\vec{\omega} \times \vec{w} - \vec{\omega} \times (\vec{\omega} \times \vec{r})]\, dV$ (T3.2) $\tau_{ij} = \mu(\partial_i w_j + \partial_j w_i) - \frac{2}{3}(\vec{\nabla}\vec{w})\,\delta_{ij}$	$\frac{\partial}{\partial t}\rho\vec{w} + \vec{\nabla}(\rho\vec{w} \otimes \vec{w} + p\bar{\bar{I}} - \bar{\bar{\tau}})$ (T3.5) $= \rho[\vec{f}_e - 2\,\vec{\omega} \times \vec{w} - \vec{\omega} \times (\vec{\omega} \times \vec{r})]$	$\rho\frac{d\vec{w}}{dt} = -\vec{\nabla}p + \vec{\nabla}\bar{\bar{\tau}} - 2\rho(\vec{\omega} \times \vec{w}) - \rho\vec{\omega} \times (\vec{\omega} \times \vec{r}) + \rho\vec{f}_e$ (T3.8) CROCCO'S FORM $\frac{\partial\vec{w}}{\partial t} - \vec{w} \times \vec{\zeta} = T\vec{\nabla}s - \vec{\nabla}I + \frac{1}{\rho}\vec{\nabla}\bar{\bar{\tau}} + \vec{f}_e$ (T3.9)
Conservation of energy	$\frac{\partial}{\partial t}\int_V \rho E^{*}\, dV + \oint_S (\rho\vec{w}I - k\vec{\nabla}T - \bar{\bar{\tau}}\vec{w})\, d\vec{S}$ $= \int_V (\rho\vec{f}_e\vec{w} + q_H)\, dV$ (T3.3)	$\frac{\partial}{\partial t}\rho E^{*} + \vec{\nabla}(\rho\vec{w}I - k\vec{\nabla}T - \bar{\bar{\tau}}.\vec{w}) = W_f + q_H$ (T3.6)	$\rho\frac{dI}{dt} = \frac{\partial p}{\partial t} + \vec{\nabla}(k\vec{\nabla}T + \bar{\bar{\tau}}.\vec{w}) + W_f + q_H$ (T3.10) ENTROPY EQUATION $\rho T\frac{ds}{dt} = \vec{\nabla}(k\vec{\nabla}T) + \varepsilon + q_H$ (T3.11)
Definitions	$E^{*} = e + \frac{\vec{w}^2}{2} - \frac{\vec{u}^2}{2} = E - \vec{u}\vec{v}$ $I = h + \frac{\vec{w}^2}{2} - \frac{\vec{u}^2}{2} = H - \vec{u}\vec{v}$ rothalpy $\vec{u} = \vec{\omega} \times \vec{r}$ $\vec{v} = \vec{u} + \vec{w}$	$W_f = \rho\vec{f}_e\vec{w}$ work of external forces $\vec{f}_e$	$\vec{\zeta} = \vec{\nabla} \times \vec{v}$ absolute vorticity $\varepsilon = \frac{1}{2\mu}(\bar{\bar{\tau}} \otimes \bar{\bar{\tau}}) = (\bar{\bar{\tau}}\vec{\nabla})\,\vec{w}$ viscous dissipation q_H heat sources

Table 3 : The system of flow equations in a steadily rotating relative frame of reference

REFERENCES

1. Chapman D.R., Computational Aerodynamics Development and Outlook, AIAA Journal, Vol 17 (1979), 1293-1313
2. Kutler P., A Perspective of Theoretical and Applied Computational Fluid Dynamics, AIAA Paper-83-0037, AIAA 21st Aerospace Sciences Meeting, 1983
3. Kline et al., Proceedings of the 1980-1981 AFOSR/Stanford Conference on Complex Turbulent Flows, Vol I,II,III, Stanford University, Stanford, California, 1982
4. Le Balleur J.C., Progres dans le calcul de l'interaction fluide parfait fluide visqueux, AGARD CP 351 on "Viscous Effects in Turbomachines, 1983
5. Thompson J.F., A Survey of Grid Generation Techniques in Computational Fluid Dynamics, AIAA Paper 83-0447, AIAA 21st Aerospace Sciences Meeting (1983), see also AIAA Journal, Vol 22 (1984), 1505-1523
6. Rogallo R.S., Moin P., Numerial simulation of turbulent flows, Ann. Rev. Fluid Mech. 1984.16:99-137
7. Moin P., Probing turbulence via large eddy simulation, AIAA Paper 84-017 1984
8. Cebeci T., Smith A.M.O., Analysis of Turbulent Boundary Layers, Academic Press, New York, 1974
9. Pulliam T.H., Steger J.L., On Implicit Finite Difference Simulations of Three Dimensional Flows, AIAA Paper 78-10 (1978), AIAA 16th Aerospace Science Meeting
10. Steger J.L., Coefficient Matrices for Implicit Finite Difference Solution of the Inviscid Fluid Conservation law Equations, Computer Methods in Appl. Mech. Eng., Vol 13, 175-188 (1978)
11. Patankar S.V., Numerical Heat Transfer and Fluid Flow, Hemisphere Publ.Co Mc Graw Hill (1980)
12. Cebeci T., Bradshaw P., Physical and Computational Aspects of Convective Heat Transfer, Springer Verlag, New York, 1984
13. Dring R.P., Joslyn H.D., Hardin L.W., An Investigation of Compressor Roto Aerodynamics, Transaction ASME, Journal of Engineering for Power, Vol 104 (1982), 84-96
14. Horlock J.H., Marsh H., Flow Models for Turbomachines, Journal of Mechanical Engineering Science, Vol 13 (1971), 358-368
15. Bosman C., Marsh H., An Improved Method for Calculating the Flow in Turbomachines, including a consistent loss Model, Journal Mechanical Engineering Science, Vol 16 (1974), 25-31
16. Lacor C., Hirsch Ch., Non-viscous three-dimensional Rotational flow Calculation in Blade Passages, Proceedings of 4th GAMM Conf. on Num, Meth. in Fluid Mechanics (1981) 150-161
17. Casal P., Principes Variationnels en Fluide Compressible et en Magnéto dynamique des Fluides, J. de Mécanique, Vol 5, n°2 (1966), 149-161
18. Hawthorne W.R., Secondary Vorticity in Stratified Compressible

Fluids in Rotating Systems, CUED/A - Turbo/TR63, 1974
19. Akay H.U., Eur A., Application of a Finite Element Algorithm for the Solution of Steady Transonic Euler Equation, AIAA Paper 82-0970 (1982)
20. Lacor C., Hirsch Ch., Rotational Flow Calculations in three-dimensional Blade Passages, ASME Paper 82-GT-316 (1982)
21. Lacor C., Hirsch Ch., Three-dimensional, inviscid, rotational flow Calculations in turbomachines, Proceedings 15th Int. Congress on Combustion Engines, CIMAC, 1983
22. Klopfer G.H., Nixon D., Non-isentropic potential formulation for Transoni Flows, AIAA Paper 83-0375 (1983)
23. Deconinck H., Hirsch Ch., Boundary Conditions for the Potential Equation in transonic internal Flow Calculations, ASME Paper 83-GT-135 (1983)
24. Habashi W.G., Hafez M.M., Kotiuga P.L., Finite Element Methods for Internal Flow Calculations, AIAA Paper 83-1404 (1983)
25. Steinhoff J., Jameson A., Multiple Solution of the Transonic Potential Flow Equation, AIAA Journal, Vol 20, n°11 (1981), 1521-1525
26. Salas M.D., Jameson A., Melnik R.E., A Comparative Study of the non-uni- queness problem of the potential equation, AIAA Paper 83-1888 (1983)

NOMENCLATURE

C_p, C_v	specific heats
$\vec{f}_e$	external forces
$\vec{F}_b$	Friction force
F	heat diffusive flux
H	stagnation enthalpy
h	enthalpy
E	total energy
e	internal energy
I	rothalpy
k	heat conductivity, turbulent kinetic energy
P	production of turbulent energy
p	pressure
q_H	heat sources
s	entropy
T	temperature
$\vec{u}$	rotational velocity of relative reference system

$\vec{v}$ absolute three-dimensional velocity vector with Cartesian components u,v,w

W_f work of the external forces

$\vec{w}$ relative velocity

γ ratio of specific heats

ε viscous dissipation

$\vec{\zeta}$ vorticity in absolute reference system

κ diffusivity coefficient

μ dynamic viscosity

ν kinematic viscosity

ρ density

$\bar{\bar{\tau}}$ viscous stress tensor

ϕ potential function

ψ vorticity function

$\vec{\omega}$ angular velocity of relative reference system

MODELLING VISCOUS FLOWS IN TURBOMACHINERY

Peter Stow

Chief Aerodynamic Scientist
Rolls-Royce Ltd.
Derby
England

1. INTRODUCTION

A turbomachinery blade designer's objective is to produce blades that satisfy a specific duty with acceptable levels of efficiency and component life. It is therefore essential that viscous phenomena are modelled in the blade design system that he uses from aerodynamic, loss production and heat transfer points of view. Traditionally blade design systems are centred around the use of quasi-three-dimensional through-flow and blade-to-blade programs see Fig. 1. The through-flow program provides radial distributions of inlet and exit conditions, e.g. Mach number and whirl angle, to be used as boundary conditions for the design of blade sections. In a linked quasi-three-dimensional blade design system, see for example Jennions and Stow [1] and Hirsch and Warzee [2], blading information in the form of blockage, blade lean, loss and turning is needed in the through-flow analysis. As well as obviously influencing the outlet conditions produced, this information is important in determining the variations of streamline radius and stream-tube height through the blade row. The sections are designed, in isolation using a blade-to-blade program such that the inlet and exit conditions are achieved, levels of loss and, where appropriate, heat transfer are acceptable and any imposed geometric constraints are satisfied. The sections are then stacked radially and circumferentially to form a three-dimensional blade geometry taking into account further geometrical aerodynamic or stress constraints.

With a linked system the blading information is passed back to the through-flow program ready for the next iteration. This process is continued (often automatically) until an acceptable level of

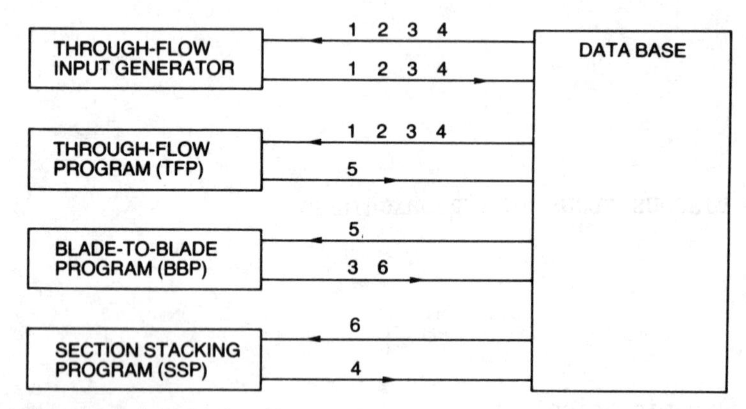

Fig. 1. Quasi - 3D blade design system

convergence between the two programs is achieved. Once a three-dimensional blade geometry has been designed then a full three-dimensional analysis can be performed in order to assess the importance of effects that cannot be modelled with a quasi-three-dimensional approach and modifications to the geometry are made on the basis of the results.

In such a blade design system it is important that each program in the system models phenomena that will affect the overall aerodynamics of the blade row. For example some of the phenomena associated with the annulus that need to be modelled are

(i) development of the annulus end-wall boundary layers
(ii) mixing of disc leakage flows
(iii) mixing of ejected end-wall cooling flows.

Some associated with the blade are

(i) development of the blade surface boundary layers and trailing edge mixing region
(ii) development and progression of the blade wake
(iii) mixing of ejected blade cooling flows e.g. film-cooling, trailing edge ejection.

Others associated with the three-dimensional nature of the flow or with blade-annulus interaction are

(i) mixing of over-tip leakage flow and of tip ejected cooling air
(ii) leading edge horse-shoe vortex
(iii) secondary flow

The following sections discuss how some of the above effects can be included in the through-flow and the blade-to-blade analyses.

2. THROUGH-FLOW

In a through-flow calculation the equations of motion to be solved are simplified in a number of ways. Firstly the flow is taken as steady in an absolute co-ordinate system for a nozzle or stator and in a relative co-ordinate system rotating with the blade speed for a rotor. In addition outside of the blade rows the flow is taken to be axisymmetric, which means that the effects of wakes from an upstream blade row are assumed to mix out to give uniform circumferential conditions. Inside the blade rows the effects of the blades are modelled using either a mean stream-sheet approach or using a passage averaging technique.

For example, for steady flow in a relative co-ordinate system the three-dimensional momentum equations are

$$\underline{W}.\nabla\underline{W} + 2\underline{\Omega}\wedge\underline{W} - \Omega^2\underline{R} = -\frac{1}{\rho}\nabla p + \underline{F}_\tau \tag{1}$$

where $\underline{W}$ is the relative velocity vector and Ω is the blade speed and $\underline{F}_\tau$ is a dissipative force introduced to simulate losses through the blade row. The passage averaged form of the equations is

$$\tilde{W}_x \frac{\partial \tilde{W}_x}{\partial x} + \tilde{W}_R \frac{\partial \tilde{W}_x}{\partial R} = -\frac{1}{\bar{\rho}}\frac{\partial \bar{p}}{\partial x} + F_{B_x} + \tilde{F}_{\tau_x} - P_x \tag{2}$$

$$\tilde{W}_x \frac{\partial \tilde{W}_R}{\partial x} + \tilde{W}_R \frac{\partial \tilde{W}_R}{\partial R} - \frac{1}{R}(\tilde{W}_\theta + \Omega R)^2 = -\frac{1}{\bar{\rho}}\frac{\partial \bar{p}}{\partial R} + F_{B_R} + \tilde{F}_{\tau_R} - P_R \tag{3}$$

and

$$\tilde{W}_x \frac{\partial \tilde{W}_\theta}{\partial x} + \tilde{W}_R \frac{\partial \tilde{W}_\theta}{\partial R} + \frac{\tilde{W}_R}{R}(\tilde{W}_\theta + 2\Omega R) = F_{B_\theta} + \tilde{F}_{\tau_\theta} - P_\theta \tag{4}$$

where "-" denotes a passage averaged value and "~" denotes a density weighted passage average; $\underline{F_B}$ is the blade force and P involves averages of products of perturbations from the means, see Jennions and Stow [1] for more details.

The continuity equation is

$$\nabla.\rho\underline{W} = 0 \tag{5}$$

which after passage averaging becomes

$$\frac{\partial}{\partial x}(BR\bar{\rho}\tilde{W}_x) + \frac{\partial}{\partial R}(BR\bar{\rho}\tilde{W}_R) = 0 \tag{6}$$

where

$$B = 2\pi - \frac{Nt}{R} \tag{7}$$

where N is the number of blades and t is the blade thickness.

The above equations (2), (3), (4) and (6) together with an energy equation can be solved using a number of techniques e.g. streamline curvature, matrix through-flow, time-marching or finite elements. However, in order to provide an adequate description of the real flow it is necessary to supplement the simplified analysis using a number of models.

2.1 Blade Loss Model

One commonly adopted procedure for accounting for viscous losses through a blade row is to include a dissipative force $\underline{F}_\tau$ in the equations of motion, as in equations (1), and relate this to entropy production.

It can be shown that

$$t\underline{W}.\nabla S = \underline{W}.\nabla I - \underline{W}.\underline{F}_\tau \tag{8}$$

where I is the rothalpy defined by

$$I = C_p t + \frac{1}{2} W^2 - \frac{1}{2} \Omega^2 R^2$$

t being the static temperature. For adiabatic flow the energy equation is taken as

$$\underline{W}.\nabla I = 0 \qquad (9)$$

so that

$$\underline{W}.\underline{F}_{\tau} = -t\ \underline{W}.\nabla S$$

If in addition the dissipative force is taken to act in the streamline direction i.e.

$$\underline{F}_{\tau} = F_{\tau}\ \underline{s} \qquad (10)$$

then it follows that

$$F_{\tau} = -\ t\frac{\partial S}{\partial s}$$

Assuming that perturbation terms are negligible this gives

$$\tilde{F}_{\tau} = -\tilde{t}\ \frac{\partial \tilde{S}}{\partial s} \qquad (11)$$

from which the three components can be found for use in equations (2), (3) and (4).

The distribution of entropy through a blade can be obtained from a blade-to-blade analysis. Alternatives are to use loss correlations or to use experimental data in the form of loss-incidence curves for particular blade designs.

An alternative procedure for modelling the effects of the blade losses is to use the blade surface boundary layer displacement thicknesses in the form of a blockage in the through-flow calculation. For example, equation (7) becomes

$$B = 2\pi - N\frac{1}{R}(t + \delta_s^* + \delta_p^*) \qquad (12)$$

where δ_s^* and δ_p^* are the suction and pressure surface displacement thicknesses. In this way the effect of the wake development downstream of the blade trailing edge can be accounted for. Note, however, that it will have to be assumed that the flow is circumferentially uniform by the leading edge of the following blade row.

Even with the blade boundary layer blockage model it is usual to account for any additional end-wall or secondary flow losses using the dissipative force model.

2.2 End-wall Boundary Layer

The usual manner in which the effects of the end-wall boundary layers are included is to calculate the boundary layer development using an integral method and represent the effects using a displacement model. In this the annulus is altered by the displacement thickness of the boundary layer and the inviscid flow calculated within the modified geometry.

The governing passage-averaged equations can be derived in a manner similar to that in Section 2.1. The boundary layer equations are first written in terms of a local meridional streamline coordinate system with the usual boundary layer assumptions that only gradients normal to the meridional direction are retained in the viscous terms. The equations can then be passage averaged by integrating from one blade to the next (skin friction effects on the blade surfaces being ignored). For example the momentum integral equations become, see Stow [3]

$$\frac{1}{BR_b}\frac{d}{dm}[BR_b\rho_e Q_e^2\theta_{mm}] + \rho_e Q_e \delta_m^* \frac{dq_{me}}{dm}$$
$$-\rho_e \frac{\sin\lambda_b}{R_b}(Q_e^2\theta_{\theta\theta} + q_{\theta_e} Q_e \delta_\theta^*) = F_m + (\tau_m)_W + P_m \qquad (13)$$

$$\frac{1}{BR_b}\frac{d}{dm}[BR_b\rho_e Q_e^2\theta_{\theta_m}] + \rho_e Q_e \delta_m^* \frac{dW_{\theta_e}}{dm}$$
$$+ \rho_e \frac{\sin\lambda_b}{R_b} Q_e (Q_e\theta_{m_\theta} + q_{m_e}\delta_\theta^*) = F_\theta + (\tau_\theta)_W + P_\theta \qquad (14)$$

where m is the meridional co-ordinate see Fig. 2 R_b is the radius of the annulus, B is given by equation (7), F_m and F_θ are the components of the blade force defect and represent integrals through the boundary layer of differences of the blade force from the value at the edge of the boundary layer; τ_w is the wall shear stress, P_m and P_θ represent perturbation terms, Q is the absolute velocity given by

$$Q^2 = q_m^2 + q_\theta^2$$

W is the relative velocity and the subscript e denotes quantities at the edge of the boundary layer; it should be noted that the "-" notation has been dropped in the above equation. Appendix A gives the definitions of the momentum and displacement thicknesses.

The entrainment equation becomes

$$\frac{d}{dm}\left[BR_b\rho_e(q_{m_e}\delta - Q_e\delta_m^*)\right] = BR_b\rho_e Q_e C_E \qquad (15)$$

It can be seen that in cases where the effects of the hade angle λ can be ignored the two momentum integral equations become uncoupled. If in addition the effects of averages of products of perturbations are ignored i.e. $P_m = 0 = P_\theta$, and the blade force is taken as constant through the boundary layer i.e. $F_m = 0 = F_\theta$, then particularly simple forms result. For example equation (13) becomes

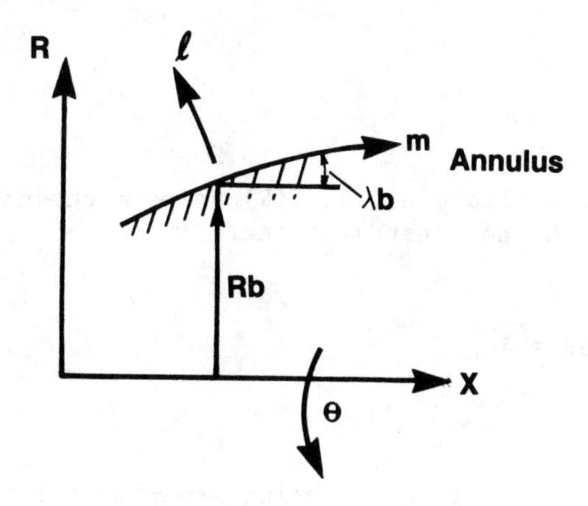

Fig. 2. Meridional coordinate system

$$\frac{1}{B}\frac{d}{dm}[B\rho_e Q_e^2 \theta_{mm}] + \rho_e Q_e \delta_m^* \frac{dq_{m_e}}{dm} = (\tau_m)_w \qquad (16)$$

see Marsh and Horlock [4]. This equation can be solved together with the entrainment equation (15) assuming the usual two-dimensional boundary layer correlations to apply in the streamwise direction.

More sophisticated treatments including the effects of the blade force defect and cross-flows in the boundary layers are given for example by De Ruyck and Hirsch [5] and De Ruyck, Hirsch and Kool [6]; see also Le Boeuf [7].

2.3 Mixing Models

The effects of the mixing of various leakage flows, e.g. disc leakage, over-tip leakage, and of various ejected cooling flows needs to be taken into account in the through-flow analysis. As well as being a source of loss they can also affect the distribution of flow quantities, e.g. total temperature, pressure, whirl velocity, which are obviously important in determining the inlet flow conditions for the following blade row.

One approximate manner of accounting for the effects is to perform a simple mixing calculation, for example one-dimensional mixing in stream-tubes assuming the ejected flow, in terms of mass, momentum and energy, to be distributed radially between the stream-tubes. So for example the distribution of ejected mass flow may be taken as

$$m(R) = M_c . f(R)$$

where M_c is the total mass flow ejected, f(R) being a chosen distribution function with the constraint that

$$2\pi \int_{R_{hub}}^{R_{tip}} R.f(R)\, dR = 1$$

The distribution function may be chosen using experimental information or using a more sophisticated viscous calculation to analyse the configuration. In theory the model can also account for axial mixing effects by making f a function of axial distance.

An alternative model to account for the effects of over-tip leakage is given by Adkins and Smith [8]. In the case of an unshrouded blade they calculate secondary flow vorticity based on the model of Lakshminarayan and Horlock [9] and use this in a secondary flow calculation to determine secondary flow velocities, see later.

The effects of film cooling flows and trailing edge ejected cooling flows are best handled in a blade-to-blade analysis, see later, and the information passed back to the through-flow calculation in terms of for example mass flow and blockage; three-dimensional mixing effects are obviously ignored with this approach.

2.4 Secondary Flow Model

A quasi-three-dimensional blade design system assumes that stream-surfaces through a blade row are axisymmetric. In addition in calculating blade profile losses it is assumed that the boundary layer develops on the blade surface in a two-dimensional manner i.e. cross-flows in the boundary layer are ignored. In practice the stream-surfaces twist as they pass through the blade row so that there is flow through the blade-to-blade surfaces used, also boundary layers will be three-dimensional in nature. In addition the annulus end-wall boundary layer will separate near the blade leading edge to produce a horse-shoe vortex. The pressure side leg of this vortex tends to move across the passage and interact with the suction side leg; the blade surface boundary layers are also affected by the end-wall interaction, the extent being determined by the aspect ratio of the blade.

Secondary flow is a term often used to encompass all the three-dimensional effects mentioned which are not included in a quasi-three-dimensional approach. These effects give rise to additional or secondary losses as well as affecting the distribution of parameters such as total pressure, whirl angle etc., at exit to the blade row.

Secondary losses are usually accounted for in the through-flow analysis using a dissipative force model together with correlations or experimental experience. One simple way of attempting to account for the effects on the mean passage whirl angle is to use a distributed deviation on the predicted blade-to-blade flow. An alternative is to use inviscid secondary flow theory, see for example Lakshminarayan and Horlock [10], Came and Marsh [11] and Smith [12]. In this the quasi-three-dimensional through-flow is taken as the primary flow to convect inlet vorticity in the end-wall boundary layers giving rise to streamwise vorticity at the exit to the blade row; the effects of the secondary flow on the primary flow are assumed small. For example with the Came and Marsh theory the streamwise vorticity ξ_{sec} at exit to a blade row is given by

$$\xi_{sec} = \xi_{s_1} \frac{\cos\alpha_1}{\cos\alpha_2} + \frac{\xi_{n_1}}{\cos\alpha_1 . \cos\alpha_2} [\alpha_2 - \alpha_1 + \frac{1}{2} (\sin 2\alpha_2 - \sin 2\alpha_1)] \quad (17)$$

where ξ_{s_1} and ξ_{n1} are the inlet streamwise and normal vorticities and α_1, and α_2 are the inlet and exit whirl angles. It should be noted that the above model is derived for incompressible flow in a linear cascade, where the primary flow is irrotational, but can be extended, as can that of Smith, to include the effects of compressibility, radius change and blade rotation, see James [13].

Once the distribution of exit streamwise vorticity is known the secondary flow velocities can be determined by solving the continuity equation in terms of a secondary flow stream-function. For example for incompressible flow the stream-function ψ satisfies

$$\frac{\partial^2\psi}{\partial n^2} \quad \frac{\partial^2\psi}{\partial z^2} = \xi_{sec} \quad (18)$$

see Fig. 3. for the co-ordinate system and Appendix B for details of the analysis.

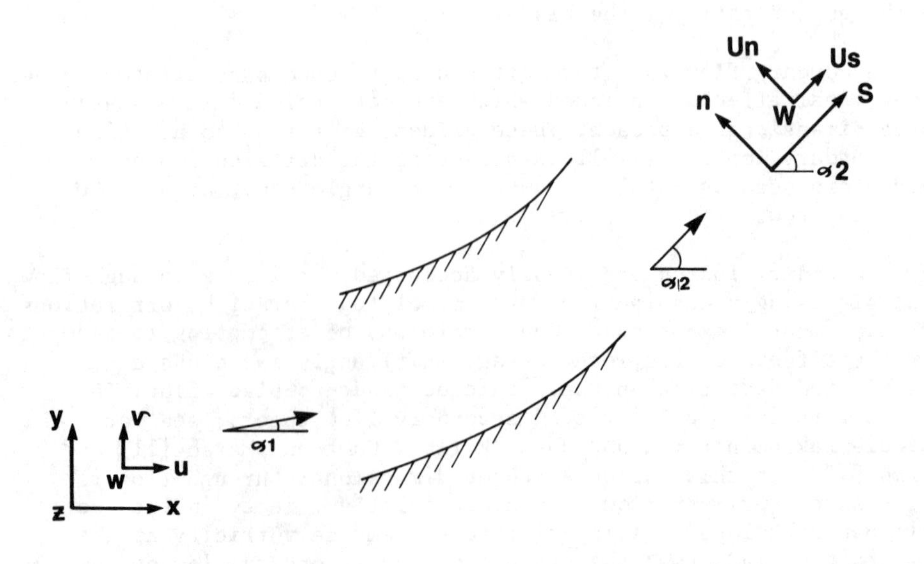

Fig. 3. Streamline coordinate system

The results produced from inviscid secondary flow models are very dependent on the inlet conditions assumed, especially the inlet streamwise vorticity. This is particularly important when considering stage calculations as illustrated by the following example. The case considered is a compressor stage described by Freeman and Dawson [14] consisting of an inlet guide vane, rotor and stator all double circular arc blades. Fig. 4 shows results for the inlet guide vane from James [15] using a method based on that of Came and Marsh. It can be seen that the radial distribution of the pitchwise mass averaged whirl angle is well predicted. In this case the inlet boundary layer was taken as collateral. We consider now predictions for the stator. It can be seen from Fig. 5 that even for the case of a collateral hub boundary layer out of the upstream rotor the absolute inlet velocity will exhibit skew because of the change from a rotating to a stationary hub. This gives rise to streamwise vorticity at inlet to the stator; the effects of secondary flow in the upstream rotor will add to this effect. If the streamwise vorticity is included in the analysis, as in equation (17), then it can be seen from Fig. 6 that poor predictions result near the hub and tip where considerable underturning is predicted and the sense of the inlet skew is maintained. The reason for this is the effects of viscosity. Considering again Fig. 5 it can be seen that since the velocity on the stationary hub is zero then intense shearing will take place in the region near to the hub which will tend to

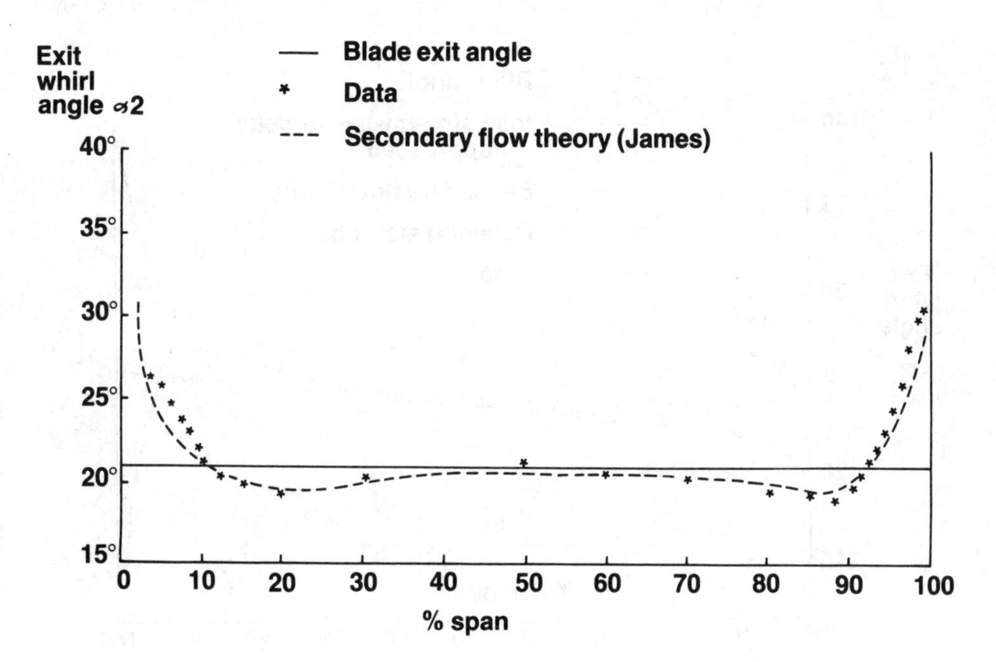

Fig. 4. Compressor inlet guide vane

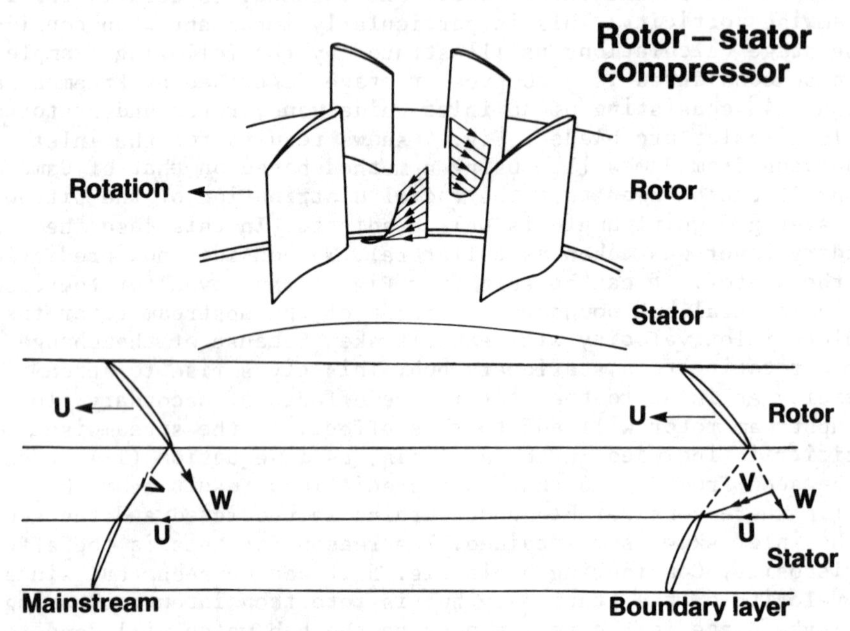

Fig. 5. Boundary layer skewing

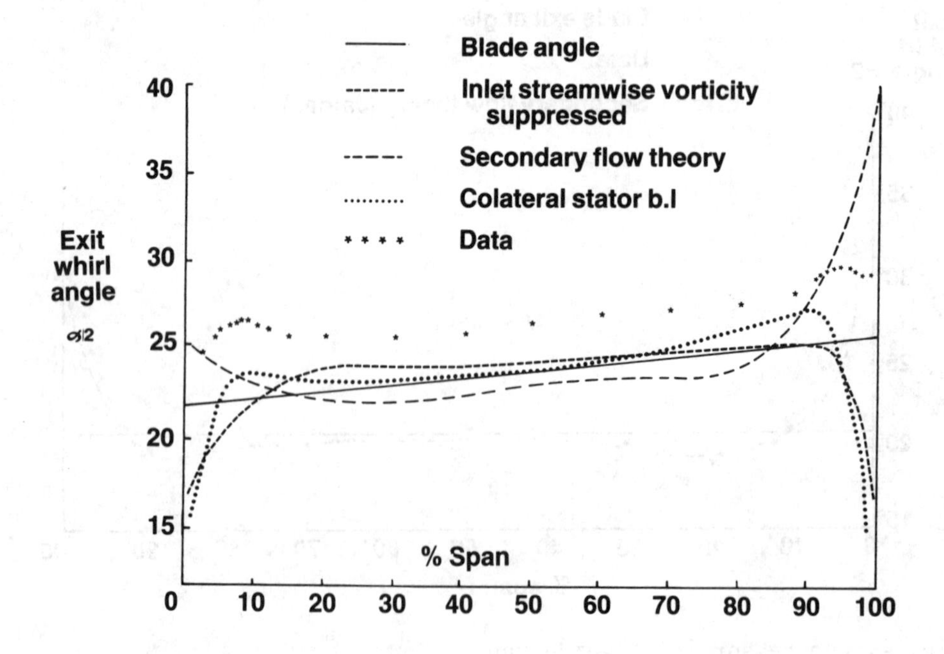

Fig. 6. Compressor stator

unskew the inlet boundary layer reducing the streamwise vorticity. Even if the skew is maintained up to the leading edge the process will continue through the blade row assisted in the case of a compressor rotor by the blade force. Fig. 6 shows results from the calculation where the inlet streamwise vorticity is suppressed. It can be seen that the trend is now improved even though the level is wrong in the middle of the stator (it is felt that this is caused either by errors in the experimental measurements or deviation that is not taken into account in the analysis). Also shown in Fig. 6 are results assuming that the inlet boundary layer is collateral in the stator frame, the inlet skew being completely destroyed by viscosity. Again even though the overturning is exagerated and the mid-passage level is incorrect the trends are encouraging. The results shown and the mechanisms described above have been substantiated by Birch [16] using a three-dimensional viscous program for this example. The conclusion is that care must be exercised in using inviscid secondary flow theory for stage calculations. In fact this is true for any inviscid calculation where the effects of an inlet boundary layer are being simulated.

In the case of a turbine nozzle or rotor the effect of viscosity is again to tend to destroy the inlet skew from an upstream blade row but now the effect of the blade force is opposite to that of the compressor stator or rotor and tends to increase the skew. It is found that the latter effect tends to dominate and as a consequence good predictions can be obtained using an inviscid program, see for example the work of Boletis, Sieverding and Van Hove [17] and Birch [18] using three-dimensional programs. It should be mentioned that for a turbine the effects of the secondary flow on the "primary" flow can be large and that because these are ignored in an inviscid secondary flow analysis the results must be used with care (these effects are obviously included in a full three-dimensional analysis).

2.5 Spanwise Mixing Model

A spanwise mixing model is aimed at accounting for two effects,

(i) turbulent diffusion in the radial direction not accounted for in a blade-to-blade analysis,

(ii) an inviscid secondary flow effect arising from the fact that in general there will be a component of velocity normal to the axisymmetric blade-to-blade surfaces used in the design of the blade sections.

Consider firstly the second effect mentioned above. It is shown in Appendix C that, using the analysis of James [19], the passage averaged energy equation may be written as

$$\tilde{W}_m \frac{\partial \hat{I}}{\partial m} = - \frac{1}{BR\bar{\rho}} \frac{\partial}{\partial R} \left[\frac{BR}{\cos\tilde{\lambda}} \overline{\rho V_\ell I} \right] \tag{19}$$

where m is the meridional direction along the axisymmetric blade-to-blade surface see Fig. 2 and V_ℓ is the velocity normal to this surface. In the usual quasi-three-dimensional design system the blade-to-blade surface is assumed to be a streamsurface and V_ℓ is taken as zero, leading to the familiar energy equation.

In a similar manner it can be shown that the passage-averaged circumferential momentum equation becomes

$$\frac{\tilde{W}_m}{R} \frac{\partial}{\partial m} (R\hat{q}_\theta) = - \frac{1}{BR^2\bar{\rho}} \frac{\partial}{\partial R} \left[\frac{BR}{\cos\tilde{\lambda}} \overline{\rho V_\ell R q_\theta} \right] - F_{B_\theta} + \tilde{F}_{\tau_\theta} \tag{20}$$

It would be possible to determine V_ℓ using an inviscid secondary flow analysis but it should be noted that the distribution through a blade row is needed, at least in theory. An alternative approach is given in Appendix C where it is shown that equation (19) may be written as

$$\tilde{W}_m \frac{\partial \hat{I}}{\partial m} \simeq \frac{1}{BR\bar{\rho}} \frac{\partial}{\partial \ell} \left[BR \bar{\rho} \varepsilon \frac{\partial \hat{I}}{\partial \ell} \right] \tag{21}$$

where ε is related to the velocity normal to the blade-to-blade surface, being determined by the secondary flow through the blade row.

It is possible to include the effects of turbulent diffusion in the above model by considering ε to be an effective diffusion coefficient. In order to determine this using experimental data a more appropriate form is

$$\frac{\partial \hat{I}}{\partial m} \simeq \varepsilon^* \frac{\partial^2 \hat{I}}{\partial \ell^2} \tag{22}$$

Adkins and Smith [8] derive equation (22) using a Taylor expansion analysis based effectively on tracing streamlines through the blade row taking into account secondary flow. They use diffusion equations for total temperature, total pressure and Rq_θ, determining the mixing coefficients empirically to account also for the effects of viscosity on the secondary flow and interaction with downstream blade rows. They show that the model has a significant effect on predictions, see Fig. 7 .

3. BLADE-TO-BLADE ANALYSIS

3.1 Boundary Layer Model

In a quasi-three-dimensional blade-to-blade analysis the flow on an isolated axisymmetric stream-surface is considered. It is usual to adopt an inviscid model with a coupled boundary layer analysis to account for the effects of viscosity on the mainstream flow (blockage, deviation etc.) and to calculate profile losses although fully viscous methods based on the solution of the Navier-Stokes equations are under development.

Integral and finite difference methods are available for boundary layer analysis the former being ideal for inclusion in interactive design systems because of the speed. The main features that need to be included in any analysis are as follows

(i) Laminar flow
(ii) Laminar separation and re-attachment
(iii) Transition - start and end
(iv) Turbulent flow
(v) Turbulent separation
(vi) Re-laminarization.

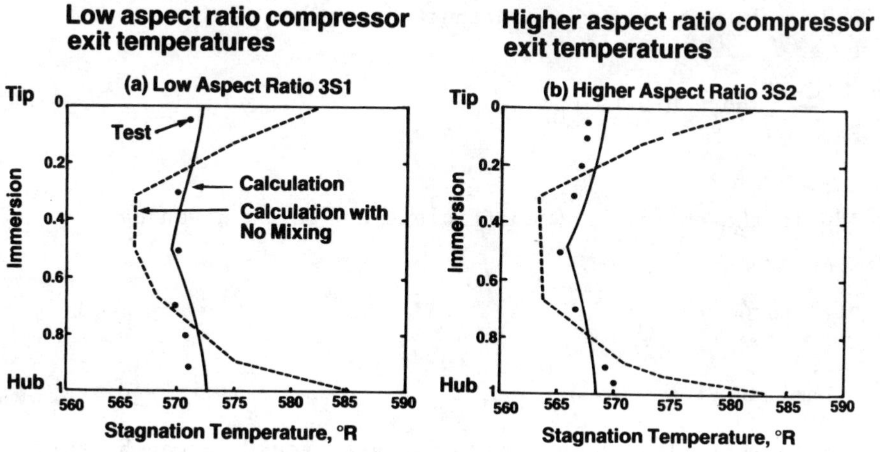

Fig. 7 . Exit temperature distributions from 3-stage compressors having different aspect ratios

The following discussions are centred around the integral method presented in Williams and Stow [20] in which direct mode coupling is adopted, see later.

3.1.1 Integral Equations

The governing integral equations are formed by integrating the boundary layer equations through the boundary layer. It is usual to assume that the blade surface boundary layers are two-dimensional i.e. boundary layer velocities normal to the blade-to-blade surface are ignored. In the case where the effects of blade surface curvature and of blade rotation can be ignored the momentum integral equation may be written as

$$\frac{1}{Rh}\frac{d}{ds}[Rh\rho_d u_d^2\theta] = \tau_w - (\rho u)_\delta \frac{du_\delta}{ds}.H.\theta \qquad (23)$$

where s is measured along the blade, u is the streamwise velocity, R is the streamline radius, h is the stream-tube height, τ_w is the wall shear stress, the subscript δ denotes conditions at the edge of the boundary layer, θ is the momentum thickness and H the form factor given by

$$H = \frac{\delta^*}{\theta}$$

δ^* being the displacement thickness. The entrainment equation, obtained by integrating the continuity equation, is

$$\frac{1}{Rh}\frac{d}{ds}[Rh\rho_\delta u_\delta H_1\theta] = C_E \qquad (24)$$

where C_E is the entrainment coefficient and H_1 is given by

$$H_1 = (\delta-\delta^*)/\theta$$

3.1.2 Laminar Flow

For laminar flow the momentum integral equation can be solved using the correlations of Luxton and Young [21] in which skin friction τ_w and form factor H are related to a single parameter dependent on du_δ/ds .

Suitable stagnation point starting conditions can be obtained by considering incompressible flow round a cylinder with circulation in which case θ is given by

$$\theta = 0.29234 \sqrt{\frac{\nu}{\frac{du_\delta}{ds}}} \tag{25}$$

where ν is the kinematic viscosity.

Equation (23) is solved using the supplied conditions at the edge of the boundary layer until either separation of transition is predicted.

3.1.3 Laminar Separation

Once laminar separation is predicted, either by zero skin friction or based on the Pohlhaussen parameter, then a laminar separation bubble is calculated using an empirical method due to Horton [22] and Ntim [23] with improved correlations by Roberts [24]. The calculation is in two parts. Firstly the length ℓ_1 from separation to transition in the separated shear shear layer is given by

$$\frac{\ell_1}{\theta_{sep}} = 2.5.10^4 \log_{10} [\coth (10.TF)] / R_e\theta_{sep} \tag{26}$$

where TF is the local fractional turbulence intensity. Secondly the length ℓ_2 from transition to re-attachment is given by

$$\frac{\ell_2}{\theta_{sep}} = \frac{B(1-q)}{q^4-C} \tag{27}$$

where

$$q = \frac{u_R}{u_s}$$

u_s being the mainstream velocity at the separation point, u_R being the value at the re-attachment point (determined by interpolation using $\ell_1+\ell_2$) and B and C are constants given by Roberts. Equation

(27) is solved iteratively for ℓ_2. The momentum thickness of re-attachment can be found once q is determined, see Roberts [24]. If the iteration for ℓ_2 fails to converge it is assumed that re-attachment does not occur. There is no really adequate treatment that can be adopted in this case and in order to continue the calculation it is assumed that immediate transition takes place at the separation point; the final solution must however be used with caution.

3.1.4 Transition

The start of transition is determined using modifications to the correlations given by Abu-Ghannam and Shaw [25] in which the critical Reynolds number is determined by the Pohlhaussen parameter λ where

$$\lambda = \frac{\theta^2}{\nu} \frac{du_\delta}{ds}$$

and the level of free-stream turbulence. Flow history effects are included by the way that λ is chosen for use in the correlation. The length of transition and the conditions at the end of transition (i.e. θ and H) are also given by Abu-Ghannam and Shaw. It is found that in many cases the transitional nature of the boundary layer is important in terms of loss prediction and must be included in the analysis.

3.1.5 Turbulent Flow

Turbulent flow is calculated using the lag-entrainment method of Green, Weeks and Brooman [26]. This method uses the momentum integral equation (23), together with the entrainment equation (24) and a lag equation for the entrainment coefficient C_E of the form

$$\theta \frac{dC_E}{ds} = C \tag{28}$$

where C is a function of θ, H and conditions at the edge of the boundary layer.

3.1.6 Turbulent Separation

The lag-entrainment method has been shown to give good predictions for attached flows in a direct mode of operation but for separated flows an inverse mode, must be used to avoid convergence problems with the inviscid calculation, see East, Smith and Merryman [27]; this point will be covered later. As boundary layer separation is approached the direct mode coupling becomes ill-conditioned and in order to be able to continue the solution scheme it is necessary to limit the growth of the displacement thickness usually by placing an upper limit on the form factor H or modified form factor $\bar{H}$. In a design situation where one can modify the blade shape in order to avoid separation this treatment is perfectly adequate. In an analysis situation, however, the results must be used with caution when such a measure has been adopted.

3.2 Boundary Layer Coupling

There are a number of ways of coupling the boundary layer calculation with the inviscid mainstream calculation and these will be covered in later lectures. Only direct mode coupling is considered here, see Fig. 8. The effect of the boundary layer on the mainstream is usually modelled using either a displacement or transpiration model. In the former the blade is thickened by the boundary layer displacement thickness and the inviscid flow calculated within the modified blade geometry. In the latter model mass, momentum and energy are transpired through the blade surface to simulate the effects of the boundary layer on the mainstream. As the displacement model usually entails re-calculating the grid as the displacement thickness changes it is only useful where grid

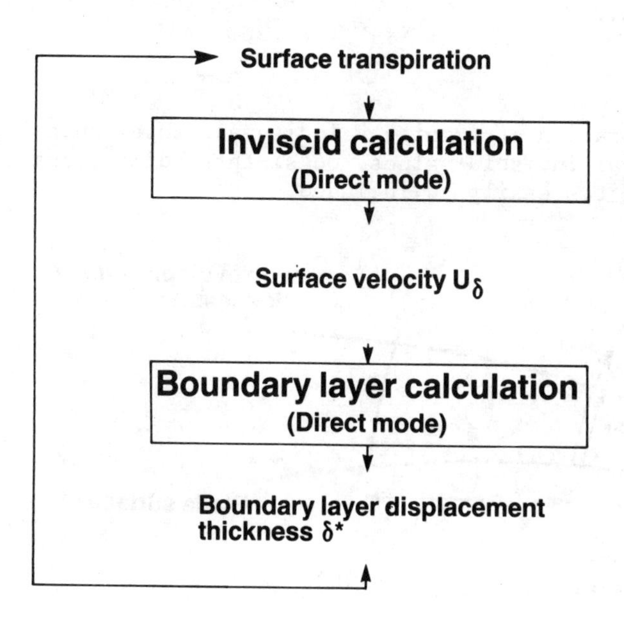

Fig. 8. Direct mode boundary layer coupling

generation is fast or where the use of an effective blade streamline is needed in the inviscid method e.g., streamline curvature. The transpiration model has wider applicability and can be used in methods which solve the potential flow or Euler equations whether they be finite volume or finite element; with finite element methods the model avoids mesh re-generation which can be costly. In the case of two-dimensional flow, where the effects of changes in streamline radius and stream-tube height and of blade rotation are ignored it can be shown that the transpiration velocity u_n normal to the blade surface is given by

$$(\rho u_n)_o = \frac{d}{ds}[(\rho u_s)_\delta \, \delta^*] \tag{29}$$

see Fig. 9 , where the subscript '0' refers to conditions at the blade surface. From consideration of streamwise momentum and energy it can be shown that

$$(u_s)_o = (u_s)_\delta \tag{30}$$

and

$$H_o = H_\delta \tag{31}$$

i.e. the transpiration streamwise velocity and total enthalpy are taken as the local inviscid values. Quasi-three-dimensional and rotation effects can easily be included.

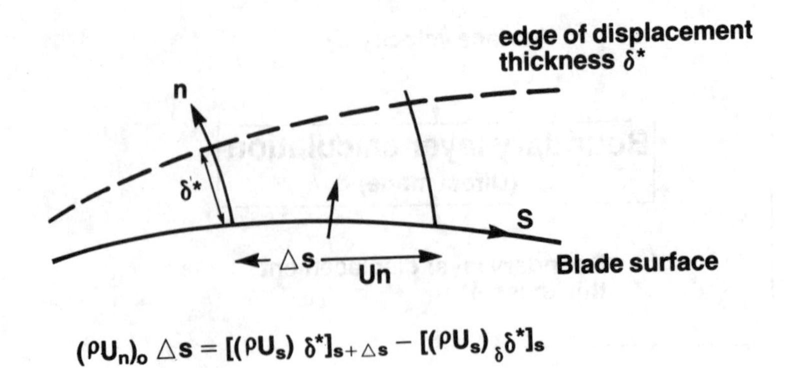

Fig. 9 . Transpiration boundary layer model

It is shown in Appendix D how the transpiration model can be included in a finite element method solving the potential flow equations and in Appendix E how it can be included in a finite volume method solving the Euler equations.

3.3 Film Cooling and Trailing Edge Ejection

A detailed analysis of the effects of ejected film cooling flows and of trailing edge ejected flows necessitates the use of a fully viscous method including elliptic effects. As this is not a practical proposition for use in a blade design system simple models based on extensions to the boundary layer analysis are usually adopted. For example, in the case of film cooling, with an integral boundary layer method one-dimensional considerations of conservation of mass, momentum and energy can be used to provide starting values of θ, H etc. for use downstream of the ejection point. With finite difference methods a slightly more sophisticated model can be adopted for example using a distribution of mass, momentum and energy of the ejected flow through the boundary layer together with the conservation laws to calculate conditions for continuing downstream of the ejection point.

REFERENCES

1. Jennions, I.K., and Stow, P. A Quasi Three-Dimensional Turbomachinery Blade Design System, Part I - Through-Flow Analysis. ASME 84-GT-26. Part II - Computerised System. ASME 84-GT-27.
2. Hirsch, C.H., and Warzee, G. An Integrated Quasi-Three-Dimensional Finite Element Calculation Program for Turbomachinery Flows. Trans. ASME, Journal of Engineering for Power. Paper No. 78-GT-56, pp. 1-8, 1978.
3. Stow, P. Derivation of the Entrainment Equation and Momentum Integral Equations for an Annulus Wall Boundary Layer in a Rotating co-ordinate System. Rolls-Royce Report.
4. Marsh, H.,and Horlock, J.H. Wall Boundary Layers in Turbomachinery. Journal of Mechanical Engineering Science. Vol.14, No.6, pp. 411-422, 1972.
5. De.Ruyck, J., and Hirsch, C. Investigations of an Axial Compressor End-wall Boundary Layer Prediction Method. ASME Journal of Engineering for Power. Vol.103, January 1980.
6. De.Ruyck, J., Hirsch, C., and Kool, P. An Axial Compressor End-wall Boundary Layer Calculation Method. ASME Journal of Engineering for Power. Vol.101, April 1979.
7. Le.Boeuf, F. Annulus End-wall Boundary Layer Theory. VKI Lecture Series 05, 1984. Secondary Flows in End-wall Boundary Layers in Axial Turbomachines.
8. Adkins, G.G., and Smith, L.H. Spanwise Mixing in Axial-Flow Turbomachines. ASME 81-GT-57, 1981.

9. Lakshminarayan, B., and Horlock, J.H. Leakage and Secondary Flows in Compressor Cascades. Aeronautical Research Council R&M No.3483, 1965.
10. Lakshminarayan, B., and Horlock, J.H. Generalised Expressions for Secondary Vorticity Using Intrinsic Co-ordinates. Journal of Fluid Mechanics, Vol.59, Part 1, pp. 97-115, 1973.
11. Came, P.M., and Marsh, H. Secondary Flow in Cascades - Two Simple Derivations for the Components of Vorticity. Journal of Mechanical Engineering Sciences, Vol.16, 1974.
12. Smith, L.H. Secondary Flow in Axial Flow Turbomachinery. Trans. ASME, Vol.77, No.7, October, 1955.
13. James, P.W. Derivation of Expressions for Secondary Streamwise Vorticity Which Allow for Blade Rotation and Streamwise Entropy Gradients. Rolls-Royce Report.
14. Freeman, C.R., and Dawson, R.E. Core Compressor Developments for Large Civil Jet Engines. International Gas Turbine Conference, Tokyo 1983.
15. James, P. Rolls-Royce private communication.
16. Birch, N.T. The effects of Viscosity on Inlet Skew in Axial Flow Turbomachinery Calculations. Institute of Mechanical Engineers Conference on Computational Methods in Turbomachinery. University of Birmingham, England, 1984.
17. Boletis, E., Sieverding, C.H., and Van Hove, W. Effects of a Shewed Inlet End-wall Boundary Layer on the 3-D Flow Field in an Annular Turbine Cascase. Agard Pep "Viscous Effects in Turbomachinery", 1983.
18. Birch, N.T. Viscous and Inviscid Predictions for VKI Annular Turbine with a Skewed Inlet End-wall Boundary Layer. Rolls-Royce report.
19. James, P.W. Spanwise Mixing Effects. Rolls-Royce report.
20. Williams, C.R., and Stow, P. Boundary Layer Inclusion in a Finite Element Blade-to-Blade Method. To be published 1984.
21. Luxton, R.E., and Young, A.D. Skin Friction in the Compressible Laminar Boundary Layer with Heat Transfer and Pressure Gradient. Aeronautical Research Council Report 20336, England.
22. Horton, H.P. A Semi-empirical Theory for the Growth and Bursting of Laminar Separation Bubbles. Aeronautical Research Council. CP 1073, 1969, England.
23. Ntim, B.A. A Theoretical and Experimental Investigation of Separation Bubbles. Ph.D. Thesis, Queen Mary College, London, 1969.
24. Roberts, W.B. Calculation of Laminar Separation Bubbles and their Effect on Airfoil Performance. AIAA Journal, Vol.18, No.1, 1980.
25. Abu-GHannam, B.J., and Shaw, R. Natural Transition of Boundary Layers - The Effects of Turbulence, Pressure Gradient and Flow History. Journal of Mechanical Engineering Sciences, Vol.22, 1980.

26. Green, J.E.,Weeks, D.J., and Brooman, J.W.F. Prediction of Turbulent Boundary Layers and Wakes in Compressible Flow by a Log-entrainment Method. Royal Aircraft Establishment, TR 72231.
27. East, L.F., Smith, P.D., and Merryman, P.J. Prediction of the Development of Separated Turbulent Boundary Layers by the Log-entrainment Method. Royal Aircraft Establishment TR 77046, 1977.
28. Cedar, R.D., and Stow, P., The Addition of Quasi-three-dimensional Terms in a Finite Element Method for Transonic Turbomachinery Blade-to-Blade Flows. To be Published in International Journal of Numerical Methods in Fluids, 1984.
29. Denton, J.D. An Improved Time-Marching Method for Turbomachinery Calculations. ASME Paper 82-GT-239, 1982.

Appendix A. Definitions of Momentum and Displacement Thicknesses

The momentum thickness in the meridional and circumferential directions are defined as

$$\theta_{mm} = \frac{1}{\rho_e Q_e^2} \int_o^\delta \rho q_m (q_{m_e} - q_m)\, d\ell$$

$$\theta_{\theta\theta} = \frac{1}{\rho_e Q_e^2} \int_o^\delta \rho q_\theta (q_{\theta_e} - q_\theta)\, d\ell$$

and the two coupled thickness as

$$\theta_{m\theta} = \frac{1}{\rho_e Q_e^2} \int_o^\delta \rho q_\theta (q_{m_e} - q_m)\, d\ell$$

$$\theta_{\theta m} \doteq \frac{1}{\rho_e Q_e^2} \int_o^\delta \rho q_m (q_{\theta_e} - q_\theta)\, d\ell$$

The displacement thicknesses are defined as

$$\delta_m^* = \frac{1}{\rho_e Q_e} \int_o^\delta (\rho_e q_{m_e} - \rho q_m)\, d\ell$$

and

$$\delta_\theta^{*} = \frac{1}{\rho_e Q_e} \int_o^\delta (\rho_e q_{\theta_e} - \rho q_\theta)\, d\ell$$

Appendix B. Secondary Flow Stream-Function

For incompressible flow the continuity equation is

$$\frac{\partial u_s}{\partial s} + \frac{\partial u_n}{\partial n} + \frac{\partial W}{\partial z} = 0 \tag{B.1}$$

where the co-ordinate system is shown in Fig. 3. It is usually assumed that

$$\frac{\partial}{\partial s} \equiv 0 \tag{B.2}$$

so that from equation (B.1) a stream-function ψ can be introduced such that

$$u_n = -\frac{\partial \psi}{\partial z}, \qquad W = \frac{\partial \psi}{\partial n} \tag{B.3}$$

The secondary streamwise vorticity ξ_{sec} is given by

$$\xi_{sec} = \frac{\partial W}{\partial n} - \frac{\partial u_n}{\partial z} \tag{B.4}$$

so that from equation (B.3) ψ satisfies.

$$\frac{\partial^2 \psi}{\partial n^2} + \frac{\partial^2 \psi}{\partial z^2} = \xi_{sec}$$

Knowing the distribution of ξ_{sec}, for example from equation (17), ψ can be determined and the secondary velocities found from equation (B.3). The boundary conditions are

$$\psi = 0$$

on the boundaries of the domain.

The above analysis can be extended to cylindrical polar co-ordinates and to include the effects of compressibility.

Appendix C. Spanwise Mixing Model

Much of the following analysis is taken by James [19]. For steady inviscid flow in a relative co-ordinate system the energy equation is

$$\nabla.\rho\underline{W}\ I = 0 \tag{C.1}$$

Where $\underline{W}$ is the relative velocity and I is the rothalpy defined as

$$I = C_p t + \frac{1}{2} W^2 - \frac{1}{2} \Omega^2 R^2 \tag{C.2}$$

Ω being the blade speed.

By integrating the equation from one blade to the next the passage averaged energy equation becomes

$$\frac{\partial}{\partial X} (BR\ \overline{\rho W_X I}) + \frac{\partial}{\partial R} (BR\ \overline{\rho W_R I}) = 0 \tag{C.3}$$

Where X and R are the usual cylindrical polar co-ordinates, B is given by equation (7) and where the "-" denotes a passage averaged value, see Jennions and Stow [1] for more details. We can write the equation in the form

$$\frac{\partial}{\partial X}(BR\ \overline{\rho W_X \hat{I}}) + \frac{\partial}{\partial R}(BR\ \overline{\rho W_R \hat{I}})$$

$$= \frac{\partial}{\partial R}(BR\ \overline{\rho W_R}\hat{I}) - \frac{\partial}{\partial R}(BR\ \overline{\rho W_R I}) \quad (C.4)$$

where $\hat{I}$ is defined as

$$\hat{I} = \frac{\overline{\rho W_X I}}{\overline{\rho W_X}} = \frac{\overline{\rho W_X I}}{\bar{\rho}\tilde{W}_X} \quad (C.5)$$

where "~" denotes a density averaged value e.g.

$$\tilde{W}_X = \frac{\overline{\rho W_X}}{\bar{\rho}}$$

Using the passage averaged continuity equation, equation (C.4) may be written as

$$BR\bar{\rho}\left[\tilde{W}_X \frac{\partial}{\partial X} + \tilde{W}_R \frac{\partial}{\partial R}\right]\hat{I} = \frac{\partial}{\partial R}\left[BR\ (\frac{\tilde{W}_R}{\tilde{W}_X}\ \overline{\rho W_X I} - \overline{\rho W_R I})\right] \quad (C.6)$$

We introduce a passage averaged hade angle $\tilde{\lambda}$ defined as

$$\tan\tilde{\lambda} = \frac{\tilde{W}_R}{\tilde{W}_X} \quad (C.7)$$

which defines the slope of the axisymmetric blade-to-blade surfaces used in blade design. With this equation (C.6) becomes

$$\tilde{W}_m \frac{\partial \hat{I}}{\partial m} = \frac{1}{BR\bar{\rho}}\frac{\partial}{\partial R}\left[BR\ \overline{\rho(W_X \tan\tilde{\lambda} - W_R)I}\right]$$

or

$$\tilde{W}_m \frac{\partial \hat{I}}{\partial m} = - \frac{1}{BR\bar{\rho}} \frac{\partial}{\partial R} \left[\frac{BR}{\cos\tilde{\lambda}} \overline{\rho V_\ell I} \right] \tag{C.8}$$

where M denotes the meridional direction see Fig. 2 and V_ℓ is the velocity normal to the axisymmetric blade-to-blade surface. In a standard quasi-three-dimensional design system V_ℓ is taken as zero, i.e. the axisymmetric blade-to-blade surfaces are assumed to be stream-surfaces, and the familiar energy equation results.

We can write

$$\overline{\rho V_\ell I} = \bar{\rho}\ \widetilde{V_\ell I} = \bar{\rho}\ \widetilde{V_\ell I'''}$$

where I''' represents a perturbation from the mean $\hat{I}$; note that

$$\tilde{V}_\ell = 0$$

Rothalpy perturbations can be determined as follows. Fig. C1 indicates a blade-to-blade calculating surface and twisted stream-surface in a plane M = constant; it is assumed that these start out coincident at the blade leading edge. We can write

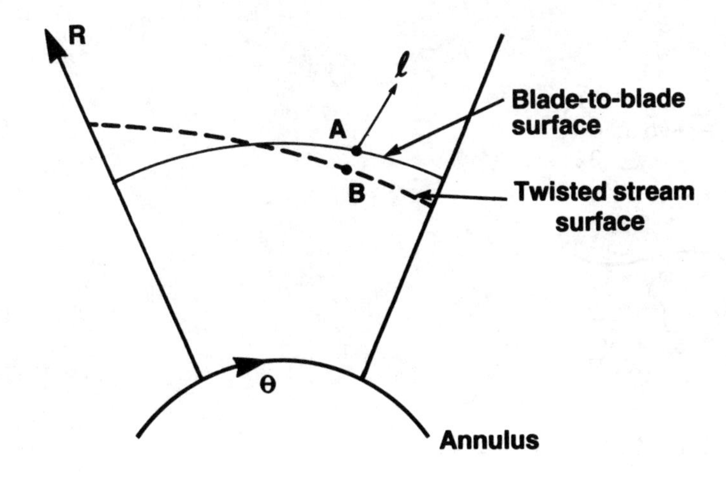

Fig. C1

$$I_A(\theta) \simeq I_B - (\ell_A - \ell_B)\frac{\partial I_A}{\partial \ell}$$

$$\simeq I_B - (\ell_A - \ell_B)\frac{\partial \hat{I}_A}{\partial \ell}$$

$$= I_{in}(R_A) - (\ell_A - \ell_B)\frac{\partial \hat{I}_A}{\partial \ell} \qquad (C.10)$$

where I_{in} is the inlet value to the blade row which is a function of radius. It follows then that

$$I_A'''(\theta) = -(\ell_A - \ell_B)\frac{\partial \hat{I}_A}{\partial \ell} \qquad (C.11)$$

In addition

$$\ell_A - \ell_B = \int_o^{m_A} \frac{V_\ell}{V_m} d_m \qquad (C.12)$$

so that

$$\overline{\rho V_\ell I} = -\bar{\rho}\,\varepsilon \frac{\partial \hat{I}_A}{\partial \ell} \qquad (C.13)$$

where

$$\varepsilon = \widetilde{V_\ell \int_o^{m_A} \frac{V_\ell}{V_m} d_m} \qquad (C.14)$$

Equation (C.8) now becomes

$$\tilde{W}_m \frac{\partial \hat{I}}{\partial m} = \frac{1}{BR\bar{\rho}} \frac{\partial}{\partial R} \left[\frac{BR\bar{\rho}}{\cos\tilde{\lambda}} \varepsilon \frac{\partial \hat{I}}{\partial \ell} \right]$$

$$\simeq \frac{1}{BR\bar{\rho}} \frac{\partial}{\partial \ell} \left[BR\bar{\rho}\, \varepsilon \frac{\partial \hat{I}}{\partial \ell} \right] \tag{C.15}$$

Appendix D. Transpiration in a Finite Element Method

A brief introduction is given here of how to incorporate a transpiration boundary layer model into a velocity potential finite element method described by Cedar and Stow [28]; for more details see Williams and Stow [20].

In the case of two dimensional flow the method solves the continuity equation

$$\nabla . \rho \underline{q} = 0 \tag{D.1}$$

together with the energy equation in the form

$$\rho = \rho_{stag} \left[1 - \frac{q^2}{2H} \right]^{\frac{1}{\gamma - 1}} \tag{D.2}$$

where ρ stag is the stagnation density and H the stagnation enthalpy. For isentropic flow a velocity potential ϕ is introduced given by

$$\underline{q} = \nabla \phi \tag{D.3}$$

With the finite element method the flow domain is divided into elements, see Fig. D1 . Equation (D.1) is solved using a Galerkin weighted residual approach leading to a system of equations of the form

$$\sum_{e(I)} \iint (\nabla . \rho \underline{q})\, N_I \, dA = 0 \tag{D.4}$$

for each node I, where N_I represents the shape function corresponding to node I and where the summation is over all elements containing I as a node.

We can write equation (D.4) as

$$\sum_{e(I)} \iint \{ \nabla . \rho \underline{q}\, N_I - \rho \underline{q} . \nabla N_I \} \, dA = 0$$

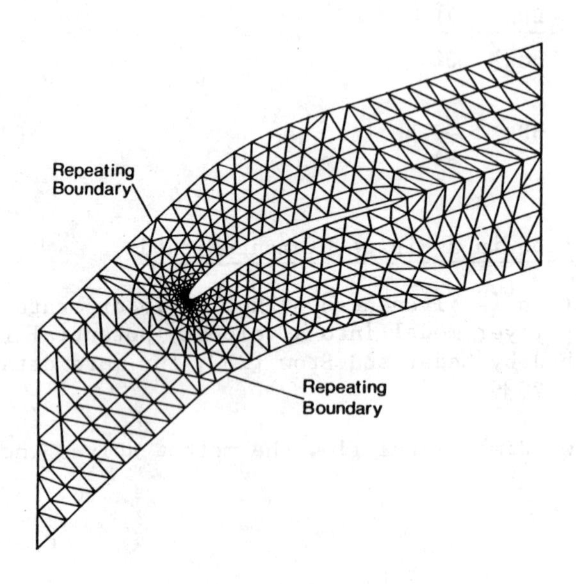

Fig. D1. Typical finite element mesh

or using the divergence theorem

$$\sum_{e(I)} \iint \rho \underline{q} . \nabla N_I \, dA = \sum_{e(I)} \int \rho \underline{q} . \underline{n} \, N_I \, dS \tag{D.5}$$

where S represents the perimeter of the domain of integration. The right-hand side of equation (D.5) is zero for all internal nodes of the domain. For nodes on the inlet and outlet boundaries it can be determined in terms of inlet and outlet boundary conditions. For nodes on the blade surface we have

$$RHS = - \sum_{e(I)} \int (\rho u_n)_o \, N_I \, dS$$

which using equation (29) gives

$$RHS = - \sum_{e(I)} \int \frac{d}{dS} \left[(\rho u)_\delta \, \delta^* \right] N_I \, dS$$

$$= \sum_{e(I)} \int (\rho u)_\delta \, \delta^* \frac{dN_I}{dS} \, dS \tag{D.6}$$

In the method three node straight sided triangles are used so that the integrals involved in equation (D.6) and the left-hand side of equation (D.5) are easily evaluated. The final system of equations may be written as

$$\underline{A}(\underline{\phi}) = \underline{C} - B\,\underline{\delta}^* \tag{D.7}$$

where $\underline{\phi}$ is the vector of unknown potentials, $\underline{A}$ is a Nx1 column vector, N being the total number of nodes, $\underline{C}$ is formed from the inlet and exit boundary conditions (being non-zero only for nodes on the inlet and exit boundaries), $\underline{\delta}^*$ is the vector of displacement thicknesses and B is an NxNs matrix where Ns is the number of surface nodes.

Both B and $\underline{\delta}^*$ depend on the surface velocities and with direct mode boundary layer coupling are evaluated using the last cycle values. This scheme may be written as

$$\underline{A}\,(\underline{\phi}^{n+1}) = \underline{C} - [B\underline{\delta}^*]^n \tag{D.8}$$

for the mainstream and

$$\underline{\delta}^{*\,n+1} = \underline{\delta}^*\,(\underline{u}_s^{n+1}) \tag{D.9}$$

for the boundary layer, where the superscript denotes the iteration number.

The system of equations (D.8) are solved using a Newton-Raphson linearisation technique which putting

$$\underline{\phi}^{n+1} = \underline{\bar{\phi}} + \underline{\phi}'$$

may be written as

$$J\underline{\phi}' = -\underline{A}(\underline{\bar{\phi}}) + \underline{C} - (B\underline{\delta}^*)^n \tag{D.10}$$

where J is the Jacobian matrix

$$J_{ij} = \frac{\partial A_i}{\partial \phi_j} \tag{D.11}$$

and can be obtained analytically. The iterative procedure is repeated until convergence has been obtained whereupon, after calculating the new approximation to the surface velocities, $\underline{\delta}^*$ can be calculated using the boundary layer method and the right-hand side of equation (D.8) up-dated.

It is found that the whole process is fast to converge leading to a fully interactive program; results from this will be shown in a later lecture.

Appendix E. Transpiration in a Finite Volume Method

A brief introduction is given here of how to incorporate a transpiration boundary layer model into a finite volume time-marching method. For two-dimensional flow the unsteady continuity and momentum equations may be written as

$$\frac{\partial \rho}{\partial t} + \nabla.\rho\underline{q} = 0 \tag{E.1}$$

$$\frac{\partial}{\partial t}(\rho u_x) + \nabla.\rho\underline{q}u_x = -\frac{\partial p}{\partial x} \tag{E.2}$$

$$\frac{\partial}{\partial t}(\rho u_y) + \nabla.\rho\underline{q}u_y = -\frac{\partial p}{\partial y} \tag{E.3}$$

Integrating equation (E.1) over an element or control volume gives

$$\iint \frac{\partial \rho}{\partial t}\, dA + \int \rho\underline{q}.\underline{n}\, dS = 0$$

where A represents the area and S the perimeter of the element. We can write this as

$$\frac{\partial \bar{\rho}}{\partial t}.A = -\Sigma \int \rho\underline{q}.\underline{n}\, dS \tag{E.4}$$

where the summation is over the sides of the element and "-" is a representative value for the element. In a similar fashion equations (E.2) and (E.3) may be written as

$$\frac{\partial}{\partial t}(\overline{\rho u_x})\, A = -\Sigma \int \rho\underline{q}.\underline{n}\, u_x\, dS - \iint \frac{\partial p}{\partial x}\, dA \tag{E.5}$$

and

$$\frac{\partial}{\partial t}(\overline{\rho u_y})A = -\Sigma\int \rho\underline{q}.\underline{n}\, u_y\, dS - \iint \frac{\partial p}{\partial y}\, dA \tag{E.6}$$

How the element values of

$$\frac{\partial \rho}{\partial t}, \quad \frac{\partial \rho u_x}{\partial t}, \quad \frac{\partial \rho u_y}{\partial t}$$

are used to up-date the nodal values depends on the method used, see for example Denton [29]. The only change to the inviscid method comes from element faces on the blade surfaces.

For example the contribution to the continuity equation is

$$\int \rho \underline{q}.\underline{n} \, dS = -\int (\rho u_n)_o \, dS$$

$$= -\Delta[(\rho u_s)_d \, \delta^*] \qquad (E.7)$$

where Δ represents the change over the side.

Similarly the contributions to the x and y momentum equations are

$$\int \rho \underline{q}.\underline{n} \, u_x \, dS = -\int (\rho u_n u_x)_o \, dS \qquad (E.8)$$

$$\int \rho \underline{q}.\underline{n} \, u_y \, dS = -\int (\rho u_n u_y)_o \, dS \qquad (E.9)$$

where from equation (30)

$$(u_x)_o = (u_s)_o.\cos\beta - (u_n)_o.\sin\beta$$

$$\simeq (u_x)_\delta$$

and

$$(u_y)_o \simeq (u_y)_\delta \qquad (E.10)$$

The fluxes of mass, momentum and energy through faces on the blade surface can be calculated using the latest boundary layer information. The flow parameters at the nodes can then be up-dated using the usual time-marching method. The new surface velocities are then used in the boundary layer analysis to re-calculate the displacement thickness and transpiration quantities and the procedure repeated.

List of Symbols

B blade blockage
C_E entrainment coefficient
f mixing function
F_B Blade Force
F_τ Dissipative force
h stream-tube height
H form factor
H_1 modified form factor
I rothalpy
ℓ direction normal to m
m meridional direction
$\dot{m}$ local mass flow
Mc ejected mass flow
n normal to blade surface
N number of blades
N shape function
P perturbation term
p static pressure
Q absolute total velocity
q absolute velocity component
R radius
s streamline coordinate
S entropy
t blade thickness, time, static temperature
u velocity component
V_ℓ spanwise velocity
W relative velocity
α absolute whirl angle
β relative whirl angle
γ ratio of specific heats
δ boundary layer thickness
δ^* displacement thickness
ε effective diffusion coefficient
λ hade angle
θ momentum thickness, circumferential coordinate
ρ density
ϕ velocity potential
ν kinematic viscosity
τ skin friction
ξ vorticity
ψ stream function

Subscripts

b blade surface
d edge of boundary layer
m meridional

n	normal to blade
p	pressure surface
R	radial
S	suction surface, streamwise direction
sec	secondary
stag	stagnation
W	wall value
0	blade surface value
1	inlet
2	exit

THROUGH FLOW MODELS FOR TURBOMACHINES : STREAM SURFACE AND PASSAGE AVERAGED REPRESENTATIONS

Ch. Hirsch, H. Deconinck

Vrije Universiteit Brussel, Department of Fluid Mechanics

INTRODUCTION

Although fully three-dimensional computations are and will be applied in increasing measure in the coming years, many physical situations can still be described with reasonable accuracy by a two-dimensional approach if the geometry of the system allows for this possibility.

In practical situations, many configurations can be considered as not "strongly" three-dimensional in the sense that one expects that the variations of the third velocity component are less important than the variations of the other two. For example, the twisted blade of a turbomachinery blade row will induce three-dimensional effects which in some cases are of lower order of magnitude and a description based on two-dimensional approaches might in these cases be a valid and acceptable approximation.

In order to treat these flow situations with limited threedimensional effects an intermediate description between the fully three-dimensional and the two-dimensional can be introduced, which is called Quasi-Three-Dimensional.

A first type of Quasi-three-dimensional descriptions approaches the flow by a succession of properly chosen, interacting families of two-dimensional flows along intersecting surfaces. Appropriate families of surfaces are defined, which can be defined as streamsurfaces, along which a two-dimensional velocity field projection is determined. Obviously a fully three-dimensional flow field will require three families of intersecting surfaces in order to be determined completely, corresponding to three interdependent scalar functions of three coordinates necessary to describe the three component velocity field restricted by the continuity equation. This can be reduced to two surfaces for steady flows. Such a representation is completely equivalent to a complete three-dimensional description and no spatial approximations are involved in this case. The Quasi-Three-Dimensional approximation comes in when the number of surfaces is reduced, that is when it is considered that a valid description of the flow behaviour is obtained by neglecting the flow contributions along either certain families of surfaces or along certain members of a given family of surfaces.

An alternative to the stream- or pseudo streamsurface method for obtaining a quasi-three-dimensional approximation is offered by the averaging method. This approach consists in averaging out the conservation equations with respect to a chosen coordinate, for instance the ξ^3 direction, leading to equations with the remaining coordinates as independent variables.

Two-dimensional equations are obtained in this way which are representative of the average flow with respect to the ξ^3 coordinate, but containing eventual terms of geometrical and kinematical nature function of the average space direction. Clearly, limits of integration have to be defined and hence this procedure is best suited for internal flows with eventual varying cross-sections.

Both approaches are presented and discussed in sections 1 and 2. Section 3 and 4 discuss the particular application in a turbomachine where the streamsurface is chosen to be normal to the circumferential direction in the first approach or alternatively where the averaging is carried out along the circumferential direction in the second approach. This choice corresponds to the so-called through-flow descriptions. Section 5 summarizes the through flow equations and discusses the interaction with the end wall boundary layers. In section 6, the computational aspects of the through flow methods are discussed.

1. THE QUASI-THREE-DIMENSIONAL REPRESENTATION (Q3D) FOR INTERNAL FLOWS : C.H. WU'S STREAMSURFACE METHOD (WU [1], WU [2]).

In general curvilinear coordinates a time independent streamsurface is defined by an equation

$$S(\xi^1,\xi^2,\xi^3)=0 \qquad \text{or} \qquad \xi^3-\xi^3(\xi^1,\xi^2)=0 \tag{1.1}$$

and the normal to the streamsurface by

$$\vec{n} = \vec{\nabla}S = \partial_\alpha S.\vec{\varepsilon}^\alpha = n_\alpha \vec{\varepsilon}^\alpha \tag{1.2}$$

where the $\vec{\varepsilon}^\alpha$ are the contravariant base vectors of the coordinate system. Two possibilities are open for the description of the considered surfaces.

Case 1. The surface is defined by a ξ^3=constant surface. In other words, the surface S contains the lines ξ^1,ξ^2, figure 1. In this case the normal vector $\vec{n}$ is equal to $\vec{\varepsilon}^3$, and one has $\vec{\varepsilon}^3\vec{v}=0$ or $v^3=0$. If the coordinate system is chosen to be orthogonal, then the length of the vector $\vec{n}$ is related to the third metric coefficient h_3 of the coordinates $\xi^1\xi^2\xi^3$. Indeed, one has

$$1/h_3^2 = \vec{\nabla}\xi^3.\vec{\nabla}\xi^3 = \vec{\varepsilon}^3.\vec{\varepsilon}^3 = \vec{n}^2 \quad \text{or} \quad \vec{1}_n = \frac{\vec{n}}{|\vec{n}|} = h_3.\vec{n} \tag{1.3}$$

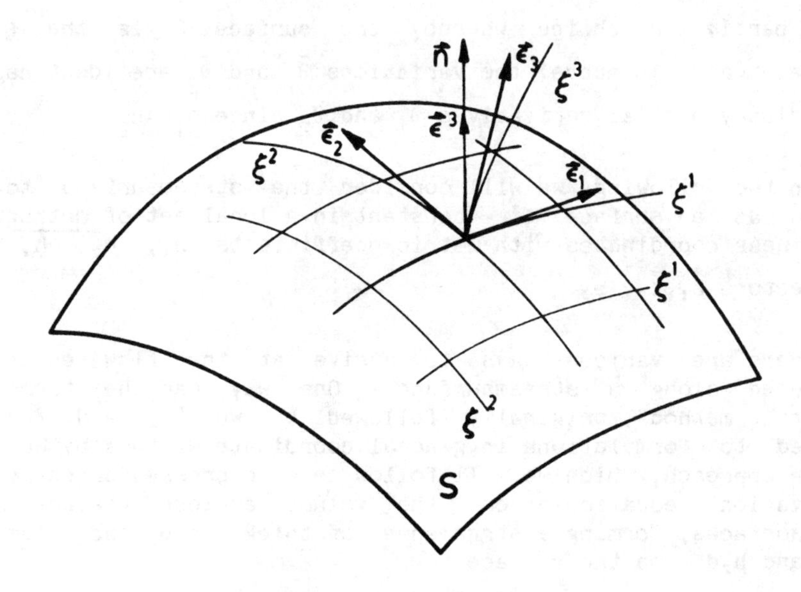

figure 1 : Streamface ξ^3 = constant

Case 2. The surface S does not contain the ξ^1 and ξ^2 lines and is situated arbitrarily with respect to the local coordinate system ξ^α. The normal $\vec{n}$ is then given by

$$\vec{n} = -\frac{\partial \xi^3}{\partial \xi^1}\vec{\varepsilon}^1 - \frac{\partial \xi^3}{\partial \xi^2}\vec{\varepsilon}^2 + \vec{\varepsilon}^3 = n_\alpha \vec{\varepsilon}^\alpha \qquad (1.4)$$

leading to

$$\frac{\partial \xi^3}{\partial \xi^1} \triangleq \partial_1 \xi^3 = -\frac{n_1}{n_3} \quad \text{and} \quad \frac{\partial \xi^3}{\partial \xi^2} \triangleq \partial_2 \xi^3 = -\frac{n_2}{n_3} \quad (n_3=1) \qquad (1.5)$$

This allows to define the variations of any flow quantity along the streamsurface S by the following relations, where the overbars indicate variations along the streamsurface,

$$\frac{\bar{\partial}}{\partial \xi^1} \triangleq \frac{\partial}{\partial \xi^1} + \partial_1 \xi^3 . \frac{\partial}{\partial \xi^3} \quad \text{and} \quad \frac{\bar{\partial}}{\partial \xi^2} \triangleq \frac{\partial}{\partial \xi^2} + \partial_2 \xi^3 . \frac{\partial}{\partial \xi^3} \qquad (1.6)$$

or in condensed notation

$$\bar{\partial}_1 = \partial_1 - \frac{n_1}{n_3}\partial_3 \quad \text{and} \quad \bar{\partial}_2 = \partial_2 - \frac{n_2}{n_3}\partial_3 \quad \bar{\partial}_3 = 0 \qquad (1.7)$$

In the particular choice whereby the surface S is the ξ^1-ξ^2 surface, case 1 above, the variations $\bar{\partial}_1$ and $\bar{\partial}_2$ are identical to the ordinary partial derivatives ∂_1 and ∂_2 since $n_1 = n_2 = 0$.

In the following, we will consider the streamsurface to be defined as a surface ξ^3 = constant in a local set of orthogonal curvilinear coordinates with metric coefficients h_1, h_2, h_3 and basisvectors $\vec{\varepsilon}_1, \vec{\varepsilon}_2, \vec{\varepsilon}_3$.

There are various ways to arrive at the flow equations considered along a streamsurface. One way can be termed as "algebraic method", originally followed by Wu [1] and further extended to formulations in general coordinate systems by Wu [2]. Another approach, which we will follow here, expresses directly the conservation equations on the volume enclosed between two streamsurfaces, forming a streamsheet of thickness b and elements $h_1 d\xi_1$ and $h_2 d\xi_2$ on the surface.

The streamsheet thickness b is defined in the direction of ξ^3 in units of the variable ξ^3. That is, the physical streamsheet thickness is $(b.h_3)$, while the volume element is given by $d\Omega = b.h_1 h_2 h_3 \, d\xi^1 d\xi^2$

Applying the general form of a conservation law for the scalar quantity ρc

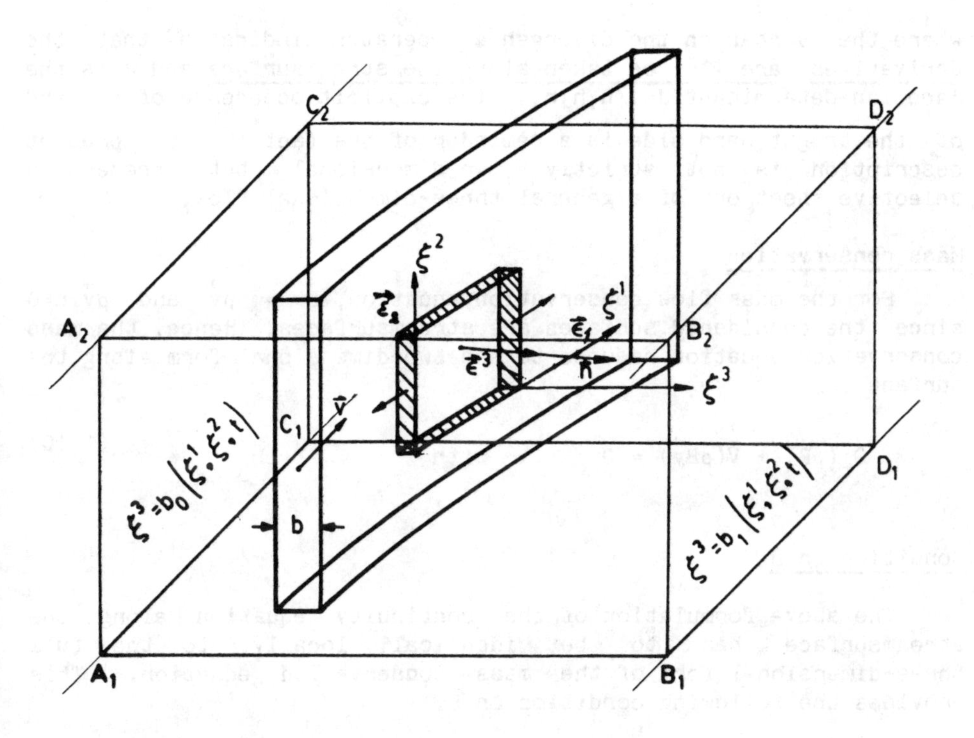

Figure 2 : Elementary volume of a streamsheet of thickness b

$$\partial_t \int_\Omega \rho c d\Omega + \oint_S \vec{F}\, \vec{dS} = 0$$

to the infinitesimal volume of fig. 2 one obtains

$$\partial_t(\rho c b h_3) + \frac{1}{h_1 h_2} [\partial_1(b h_3 h_2 \tilde{f}^1) + \partial_2(b h_3 h_1 \tilde{f}^2)]$$
$$= - \frac{b}{h_1 h_2} \partial_3(\vec{n}.\vec{F} h_1 h_2 h_3) \tag{1.8}$$

where $\tilde{f}^\alpha = \vec{F}.\vec{e}^\alpha$ is the projection of the flux vector on the direction ξ^α with unit vector $\vec{e}^\alpha$.

In the left hand side one recognizes the expression of the two-dimensional divergence operator, in the coordinates ξ^1 and ξ^2 along the surface, and the conservation equation along the streamsurface becomes

$$\partial_t\ (\rho c b h_3) + \vec{\tilde{\nabla}}\ (b h_3 . \vec{F}) = -\ b h_3 J\ \partial_3(\vec{n}\ \vec{F}/J) \tag{1.9}$$

where the overbar on the divergence operator indicates that the derivatives are to be taken along the streamsurface and J is the Jacobian determinant $J=1/h_1h_2h_3$. The explicit occurence of h_3 and of the right hand side is a reminder of the fact that the present description is not strictly two-dimensional but presents a selective sheet out of a general three-dimensional flow.

Mass conservation

For the mass flow conservation equation, $\vec{F} = \rho\vec{v}$ and $\rho\vec{v}.\vec{n}=0$ since the considered surfaces are streamsurfaces. Hence, the mass conservation equation reduces to the two-dimensional form along the surface

$$\partial_t(\rho B) + \vec{\bar{\nabla}}(\rho B\vec{v}) = 0 \qquad \text{with} \qquad B=bh_3 \tag{1.10}$$

Condition on B

The above formulation of the continuity equation along the streamsurface has to be identical, locally, to the full three-dimensional form of the mass conservation equation. This provides the following condition on B.

$$\frac{1}{B}(\partial_t + \vec{v}\,\vec{\bar{\nabla}})\,B = \vec{n}\,\partial_3\,\vec{v} \tag{1.11}$$

for time independent metric coefficients.

Equation (1.11) allows to determine the variations of the streamtube thickness b or B from the knowledge of the complete three-dimensional flow field. This equation plays therefore an important role in a quasi-three dimensional interactive procedure.

Quasi-Three-Dimensional form for the Momentum Conservation Equation

Following the same procedure one obtains for the momentum equation

$$\partial_t\,(\rho\vec{v}bh_3) + \frac{1}{h_1h_2}\,[\bar{\partial}_1\,(\bar{\bar{F}}.\vec{e}^1.bh_3h_2) + \bar{\partial}_2\,(\bar{\bar{F}}.\vec{e}^2.bh_3h_1)]$$

$$= \rho\vec{f}_e\,.\,bh_3 - bh_3J\,\partial_3(\bar{\bar{F}}.\vec{n}/J) \tag{1.12}$$

where $\bar{\bar{F}} = \rho\,\vec{v}\otimes\vec{v} + pI - \bar{\bar{\tau}}$ is the momentum flux tensor. The term $J\partial_3(\bar{\bar{F}}.\vec{n}/J)$ reduces to the contribution from the pressure and shear stresses since $\vec{v}.\vec{n} = 0$. Hence, a body force $\vec{f}$ is defined by

$$\rho\vec{f} \triangleq -J\ \partial_3(\bar{\bar{F}}\ .\ \vec{n}/J) = -J\partial_3[(p\bar{\bar{I}} - \bar{\bar{\tau}})\vec{n}/J] \qquad (1.13)$$

The derivative terms on the left hand side, with an overbar to indicate that they are to be considered along the streamsurface, have the form of the divergence of a tensor in the <u>two-dimensional</u> space $\xi^1 - \xi^2$. Hence the quasi-three-dimensional momentum equation becomes

$$\frac{1}{B}\ [\ \partial_t\ (\rho\ B\vec{v}) + \bar{\vec{\nabla}}(\rho\ B\vec{v} \otimes \vec{v})\ ] + \frac{1}{B}\bar{\vec{\nabla}}\ [(p\bar{\bar{I}} - \bar{\bar{\tau}})\ B] = \rho\ \vec{f}_e + \rho\vec{f} \qquad (1.14)$$

written completely in the two-dimensional space $\xi^1 - \xi^2$.

Compared to the three-dimensional momentum equation, the quasi-three dimensional equation (1.14) maintains the same general form with the "streamsurface" derivatives $\bar{\vec{\nabla}}$ instead of the ordinary derivatives and with the addition of the streamtube thickness in the flux term. More importantly, an extra force term $\vec{f}$ in the right hand side, appears in this equation.

The additional forces appear as a consequence of the confinement of the conservation equation and of the flow description to the streamtube of thickness b around the "streamsurface" S. This force $\vec{f}$ is defined by equ. (1.13). Substracting the quasi-three-dimensional continuity equation, equ. (1.10), multiplied by $\vec{v}$, from the momentum equation (1.14), one obtains the alternative form

$$\partial_t\ \vec{v} + (\vec{v}\ \bar{\vec{\nabla}})\ \vec{v} = \frac{-1}{\rho}\ \bar{\vec{\nabla}}\ (p\bar{\bar{I}} - \bar{\bar{\tau}}) + \vec{f}_e + \vec{f} - \frac{1}{\rho B}\ (p\bar{\bar{I}} - \bar{\bar{\tau}})\ \bar{\vec{\nabla}}B \qquad (1.15)$$

A new distributed force $\vec{f}_B$ is defined which includes the last term together with the force $\rho\vec{f}$

$$\vec{f}_B = \vec{f} - \frac{1}{\rho B}\ (p\bar{\bar{I}} - \bar{\bar{\tau}})\ \bar{\vec{\nabla}}B \qquad (1.16)$$

This can be simplified for stationary flow to

$$\vec{f}_B = \frac{-1}{h_1 h_2}\ \frac{\vec{n}}{\rho}\ \ \partial_3\ [\ (p\bar{\bar{I}} - \bar{\bar{\tau}})\ h_1 h_2\ ] \qquad (1.17)$$

and the obtained momentum equation reduces to the usual,

two-dimensional momentum equation written along the streamsurface, with however, the addition of the distributed force $\vec{f}_B$ acting on the flow. Hence, the whole effect of the three-dimensionality on the two-dimensional description is contained in the force term $\vec{f}_B$.

The results remain valid for other forms of the momentum equation, in particular for Crocco's form, or for the equations in a relative system.

Quasi-Three-Dimensional form for the energy conservation law

The energy equation can be treated exactly in the same way as the mass conservation equation. The details are omitted and the resulting form is given in table 1.

In absence of shear stresses and heat conduction effects, the quasi-three dimensional form of the energy equation reduces to

$$\frac{\bar{d}H}{dt} = \frac{1}{\rho B} \partial_t (pB) \qquad (1.18)$$

where $\frac{\bar{d}}{dt} = \bar{\partial}_t + \vec{v}\,\bar{\vec{\nabla}}$ is the total material derivative along the streamsurface.

The equations resulting from the streamsurface approach and derived in this section are summarized in table 1.

2. THE QUASI-THREE-DIMENSIONAL REPRESENTATION (Q3D) FOR INTERNAL FLOWS : THE AVERAGING PROCEDURE

As discussed in the introduction, an alternative to the streamsurface approach in reducing the complexity of the three dimensionality of the flow, can be obtained by taking the average of the conservation equations over one of the spatial directions, Hirsch [3], [4], Jennions and Stow [5].

Let us consider therefore a channel limiting the flow region by upper and lower surfaces, as well as side surfaces, these surfaces being either material, solid walls or fluid boundaries.

Average values of any flow variable A over, say, the ξ^3 direction can be defined through

$$\bar{A} = \frac{1}{b(\xi^1,\xi^2,t)} \int_{b_0}^{b_1} A\ (\xi^1,\xi^2,\xi^3,t)\ d\xi^3 \qquad (2.1)$$

The quantities b_1 (ξ^1,ξ^2,t) and b_0 (ξ^1,ξ^2,t) define the limits of the integration region with

$$b = b_1 - b_0 \qquad (2.2)$$

and are, in general, functions of the other coordinates and eventually of time. Note that b is the width of the integration region expressed in units of ξ^3.

The results for the averaged derivatives are defined as follows : From the general properties of the derivative of an integral one obtains

$$\overline{\partial_\alpha A} = \frac{1}{b}\,\partial_\alpha\,(b\bar{A}) - \frac{1}{b}\,(A_1\partial_\alpha b_1 - A_0\partial_\alpha b_0) \qquad \alpha = 1,2 \qquad (2.3)$$

where A_1 and A_0 are the values taken by the variable A on the upper and lower limits of the integration domain. Similarly, for $\alpha = 3$, one obtains

$$\overline{\partial_3 A} = \frac{1}{b}\,(A_1 - A_0) \qquad (2.4)$$

The boundary surfaces of the integration region will be defined by

$$S_1 = \xi^3 - b_1(\xi^1,\xi^2,t) = 0 \quad \text{and} \quad S_0 = \xi^3 - b_0(\xi^1,\xi^2,t) = 0 \qquad (2.5)$$

With these relations, the averaging rules can be written, grouping (2.3) and (2.4) as the average of the gradient of a scalar quantity A.

$$\overline{\vec{\nabla}\,A} = \frac{1}{b}\,\vec{\nabla}\,(b\,\bar{A}) + \frac{1}{b}\,[A.\vec{n}] \qquad (2.6)$$

where the square brackets indicate the difference over the two boundaries. The same relations apply on the divergence of a vector or a tensor quantity, if the metric coefficients are independent of the averaging coordinate ξ^3.

<u>The averaged form of the continuity equation</u>

Applying the above rules to the continuity equation leads to

$$\partial_t \bar{\rho}\, b + \vec{\bar{\nabla}} (\bar{\rho} . \vec{\bar{v}}\, b) = - \vec{\bar{\nabla}}\, \overline{\rho' \vec{v}' b} \tag{2.7}$$

If an integrating factor μ is defined, such that

$$\vec{\bar{\nabla}}(\overline{\rho' \vec{v}' b}) = \frac{1}{\mu} \rho b(\vec{\bar{v}}\, \vec{\bar{\nabla}})\mu \tag{2.8}$$

equation (2.7) becomes

$$\partial_t \bar{\rho}\, B + \vec{\bar{\nabla}} (\bar{\rho} . \vec{v}\, B) = 0 \qquad \text{with} \qquad B=b\mu \tag{2.9}$$

Comparing this with equation (1.10), a two-dimensional mass conservation equation is obtained where the B-factor represents a varying streamtube thickness which is equal to the product of the geometrical thickness of the fluid layer over which the averaging is performed times a correction factor μ which represents the effect of the mass flux fluctuations over the flow domain. Hence, this B-factor is dependent on the three-dimensional flow properties. This can be avoided by introducing density averages, see for instance Hirsch and Warzee [4].

Denoting, in the following, the ordinary average as defined by equation (2.1) by an overbar and the density averaged values by a tilda, one has

$$\tilde{A} = \frac{\overline{\rho A}}{\bar{\rho}} = \frac{1}{\bar{\rho} b} \int_{b_0}^{b_1} \rho A \, d\xi^3 \tag{2.10}$$

The fluctuation of the variable A with respect to its usual average $\bar{A}$ will be indicated by A' while A" will indicate the fluctuations with respect to the density average $\tilde{A}$. One has

$$\tilde{A} - \bar{A} = - \overline{A''} = \frac{\overline{\rho' A'}}{\bar{\rho}} \qquad \text{and} \qquad \overline{\rho A B} = \bar{\rho}\, \tilde{A}\, \tilde{B} + \overline{\rho A'' B''} \tag{2.11}$$

while the above relations for the average of a derivative remain unchanged.

Application of these various properties to the averaged mass conservation leads to the equation

$$\partial_t \bar{\rho}\, b + \vec{\nabla} (\bar{\rho}\, \vec{V}\, b) = 0 \tag{2.12}$$

writing $\vec{V}$ for the averaged velocity vector $\tilde{\vec{v}}$. This exact relation has the advantage that the flow dependent streamtube thickness b in equation (2.9) reduces here, in all cases, to the geometrical thickness b of the flow region in ξ^3 direction, since the influence of the density fluctuations is included in the definition of $\vec{V}$.

The averaged momentum conservation equation

The application of the averaging rules to the momentum equation leads to the following averaged momentum equation

$$\partial_t \, (\overline{\rho\vec{v}b}) + \vec{\nabla} \, (\overline{\rho\vec{v} \otimes \vec{v}b} + \bar{p}b - \bar{T}b) = \overline{\rho \, \vec{f}_e \, . \, b} - [p\vec{n}] + [\bar{\bar{\tau}}\vec{n}] \tag{2.13}$$

writing $\bar{T}$ for the averaged shear stresses. The last two terms represent the components of an additional force term which appears as an outcome of the averaging procedure in a way that completely parallels the way the additional force $\vec{f}$ appears in the streamsurface approach equation (1.13). This force term, defined by

$$\bar{\rho} \, \vec{f} = - \frac{1}{b} \, [(p\bar{\bar{I}} - \bar{\bar{\tau}})\vec{n}] \tag{2.14}$$

is resulting from the difference in the normal components of the pressure and shear stresses acting on both limiting surfaces b_1 and b_0. Hence, equation (2.13) becomes

$$\partial_t \, (\overline{\rho\vec{v}b}) + \vec{\nabla} \, (\overline{\rho\vec{v} \otimes \vec{v} \, b}) = -\vec{\nabla} \, (\bar{p}b\bar{\bar{I}} - \bar{T}b) + \overline{\rho\vec{f}_e b} + \bar{\rho}\vec{f}b \tag{2.15}$$

If the averaging rule of products, equ. (2.11), is introduced with density weighted averages, one obtains the following equation for the density averaged velocity vector $\tilde{\vec{v}} = \vec{V}$.

$$\begin{aligned} &\partial_t \, \bar{\rho}\vec{V}b + \vec{\nabla} \, (\bar{\rho}b\vec{V} \otimes \vec{V}) = \\ &= - \vec{\nabla} \, (\bar{p}b - \tilde{T}b) + \overline{\rho\vec{f}_e} \, b + \bar{\rho}\vec{f} \, b - \vec{\nabla} \, (\overline{\rho\vec{v}'' \otimes \vec{v}'' \, b}) \end{aligned} \tag{2.16}$$

As before, for the non conservative form an equation can be made formally identical to the standard two-dimensional momentum equation, by introducing the force $\vec{f}_B$ instead of $\vec{f}$, with

$$\bar{\rho}\ b\ \vec{f}_B = -(\bar{p}\bar{\bar{I}} - \bar{T})\vec{\nabla} b + \bar{\rho}\ b\ \vec{f} \tag{2.17}$$

leading to

$$\partial_t\ \vec{V} + (\vec{V}\ \vec{\nabla})\ \vec{V} = \frac{-1}{\bar{\rho}} \cdot \vec{\nabla}\ (\bar{p}\bar{\bar{I}} - \bar{T}) + \vec{\vec{f}}_e + \vec{f}_B - \frac{1}{b\bar{\rho}} \vec{\nabla}\ \overline{(\rho\ \vec{v}'' \otimes \vec{v}''\ b)} \tag{2.18}$$

With the exception of the last term, this equation and equation (1.15) obtained by following a streamsurface, have very similar forms, although the interpretations of each term differ by the fact that here <u>local derivatives of averaged quantities appear</u> while in the former approach one has to do with <u>streamsurface derivatives of the local flow quantities</u>. In both cases however, the quasi-three-dimensional approximation leads to the introduction of a streamtube thickness (B or b) and to additional forces $\vec{f}$ whose non-viscous components act in the direction normal to the considered surfaces.

In the present approach an extra term appears on the right hand side of the momentum equation. This term, which has the same structure as the Reynolds stress tensor in the turbulent averaged Navier-Stokes equations, contains the influence of the three-dimensionality of the flow on its averaged representation. If the flow is mildly non-uniform in the averaging direction these terms may become neglectable with respect to the other terms.

<u>The averaged energy conservation equation</u>

The energy equation can be derived in the same way as described above using the local form

$$\partial_t\ \rho\ E + \vec{\nabla}\ (\rho\ \vec{v}\ H - \vec{B}) = Q \tag{2.19}$$

where $\vec{B}$ is the flux of the shear stresses and heat conduction

$$\bar{B} = k\ \vec{\nabla}\ T + \bar{\bar{\tau}}.\vec{v} \tag{2.20}$$

The averaging procedure leads to the following form of the averaged energy equation

$$\partial_t\ \overline{\rho E}\ b + \vec{\nabla}\ \overline{(\rho\ \vec{v}\ H - \vec{B})}\ b = \bar{Q}.b - [p\partial_t b] + [\vec{B}.\vec{n}] \tag{2.21}$$

Two additional contributions to the averaged energy balance appear in the right hand side as source terms. The first one - $[p\ \partial_t\ b]$ is the work of the pressure forces against the moving boundaries, while the second term $[\vec{B}.\vec{n}]$ is the transfer of heat flux and work of the viscous stresses accross the boundaries of the integration region. These two terms will be grouped in the source term

$$b\, Q_b = [\vec{B}\, \vec{n}] - [p\, \partial_t\, b] \tag{2.22}$$

When the averaged products on the left hand side are worked out, one obtains

$$\partial_t(\bar{\rho}\, \tilde{E}\, b) + \vec{\nabla}\, (\bar{\rho}\, \tilde{\vec{V}}\, \tilde{H} - \bar{\vec{B}})b = \bar{Q}.b + Q_b.b - \vec{\nabla}\, (\overline{b\rho\, \vec{v}''\, H''}) \tag{2.23}$$

or by application of the averaged continuity equation

$$\rho\, (\partial_t\, \tilde{H} + (\vec{V}\, \vec{\nabla})\, \tilde{H}) = \frac{1}{b}\, \partial_t\, \bar{p}\, b + \frac{1}{b}\, \vec{\nabla}\, (\bar{\vec{B}}\, b) + \bar{Q} + Q_b - \frac{1}{b}\, \vec{\nabla}\, (\overline{b\, \rho\, \vec{v}''\, H''}) \tag{2.24}$$

This form of the averaged energy conservation equation parallels completely the form obtained by Wu's streamsurface method with the difference in interpretation already discussed above for the momentum equation. An additional energy flux term appears $\overline{\rho\, \vec{v}''\, H''}$ which represents the total enthalpy fluctuations with respect to the average value as convected by the large scale velocity fluctuations.

Some care has to be taken in the treatment of the averaged total energy $\tilde{E}$ and total enthalpy $\tilde{H}$, with respect to the way the kinetic energy is accounted for.

Indeed, the averaged total energy $\tilde{E}$ or $\tilde{H}$ are not equal to the total energy of the averaged flow which we will denote by $\hat{E}$. The latter is defined by

$$\hat{E} = \tilde{e} + \frac{\vec{V}^2}{2} \tag{2.25}$$

while the averaged total energy is given by

$$\tilde{E} = \tilde{e} + \frac{\widetilde{\vec{v}^2}}{2} = \tilde{e} + \frac{\vec{V}^2}{2} + \frac{\overline{\rho\, \vec{v}''\vec{v}''}}{2\bar{\rho}} \tag{2.26}$$

Identical relations exist for $\tilde{H}$ and $\hat{H}$. In practical computations, one disposes of $\vec{V}$ through the momentum equations, for instance and

STREAMSURFACE APPROACH	AVERAGING APPROACH
CONTINUITY EQUATION	
$\partial_t \rho B + \vec{\bar{\nabla}}_\rho \vec{v}B = 0$ (T4.1)	$\partial_t \bar{\rho} b + \vec{\nabla}\bar{\rho} \vec{V}b = 0$ (T4.2)
MOMENTUM CONSERVATION EQUATION	
$\partial_t (\rho B\vec{v}) + \vec{\bar{\nabla}}(\rho\vec{v} \otimes \vec{v}) + p\bar{\bar{I}} - \bar{\bar{\tau}})B = \rho\vec{f}_e B + \rho\vec{f}B$ (T4.3)	$\partial_t(\bar{\rho} b\vec{\tilde{V}}) + \vec{\nabla}(\bar{\rho}\vec{V} \otimes \vec{V} + \bar{p} - \bar{\bar{T}})b = \bar{\rho}\tilde{f}_e b + \bar{\rho}\vec{f}b - \vec{\nabla}(\overline{\rho\vec{v}'' \otimes \vec{v}''b})$ (T4.4)
$\rho\vec{f} = J\partial_3 \{(p\bar{\bar{I}} - \bar{\bar{\tau}}) \frac{\vec{n}}{J}\}$ (T4.5)	$\bar{\rho}\bar{f} = -\frac{1}{b}[(pI - \bar{\bar{\tau}})\vec{n}]$ (T4.6)
ALTERNATIVE FORMULATION - NON-CONSERVATIVE FORM	
$\partial_t\vec{v} + (\vec{v}\,\vec{\bar{\nabla}})\vec{v} = -\frac{1}{\rho}\vec{\bar{\nabla}}(p\bar{\bar{I}} - \bar{\bar{\tau}}) + \vec{f}_e + \vec{f}_B$	$\partial_t\vec{V} + (\vec{V}\vec{\nabla})\vec{V} = -\frac{1}{\bar{\rho}}\vec{\nabla}(\bar{p}\bar{\bar{I}} - \bar{\bar{T}}) + \vec{\tilde{f}}_e + \vec{f}_B - \frac{1}{b\bar{\rho}}\vec{\nabla}(\overline{\rho\vec{v}'' \otimes \vec{v}''b})$
$\vec{f}_B = \vec{f} - \frac{1}{\rho B}(p\bar{\bar{I}} - \bar{\bar{\tau}})\,\vec{\bar{\nabla}}B = \frac{\vec{n}}{\rho h_1 h_2}\partial_3[(p\bar{\bar{I}} - \bar{\bar{\tau}})\,h_1 h_2]$ (T4.7)	$\vec{f}_B = \bar{f} - \frac{1}{\bar{\rho}b}(\bar{p}\bar{\bar{I}} - \bar{\bar{T}})\,\vec{\nabla}b$ (T4.8)
CROCCO'S FORM	
$\partial_t\vec{v} - \vec{v} \times \vec{\bar{\zeta}} = T\vec{\bar{\nabla}}S - \vec{\bar{\nabla}}H + \frac{1}{\rho}\vec{\bar{\nabla}}\bar{\bar{\tau}} + \vec{f}_e + \vec{f}_B$ (T4.9) $\vec{\bar{\zeta}} = \vec{\bar{\nabla}} \times \vec{v}$	
ENERGY CONSERVATION LAW	
$\partial_t(\rho EB) + \vec{\bar{\nabla}}(\rho\vec{v}H - k\vec{\nabla}T - \bar{\bar{\tau}}.\vec{v})B = Q.B + Q_b B$ (T4.11)	$\partial_t(\bar{\rho}\tilde{E}b) + \vec{\nabla}(\bar{\rho}\vec{V}\tilde{H} - \vec{\bar{B}})b = \bar{Q}.b + Q_b.b - \vec{\nabla}(\overline{\rho\vec{v}''H''b})$ (T4.12)
$Q_b = +\,\partial_3\,(k\vec{\nabla}T + \bar{\bar{\tau}}\vec{v})\vec{n}$	$Q_b = \frac{1}{b}[\,\vec{B}.\vec{n}\,] - \frac{1}{b}[p\;\partial_t b\,]$ $\vec{B} = k\vec{\nabla}T + \bar{\bar{\tau}}.\vec{v}$
NON-CONSERVATIVE FORM	
$\rho(\partial_t H + \vec{v}\vec{\bar{\nabla}}H) = \frac{1}{B}\partial_t\, pB + \frac{1}{B}\vec{\bar{\nabla}}(k\vec{\nabla}T + \bar{\bar{\tau}}\vec{v})B + Q + Q_b$ (T4.13)	$\bar{\rho}\,[\,\partial_t\tilde{H} + (\vec{V}\vec{\nabla})\tilde{H}\,] = \frac{1}{b}\partial_t\;\bar{p}b + \frac{1}{b}\vec{\nabla}(\vec{\bar{B}}b) + \bar{Q} + Q_b - \frac{1}{b}\vec{\nabla}(\overline{\rho\vec{v}''H''b})$ (T4.14)
NOTATIONS	
The divergence operator $\vec{\bar{\nabla}}\vec{F}$ is defined by $\frac{1}{h_1 h_2}[\partial_1(h_1 h_2 F^1) + \partial_2(h_1 h_2\,F^2)] = \vec{\bar{\nabla}}\vec{F}$ where $\vec{n}$ is the normal to the surface $\xi^3 = \xi^3(\xi^1, \xi^2)$ defined by the components $\vec{n} = (-\frac{\partial\xi^3}{\partial\xi^1}, -\frac{\partial\xi^3}{\partial\xi^2}, 1)$ and $J = \frac{1}{h_1 h_2 h_3}$	The averaged quantities are defined by $\bar{\rho}\tilde{A} = \frac{1}{b}\int_{b_o}^{b_1} \rho A\, d\xi^3$ where ξ^3 is the averaged coordinate direction $b = (b_1 - b_o)$ $\vec{V}$ is the averaged velocity

Table 1 : Quasi-Three Dimensional Formulation. Streamsurface and averaging approach.

the average kinematic energy $\bar{k}$ of the fluctuating flow field $\vec{v}''$,

$$\bar{k} = \frac{\overline{\rho\, \vec{v}''\vec{v}''}}{2\bar{\rho}} \tag{2.27}$$

is unknown. Note also the definition of the stagnation enthalpy fluctuation

$$H'' = H - \tilde{H} = h'' + \vec{v}''.\vec{V} + \frac{\vec{v}''^2}{2} - \bar{k} \tag{2.28}$$

3. THE STREAMSURFACE THROUGH FLOW APPROXIMATION FOR TURBOMACHINERY FLOWS

Two families of streamsurfaces are generally considered of the form illustrated in fig. 3. The S1 family is of the blade-to-blade type and can be considered as generated by the particles situated initially on a line at a constant radius r. The S2 family is generated by particles located initially on a radial line and is designated as a hub-to-shroud surface and the corresponding flow as the through flow.

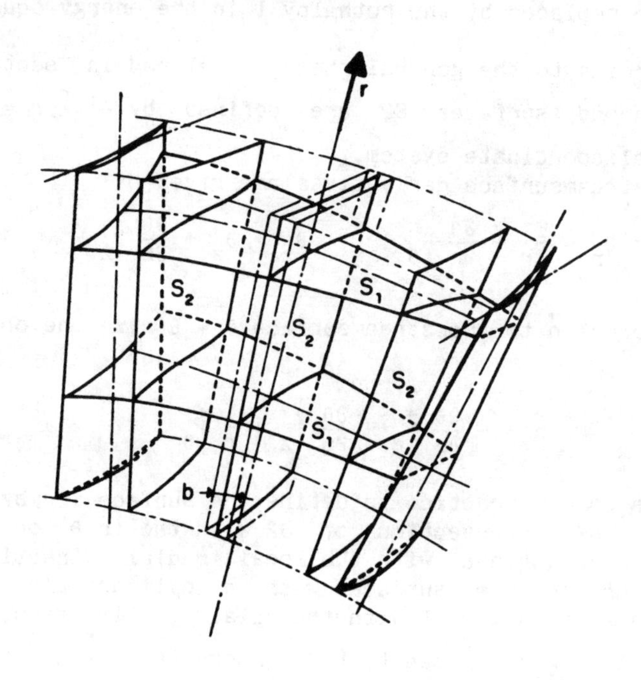

Figure 3 : Families of streamsurfaces

The streamsurface Q-3-D approximation has been introduced by Wu [1] and applied by a large number of authors, among which Katsanis [6],[7], Novak [8], Marsh [9], Adler & Krimerman [10], Biniaris [11], Bosman & El-Shaarawi [12]. A more complete list of references, including program development descriptions which have been published outside the journal literature can be found in the review article by S. Serovy in a recent AGARD report, Hirsch & Denton [13].

In the following, we will concentrate on the flow in the S2 surface, namely the through flow. Some of its iteraction aspects with the blade to blade flow in the Q-3-D framework will also be considered.

In accordance with the distributed loss model, the heat conduction and shear stresses are neglected, but a function force $\vec{F}$ is added to the external force.

In addition, we will consider the flow relative to a rotating frame of reference, rotating with a constant angular velocity ω. As a consequence centrifugal and Coriolis forces will have to be added to the momentum equations and the stagnation enthalpy H will have to be replaced by the rothalpy I in the energy equation.

Referring to the general theory developed in section 1, the hub-to-schroud surfaces S2 are defined by $\xi^3 \triangleq r = \theta(r,z)$ in a cylindrical coordinate system.

The streamsurface derivatives are given by

$$\bar{\partial}_r = \partial_r + \frac{\partial\theta}{\partial r}\,\partial_\theta \qquad \bar{\partial}_z = \partial_z + \frac{\partial\theta}{\partial z}\,\partial_z \tag{3.1}$$

For the normal $\vec{n}$ to the streamsurface $\theta = \theta(r,z)$ one obtains, equ. (1.4)

$$\vec{n} = -\frac{\partial\theta}{\partial r}\,\vec{1}_r - \frac{\partial\theta}{\partial z}\,\vec{1}_z + \frac{1}{r}\,\vec{1}_\theta \tag{3.2}$$

It is common practice to define the surface S2 by angles β' and ε'. The intersection of S2 with the (r.θ) or z = constant plane makes an angle ε' with the local radial direction and the intersection of the surface with a cylindrical, r = constant surface is at an angle β' with the axial z direction, figure 2. These angles are related to the components of the normal vector $\vec{n}$ since they define uniquely the local orientation of the streamsurface. They are defined by

$$r \frac{\partial\theta}{\partial r} = \tan \varepsilon' \qquad r \frac{\partial\theta}{\partial z} = \tan \beta' \qquad (3.3)$$

and the normal $\vec{n}$ is given by

$$r.\vec{n} = - \tan \varepsilon' \vec{1}_r - \tan \beta' \vec{1}_z + \vec{1}_\theta \qquad (3.4)$$

In the following it is assumed that the streamsurface is time independent, hence one has

$$\vec{w}.\vec{n} = 0 \qquad \text{or} \qquad w_\theta = w_z \tan \beta' + w_r \tan \varepsilon' \qquad (3.5)$$

Further, an axial flow angle $\hat{\beta}$ in the cylindrical surface r = constant and a meridional flow angle σ in the meridional plane (r-z) are introduced by (figure 4 and 5)

$$\text{tg } \hat{\beta} = w_\theta / w_z \qquad \text{and} \qquad \text{tg } \sigma = w_r / w_z \qquad (3.6)$$

The flow angle β is defined with respect to the meridional velocity component w_m in the meridional plane and having the direction σ with the axial direction

$$w_m = \sqrt{w_r^2 + w_z^2} \qquad \tan \beta = \frac{w_\theta}{w_m} \qquad (3.7)$$

The streamtube thickness, also called the <u>tangential blockage</u> factor b is related to the tangential variations of the flow by equation (1.11)

$$\frac{1}{b}\frac{\bar{d}b}{dt} = \frac{1}{r} (-\tan \beta' . \partial_z w_\theta - \text{tang } \varepsilon' . \partial_r w_\theta + \partial_\theta w_\theta)$$
$$= \frac{1}{b} (\bar{\partial}_t b + w_r \bar{\partial}_r b + w_z \partial_z b) \qquad (3.8)$$

For an axisymmetric flow the tangential blockage factor b will be constant and can be set equal to one. But inside blade rows, the blade-to-blade flow creates a tangential variation of the flow field and hence variations of b. In absence of any other information the blockage factor b may be approximated by the tangential spacing variation between two adjacent blades. A better approximation can be obtained from the knowledge of the flow in the θ-direction that is from the flow along the S1-blade-to-blade surfaces, in this case b can be represented by the distance variation between two adjacent streamlines. The body force $\vec{f}_B$, appearing in Crocco's form of the streamsurface momentum equation, and defined by equation (1.17) reduces to

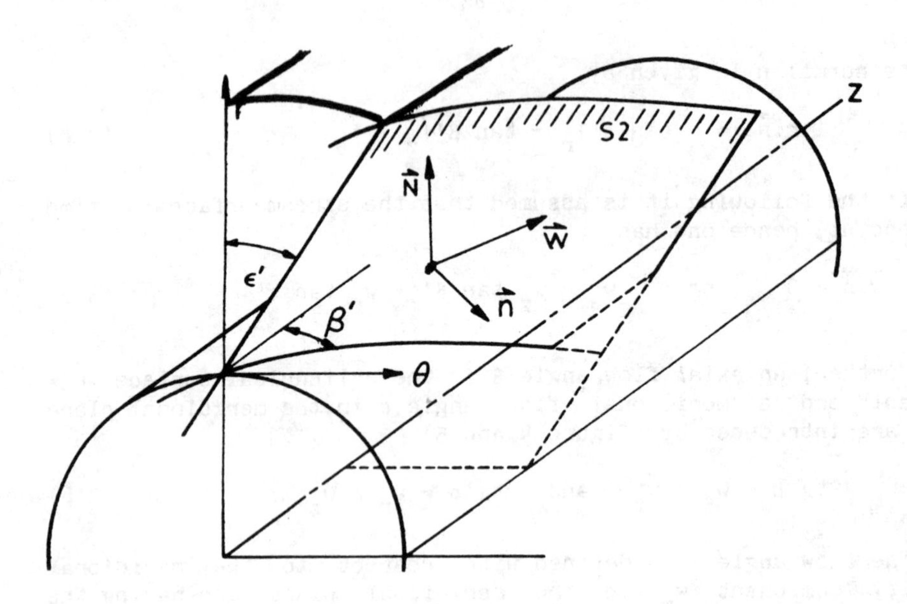

Figure 4 : Angles of the S^2 surface with radial and axial direction.

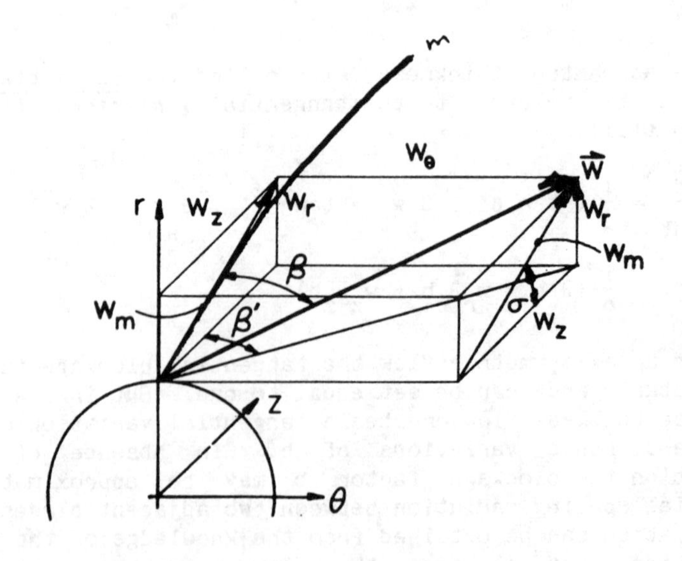

Figure 5 : Definition of flow angles and velocity components

$$\vec{f}_B = -\frac{\vec{n}}{\rho}\frac{\partial p}{\partial \theta} \tag{3.9}$$

This force, which results from the confinement of the flow description to a two-dimensional surface, is directed along the normal, and hence is orthogonal to the local velocity vector ($\bar{w}.\vec{f}_B = 0$).

Combining equ. (3.9) and (3.4) allows the determination of two force components f_Br and f_Bz in function of the third one f_θ.

$$f_{Br} / f_{B\theta} = -\tan \varepsilon' \quad f_{Bz} / f_{B\theta} = -\tan \beta' \quad f_{B\theta} = \frac{-1}{\rho r}\frac{\partial p}{\partial \theta} \tag{3.10}$$

The direction of the friction force $\vec{F}_f$ is defined as being opposed to the relative velocity vector $\vec{w}$. Hence, two components of $\vec{F}_f$ are determined by the flow angles in function of the third one, or alternatively in function of the magnitude of $\vec{F}_f$.

$$F_{f\theta} = -F_f.\frac{w_\theta}{w} \quad F_{fr} = -F_f.\frac{w_r}{w} \quad F_{fz} = -F_f.\frac{w_z}{w} \tag{3.11}$$

The through flow equations, in cylindrical coordinates, take the following form within the streamsurface approach, when Crocco's form is selected for the momentum equation.

$$\partial_t (\rho br) + \bar{\partial}_r (\rho brw_r) + \bar{\partial}_z (\rho brw_z) = 0$$

$$\partial_t w_r - w_z (\bar{\partial}_r v_z - \bar{\partial}_z v_r) - \frac{w_\theta}{r} \bar{\partial}_r (rv_\theta) = T\bar{\partial}_r s - \bar{\partial}_r I + F_{fr} + f_{Br}$$

$$\partial_t w_z + w_r (\bar{\partial}_r v_z - \bar{\partial}_z v_r) - \frac{w_\theta}{r} \bar{\partial}_z (rv_\theta) = T\bar{\partial}_z s - \bar{\partial}_z I + F_{fz} + f_{Bz} \tag{3.12}$$

$$\partial_t w_\theta + \frac{1}{r} (w_r\bar{\partial}_r + w_z\bar{\partial}_z)(rv_\theta) = F_{f\theta} + f_{B\theta}$$

$$\bar{\partial}_t s + w_m\bar{\partial}_m s = \rho \frac{W}{T} F_f$$

$$\bar{\partial}_t I + w_m\bar{\partial}_m I = \frac{1}{\rho} \bar{\partial}_t p$$

The latter two equations are the entropy and energy equation, where m is a coordinate along a meridional streamline, hence $w_m\bar{\partial}_m = w_r\bar{\partial}_r + w_z\bar{\partial}_z$. Actually, the whole system (3.12) is frequently written in a (θ,m,n) coordinate system. The quasi-three-dimensional approximation coupled to the distributed loss model, neglects the details of the viscous phenomena and in particular the effects of the individual wakes of the blades. As mentioned earlier one considers instead a mixed out, uniformized flow entering a given blade row. Therefore it is fully consistent with the already assumed approximations to consider that, for steady inlet flow conditions to the machine and steady rotation, the <u>relative flow</u> remains <u>steady</u>. In this case, the set of S2-streamsurface equations simplify to the following form, in cylindrical coordinates.

$$\bar{\partial}_r (\rho r b w_r) + \bar{\partial}_z (\rho r b w_z) = 0 \tag{3.13a}$$

$$- w_z (\bar{\partial}_r w_z - \bar{\partial}_z w_r) = T\bar{\partial}_r s - \bar{\partial}_r I + \frac{w_\theta}{r} \bar{\partial}_z (r v_\theta) + F_{fr} + f_{Br} \tag{3.13b}$$

$$w_r (\bar{\partial}_r w_z - \bar{\partial}_z w_r) = T\bar{\partial}_z s - \bar{\partial}_z I + \frac{w_\theta}{r} \bar{\partial}_z (r v_\theta) + F_{fz} + f_{Bz} \tag{3.13c}$$

$$\frac{1}{r} w_m \bar{\partial}_m (r v_\theta) = F_{f\theta} + f_{B\theta} \tag{3.13d}$$

$$w_m \bar{\partial}_m s = \frac{w}{T} F_f \tag{3.13e}$$

$$w_m \bar{\partial}_m I = 0 \tag{3.13f}$$

Note that one of the above equations is not independent from the others.

The last equation (3.13f) states that the total energy or rothalpy I is constant along a meridional projection of the streamline along the S2-surface.

Equation (3.13d) defines the meridional variations of the angular momentum rv_θ as determined by the component of the friction force and the body force. In practical computations equ. (3.13.e) determines the magnitude of the friction force F_f and its components are given by equation (3.11), while the necessary information is provided for the variations of rv_θ - see next section.

The remaining equations (3.13a) to (3.13c) form a system for the velocity components w_r, w_z. Since the entropy equation has been used explicitely, the two momentum equations are not independent and one of them can be applied indifferently, while the other is discarded.

The radial component of the momentum equation (3.13b) is known in the turbomachinery literature as the radial <u>equilibrium equation</u>.

Principles of Through-flow computations

Three sets of external information have to be provided to the system of equations (3.13), expressing the missing information with regard to viscous effects ($\vec{F}_f$) as well as the properties of the flow in the omitted direction, that is the θ or tangential direction (b, rv_θ).

i) The tangential blockage or streamsheet thickness b.

This information has to be defined by the flow along the other family or surfaces, namely the S1 or blade-to-blade surfaces according to equ. (3.8). In absence of this information, the following approximation is usually applied, whereby b is made flow independent

$$b = \frac{\text{geometrical width of channel}}{\text{inlet width of channel}} = 1 - \frac{d}{s} \tag{3.14}$$

where d is the blade thickness in the tangential direction and s the blade spacing or pitch. In duct regions, outside blade rows, b can be taken equal to 1.

ii) The angular momentum rv_θ.

Since equations (3.13.a) and (3.13.b) or (3.13.c) solve for the velocity components w_r and w_z in the meridional plane, the tangential components w_θ or $v_\theta = w_\theta + w_r$ have to be known from another source. Here again, this source is the flow in the other family of surfaces, namely the blade-to-blade or S1 surfaces. When this information is not available, empirical correlations between the flow angle β and some geometrical as well as aerodynamic parameters of the blade-to-blade flow have to be provided, replacing the knowledge of the detailed blade-to-blade flow.

These turning or deviation correlations, will approximate the flow angle β_2 at outlet of a blade row. When calculations are performed inside blade rows, the streamwise variation of β from its inlet value to the outlet value β_2, will have to be assumed. Hence a certain law of evolution of the flow angle

$$\beta = \beta(z) \qquad \text{or} \qquad \beta = \beta(m) \tag{3.15}$$

has to be given, next to the empirically determined outlet angles β_2.

Therefore, one will define, in a blade row, the tangential velocity by

$$w_\theta = w_m \tan \beta \tag{3.16}$$

Outside blade rows, the flow can be assumed as axisymmetric because of the absence of boundary constraints and of the assumed mixed-out flow conditions attached to the distributed loss model assumptions. In this case, the body force as well as the friction force vanish and equation (3.13.d) shows that the angular momentum is constant along a meridional projection of the streamlines. Hence,

rv_θ = constant along a m-line outside blade rows

The above considerations can be applied to the analysis problem whereby the blade row geometry is known. However, the radial equilibrium equations are also used in the design procedures whereby the geometry is undefined but whereby a given distribution of energy H in the spanwise direction of the blades is desired as well as a given distribution of (rv_θ). Hence the design problem, the tangential velocity follows an imposed distribution $v_\theta(r)$ or $v_\theta(z)$ which will have to be satisfied by the selected blading.

iii) The entropy production.

From equation (3.13.e), the entropy is generated by the distributed friction force F_f which represents the overall effect of the viscous and turbulent shear stresses on the flow. The contributions of the friction force F_f are solely connected to the energy dissipation, that is to the energy losses within the blade rows and the machine components.

It is customary in turbomachinery practice, to represent the energy losses by the stagnation pressure variations for compressible flow, or by the variations of the rotary stagnation pressure p* for incompressible flows.

In the first case, one has for the perfect gas under consideration, the variation of entropy from a point A to the current point

$$e^{-(s - s_A)/r} = \frac{p_o}{p_{o_A}} \left(\frac{T_{o_A}}{T_o}\right)^{\gamma/(\gamma-1)} \tag{3.17}$$

considered in an absolute or relative system. From the constancy of I along a streamline, one has with T_o' denoting relative stagnation temperatures ($T_o' = T + \frac{\vec{w}^2}{2c_p}$) :

$$c_p T'_o - \frac{u^2}{2} = c_p\, T'_{o_A} - \frac{u_A^2}{2} \qquad (3.18)$$

The stagnation pressure loss between an inlet point A and the current position is defined by a coefficient Π

$$\Pi = \frac{p_o^{(is)'} - p'_o}{p_d} \qquad (3.19)$$

where p'_o is the relative stagnation pressure and $p^{(is)'}$ the relative stagnation pressure at the current point obtained by an isentropic transformation from A

$$\frac{p_o^{(is)'}}{p_{o_A}'} = \left(\frac{T'_o}{T'_{o_A}}\right)^{\gamma/\gamma-1} = \left(1 - \frac{u_A^2 - u^2}{2c_p\, T'_{o_A}}\right)^{\gamma/\gamma-1} \qquad (3.20)$$

In an absolute, nonrotating system

$$p_o^{(is)} = p_{o_A} \qquad \text{since} \qquad T_o = T_{o_A}$$

The stagnation pressure loss is non-dimensionalized by a reference dynamic pressure p_d defined either at inlet or at outlet of a blade row. From the knowledge of the loss coefficient Π, one obtains the entropy variation

$$\frac{s - s_A}{r} = -\ln\left(1 - \frac{\Pi p_d}{p_o^{(is)'}}\right) \qquad (3.21)$$

from which the friction force F_f can be defined.

The determination of the stagnation pressure losses is strictly outside the present level of approximation and the losses have to be provided by external sources such as empirical informations.

The boundary conditions are the inviscid conditions, namely the normal relative velocity is zero on a solid wall, since F_f is treated as an external force.

The density is computed from the knowledge of the flow velocities and the local stagnation conditions, for instance

$$\frac{\rho}{\rho_0'} = \left(1 - \frac{\vec{w}^2}{2H_0'}\right)^{1/(\gamma-1)} = \left(1 - \frac{w_m^2 + w_\theta^2}{2H_0'}\right)^{1/(\gamma-1)} \tag{3.22}$$

where $H_0' = c_p\, T_0'$.

Alternative forms of through-flow equations

The choice between equation (3.13b) and (3.13c) depends on the general flow configuration. Equation (3.13b), the radial component of the momentum equation, has a coefficient w_z in the left hand side and will not be suitable for a machine with a predominant radial flow, that is for radial or centrifugal machines. Similarly, the axial component of the momentum equation, equation (3.13c) cannot be applied for axial machines where w_r might vanish in all or part of the flow domain. For mixed flow machines one would have to use one or the other equation according to the flow region. An elegant way to avoid this has been introduced by Bosman and Marsh [14], and consists in using the projection of the momentum equation on a direction $\vec{N}$ situated in the S2 surface and perpendicular to the velocity vector $\vec{w}$. We refer to this reference for a detailed description, see also Hirsch [15].

Mathematical Properties of the through-flow equations

It is important to analyze the mathematical properties of the system formed by the continuity equation and the momentum equation, for instance the radial component for an axial machine, with regard to the characteristics and their elliptic or hyperbolic properties. The radial momentum equation is best written, for this analysis, in the following form, obtained after introduction of the continuity equation, for steady flows

$$w_r \bar{\partial}_r w_r + w_z \bar{\partial}_z w_r + \frac{1}{\rho} \bar{\partial}_r p = F_{fr} + f_{Br} + \omega^2 r + 2\omega . w_z \tag{3.23}$$

where the right hand side is unimportant for the properties of the system. The continuity equation is transformed to the quasi-linear form, using the relation between the variations of density and velocity. One has, from equation (3.22)

$$\delta\rho = -\frac{\rho}{a^2}(w_r \delta w_r + w_z \delta w_z + w_\theta \delta w_\theta) \quad \text{and} \quad \partial_\theta p = \left(\frac{\partial p}{\partial \rho}\right) \partial_r \rho = a^2 . \partial_r \rho \tag{3.24}$$

where a^2 is the square of the speed of sound.

The dependent variables are w_r and w_z and as discussed above, w_θ is supposed to be known from the blade-to-blade information. If

w_θ is defined by the flow angle $\hat{\beta}$, equation (3.6), then w w_θ will be function of w_z and contribute to the definition of the properties of the equation. In the other case, if w_θ is given as a function of coordinates, as in the design option, w_θ will be considered independent of the velocity component w_z. We will combine these two cases by writing formally

$$\delta w_\theta = \nu . \tan \hat{\beta} . \delta w_z \tag{3.25}$$

where $\nu = 1$ when the flow angle $\hat{\beta}$ is given and $\nu = 0$, when w_θ is imposed.

Hence, the continuity and radial momentum equations can be written under the system form as follows

$$A \partial_r \begin{vmatrix} w_r \\ w_z \end{vmatrix} + B \partial_z \begin{vmatrix} w_r \\ w_z \end{vmatrix} = \begin{vmatrix} q_1 \\ q_2 \end{vmatrix}$$

or explicitly

$$\begin{vmatrix} (1-M_r^2 & -M_r M_z (1+\nu \tan^2\hat{\beta}) \\ 0 & -w_z (1+\nu \tan^2\hat{\beta}) \end{vmatrix} \partial_r \begin{vmatrix} w_r \\ w_z \end{vmatrix}$$

$$+ \begin{vmatrix} -M_r M_z & 1-M^2(1+\nu \tan^2\hat{\beta}) \\ w_z & 0 \end{vmatrix} \partial_z \begin{vmatrix} w_r \\ w_z \end{vmatrix} = \begin{vmatrix} q_1 \\ q_2 \end{vmatrix} \tag{3.26}$$

where M_r and M_z are the radial and axial Mach number components and the right hand sides represent the terms independent of the velocity derivatives. The above system is elliptic or hyperbolic if the characteristics are imaginary or real. They are determined by the solutions of the determinant $|A + \lambda B| = 0$. One obtains the characteristic directions

$$\lambda = \frac{M_r M_z (1+\nu\tan^2\hat{\beta}) \pm \sqrt{[M_r^2 + M_z^2 (1+\nu\tan^2\hat{\beta}) - 1](1+\nu\tan^2\hat{\beta})}}{1 - M_z^2 (1 + \nu \tan^2\hat{\beta})} \tag{3.27}$$

and the important following properties, valid for all quasi-three-dimensional descriptions along the S2 type of surfaces : The system of equ. (3.13) is <u>elliptic</u> for

$$M_m^2 + \nu M_\theta^2 < 1 \tag{3.28}$$

that is, for <u>subsonic meridional Mach numbers</u> when w_θ is imposed, while the system is elliptic for <u>subsonic relative Mach numbers</u> $M^2 = M_m^2 + M_\theta^2 < 1$, when the flow angle $\hat{\beta}$ is given. Any solution method will have to take these properties into account. The first case, namely the design type, whereby w_θ is considered as a known function, is of particular interest and deserves some comments. The condition (3.28) indicates that the through-flow problem can be treated in an elliptic way for subsonic meridional Mach numbers, although the relative Mach number can be supersonic. Hence, for given w_θ distributions, one can handle relative supersonic flows inside the blade rows in the same way as the subsonic relative flows and compute the meridional velocity components within the blade rows by implementing in the appropriate way, the information about w_θ to be transfered from the blade-to-blade surfaces.

However if this information is transmitted under the form of the flow angles β or $\hat{\beta}$, no computational method of elliptic type will be able to handle correctly supersonic relative flows at calculation points situated within the blade rows.

Note that outside blade rows, that is in the duct region, the above conditions automatically apply since v_θ is always given (rv_θ = constant). Hence in duct regions the problem will always be elliptic for subsonic meridional Mach numbers.

4. THE PASSAGE-AVERAGED THROUGH FLOW APPROXIMATION

The averaging procedure for obtaining a quasi-three dimensional approximation to the turbomachinery flow consists in defining an average meridional cross-section of the machine representative of the average flow over the pitch of the blade row.

The averaged flow at a given point P of the meridional cross-section is obtained by performing an integration over θ of the flow properties from pressure surface to suction of the next blade along the axisymmetric surface generated by the rotation of the meridional streamline m.

The density weighted average of a quantity A is defined by

$$\bar{\rho}\tilde{A} = \frac{1}{\frac{2\pi b}{N}} \int_{\theta_p}^{\theta_s} \rho A d\theta \tag{4.1}$$

where $\bar{b} = \frac{2\pi b}{N} = \frac{2\pi}{N}(1 - d/s)$ is the angular difference between pressure and suction surface and b a streamtube thickness, only function of the geometry.

Applying the general procedure of section 2, one obtains the averaged equations for the through flow in a meridional cross section of the machine. With the cylindrical coordinates (θ, r, z), the jacobian $J = \frac{1}{r}$ is independent of the averaging direction θ. Hence, equation (2.13) for the averaged continuity equation in the relative system becomes

$$\partial_t (\bar{\rho} b r) + \partial_r (\bar{\rho} b r W_r) + \partial_z (\bar{\rho} b W_z) = 0 \qquad (4.2)$$

where W_r and W_z are the density weighted averages of the relative velocity components with $b = 1 - d/s$. This is an exact equation to be compared with equation (3.12a) where b is given by equ. (3.8). Formally, both equations are identical, but in the streamsurface approach b is flow dependent and is only approximately equal to the value given by equation (3.14). On the other hand, equation (3.12a) is the mass conservation law for the local velocities along an S2 streamsurface which has to be defined, while equation (4.2) is the mass conservation law for the averaged meridional flow considered in a true meridional cross section of the machine. For an axisymmetric flow, however, both approaches lead to exactly the same equations.

<u>The passage averaged momentum equation</u>

Applying the results summarized in equation (2.11) with the body force $\vec{f}$ defined by equation (2.14) to the momentum equation of the distributed loss model, in the relative system gives

$$\frac{1}{b}\partial_t (\rho \vec{W} b) + \frac{1}{b} \vec{\nabla} (\overline{\rho \vec{W}} \otimes \vec{W} b) = \frac{-1}{b} \vec{\nabla} \bar{p} b - 2\bar{\rho} (\vec{\omega} \times \vec{W}) - \bar{\rho}\omega^2 \vec{r} \qquad (4.3)$$

$$+ \overline{\rho \vec{F}_f} + \overline{\rho \vec{f}} - \frac{1}{b} \vec{\nabla} (\overline{\rho \vec{w}'' \otimes \vec{w}'' b})$$

where the gradient is operating on the r and z coordinate space.

The body force $\vec{f}$ is defined by

$$\overline{\rho \vec{f}} = \frac{-1}{\bar{b}} [p\vec{n}] = \frac{1}{\frac{2\pi b}{N}} (p_p \vec{n}^{(p)} - p_s \vec{n}^{(s)}) \qquad (4.4)$$

The normal $\vec{n}$ to the blade surfaces are defined by the angles β' and ε' as in equation (3.4) and are given for the blade pressure surfaces, by equation (3.2)

$$n_r^{(p)} = -\frac{1}{r}\,\text{tg}\,\varepsilon'_p \qquad n_z^{(p)} = -\frac{1}{r}\,\text{tg}\,\beta'_p \tag{4.5}$$

Similar equations hold for the suction surface, defining the angles β'_s, ε'_s. Hence, one obtains for the body force $\vec{f}$:

$$\bar{\rho} f_z = -\frac{1}{bs}(p_p \tan\beta'_p - p_s \tan\beta'_s)$$

$$\bar{\rho} f_r = -\frac{1}{bs}(p_p \tan\varepsilon'_p - p_s \tan\varepsilon'_s) \tag{4.6}$$

$$\bar{\rho} f_\theta = \frac{1}{bs}(p_p - p_s)$$

This last relation shows that the body force originates from the presence of the blades in the flow field as solid surfaces able to sustain pressure forces. Actually f_θ is nothing else than the tangential projection of the local lift force on the blade. Hence, outside the blade region, $\vec{f} = 0$ and the flow is considered as axisymmetric. The last term of equation (4.3), which has the same structure as the Reynolds stresses in turbulent averaged Navier-Stokes equations, represents the influence of the full three dimensional flow on the averaged flow components. The velocity fluctuations $\vec{w}''$ are the velocity components generated by the deviation from axisymmetry such as the secondary flow contributions on the blade-to-blade flow variation and we will designate these terms as the <u>interaction terms</u> or <u>secondary stresses</u> to refer to the fact that their contribution arises essentially from secondary flows.

An alternative form to equation (4.3) is obtained by taking into account the continuity equation (4.2), leading to

$$\partial_t \vec{W} + (\vec{W}\vec{\nabla})\,\vec{W}$$

$$= -\frac{1}{\bar{\rho} b}\vec{\nabla}\bar{p}b - 2\vec{\omega}\times\vec{W} + \omega^2\vec{r} + \vec{\vec{F}}_f + \vec{f} - \frac{1}{b\bar{\rho}}\vec{\nabla}\,\overline{(\rho\vec{w}''\otimes\vec{w}''b)} \tag{4.7}$$

and the projections in cylindrical coordinates :

<u>Radial momentum equation</u>

$$\partial_t W_r + (W_r\partial_r W_r + W_z\partial_z W_r - \frac{W_\theta^{\,2}}{r}) = -\frac{1}{\bar{\rho} b}\partial_r b\bar{p} + \omega^2 r + 2\omega W_\theta + F_{f,r} + f_r - \sum_{i=1}^{3} N_i^{(r)} \tag{4.8}$$

where the secondary stresses are represented by the following three contributions

$$N_1^{(r)} = \frac{1}{\bar{\rho} br} \partial_r \left(\overline{br\rho w''_r w''_r}\right)$$

$$N_2^{(r)} = \frac{1}{\bar{\rho} br} \partial_z \left(\overline{br\rho w''_r w''_z}\right) \quad N_3^{(r)} = \frac{-1}{\bar{\rho} r} \left(\overline{\rho w''_\theta w''_\theta}\right) \tag{4.9}$$

Similarly the axial component becomes

$$\partial_t W_z + W_r \partial_r W_z + W_z \partial_z W_z = -\frac{1}{\bar{\rho} b} \partial_z (\bar{p} b) + F_{f,z} + f_z - \sum_{i=1}^{2} N_i^{(z)} \tag{4.10}$$

where

$$N_1{}^{(z)} = \frac{1}{\bar{\rho} br} \partial_r \left(\overline{br\rho w''_r w''_z}\right) \quad N_2{}^{(z)} = \frac{1}{\bar{\rho} br} \partial_z \left(\overline{br\rho w''_z w''_z}\right) \tag{4.11}$$

while the tangential component becomes

$$\partial_t W_\theta + W_r \partial_r W_\theta + W_z \partial_z W_\theta + \frac{1}{r} W_r W_\theta$$
$$= -2\omega W_r + F_{f,\theta} + f_\theta - \sum_{i=1}^{3} N_i^{(\theta)} \tag{4.12}$$

or

$$\partial_t W_\theta + \frac{W_m}{r} \partial_m (rV_\theta) = F_{f,\theta} + f_\theta - \sum_{i=1}^{3} N_i^{(\theta)}$$

where

$$N_1^{(\theta)} = \frac{1}{\bar{\rho} br} \partial_r \left(\overline{br\rho w''_r w''_\theta}\right)$$

$$N_2^{(\theta)} = \frac{1}{\bar{\rho} br} \partial_z \left(\overline{br\rho w''_z w''_\theta}\right) \quad N_3{}^{(\theta)} = \frac{1}{\bar{\rho} r} \overline{\rho w''_\theta w''_r} \tag{4.13}$$

A detailed investigation of the influence and order of magnitude of the interaction terms originating from the blade-to-blade flow variations inside a compressor blade row has been presented by Smith [16]. The main conclusion is that the interaction terms are proportional to the blade loading or lift coefficient but that this effect is rather small in the considered case, although their influence might be non-negligible in certain configurations with high loadings and secondary flows, for instance in turbine passages.

A similar analysis, performed by Hirsch [3], for the contributions arising from the two-dimensional wakes of the blades, shows that the interaction terms are proportional, outside the blade passage in the vicinity of the trailing edge, to the loss coefficients of the corresponding blade section. The influence of these terms has been found to be generally negligible although in regions of high losses such as the sections close to the end-walls, their effect might need to be accounted for.

In addition to the two-dimensional effects, three-dimensional effects due to secondary flows and local vorticity production can strongly influence the averaged flow, through significant contributions to the interaction terms.

An experimental investigation by Sehra [17] in a transonic compressor rotor showed that the magnitude of these terms have a non-negligible effect on the averaged flow and that the omission of their contribution led to an inaccurate prediction of the radial flow distribution, see also Sehra and Kerrebroeck [18].

Hence in general, the contributions of the secondary stress will influence the averaged flow unless the blade rows are formed of lightly loaded blades generating small secondary flows.

The passage averaged energy equation

The averaged energy equation is obtained from the general form given by equation (2.22) and (2.23) written for a relative system. Writing I instead of H and W instead of V, one obtains, within the assumptions of the distributed loss model,

$$\partial_t\,(\bar{\rho}\tilde{I}b) + \vec{\nabla}\,(\bar{\rho}\tilde{W}\tilde{I}b) = \partial_t\bar{p}b - \vec{\nabla}\,\overline{(\rho\vec{w}''I''b)} - [p.\partial_t\theta] \tag{4.14}$$

or with the continuity equation

$$\bar{\rho}\partial_t\tilde{I} + \rho\vec{W}\vec{\nabla}\tilde{I} = \frac{1}{b}\,\partial_t\bar{p}b - \frac{1}{b}\,\vec{\nabla}\,\overline{(\rho\vec{w}''I''b)} - \frac{1}{b}\,[p.\partial_t\theta] \tag{4.15}$$

where $[p\partial_t\theta] = (p_s.\partial_t\theta_s - p_p\partial_t\theta_p) = [p\vec{n}\vec{w}]$. It is seen from the above equations that the averaged total energy represented by the averaged rothalpy $\tilde{I}$

$$\tilde{I} = \tilde{h} + \frac{\widetilde{\vec{w}^2}}{2} - \frac{\vec{u}^2}{2} = \tilde{h} + \frac{\vec{W}^2}{2} - \frac{\vec{u}^2}{2} + \frac{\overline{\rho\vec{w}''\vec{w}''}}{2\bar{\rho}} = \hat{I} + \frac{\overline{\rho\vec{w}''\vec{w}''}}{2\bar{\rho}} \tag{4.16}$$

cannot be considered as constant anymore along a streamline for the steady relative flow. Indeed, in the relative steady state situation, equ. (4.15) becomes

$$\rho\vec{W}\vec{\nabla}\tilde{I} = -\frac{1}{b}\,\vec{\nabla}\,\overline{(\rho\vec{w}''I''b)} \tag{4.17}$$

showing that the averaged total energy varies along a streamline, as a consequence of the non-axisymmetric energy fluxes $\overline{\rho\vec{w}''I''}$.

The energy conservation law can also be written for the total energy of the averaged flow $\hat{I}$ defined by equ. (4.16). One has instead of equation (4.14)

$$\partial_t(\bar{\rho}\hat{I}b) + \vec{\nabla}(\bar{\rho}\vec{\tilde{W}}\hat{I}b) = \partial_t\bar{p}b - \vec{\nabla}(\overline{\rho\vec{w}''I''b}) - \vec{\nabla}(\bar{\rho}\vec{\tilde{W}}\tilde{k}b) - [p\partial_t\theta] \tag{4.18}$$

where $\tilde{k}$ represents the average kinetic energy of the large scale fluctuations.

Written out explicitly in cylindrical coordinates, the energy equation takes the form

$$\begin{aligned}\partial_t\bar{\rho}\tilde{I}b + \frac{1}{r}\partial_r(b\bar{\rho}rW_r\tilde{I}) + \frac{1}{r}\partial_z(b\bar{\rho}rW_z\tilde{I}) &= \partial_t(\bar{p}b) - [p\partial_t\theta] \\ &- \frac{1}{r}\partial_r(\overline{b\rho w''_r I''r}) - \frac{1}{r}\partial_z(\overline{b\rho w_z''I''r})\end{aligned} \tag{4.20}$$

and with the introduction of the continuity equation, one obtains the alternative form

$$\begin{aligned}&\bar{\rho}\partial_t\tilde{I} + \bar{\rho}W_m\partial_m\tilde{I} \\ &= \frac{1}{b}\partial_t(\bar{p}b) - \frac{1}{b}[p\partial_t\theta] - \frac{1}{br}\partial_r(\overline{b\rho w''_r I''r}) - \frac{1}{br}\partial_z(\overline{b\rho w''_z I''r})\end{aligned} \tag{4.21}$$

Equation (4.18) shows that the total energy of the averaged flow does not remain constant along an averaged streamline following $\vec{W}$ in the steady state limit. This is an essential limitation of the averaged flow model for turbomachinery computation, since the general procedure in practical calculations is based on the constancy of energy, together with the entropy equation, which are explicitly used in order to replace one of the two momentum equations.

In the averaged formulation, Crocco's form can be derived either directly by averaging the three-dimensional form, or bij introducing the entropy and the enthalpy in the averaged momentum equation (4.7), Hirsch & Warzee [4].

One obtains the following averaged form of Crocco's momentum equation

$$\partial_t \vec{W} - \vec{W} \times \vec{\vec{\zeta}} = \tilde{T}\vec{\nabla}s - \vec{\nabla}\hat{I} + \vec{F}_f + \vec{f}^{(h)} - \frac{1}{b\rho}\vec{\nabla}\,\overline{(\rho\vec{w}'' \otimes \vec{w}''b)} + \frac{\overline{h''\vec{\nabla}\rho}}{\rho} \quad (4.22)$$

where the averaged absolute vorticity is defined by $\vec{\vec{\zeta}} = \vec{\nabla} \times \vec{V}$, and where $\vec{f}^{(h)}$ is a blade force expressed in function of the enthalpy variations.

The total energy gradient term appearing in this equation contains the total energy of the averaged flow $\hat{I}$. In practical computations, where the averaged quantities $\tilde{h}$, $\vec{W}$ are calculated, this quantity is the obvious one to consider, since it is only dependent on the averaged flow variables. This is not the case for the averaged total energy $\tilde{I}$ which depends also on the large scale flow deviations from axisymmetry $\vec{w}''$. However, neither of these quantities are conserved along a streamline of the averaged flow in steady state conditions. The implications of this situation have been discussed by Sehra [17] with respect to the interpretations of overall radial equilibrium as described by equation (4.22).

Compared to the corresponding equation obtained in the streamsurface approach, one obtains here the same terms with the addition of the already discussed interaction terms.

For the passage averaged form of the entropy equation one obtains

$$\bar{\rho}\,\tilde{T}\,(\partial_t s + \vec{W}\,\vec{\nabla}\,s) + \overline{\rho\,T''\vec{w}''\vec{\nabla}s} = -\,\bar{\rho}\,\vec{W}\,\vec{F}_f \quad (4.23)$$

It is important to notice that the averaged entropy equation does not contain contributions from the secondary stresses $\bar{\bar{\tau}}^{s}$, nor from the secondary flow enthalpy flux $-\overline{\rho\vec{w}''h}$.

Quasi-Three-Dimensional Interaction for the Averaged Flow Model

The only quasi-three-dimensional interaction possible here is the simplified interaction whereby the role of the mean streamsheet is played by the averaged flow considered in a true meridional plane.

The blade-to-blade streamsheet thickness B as well as the location of the meridional streamlines m are an output of the through flow computation of the averaged flow. With an initial approximation whereby all the secondary stresses are set to zero, that is, an axisymmetric flow approximation, the axisymmetric averaged flow is computed.

After the computation of the blade-to-blade flows an estimation of the interaction terms can be obtained, from the blade-to-blade flows, but also from additional, direct estimations of secondary flows and stream-wise vorticity, Hawthorne [19].

A new iteration cycle can then be initiated. The question arises actually in the description of turbomachinery flows as to the equivalence of the averaged flow representation and the mean streamsheet representation used in the simplified quasi-three-dimensional flow description discussed in the previous section. This topic has been discussed in particular by Horlock and Marsh (1971).

If a mean streamsheet is defined by angles β' and ε', is there a choice of values of β' and ε' which will render the two representations identical ?

Comparing the equations of the two models, a first conclusion arises from the energy equations (3.12f) and (4.17). It is clear that neither the total energy of the averaged flow $\hat{I}$, nor the averaged total energy $\tilde{I}$, are able to represent the total energy along the mean S2 streamsurface. For a steady state flow, equation (3.12f) gives

$$w_m \bar{\partial}_m I = 0$$

or

$$w_m \partial_m I = -(w_\theta \partial_\theta I)$$

while equation (4.17) leads to

$$\bar{\rho} W_m \partial_m \tilde{I} = - \frac{1}{br} [\partial_r (\overline{r\rho w''_r I'' b}) + \partial_z (\overline{\rho w''_z I'' br})]$$

Unless the right hand sides of both equations are zero, which is the case in an axisymmetric flow, no general choice of $\tilde{I}$ would render the energy fluxes of the interaction terms zero. With regard to the body forces in the two models, one obtains the following conclusion : If the pressure variation between pressure and suction surface, at a constant radial and axial position, is linear, the body forces of both models are identical, if the mean S2 streamsurface is selected to be identical with the camberline.

However for a more general pressure variation law, the two body forces will not be equal, unless the blade thickness is constant or zero. This corresponds to a infinitely thin blade and a mean streamsheet following the blade surface.

In addition, as discussed previously the interaction terms in the momentum equations are non-negligible except for lightly loaded blades generating negligible secondary flow effects. This would be the case for instance for highly spaced, low cambered cascade blades or with many, thin blade cascades.

Some direct estimations in simplified cases of the different contributions to various flow representations can be found in Horlock and Marsh [20].

It is to be concluded from these considerations, that in general, it is not possible to define a mean S2 streamsurface on which the flow variations are identical to the averaged flow variations. Considering that the averaged flow model describes correctly the behaviour of the real physical average flow over the blade spacing, the flow on a mean streamsurface will not be able to represent correctly the averaged flow.

Only in the hypothetical case of an axisymmetric flow, will both models be identical. Therefore, the interpretation of the calculated local behaviour of the flow variables along a mean S2 streamsurface should be taken with caution when considering them as representative of the averaged actual flow.

5. PASSAGE AVERAGED THROUGH FLOW EQUATIONS. CONNECTION WITH END WALL BOUNDARY LAYER EQUATIONS

Several methods have been developed for the numerical solution of the through-flow equations in an arbitrary, multistage turbomachine, which can be classified in two broad families. The first family of methods is based on the solution of the momentum and continuity equations for the physical variables, while the second family is based on the introduction of a streamfuntion in the meridional coordinates (r,z) leading to a second order partial differential equation for the streamfunction.

Within the first family, the streamline curvature method is based on an equation for the meridional velocity w_m ($=v_m$), along a given calculation station, obtained from the stationary momentum equation. It will be treated in detail in several lectures.

The streamfunction approach was introduced by Wu [1] in a finite difference discretization and renamed matrix through-flow method by Marsh [9]. The same approach has been followed by Davis & Millar [21],[22]. Davis [23], Biniaris [11], Bosman & Marsh [14], Bosman & El Shaarawi [12].

The discretization of the same streamfunction equations by Finite Elements has been introduced independently by Adler & Krimerman [10] and Hirsch & Warzee [24], [25].

The current approximations in through-flow computations assume relative steady flow situations at constant wheel speed ωr and some form of axisymmetry of the flow configuration. Actually, within the streamsurface approach, the equations along the S2 streamsheet are formally identical to the axisymmetric flow equations since the replacement of the streamsurface derivatives $\bar{\partial}$ by ordinary derivatives leads to the axisymmetric equations. Hence, within the streamsurface approach, the distinction between the flow along an arbitrary S2 streamsheet and an axisymmetric flow is a matter of interpretation, the numerical resolution techniques being identical in both cases.

This is not the case within the passage averaged model, since the extra terms represented by the secondary stresses $\vec{\nabla}\bar{\bar{\tau}}$ in the momentum equations and by the additional energy exchange and dissipation terms in the energy equations, vanish only in the axisymmetric assumption. Within a complete quasi-three-dimensional computation, the first step of the calculation will be based on an axisymmetric through-flow, while the next iterations will introduce some estimations of the interaction terms between the averaged flow and the non-axisymmetric contributions, and take into account their contributions to the momentum equations as well as on the energy and entropy equations.

The passage averaged equations allow in addition to define a coherent interaction between the meridional through-flow and the end-wall boundary layers developing along the hub and shroud walls of a turbomachine, fig. 6

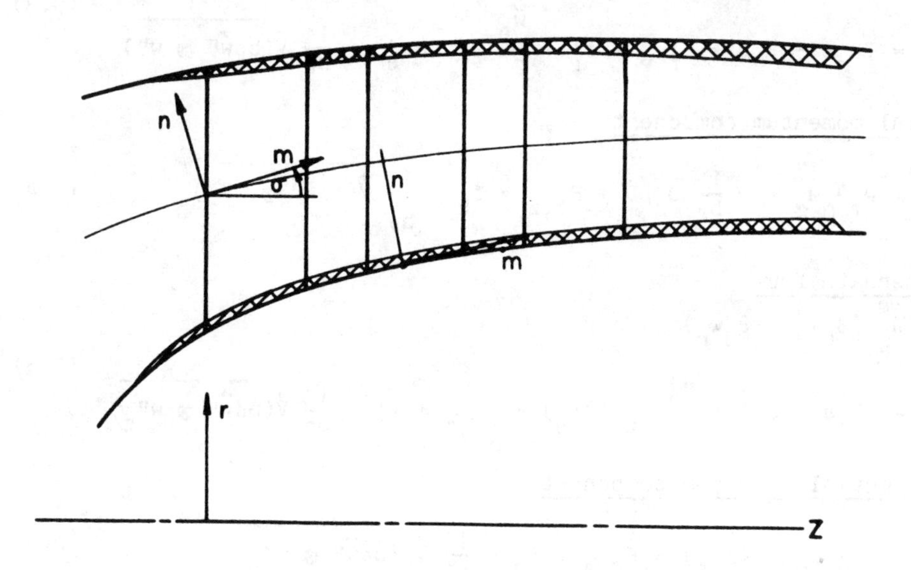

Figure 6 : Meridional cross section of multibladed turbomachine.

As shown in the previous sections, the through-flow equations consist of the steady state form of the mass conservation equation, one of the meridional (axial, radial or any combination such as the N-component) projections of the momentum conservation laws, the energy conservation and the entropy equation connecting the stagnation pressure variation to the empirical loss coefficients. In order to clarify the notation, we will remove the overbars on the derivatives in the streamsurface approach and the overbars and tilda on the flow variables in the averaged formulation of the conservation equations. Furthermore, since the averaged equations become identical to the streamsurface equations in the axisymmetric limit, we will consider only the averaged form implying that there is a single axisymmetric streamsurface through-flow model, obtained by setting all the interaction terms to zero.

The through-flow equations can be summarized, relying on section 4, as follows in cylindrical coordinates

Continuity equation

$$\partial_r (\rho b r w_r) + \partial_z (\rho b r w_z) = 0 \tag{5.1}$$

Radial equilibrium equation

$$w_m \partial_m w_r = -\frac{1}{b\rho} \partial_r pb + \frac{v_\theta^2}{r} + F_{f,r} + f_r - \frac{1}{b\rho} \vec{\nabla}\overline{(b\rho \vec{w}'' \otimes w_r'')} \tag{5.2}$$

Alternate form

$$w_z (\partial_z w_r - \partial_r w_z)$$

$$= T\partial_r s - \partial_r I + F_{f,r} + f_r' + \frac{w_\theta}{r} \partial_r (r v_\theta) - \frac{1}{b\rho} \vec{\nabla}\overline{(b\rho \vec{w}'' \otimes w'')} \tag{5.3}$$

Axial momentum component

$$w_m \partial_m w_z = -\frac{1}{b\rho} \partial_z pb + F_{f,z} + f_z - \frac{1}{b\bar{\rho}} \vec{\nabla}\overline{(b\rho w'' \otimes w''_z)} \tag{5.4}$$

Alternate form

$$w_r (\partial_r w_z - \partial_z w_r)$$

$$= T\partial_z s - \partial_z I + \frac{w_\theta}{r} \partial_z (r v_\theta) + F_{f,z} + f_z' - \frac{1}{b\rho} \vec{\nabla}\overline{(b\rho \vec{w}'' \otimes w''_z)} \tag{5.5}$$

Tangential momentum component

$$\frac{1}{r} w_m \partial_m (r v_\theta) = F_{f,\theta} + f_\theta - \frac{1}{\rho b} \vec{\nabla}\, \overline{(b\rho \vec{w}'' \otimes w''_\theta)} \tag{5.6}$$

Energy equation

$$\rho w_m \partial_m I = -\frac{1}{b}\vec{\nabla}(\overline{b\rho\vec{w}''h''}) - \frac{1}{b}\vec{\nabla}(\overline{b\rho\vec{w}k''}) + \frac{1}{b}\vec{\nabla}(b\bar{\bar{\tau}}^{s}\vec{w}) \tag{5.7}$$

Entropy equation

$$\rho T w_m \partial_m s = \rho w.F_f - \overline{\rho T''\vec{w}''\vec{\nabla}s} \tag{5.8}$$

The friction force $\vec{F}_f$ is defined by

$$\vec{F}_f = -F_f.\vec{1}w \tag{5.9}$$

and the body force is normal to the streamsurface S2 in the streamsurface approach, that is

$$\vec{f} = f_\theta.\frac{\vec{n}}{n_\theta} \tag{5.10}$$

with

$$\frac{n_r}{n_\theta} = -\tan\varepsilon' \qquad \frac{n_z}{n_\theta} = -\tan\beta' \tag{5.11}$$

Similar relations will be used for the body forces $\vec{f}$ of $\vec{f}'$ as defined in the averaged flow model.

The above equations are valid outside the boundary layer regions, since the friction force $\vec{F}_f$ is not resolved and is related to overall empirical loss coefficients. However the complete through-flow is influenced by the end wall boundary layers, through the blockage effect introduced by their displacement thickness and through their effect on the overall energy exchange process, Smith [16]. A consistent representation and description of the influence and interaction of the end wall regions and the main flow has been presented by Mellor & Wood [26], and further extended by Hirsch [27], [28], Horlock & Perkins [29], De Ruyck, Hirsch & Kool [30], De Ruyck & Hirsch [31], [32].

Within the pitch-averaged flow representation, the main stream through flow will be described by the radial component of the momentum equation for instance, in addition to the continuity equation and the energy and entropy laws. Within the same representation, the end wall boundary layer equations will make use of the other two components of the momentum equations. In an axisymmetric coordinate system (m,n,θ), the mainstream through-flow will solve the n-component while the end wall flow description will be based on the m and θ-components of the momentum equations.

End-Wall Boundary Layer equations

Due to the viscous nature of the end wall regions, the friction force $\vec{F}_f$ has to be written in its local, original form, namely

$$\rho\vec{F}_f = \vec{\nabla}\bar{\bar{\tau}} \tag{5.12}$$

where $\bar{\bar{\tau}}$ represents the viscous plus turbulent shear stress tensor. The local shear gradients have to be averaged over the blade spacing, following the rules of section 2. The axial component of the shear stress term becomes in cylindrical coordinates,

$$\vec{\nabla}\bar{\bar{\tau}}\big|_z = \frac{1}{r}\,\partial_r\,(r\tau_{zr}) + \frac{1}{r}\,\partial_\theta\tau_{z\theta} + \frac{1}{r}\,\partial_z\tau_{zz} \tag{5.13}$$

and applying the averaging rules, one obtains

$$\overline{\vec{\nabla}\bar{\bar{\tau}}\big|_z} = \frac{1}{br}\,\partial_r\,(br\bar{\tau}_{zr}) + \frac{1}{br}\,\partial_z\,(rb\bar{\tau}_{zz}) + \frac{1}{bs}\,\left[\frac{\bar{\bar{\tau}}.\vec{n}}{n_\theta}\right]_z \tag{5.14}$$

where $\vec{n}$ is the vector normal to the blade surfaces. The last term is the difference of the values of the shear stress along the suction and pressure surfaces, at a given radius. Consistently with channel flows, this difference can be set equal to twice the average of the blade shear stress $\vec{\tau}_b$ and has to be added to the inviscid body force $\vec{f}$ (or $\vec{f}'$).

$$\vec{\tau}_b = \frac{-1}{2bs}\,[\bar{\bar{\tau}}.\,\frac{\vec{n}}{n_z}] \tag{5.15}$$

The classical boundary approximations imply that the radial derivatives dominate the other derivatives of the stress tensor components and that the radial velocity is negligible compared to the axial and tangential components.

Hence, one obtains the following averaged momentum equations valid in the end wall regions

Axial momentum

$$\begin{aligned} w_m\partial_m w_z = &-\frac{1}{b\rho}\partial_z pb + f_z - \frac{1}{2\rho}\tau_{bz} + \frac{1}{br\rho}\partial_r\,(rb\bar{\tau}_{rz} - \overline{rb\rho w''_r w''_z}) \\ &+ \frac{1}{br\rho}\partial_z\,(b\bar{\tau}_{zz} - \overline{b\rho w''_z\,w''_z}) \end{aligned} \tag{5.16}$$

or

$$w_m \partial_m w_z = -\frac{1}{b\rho}\,\partial_z pb + f_z^{(w)} + \frac{1}{br\bar{\rho}}\,\partial_r\,(br\bar{\tau}_{rz} - br\overline{\rho w''_r w''_z}) \tag{5.17}$$

where the axial component of a wall region blade force $\vec{f}^{(w)}$ has been introduced, with

$$f_z^{(w)} = f_z - \frac{1}{2\bar{\rho}}\,\tau_{bz} + \frac{1}{b\bar{\rho}}\,\partial_z\,(b\bar{\tau}_{zz} - b\,\overline{\rho w''_z w''_z}) \tag{5.18}$$

The <u>tangential momentum equation</u> is treated similarly, leading to

$$\frac{1}{r}\,w_m\partial_m\,(rv_\theta) = f_\theta - \frac{1}{2\bar{\rho}}\,\tau_{b\theta} + \frac{1}{b\,\bar{\rho}}\,\partial_z\,(b\bar{\tau}_{z\theta} - b\,\overline{\rho w''_z w''_\theta})$$
$$-\frac{1}{\bar{\rho} r}\,\overline{\rho w''_\theta w''_r} + \frac{\bar{\tau}_{\theta r}}{r\bar{\rho}} + \frac{1}{br\bar{\rho}}\,\partial_r\,(br\bar{\tau}_{r\theta} - br\,\overline{\rho w''_r w''_\theta}) \tag{5.19}$$

or

$$\frac{1}{r}\,w_m\partial_m\,(rv_\theta) = f_\theta^{(w)} + \frac{1}{br\bar{\rho}}\,\partial_r\,(br\bar{\tau}_{r\theta} - br\,\overline{\rho w''_r w''_\theta}) \tag{5.20}$$

with

$$f_\theta^{(w)} = f_\theta - \frac{1}{2\bar{\rho}}\,\tau_{b\theta} + \frac{1}{b\bar{\rho}}\,\partial_z\,(b\bar{\tau}_{z\theta} - br\,\overline{w''_z w''_\theta}) + \frac{\bar{\tau}_{\theta r}}{r\bar{\rho}} - \frac{1}{r\bar{\rho}}\,\overline{\rho\, w''_\theta w''_r} \tag{5.21}$$

It is important to notice that the wall body force $\vec{f}^{(w)}$ is distinct from the main stream body force $\vec{f}$ and hence, that there will be a variation of the body force within the end wall boundary layer regions.

When an integral boundary layer representation is applied to the wall regions regions the integral of the force difference term will not vanish and a body force "defect" will remain

$$\Delta\vec{f} = \frac{1}{\delta}\int_0^\delta (\vec{f} - \vec{f}^{(w)})dr \tag{5.22}$$

where δ is the boundary layer thickness, acting on the end wall flow. This "force defect" has been known to play a fundamental role in the development of the end wall layers in multistage axial compressors, Smith [16], Mellor & Wood [26].

Note also that the energy equation in the boundary layers is given by any of the forms developed in the previous section but with the addition of the contributions from the work of the shear stresses $1/b\ \vec{\nabla}^{(2)}(b\bar{\bar{\tau}}.\vec{w})$ added to the right hand side of the equations, as well as the contributions from the work of the blade surface shear stress $\vec{\tau}_b$.

$$Q_b = -\frac{1}{2\bar{\rho}}\ \vec{\tau}_b\ \vec{v}_s \qquad (5.23)$$

where $\vec{v}_s$ is the blade velocity in the considered reference system.

6. THROUGH-FLOW COMPUTATIONAL METHODS

Several methods have been developed for the numerical solution of the throug-flow equations in an arbitrary, multistage turbomachine, which can be classified in two broad families. The first family of methods is based on the solution of the momentum and continuity equations for the physical variables, while the second family is based on the introduction of a stream function in the meridional coordinates (r,z) leading to a second order partial differential equation for the stream function.

Within the first family, the streamline curvature method is based on an equation for the meridional velocity $w_m(=v_m)$, along a given calculation station, obtained from the stationary momentum equation. The dependence of the flow on the other coordinate is expressed through the curvature of the meridional projection of the streamline ; hence the name of the method. The streamline curvature method was developed for through-flow computations in its contemporary form in the early sixties, Katsanis [6], [7], Smith [16], Novak [8], Jansen & Moffatt [33] and is still widely used. It is discussed elsewhere in this book.

The streamfunction methods will be discussed in this section from the point of view of the numerical approach, leaving aside any detailed considerations with regard to the necessary empirical input under the form of loss coefficients and turning angles.

6.1. Streamfunction equations

An alternative to the streamline curvature method is provided by the introduction of a streamfunction ψ, in order to satisfy the continuity equation. This leads to a second order, quasi-linear, partial differential equation for the streamfunction, which can be discretized on a fixed mesh, either by finite difference or by finite element methods.

The continuity equation, (5.1) can be satisfied for any

continuous function ψ, by setting

$$w_z = \frac{1}{\rho b r} \partial_r \psi$$
$$w_r = - \frac{1}{\rho b r} \partial_z \psi \tag{6.1}$$

Inserting these definitions in equation (5.3), leads to the following streamfunction equation, considering an axisymmetric approximation.

$$\partial_r(\frac{1}{\rho b r} \partial_r \psi) + \partial_z(\frac{1}{\rho b r} \partial_z \psi) = \frac{-1}{w_z} [T \partial_r s - \partial_r I + \frac{w_\theta}{r} \partial_r r v_\theta + F_r^T] \tag{6.2}$$

where F_r^T stands for the sum of the blade force and friction force radial components.

This equation can be applied as long as w_z does not vanish, that is for axial flow configurations. For radial flow configurations, the axial momentum equation (5.5) can be used, leading to an equivàlent streamfunction equation

$$\partial_r(\frac{1}{\rho b r} \partial_r \psi) + \partial_z(\frac{1}{\rho b r} \partial_z \psi) = \frac{1}{w_r} [T \partial_z s - \partial_z I + \frac{w_\theta}{r} \partial_r r v_\theta + F_z^T] \tag{6.3}$$

This equation in turn ceases to be valid when the radial velocity component vanishes.

A general form, valid vor any flow configuration and which can therefore be applied in mixed flow machines with axial-radial transitions is obtained by introducing the component of the momentum equation, in the direction defined by $\vec{w} \times \vec{f}^{(h)}$, Bosman & Marsh [14]. The streamfunction equation becomes in this case

Blade region

$$\begin{aligned}\partial_r(\frac{1}{\rho b r} \partial_r \psi) + \partial_z(\frac{1}{\rho b r} \partial_z \psi) = & - \frac{1}{w^2} [(w_z + w_\theta \tan \beta')(T \partial_r s - \partial_r I) \\ & -(w_r + w_\theta \tan \varepsilon')(T \partial_z s - \partial_z I)] \\ & - \frac{1}{r} [\tan \beta' \partial_r(r v_\theta) - \tan \varepsilon' \partial_z(r v_\theta)]\end{aligned} \tag{6.4}$$

β' and ε' can be considered equal to the flow angles with respect to the axial and radial directions respectively.

Duct region

$$\partial_r\left(\frac{1}{\rho br}\,\partial_r\psi\right) + \partial_z\left(\frac{1}{\rho br}\,\partial_z\psi\right) = \frac{-1}{W_m^2}\,[w_z(T\,\partial_r s - \partial_r H) - w_r(T\,\partial_z s - \partial_z H) + \frac{w_\theta}{r}\,[w_z\partial_r(rv_\theta) - w_r\partial_z(rv_\theta)]] \tag{6.5}$$

For any of the above representations, the streamfunction equation will be written as

$$\partial_r\left(\frac{1}{\rho br}\,\partial_r\psi\right) + \partial_z\left(\frac{1}{\rho br}\,\partial_z\psi\right) = q \tag{6.6}$$

An alternative form is obtained by working out the left-hand side into a Laplace operator

$$\Delta\psi = [-\partial_r\left(\frac{1}{\rho br}\right).\partial_r\psi - \partial_z\left(\frac{1}{\rho br}\right).\partial_z\psi + q]\ (\rho br) \tag{6.7}$$

or

$$\Delta\psi = q.\rho br - w_r\partial_z(\rho br) + w_z\partial_r(\rho br)$$

Equation (6.6) is elliptic for subsonic relative Mach numbers in the analysis case and for subsonic meridional Mach numbers in the design option when rv_θ is specified. This results from the general analysis of section 3 but can also be shown directly on equation (6.6).

6.2. The determination of density

One of the central problems with the application of streamfunction methods for compressible flows is connected to the non-unique relation between the streamfunction and the density. From the isentropic relation between static and stagnation variables one has, from equation (3.22)

$$\rho/\rho_o' = \left(1 - \frac{\vec{w}^2}{2H_o'}\right)^{1/(\gamma-1)} = \left(1 - \frac{|\vec{\nabla}\psi|^2}{2(\rho br)^2 H_o'} - \frac{w_\theta^2}{2H_o'}\right)^{1/(\gamma-1)} \tag{6.8}$$

where

$$|\vec{\nabla}\psi|^2 = |\partial_r\psi|^2 + |\partial_z\psi|^2 \tag{6.9}$$

and where the <u>stagnation quantities are considered in the relative system</u>.

For fixed values of streamfunction gradient $|\vec{\nabla}\psi|^2$, stagnation enthalpy H_o' and stagnation density ρ_o', equation (6.8) has generally two solutions, one corresponding to a supersonic flow, the other to a subsonic relative flow.

Writing equation (6.8) under the form

$$\rho/\rho_o' = \left(1 - \frac{w_\theta^2}{2H_o'} - \frac{A}{(\rho/\rho_o')^2}\right)^{1/\gamma-1} \triangleq F(\rho/\rho_o) \tag{6.10}$$

$$A = \frac{|\vec{\nabla}\psi|^2}{2(\rho_o' br)^2 H_o'} = \frac{w_m^2}{2H_o'(\rho_o'/\rho)^2} \tag{6.11}$$

for the design option (rv_θ or w_θ fixed) and

$$\rho/\rho_o = \left(1 - \frac{|\vec{\nabla}\psi|^2(1 + \tan^2\beta)}{2H_o'(\rho br)^2}\right)^{1/\gamma-1}$$

$$\triangleq \left(1 - \frac{A}{(\rho/\rho_o')^2}\right)^{1/\gamma-1} \triangleq F(\rho/\rho_o) \tag{6.12}$$

$$A = \frac{|\vec{\nabla}\psi|^2(1 + \tan^2\beta)}{2H_o'(\rho_o' br)^2} = \frac{\vec{w}^2}{2H_o'(\rho_o'/\rho)^2} = \frac{(\rho w_m)^2(1 + \tan^2\beta)}{2H_o'\rho_o'^2} \tag{6.13}$$

for the analysis option ($\tan\beta$ fixed), one obtains the following graphical representation, fig. 7.

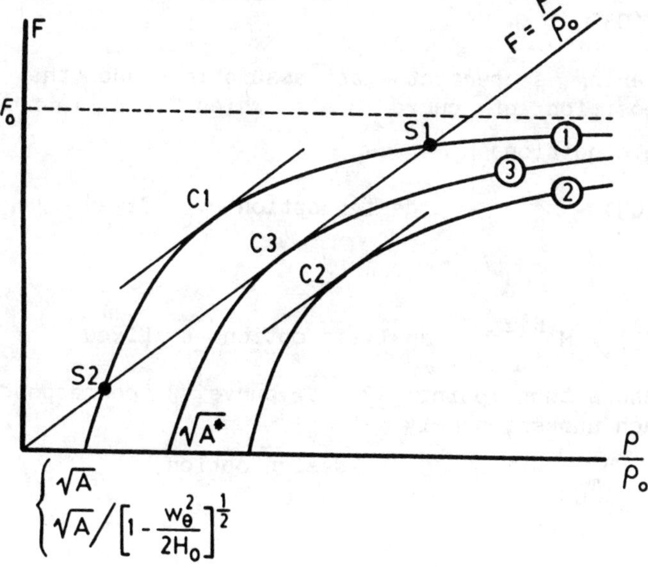

Figure 7. : Relation between density and streamfunction gradients

The asymptotic limit for large values of ρ is given by (omitting the accent for the stagnation quantities)

$$F_0 = (1 - w_\theta^2/2H_0)^{1/\gamma-1} \qquad \text{if } rv_\theta \text{ fixed} \qquad (6.14)$$
$$F_0 = 1 \qquad \text{if } \beta \text{ fixed}$$

The number of solutions for the density depends on the value of the constants A. Generally one will have two values of the density, given by the points S1 and S2 of figure 7., for a given value of the constant A. Point S1 is the subsonic solution while point S2 corresponds to a supersonic velocity. However, if point C lies below the line $F=\rho/\rho_0$, for instance curve 2, there will be no solutions for the density corresponding to the current values of the constants A.

The condition to be imposed in order to have solutions to equation (6.8) is that

$$F(C) > (\rho/\rho_0)_C \qquad (6.15)$$

where $(\rho/\rho_0)_C$ is defined by the condition

$$F'(C) = 1 \qquad (6.16)$$

The limiting case being reached for curve 3 where $F(C3)=(\rho/\rho_0)_{C3}$.

Considering a perfect gas assumption and the particular limiting position of curve 3, for which $F=\rho/\rho_0$ in point C3, one obtains the conditions

$$F'(C3) = M_m^2 \qquad \text{design option } rv_\theta\text{- fixed} \qquad (6.17)$$

or

$$F'(C3) = M^{(R)2} \qquad \text{analysis option } \beta \text{ fixed}$$

This shows that point C3 of curve 3 corresponds to the critical Mach number, namely

$$M_{m_{C3}} = 1 \qquad \text{design option} \qquad (6.18)$$

or

$$M^R_{C3} = 1 \qquad \text{analysis option}$$

and that the critical density ratio is reached

$$(\rho/\rho_0)^*_{C3} = (\frac{\gamma+1}{2})^{-1/\gamma-1} \cdot (1 + \frac{\gamma-1}{\gamma+1} M^2_\theta)^{-1/\gamma-1} \quad \text{design option} \qquad (6.19)$$

or

$$(\rho/\rho_0)^*_{C3} = (\frac{\gamma+1}{2})^{-1/\gamma-1} \quad \text{analysis option}$$

The lowest value of (ρ/ρ_0) is obtained by the intersection of the F curve with the horizontal axis, that is by F=0. Hence

$$(\rho/\rho_0)_{min} = \sqrt{A} \,/\, \sqrt{1 - w^2_\theta/2H_0} \quad rv_\theta \text{ fixed - design option}$$
$$= \sqrt{A} \quad \beta \text{ fixed - analysis option} \qquad (6.20)$$

The critical value of A,A*, corresponds to curve C3, with the condition

$$(\frac{2}{\gamma-1} \cdot \frac{A}{(\rho/\rho'_0)^2} \cdot \frac{H_0}{L})^* = 1 \qquad (6.21)$$

or

$$A^* = \frac{\gamma-1}{2} \cdot [(\rho/\rho'_0)^*]^{(\gamma+1)}$$

with $(\rho/\rho'_0)^*$ given by equation (6.19).

The condition for the density equation (6.8) to have solutions, can be expressed by the condition that the values of A corresponding to curve C1 should be larger than A*. This is actually a condition on the mass flow for unit area, ρw_m, since the critical mass flow is given by

$$(\rho w_m)^* = \rho_0 \sqrt{(\gamma-1)H'_0}\, (\frac{\rho^*}{\rho'_0})^{\gamma+1/2} = \sqrt{A^*}\, \rho'_0 \sqrt{2H'_0} \qquad (6.22)$$

Hence, the condition A < A* becomes in all cases with either w_θ or tan β fixed,

$$\frac{(\rho w_m)^2}{2H_0\rho_0{}^2} < \frac{(\rho w_m)^{*2}}{2H_0\rho_0{}^2}$$

or

$$\rho w_m < (\rho w_m)^* \tag{6.23}$$

In practical computations, when the computed streamfunction is far from the converged value, a careful check has to be performed in order to ensure that the current approximation of $|\vec{\nabla}\psi|^2$ is not too high such as to violate the condition (6.23).

When (6.23) is satisfied, two solutions occur corresponding to the subsonic branch, point S1 or to the supersonic branch of the mass flow relation versus density or Mach number.

In order to solve the density equation and obtain a unique solution, one can apply a Newton method or a fixed point method. The former will allow to obtain either of the two solutions, if the user knows which one to select, while the latter will always lead to the subsonic solution.

Graphically, it is easy to see that the Newton iteration will converge to the subsonic or the supersonic solution depending on the value chosen to initiate the iteration.

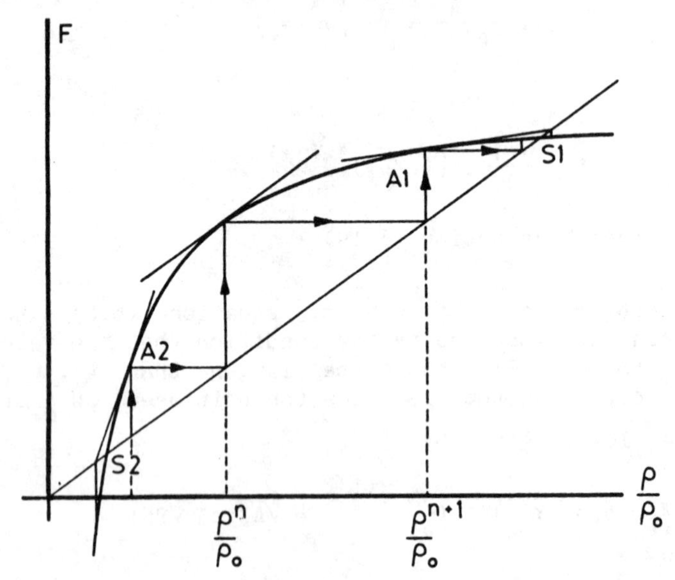

Figure 8. : Iterative solution procedure for the density equation

If the iteration starts at a subsonic velocity point A1, the Newton procedure will converge to S1, while starting with A2 at a supersonic point will lead to the supersonic solution S2.

On the other hand, the fixed point iteration

$$(\rho/\rho_o)^{n+1} = F(\rho^n/\rho_o) \qquad (6.24)$$

converges always to the subsonic solution S1 as seen on fig. 8., except when the initial value is left of the supersonic solution. In this case the iteration diverges.

Boundary conditions for the streamfunction equation

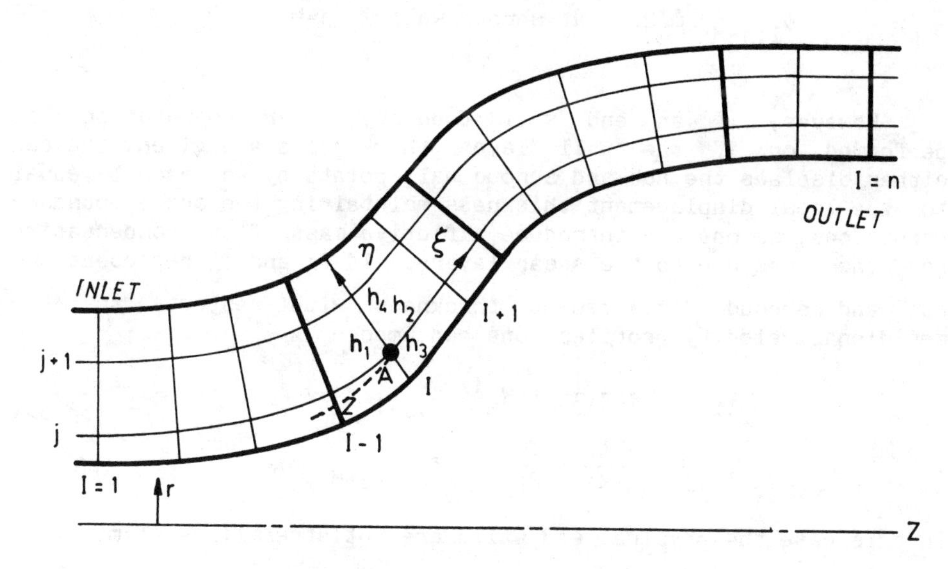

Figure 9. : Meridional section of mixed flow machine and grid layout

The boundary conditions for the solution of the streamfunction equation are easily defined and applied.

At the inlet station, the flow is assumed to be known and therefore it is easy to determine the streamfunction distribution.

For

$$I = 1 \quad : \quad \psi(x_{1,j}\ ,\ y_{1,j}) = \psi_{1,j} \quad \text{given} \qquad (6.25)$$

At the outlet station it is nearly always possible to select a station which is situated in a region of uniform meridional flow direction. Hence, if the last station is set perpendicularly to this direction, one can assume the normal derivative to be zero. That is

$$\frac{\partial \psi}{\partial n} = 0 \quad \text{at the last station I=n} \tag{6.26}$$

where n is the normal direction.

The hub and shroud walls can be taken as streamlines if no end wall boundary layer is to be considered. In this case one can select

$$\begin{aligned} \psi_{I,j=1} &= 0 \qquad \text{at hub wall} \qquad j=1 \\ \psi_{I,j=M} &= \dot{m}/2\pi \quad \text{at shroud wall} \quad j=M \end{aligned} \tag{6.27}$$

However, if an end wall boundary layer computation is performed or if the wall layer thicknesses are given, one can either displace the hub and shroud wall points by an amount equal to the local displacement thickness maintaining the above boundary conditions, or one can introduce a fictive mass flux compensating for the loss due to the shear layers. If δ^*_H and δ^*_S represent the hub and shroud displacement thickness with respect to the meridional velocity profiles, one can impose

$$\begin{aligned} \psi_{I,j=1} &= -\frac{1}{2\pi}(\rho w_m \delta^*_H)_{I,j=1} \\ \psi_{I,j=M} &= \frac{\dot{m}}{2\pi} + \frac{1}{2\pi}(\rho w_m \delta^*_S)_{I,j=M} \end{aligned} \tag{6.28}$$

In this case the physical end walls are not streamlines anymore.

6.3. Finite Difference solutions of the streamfunction equation

The discretization of the streamfunction with finite differences has been initially proposed by Wu [1]. The approach of Wu was closely followed and developed to practical computer programs by Marsh [9], Davis [23], Katsanis & McNally [34], Biniaris [11], Bosman & El-Shaarawi [12].

All these authors use central difference schemes and hence the methods are limited to subsonic flow regions, either to meridional subsonic velocities for the design option, or to relative subsonic velocities in the analysis case. The densities are obtained by the fixed point iteration scheme in all the presently published methods. Of the above mentioned methods, Davis [35] and Katsanis & McNally [34] contain full documented codes. In particular the latter, called MERIDL, is extensively documented.

In most methods an arbitrary grid is generated, although Katsanis & McNally work with a curvilinear orthogonal grid and Biniaris uses a cartesian grid in the meridional surface (r,z).

Second order central differences are applied by these two authors, while Marsh [9] and Davis [35] apply up to third order central difference formulas on arbitrary grids.

The algebraic matrix system is mostly solved by LU decompositions with the exception of Katsanis & McNally [34] and Biniaris [11] who apply successive over relaxation techniques.

In local coordinates ξ,η, second order difference formulas are used

$$\left.\frac{\partial^2\psi}{\partial\xi^2}\right|_{I,j} = \frac{2\psi_{I-1,j}}{h_1(h_1+h_2)} + \frac{2\psi_{I+1,j}}{h_2(h_1+h_2)} - \frac{2\psi_{I=j}}{h_1 h_2} \tag{6.29}$$

Higher order difference formulas are applied by Marsh [9] and Davis [35] see also Davis & Millar [21].

Computational procedure

The mesh is swept from the inlet station $I=1$ to the outlet station $I=N$ and the determination of the flow properties is based on the conservation of flow variables, such as rv_θ in duct regions or I(H) in blade rows, along a meridional streamline.

In point A of figure 9., the streamline is for instance AZ and the location of point Z, has to be found by interpolation with the condition $\psi_Z=\psi_A$. All necessary flow variables are then computed at point Z by some interpolation formula compatible with the order of accuracy of the difference approximation.

The coefficients of the discretized matrix equations are then computed at all mesh points and a new solution for $\psi_{I,j}$ can be obtained. All the above mentioned authors apply a Laplace operator for the convergence operator, solving the streamfunction equation in the form

$$L\ \delta\psi = -\tau\ R(\psi^n) \qquad \delta\psi = \psi^{n+1} - \psi^n \tag{6.30}$$

where R is the residual.

The optimal underrelaxation factor τ is depending on the degree of non-linearity of the problem. An analysis of the influence of Mach number and of angular momentum gradients on the convergence rate has been presented by Bosman [36].

Although no numerical estimations of the underrelaxation factor can be given in general, it will be necessary to reduce τ for increasing maximum Mach numbers and increasing gradients of angular momentum. This is also confirmed by numerical experiments,

Bosman [36], Hirsch & Warzee [37].

A particular mention has to be made about Katsanis & McNally's [34] program MERIDL. This code applies finite difference discretizations in the subsonic regions $M^R < 1$ for analysis problems and shifts to a streamline curvature method if shock free supersonic flow domains appear. The slopes and curvatures of the streamlines are obtained from a streamfunction, finite difference solution at a lower mass flow such that all points have subsonic velocities. These slopes and curvatures are kept fixed in the streamline curvature calculation at the full mass flow.

6.4. Finite Element Solutions of the Streamfunction Equations

Finite element solution procedures for the through-flow problem have been introduced by Adler & Krimerman [10] and Hirsch & Warzee [24]. The first authors define triangular meshes with linear elements while Hirsch & Warzee apply biquadratic elements leading to third order accuracy for the streamfunction and hence, second order accuracy for the velocities.

The basic streamfunction equation is written under the quasi-harmonic form

$$\vec{\nabla} \frac{1}{\rho b r} \vec{\nabla}\psi - f = 0 \qquad (6.31)$$

where the gradient operator $\vec{\nabla}$ acts on the coordinates (r,z) and $(-f)$ represents the right hand side terms.

Applying a standard Galerkin method and a finite element representation

$$\psi(r,z) = \sum_J \psi_J \ \phi_J(r,z) \qquad (6.32)$$

where J is a node number and ϕ_J the corresponding interpolation function one obtains the following system

$$K_{IJ} \cdot \psi_J = f_I \qquad (6.33)$$

The stiffness matrix K is given by

$$K_{IJ} = \int_\Omega \frac{1}{\rho b r} \left(\frac{\partial \phi_I}{\partial r} \cdot \frac{\partial \phi_J}{\partial r} + \frac{\partial \phi_I}{\partial z} \cdot \frac{\partial \phi_J}{\partial z}\right) dr\, dz \qquad (6.34)$$

$$f_I = \int_\Omega f(r,z)\ \phi_I\ dr\, dz \qquad (6.35)$$

where the boundary conditions are already included with the

assumption ϕ_I=0 along the inlet station and along the hub and shroud walls where Dirichlet conditions are imposed. The Neumann boundary condition at outlet is automatically satisfied by the solutions of the matrix system-natural boundary condition-. Hirsch & Warzee use 8-node quadrilateral elements and a typical mesh is shown on figure 10.

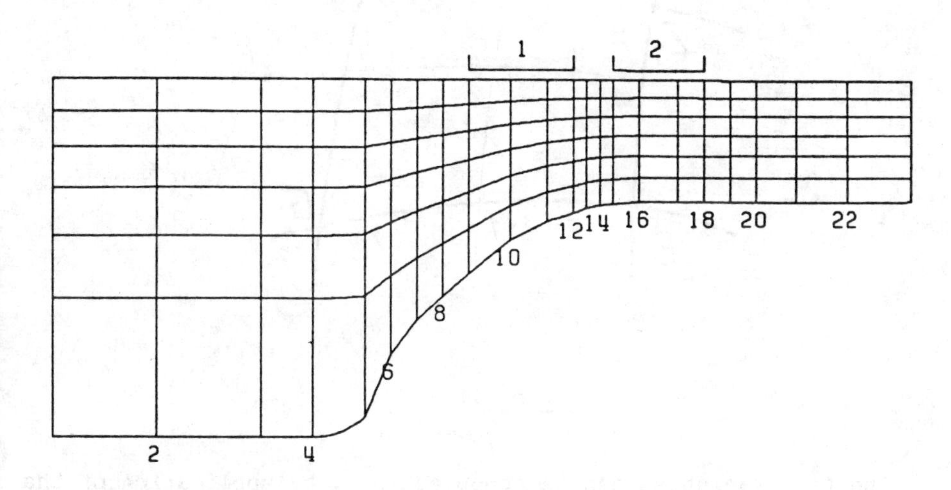

Figure 10. : Finite element meshes in meridional sections

The matrix elements are calculated, using an isoparametric transformation and Gauss points integration. Generally a (2 x 2) Gauss integration formula is sufficient.

Computational procedure

The succession of calculations follows closely the steps of the finite difference approach.

The determination of the streamline position makes use of the interpolation functions. For instance, within the element of figure 11., where EC and GA are located on two consecutive stations, the position of point Z on the streamline of node A is obtained by solving the following equation, written in the case of a biquadratic element

$$\psi_Z = \psi_A = \sum_{C}^{E} \psi_J \phi_J(\xi,\eta) \qquad (6.36)$$

leading to a second order algebraic equation in η along the side EZC which corresponds to ξ=-1.

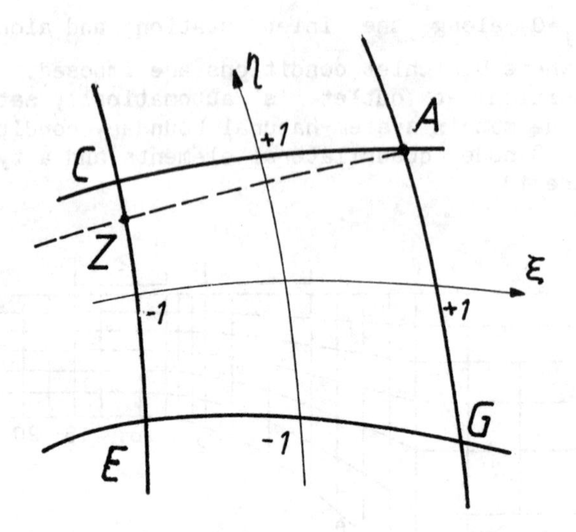

Figure 11. :

The flow variables can be computed in Z, by application of the interpolation functions. For any quantity B one can write

$$B_Z = \sum_{C}^{E} B_J \, \phi_J(\xi_Z, \eta_Z) \tag{6.37}$$

where $\xi_Z = -1$ in the case of figure 11

Adler & Krimerman [10] displace the mesh from one iteration to the other in order that the nodes of the linear triangular elements remain on a streamline. This avoids the need for the interpolations of flow quantities to the streamline position Z.

The iterative solution of the algebraic system is based on the so-called constant stiffness or variable stiffness methods

$$K_o . \delta\psi = -\tau R^n \qquad \delta\psi = \psi^{n+1} - \psi^n \tag{6.38}$$

where

$$K_o = K^m \qquad m < n-1 \qquad \text{"constant stiffness" method}$$

or

$$K_o = K^{n-1} \qquad \text{"secant stiffness" method} \tag{6.39}$$

and τ is the underrelaxation factor, $\tau < 1$.

Another choice is $K_0 = L$ where L is the Laplace operator ; this is applied by Adler & Krimerman.

An exact Newton iteration is not possible since the dependence of entropy or stagnation pressure losses and v_θ or β with ψ cannot be determined explicitly because of the empirical nature of these quantities which are external inputs to the through-flow problem.

However, a partial Newton method can be applied if these contributions are neglected in the computation of $(\partial f/\partial\psi)$. However, compared to a constant stiffness approach, whereby the K_0 matrix has only to be inverted once, the Newton method requires, like the secant stiffness method, an inversion of K_0 at each iteration. This can be a costly operation and in many cases, the reduction in the number of iterations with the Newton method does not compensate, in overall computation time, the constant stiffness iteration scheme. See Hirsch & Warzee [25], [37].

REFERENCES

1. Wu C.H., A General Theory of Three-Dimensional Flow in Subsonic and Supersonic Turbomachineries of Axial-Radial and Mixed Flow Types, NACA TN 2604 (1952)
2. Wu C.H., Three-dimensional Turbomachine Flow Equations Expressed with Respect to Non-orthogonal Curvilinear Coordinates and Methods of Solution, In : Third Int. Symp. on Air Breathing Engines, Munich (1976), 233-252
3. Hirsch Ch., Unsteady Contributions to Steady Radial Equilibrium Flow Equations, In : Unsteady Phenomena in Turbomachines, AGARD Conf. Proc. CP 195 (1975)
4. Hirsch Ch., Warzee G., An Integrated Quasi Three-dimensional Finite Element Calculation Program for Turbomachinery Flows, Trans. ASME, Journal of Engineering for Power, vol. 101 (1979), 141-148
5. Jennions K., Stow P., A Quasi-three-dimensional Blade Design System, ASME Paper 84-GI, Transaction ASME, Journal of Engineering for Power (1984)
6. Katsanis T., Use of Arbitrary Quasi-orthogonals for Calculating Flow Distributions in the Meridional Plane of a Turbomachine, NASA TN D-2546 (1964)
7. Katsanis T., Use of Arbitrary Quasi-orthogonals for Calculating Flow Distribution in a Turbomachine, Trans. ASME, Journal of Engineering for Power, vol. 88 (1966), 197-202
8. Novak R.A., Streamline Curvature Computing Procedures for Fluid Flow Problemes, Trans. ASME, Journal of Engineering for Power, vol. 89 (1967), 478-490

9. Marsch H., A Digital Computer Program for the Through-flow-fluid Mechanics in an Arbitrary Turbomachine Using a Matrix Method, Aeronautical Research Council, R&M, n° 3509 (1968)
10. Adler D., Krimerman Y., The Numerical Calculation of the Meridional Flow Field in Turbomachines using the Finite Element Method, Israel Journal of Technology, vol. 12 (1974), 268-274
11. Biniaris S., The Calculation of the Quasi-three Dimensional Flow in Axial Gas Turbine, Trans. ASME, Journal of Engineering for Power, vol. 97 (1975), 283-294
12. Bosman C., El-Shaarawi M.A.I., Quasi-three Dimensional Numerical Solution of Flow in Turbomachines, Trans. ASME, Journal of Fluids Engineering, vol. 99 (1977), 132-140
13. Serovy, Through Flow calculations in Turbomachines, AGARD AR-175 (1981)
14. Bosman C., Marsh H., An Improved Method for Calculating the Flow in Turbomachines, Including a Consistent Loss Model, Journal Mechanical Engineering Science, vol. 16 (1974), 25-31
15. Hirsch Ch., The Numerical Computation of Inernal and External Flows, J. Wiley & Sons, to be published (1985)
16. Smith L.H.Jr., The Radial Equilibrium Equation of Turbomachinery, Trans. ASME, Journal Engineering for Power 88A, 1 (1966)
17. Sehra A.K., The Effect of Blade to Blade Flow Variations on the Mean Flow Field of a Transonic Compressor, Ph.D. Thesis, MIT, Dept. Aeronautics and Astronautics, see also AFAPL-TR-79-2010 (1979)
18. Sehra A.K., Kerrebrock, Blade-to-Blade Effects on Mean Flow in Transonic Compressors", AIAA Journal, vol. 19 (1981), 476-483
19. Hawthorne W.A., Rotational Flow Through Cascades, Part 1 : The Components of Vorticity, Quarterly Journal of Mech. and Appl. Math., vol. 8 (1955), 266-279
20. Horlock J.H., Marsh H., Flow Models for Turbomachines, Journal of Mechanical Engineering Science, vol. 13 (1971), 358-368
21. Davis W.R., Millar D., A Comparison of the Matrix and Streamline Curvature Methods of Axial Flow Turbomachinery Analysis from a User's Point of View, Trans. ASME, Journal of Engineering for Power, vol. 97 (1975), 549-560
22. Davis W.R., Millar D.A.J., Through Flow Calculations based on Matrix Inversion, in : Through-flow Calculations in Axial Turbomachinery, AGARD Conf. Proc. CP-195 (1976)
23. Davis W.R., A General Finite Difference Technique for the Compressible Flow in the Meridional Plane of a Centrifugal Turbomachine, ASME Paper 75-GT-121 (1975)
24. Hirsch Ch., Warzee G., A Finite Element Method for Flow Calculations in Turbomachines, Vrije Universiteit Brussel, Dept. Fluid Mechanics, Report VUB-STR-5 (1974)
25. Hirsch Ch., Warzee G., A Finite Element Method for Through-flow Calculations in Turbomachines, Trans. ASME, Journal of Fluids Engineering, vol. 98 (1976), 403-421
26. Mellor G., Wood G., An Axial Compressor End-wall Boundary Layer

Theory, Trans. ASME, Journal of Basic Engineering, vol. 93D (1971), 300
27. Hirsch Ch., End-wall Boundary Layers in Axial Compressors, Trans. ASME, Journal of Engineering for Power, vol. 96A (1974), 413
28. Hirsch Ch., Flow Prediction in Axial Flow Compressors including End-wall Boundary Layers, ASME Paper 76-GT-72 (1976)
29. Horlock J.H., Perkins H.J., Annulus Wall Boundary Layers in Turbomachines, AGARD Report AG-185 (1974)
30. De Ruyck J., Hirsch Ch., Kool P., An Axial Compressor End-wall Boundary Layer Calculation Method, Trans. ASME, Joural of Engineering for Power, vol. 101 (1979), 233-249
31. De Ruyck J., Hirsch Ch., Investigations of an Axial Compressor End-wall Boundary Layer Prediction Method, Trans. ASME, Journal of Engineering for Power, vol. 109 (1984), 20-33
32. De Ruyck J., Hirsch Ch., End-wall Boundary Layer Calculations in Multistage Axial Compressors, in : Viscous Effects in Turbomachines, AGARD Conf. Proceedings CP 351 (1983)
33. Jansen W., Moffatt W.C., The Off-design of Axial-flow Compressors, Trans. ASME, Journal of Engineering for Power, vol. 89 (1967), 453-463
34. Katsanis T., McNally W.D., Fortran Program for Calculating Velocities and Streamlines on the Hub-shroud Mid-channel Flow Surface of an Axial or Mixed-flow Turbomachine, vol. I and II, NASA Technical Note TN-D-7343 and 7344, superseded by NASA TN-D 8430 (1977) and TN-D 8431 (1977)
35. Davis W.R., A Matrix Method Applied to the Analysis of the Flow in Turbomachinery, Report ME/A 71-6, Orleton University, Canada
36. Bosman C., The Effect of Damping Factor on the Behavior of Flow Calculations in Turbomachines, Aeronautical Research Council, Reports and Memoranda R&M, n° 3766 (1975)
37. Hirsch Ch., Warzee G., A Finite Element Method for the Axisymmetric Flow Computation in a Turbomachine, Int. Journal for Numerical Methods in Engineering, vol. 10 (1976), 99-113

NOMENCLATURE

b,B	streamsheet thickness (streamsurface approach)
b	integration interval in the ξ^3 direction (averaging approach)
E	total energy
$\vec{F}$	flux vector
$\vec{F}_f$	friction force
$\vec{f}_e$	external forces
$\vec{f}_B$	body force due to 3D effects
h	enthalpy

H	total enthalpy
h_1, h_2, h_3	metric coefficient for orthogonal curvilinear coordinate system
I	rothalpy
$\tilde{I}$	averaged rothalpy
$\hat{I}$	rothalphy of averaged flow
$\bar{\bar{I}}$	unit tensor
J	Jacobian of transformation from Cartesian to curvilinear coordinates
k	heat conduction coefficient, kinematic energy
$\dot{m}$	mass flow
M	relative Machnumber
$\vec{n}$	normal to streamsurface
p	pressure
R	residual
s	entropy
S	streamsurface
t	time
T	temperature
$\vec{v}$	absolute velocity
$\vec{V}$	density weighted average of the absolute velocity over the ξ^3 direction
$\vec{v}''$	density weighted velocity fluctuations
$\vec{w}$	relative velocity
$\vec{W}$	density weighted average of $\vec{w}$
β	relative flow angle with respect to meridional direction
β'	angle of streamsurface with axial direction in r = constant cylinder
δ^*	displacement thickness of end-wall boundary layer
ε'	angle of streamsurface with radial direction in a z = constant plane
$\vec{\zeta}$	absolute vorticity
ξ^1, ξ^2, ξ^3	curvilinear coordinates
ρ	density
σ	relative flow angle in the meridional plane with respect

	to axial direction
$\bar{\bar{\tau}}$	viscous plus turbulent stress tensor
τ	underrelaxation factor
ϕ	finite element basis function
ψ	stream function
ω	angular velocity of relative reference system

Subscripts and superscripts

$(\ )_o$	stagnation value
$(\ )'$	stagnation vaue in the relative coordinate system
$(\bar{\ })$	average over the ξ^3 direction
$(\tilde{\ })$	density weighted average over the ξ^3 direction
$(\ )_p$	pressure side of blade
$(\ )_s$	suction side of blade

Symbols

$\times$	vector product
$\otimes$	tensorial product
$\bar{\partial}$	derivative along a streamsurface

SOLUTION METHODS

THE USE OF RADIAL EQUILIBRIUM AND STREAMLINE CURVATURE METHODS FOR TURBOMACHINERY DESIGN AND PREDICTION

Ennio Macchi

Politecnico di Milano, Dipartimento di Energetica

1 INTRODUCTION

Two statements are almost invariably present in the introductory notes of papers dealing with flow in turbomachines:

- the first one points out that the real flow in a turbomachine has an extremely complex pattern, being always 1) unsteady, 2) three-dimensional, 3) viscous, and often compressible (if not transonic or supersonic), and further complicated by overtip leakage flows, cooling flows, or by peculiar thermodynamic properties, like wet steam in low pressure steam turbines;
- the second one acknowledges that the solution of the full equations of motion with the real boundary conditions existing in a turbomachine is still beyond the capabilities of modern computers.

As a consequence of the situation underlined by the above points, all methods proposed in the technical literature aiming to find the solution of flow in turbomachinery necessarily include simplifications in flow models and/or equations. Some of these methods can be classified as fully three-dimensional methods, others as two-dimensional methods, on blade-to-blade stream surfaces (S_1, according to Wu's (1) terminology) or on through flow stream surfaces (S_2).

The question may arise whether the two-dimensional methods for through flow calculation which are the subject of this lecture are nowadays made obsolete by the appearance of a large number of fully three-dimensional methods (see for instance, (3), (4), (5), (6),

(7), (8), just to quote papers presented in the last ASME International Gas Turbine Conference). Now, the great importance of fully three-dimensional methods cannot be denied. It is easy to predict that their potential will tremendously increase in a near future thanks to availability of faster and more powerful computers. However, to quote Rolls-Royce experts (2), "they will be for some years to come of such speed that only isolated blade rows, or at best single stages can be three-dimensional computed in the design times available". Moreover, according to the reasonings previously stated, it should not be underestimated the fact that the even the best fully three-dimensional methods today available are still oversimplifications of real flow. Hence, while there are no doubts that their contribution is fundamental for a deeper understanding of flow phoenomena, it is questionable whether they will be automatically a tool to design more efficient turbomachines than the ones obtained by existing design procedures, based on two-dimensional (or quasi-three-dimensional) methods and incorporating "correction factors" and/or correlations empirically derived from many years of design practice and tests.

Purpose of this lecture is to describe the methods of solution of through flow in axial flow turbomachinery based on the streamline curvature method. Other well-known techniques of solution of the same equations, i.e. methods based on matrix solution, finite elements, finite volumes and time-marching, will be the subject of other lectures in this Institute and will not be discussed here. Through flow calculations are very useful both in the design (selection of turbomachine geometry for specified thermofluid dynamic characteristics) and prediction (obtainment of turbomachine operating characteristics for specified geometry) phase for most turbomachine models, including axial and radial flow hydraulic, gas and steam turbines, compressors, fans, pumps. The streamline curvature method is a relatively old (thirty years) and well known procedure. Its application to through flow in turbomachinery has been treated in hundreds of papers (see for instance the bibliography given in (9)), and a large number of computer programs has been reported in the technical literature.

These methods can be classified in several ways. An important differentiation should be made between:

. "duct-flow" methods
. "through-flow" methods.

In the "duct-flow" methods, the flow is solved only in stations ahead of and after the blade rows (fig. 1), while in "through-flow" methods one or more computing stations are placed inside the blades (fig.2).The main advantages of the two methods are listed in tab.1. As stated in (10), the accuracy of prediction of actual test cases is similar for duct- and through-flow methods.

2 GOVERNING EQUATIONS AND SYMPLIFYING HYPOTHESES

Although similar, the duct-flow and through-flow methods use different equations, derived from different hypotheses. Hence, they will be treated separately in the following.

2.1 Duct Flow Methods

As said above, computing stations are placed outside the blade rows. The three fundamental laws of fluid dynamics, namely:

. Equation of motion
. Energy equation
. Equation of continuity

must be solved. Reference will be made in the followings to derivations given in (11) and (12).

2.1.1 Simplifying Hypotheses. The following assumptions are made:

1) Steady flow.
2) Frictionless flow at the stations where the equations of motion are solved.
3) Axisymmetric flow at the stations where the equations of motion are solved.
4) Adiabatic flow.
5) The working fluid is a perfect gas, namely a gas having the equation of state:

$$p = \rho R_G T$$

and specific heat at constant pressure is independent of temperature and pressure.

Hypotheses 1) and 3) are interrelated since the flow can be steady in both rotor and stator only if it is axisymmetric. Although they are correct only for cascades with an infinite number of blades, axisymmetry and steady conditions are necessary assumptions since any other hypothesis would require unknown quantitative information on flow inside the blade rows. One problem connected with these assumptions is that downstream conditions cannot affect the flow upstream. Therefore pressure distributions at the machine exit can deviate from those imposed by the discharge conditions.

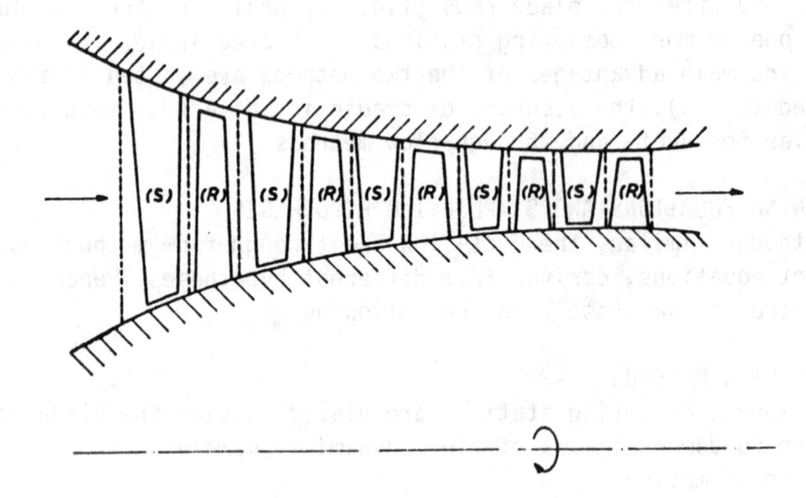

Fig. 1 Duct flow methods: computing stations for a multistage machine.

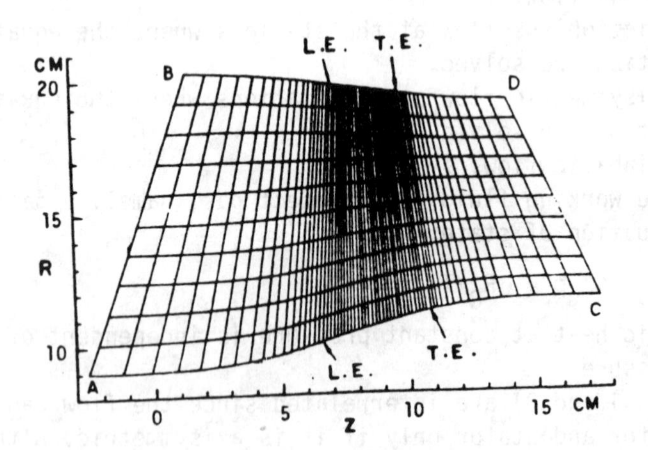

Fig. 2 Through flow methods: typical computational mesh for a single blade row.

Duct-flow calculation	Through-flow calculation
1. A lower number of stations in which the REE has to be solved.	1. It can be coupled to blade-to-blade calculations, to obtain a quasi-three-dimensional solution.
2. The flow angles and losses are to be specified only at stations were they can be more easily calculated.	2. A better understanding of three-dimensional effects is obtained.
3. The preparation of input data is greatly simplified.	3. Blade blockage effects, influence of section stacking are accounted for.
4. Blade-to-blade calculations are not required.	4. Streamline patterns are more precisely defined in machines with large radius variation and with not radial blades.
5. Streamline curvatures and slopes can be estimated by simple methods.	

Tab. 1 Comparison of advantages of duct-flow and through-flow calculation methods.

With assumption 2), all entropy changes are supposed to occur in the blade rows ahead of the considered station. This assumption can be justified by the fact that velocity gradients in the regions, where the equation of motion is applied, are in all probability smaller than those occurring in the boundary layers along the blade surfaces, and flow frictional losses between blades can therefore be neglected.
Assumption 4) is reasonable for uncooled blades, where energy changes due to heat transfer are negligible with respect to those related to velocity changes. If blades are cooled, the additional term q in the energy equation must be specified. Assumption 5) is valid in most cases. If the thermodynamic processes occur in ranges where real gas effects are not too strong, average values of gas properties give good approximations. Otherwise, the actual equation of state of the working fluid is to be specified, and the enthalpy and entropy values present in the equations must be computed by proper routines as functions of other thermodynamic parameters.

2.1.2 Equation of motion. The general equation of motion for relative flow in vectorial form, is:

$$\frac{\partial \vec{W}}{\partial t} + \nabla H_R = \vec{W} x (\nabla x \vec{W} + 2\vec{\omega}) + T \nabla S + \vec{f}_f \qquad (1)$$

$$\vec{\omega} = 0 \qquad (2)$$

For absolute flow, where

$$\{ \vec{V} = \vec{W} \qquad (3)$$

$$H = H_R \qquad (4)$$

equation (1) becomes:

$$\frac{\partial \vec{V}}{\partial R} + \nabla H = \vec{V} x (\nabla x \vec{V}) + T \nabla S + \vec{f}_f \qquad (1a)$$

With hypothesis 1) $\frac{\partial \vec{W}}{\partial t} = 0, \ \frac{\partial \vec{V}}{\partial t} = 0$

and hypothesis 2) $\vec{f}_f = 0$

equations (1) and (1a) become respectively:

$$\nabla H_R = \vec{W} x (\nabla x \vec{W} + 2\vec{\omega}) + T \nabla S \qquad (5)$$

$$\nabla H = \vec{V} x (\nabla x \vec{V}) + T \nabla S \qquad (5a)$$

The following derivations for equation (5) are valid also for equation (5a), if substitutions (2), (3), (4) are performed.

It is convenient to express equation (5) in cylindrical coordinates. Equating the three components of equation (5), the following equations are obtained:

a. tangential component:

$$\frac{1}{R}\frac{\partial H_R}{\partial \Theta} = \frac{W_A}{R}\left(\frac{\partial W_A}{\partial \Theta} - \frac{\partial (RW_u)}{\partial z}\right) - \frac{W_r}{R}\left(\frac{\partial (RW_u)}{R} - \frac{\partial W_r}{\partial \Theta}\right) - 2\omega\, W_r + \frac{T}{R}\frac{\partial S}{\partial \Theta} \qquad (6)$$

b. meridional component:

$$\frac{\partial H_R}{\partial x} = W_r\left(\frac{\partial W_r}{\partial x} - \frac{\partial W_A}{\partial R}\right) - \frac{W_u}{R}\left(\frac{\partial W_A}{\partial \Theta} - \frac{\partial (rW_u)}{\partial x}\right) + T\frac{\partial S}{\partial x} \qquad (7)$$

c. radial component:

$$\frac{\partial H_r}{\partial R} = \frac{W_u}{R}\left(\frac{\partial (rW_u)}{\partial R} - \frac{\partial W_r}{\partial \Theta}\right) - W_A\left(\frac{\partial W_r}{\partial z} - \frac{\partial W_A}{\partial R}\right) + 2\omega\, W_u + T\frac{\partial S}{\partial R} \qquad (8)$$

with hypothesis 3) $\frac{\partial}{\partial\Theta}$ ( ) = 0

and hypothesis 4) $\vec{W}\, d\, t \cdot \nabla H_R = 0$

the system of equations (6), (7), (8) reduces to equation (9):

$$\frac{\partial H_R}{\partial R} = \frac{W_u}{R}\frac{\partial (rW_u)}{\partial R} - W_A \frac{\partial W_r}{\partial z} + W_A \frac{\partial W_A}{\partial R} + 2\omega\, W_u + T\frac{\partial S}{\partial R} \tag{9}$$

By introducing:

$$H_E = H_R + \frac{u^2}{2} \tag{10}$$

with hypothesis 5):

$$H = C_p\, T \tag{11}$$

and hypothesis 4):

$$H_E = \text{const} \tag{12}$$

and by introducing the meridional velocity component:

$$V_m = W_a/\cos\lambda \tag{13}$$

one obtains the following expression for eq (9):

$$\frac{\partial V_m^2}{\partial R} + V_m^2 \{2\, k_m \cos\lambda \cos^2\beta - \frac{\partial}{\partial R}(\ln \cos^2\beta) + \frac{2}{R}\sin^2\beta - \frac{1}{C_p}\frac{\partial S}{\partial R}\}$$

$$+ V_m\ (2\omega \sin(2\beta)) +$$

$$+ \{(2\omega^2 R - \sin\lambda \frac{\partial V_m^2}{\partial m} - 2\, C_p \frac{\partial T_E}{\partial R} + 2\, T_E \frac{\partial S}{\partial R})\cos^2\beta\} = 0 \tag{14}$$

or, for absolute flow ($H_R \to H$, $T_E \to T_t$, $W \to V$, $\beta \to \alpha$, $\omega = 0$):

$$\frac{\partial V_m^2}{\partial R} + V_m^2\ \{2\, k_m \cos\lambda \cos^2\alpha - \frac{\partial}{\partial R}(\ln \cos^2\alpha + \frac{2}{R})\sin^2\alpha -$$

$$- \frac{1}{C_p}\frac{\partial S}{\partial R}\} + \{(-\sin\lambda \frac{\partial V_m^2}{\partial m} - 2\, C_p \frac{\partial T_t}{\partial R} + 2\, T_t \frac{\partial S}{\partial R})\cos^2\alpha\} = 0 \tag{14a}$$

It can be useful, for computer solutions, express the equations (14, 14a) in non-dimensional form, by introducing:

$$X = R/\overline{R} \tag{15}$$
$$Y = V_m/\overline{V}_m \tag{16}$$

$$S^* = S/C_p \tag{17}$$
$$\overline{U} = \omega\overline{R} \tag{18}$$

The following equations are obtained for relative flow:

$$\frac{\partial(\ln Y^2)}{\partial X} = -2k_m \overline{R}\cos\lambda' \cos^2\beta + \frac{\partial(\ln\cos^2\beta)}{\partial X} - 2\frac{\sin^2\beta}{X} -$$
$$-\frac{2}{Y}\frac{\overline{U}}{\overline{V}_m}\sin^2\beta - \frac{2}{Y^2}\left(\frac{\overline{U}}{\overline{V}_m}\right)^2 X\cos^2\beta +$$
$$+\left\{\sin\lambda\frac{\partial(\ln Y^2)}{\partial(m/\overline{R})} + \frac{1}{Y^2}\frac{\partial}{\partial X}\left(\frac{T_E}{\overline{V}_m^{\,2}/2C_p}\right)\right\}\cos^2\beta +$$
$$+\left\{1 - \frac{T_E\cos^2\beta}{Y^2(\overline{V}_m^{\,2}/2C_p)}\right\}\frac{\partial S^*}{\partial X} = I(X) \tag{19}$$

and, for absolute flows:

$$\frac{\partial(\ln Y^2)}{\partial X} = -2k_m \overline{R}\cos\lambda \cos^2\alpha + \frac{\partial(\ln\cos^2\alpha)}{\partial X} - \frac{2\sin^2\alpha}{X} +$$
$$+\left\{\sin\lambda\frac{\partial(\ln Y^2)}{\partial(m/\overline{R})} + \frac{1}{Y^2}\frac{\partial}{\partial X}\left(\frac{T_t}{\overline{V}_m^{\,2}/2C_p}\right)\right\}\cos^2\alpha +$$
$$+\left\{1 - \frac{T_t\cos^2\alpha}{Y^2(\overline{V}_m^{\,2}/2C_p)}\right\}\frac{\partial S^*}{\partial X} = I(X) \tag{19a}$$

2.1.3 Energy equation. The energy equation in a relative flow is given by:

$$\vec{W}dt \cdot \nabla(H_R) = \vec{W}dt \cdot \nabla(q) - \frac{\partial}{\partial t}\left(\frac{W^2}{2}\right)dt \tag{20}$$

and in an absolute flow, with (2), (3) and (4) in (20):

$$\vec{V}dt \cdot \nabla(H_R) = \vec{V}dt \cdot \nabla(q) - \frac{\partial}{\partial t}\left(\frac{V^2}{2}\right)dt \tag{20a}$$

where q is the heat added to a unit mass particle from external sources. With assumptions 1) $\partial/\partial t\ (\ ) = 0$ and 4) $q = 0$, equations (20) and (20a) become:

$$\vec{W}dt \cdot \nabla(H_R) = 0 \tag{21}$$

$$\vec{V}dt \cdot \nabla(H) = 0 \tag{21a}$$

Equations (21) and (21a) indicate that for steady, adiabatic flow the relative total enthalpy is constant along any streamline. This does not mean that $\nabla(H_R) = 0$, since relative total enthalpy can assume different values for different streamlines.

2.1.4 Equation of continuity. The mass flow through any calculating station is given by:

$$\dot{m} = 2\pi \int_{R_i}^{R_o} R\,\rho\,V_m \cos\lambda \; dR \tag{22}$$

2.2 Throughflow Analysis

Reference is made to the recent work of Jennions and Stow (2, 13), who present a very comprehensive method of through flow analysis, based on a rigorous passage-averaging technique. The derived equations are applied inside the blade row.

2.2.1 Simplifying Assumptions.

1) The flow is steady ($\partial/\partial t = 0$);
2) The flow is on surfaces of revolution (the so-called degenerated Wu's S_2 surfaces);
3) The flow properties can be represented by mean plus perturbation values with respect to the Θ coordinate;
4) There are no velocity perturbations normal to the S_1 stream surfaces;
5) The flow is adiabatic ($q = 0$);
6) The loss terms are modeled by a dissipative force which acts along a streamline (see (14));
7) The working fluid is a perfect gas ($p = \rho R_G T$).

2.2.2 Radial Equilibrium Equation. It is derived from momentum equation in radial and axial directions. With derivations found in (2), it is obtained:

$$\frac{1}{\rho}\frac{\partial \bar{p}}{\partial R} = SC + CE + PG + BF + PE \tag{23}$$

where:

$$SC = -\tilde{W}_x^{\,2} \cos^2\lambda \left(\frac{\partial \tan\lambda}{\partial X}\right)_{\Psi} \quad \text{(streamline curvature effect)} \tag{24}$$

$$CE = \frac{(\tilde{W}_x \tan \tilde{\beta} + \omega R)^2 \cos^2\lambda}{R} \text{ (centrifugal effect)} \qquad (25)$$

$$PG = \frac{\sin\lambda \cos\lambda}{\tilde{\rho}} \left(\frac{\partial \tilde{p}}{\partial x}\right)_{\Psi} \quad \text{(axial pressure gradient effect)} \qquad (26)$$

$$BF = F_{B\Theta} (\tan\lambda \tan\sigma - \tan\delta) \cos^2\lambda \text{ (blade force effect)} \qquad (27)$$

where the blade force $F_{B\Theta}$ is derived from momentum equation in Θ direction

$$F_{B\Theta} = \frac{\tilde{W}_x}{R} + \left(\frac{\partial\{(\tilde{W}_x \tan\tilde{\beta} + \omega R) R\}}{\partial x}\right)_{\Psi} - \tilde{F}_{\tau\Theta} + P_{\Theta} \qquad (28)$$

with the dissipative body force $\tilde{F}_{\tau\Theta}$ is given by:

$$\tilde{F}_{\tau\Theta} = - \frac{\tilde{T} \tan\beta}{1 + \tan^2\lambda + \tan^2\beta} \left(\frac{\partial\tilde{S}}{\partial x}\right)_{\Psi} \qquad (29)$$

and the perturbation term P_{Θ} is given:

$$P_{\Theta} = \frac{1}{B R \bar{\rho}} \left(\frac{\partial\, B R \rho\, W''_{\Theta} W''_x}{\partial x}\right)_{\Psi} + \widetilde{W''_{\Theta} W''_x} \frac{\partial \tan\lambda}{\partial R} + \frac{W''_{\Theta} W''_x}{R} \tan\lambda \qquad (30)$$

where B is the blade blockage term, defined by:

$$B = \frac{N(\Theta_s - \Theta_p)}{2\pi} \qquad (31)$$

Eventually, the last term of RHS of eq.(23) is given by:

$$PE = - \tilde{W}''_x \cos^2\lambda \left(\frac{\partial \tan\lambda}{\partial x}\right)_{\Psi} + \frac{\tilde{W}''_{\Theta}}{R} + \frac{P'_s + P'_p}{2\bar{\rho} B} \left\{\sec^2\lambda \frac{\partial B}{\partial R} - \tan\lambda \left(\frac{\partial B}{\partial x}\right)\right\}_{\Psi} \qquad (32)$$

2.2.3 Energy equation. The constancy of equivalent (for rotors) or total (for static) enthalpy along streamlines is assumed, according to hypotheses 1) and 5). With notations used in this section:

$$\frac{\partial}{\partial x} \Big\{C_p \tilde{T} + \frac{1}{2} \tilde{W}_x^2 (1 + \tan^2\lambda + \tan^2\tilde{\beta}) - \frac{1}{2} \omega^2 R^2 + \frac{1}{2} \tilde{W}_x^2 (1 + \tan^2\lambda) + \frac{1}{2} \tilde{W}''^2_{\Theta}\Big\}_{\Psi} = 0 \qquad (33)$$

2.2.4 Continuity equation. Use is made of the blade blockage term B defined by eq.(31). The constancy of the mass flow rate is expressed by:

$$\dot{m} = 2\pi \int_{R_i}^{R_o} B\, R\, \overline{\rho}\, \tilde{W}_x \, dR \tag{34}$$

3. The Streamline Curvature Method of Solution

The streamline curvature method has been described by several authors (see for instance (15, 16, 17)). It is based on an iterative procedure:

1. Initial values of streamlines locations are assumed at each computing station.
2. Initial values of some flow properties (for instance axial velocity) at a reference location (in general at mean radius) are assumed.
3. At each computing station, from inlet to output:

3.1 The radial equilibrium equation (either (19) or (23), depending on the adopted method is solved in both cases, REE is a first-order differential equation, which provides the radial distribution of the axial (or meridional) velocity, for a given boundary condition, generally the above quoted axial (or meridional) velocity value at a mean streamline. The REE solution requires the knowledge of several terms involving the flow evolution in axial direction, to be computed from quantities on preceding and following streamlines, namely:

k_m, λ, $\dfrac{\partial(\ln Y2)}{\partial(m/\overline{R})}$ in (19) or (19a) for direct flow, or

$\lambda, \left(\dfrac{\partial \tan\lambda}{\partial x}\right)_\Psi, \left(\dfrac{\partial p}{\partial x}\right)_\Psi, \left(\dfrac{\partial B}{\partial x}\right)_\Psi, \dfrac{\partial\{(\tilde{W}_x \tan\beta + \omega R)R\}_\Psi}{\partial x}, \left(\dfrac{\partial \tilde{S}}{\partial x}\right)_\Psi,$

$\dfrac{\partial(B R \overline{\rho W''_\Theta W''_x})_\Psi}{\partial x}$ in (23) for through flow.

The simplest approach is not to account for them ($k_m = 0$, $\lambda = 0$, $\partial/\partial m$ or $\partial/\partial x = 0$) in the first step of calculation.

3.2 From known velocity distribution (step 3.1), the relative location of equal mass flow points at the computing station is calculated by local continuity equation and compared with the one previously assumed (at step 1 for the first loop, or at step 3.2 of the previous loop). Corrections are applied, according to adopted numerical method.

3.3 The overall continuity equation is then checked, and corrections adopted on the previous value of the flow property at mean radius (either assumed at step 2 or at step 3.3 of the previous loop), up to the attainment of the desired flow rate.

4. Once all equal mass flow points are updated, new streamlines are determined by a proper fitting technique, and calculation is repeated from step 3, up to the attainment of convergence in all flow variables.

Although the above procedure could be in principle applied in a single computer program for all through flow problems in turbomachinery, it is most practical to use specialized programs for the various problems. Typical input/output data for a duct (or through) flow program for prediction of an axial-flow turbomachine are presented in tab. 2.

For transonic or supersonic machines, the mass flow is independent on pressure ratio, and the program structure must be modified, as discussed for instance by Cox in (18) or by Denton in (19).

For design methods, the input/output data and the computer program logic are not so easily generalized as it was done for tab. 3, since the options in the design process of a turbomachine are infinite and programs are tailored according to manufacturer experience. A possible procedure for a duct-flow program would be the following:

1. Assign a "reference" (either at inner,outer or mean radius) streamline.
2. Assign velocity triangles on the "reference" streamline.
3. Select a radial variation low for some parameter, either:
   - tangential velocity component
   - meridional velocity component
   - blade specific work
   - flow angle

   These laws can: a) completely specify the problem (like, for instance, the assumption of constant flow angle, or constant meridional velocity component, etc.), or b) leave some degree of freedom (for instance, assume a parabolic distribution of a velocity component with coefficients to be found from other conditions).
4. Solve the radial equilibrium equation; the meridional contours of the flow channel are either output data, being the result of

overall continuity equation (option a) of the proceding point), or input data specified at a previous stage.

5. For solving REE at preceding step, loss prediction routines should be incorporated in the program. Difficulties arise to obtain accurate loss coefficient predictions at this stage of calculation, without a complete knowledge of all blade geometric characteristics.
6. The procedure is obviously iterative, since streamline locations will vary during calculation, and the required operating conditions must be checked.
7. Once convergence is obtained, the machine performance (efficiency, specific work per stage, etc.) and geometry (meridional contours, blade characteristics) are determined and printed or plotted as output results.

"Optimization" routines and/or controls of various nature can be incorporated in the programs.

MAIN INPUT DATA

1. Working fluid parameters
   - 1a. Perfect gas constants (molecular mass, heat capacity ratio, viscosity).
   - 1b. Routines for definition of thermodynamic properties of fluid (equation of state, thermodynamic functions).
2. Operating conditions
   - 2.1 Number of operating points to be investigated.
   - 2.2 For each operating point:
     - mass flow rate (or expansion ratio, for supersonic turbines)
     - speed of revolution
     - thermodynamic conditions at inlet station (total pressure and temperature radial distribution or other equivalent condition).
3. Machine meridional geometry and computing grid
   - 3.1 Inner and outer channel profiles; selected option for fitting.
   - 3.2 Axial location of L.E. and T.E. of all blade rows at inner and outer radii.
   - 3.3 Radial clearances (or other details to define leakages) for each blade row extremity.
   - 3.4 Axial location of computing stations at inner and outer radii.

Tab. 2

Tab.2 (cont.)

3.5 Number of streamlines to be computed.

4. Blade definition
   For each blade row, in a specified number of radial positions:
   4a All geometrical characteristics required by loss and flow angle correlations (for duct-flow methods).
   4b 1. Blade profiles.
      2. Results of blade-to-blade calculations (for through flow methods).

5. Options in loss, flow angles and endwall boundary layer calculation

6. Selection of procedures for streamline fitting

7. Selection of tolerances for various iterative processes (on overall continuity, streamline continuity, velocity distribution, pressure ratio, etc.).

MAIN OUTPUT DATA

1. Overall machine performance
   1.1 At each computing station gap,average values of:
   - thermodynamic conditions
   - efficiency, specific work, torque, power
   - velocity triangles, degree of reaction, etc.
   1.2 Overall efficiency, pressure ratio, torque, power.

2. Detailed flow description
   2.1 Final computing grid, with tracing of streamlines.
   2.2 For a number of specified stations, at specified radial positions:
   - thermodynamic conditions
   - velocity triangles, Mach numbers, incidence, etc.
   - loss coefficients, efficiency, etc.
   - blade stresses
   - streamline curvature and hade angle.

Tab.2 Main input/output data of computer programs for prediction of axial-flow turbomachinery.

4. DIFFICULTIES IN THE METHOD OF APPLICATION

It is a common experience that computer programs of various authors based on similar through flow methods yield quite different results, when applied to an actual turbomachine. This is true both for design and prediction programs, i.e. a different machine geometry would result from the same specified operating characteristics, and/or different machine performance is predicted for the same machine geometry. This point has been demonstrated by the results of comparison on various test cases carried out by various organizations (18), (19), (20), (21): some typical results of these comparisons are given in fig. 3-6 for various types of machines. The main sources of discrepancies between the results obtained by the various methods are outlined in the followings.

4.1 Streamlines Fitting

As described in the previous section, in streamline curvature methods the streamlines are located in various computing stations by the equal mass flow condition. An analytical curve fitting these points must then be specified, to obtain the first and second order derivatives (or λ and k_m of equations (19) and (23)). The difficulties related to such a fitting and the uncertainty of achievable results, particularly for curvatures, are well known, and common to other fields of science, like calculation of thermodynamic properties of fluids. In (22), the author demonstrated the great amplification of errors occurring in the fitting process: small random errors ($\sim$ 1/1000) in interpolating points, which in our case could appear for several reasons, like inaccuracies in the numerical solutions, discontinuities in the losses or deviations routines, etc., can yield large errors ($>$ 10%) in the second derivative.

The problem presents quite different aspects for "duct" and "through" flow methods. In fact, in the first case, only few points are available for streamline definition, and it is impossible to check the streamline shape between the computing stations. Moreover, it has not physical meaning to assume that informations coming from stations located far away from the computing point influence the local shape of the curve. In this situation, the most reasonable and simplest assumption seems to be to one proposed by Vavra (15), and illustrated in fig. 7. The results obtained by Denton (19) demonstrate that the use of a coarse grid with simple interpolating (quadratic) functions is acceptable for usual axial flow machines, due to the relatively small influence of the streamline shape on flow distribution outside the blade rows.

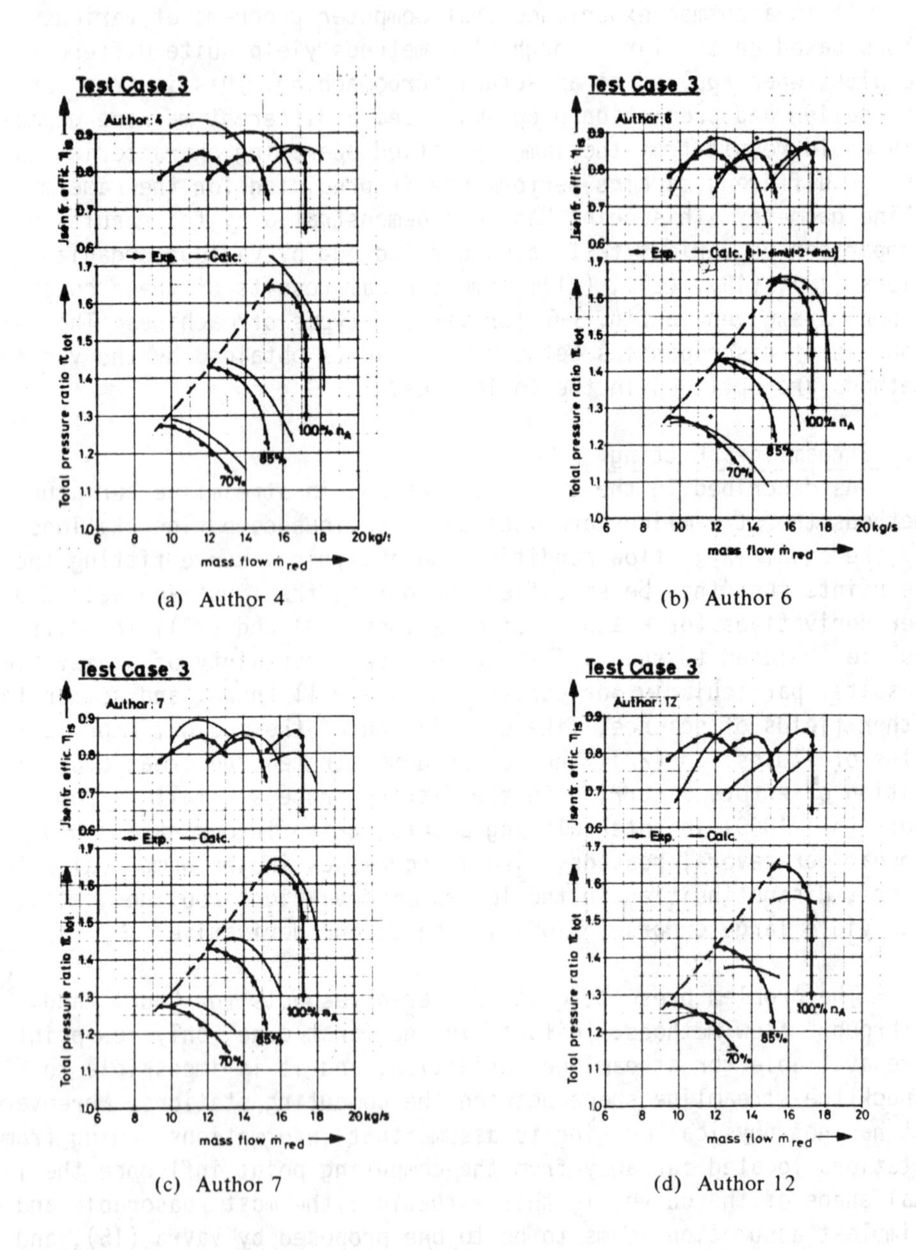

Fig. 3 Overall performance of a single stage transonic compressor compared to performance predictions (from (18)).

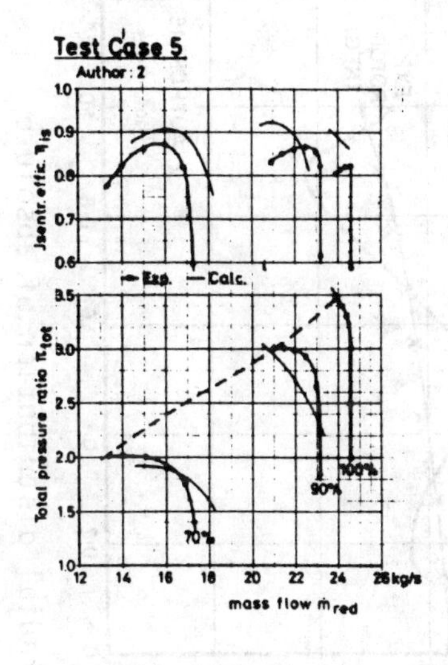

(a) Author 2

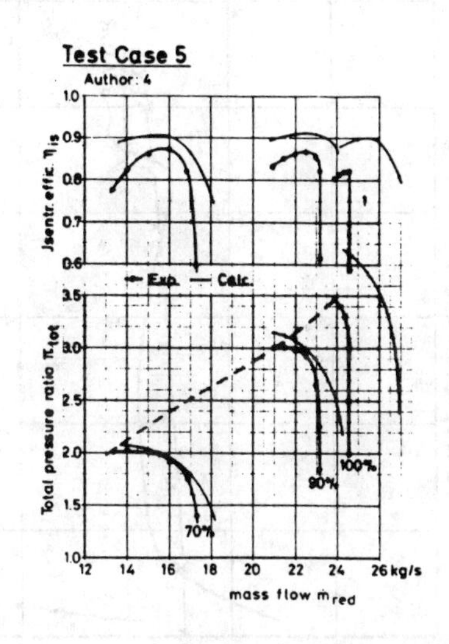

(b) Author 4

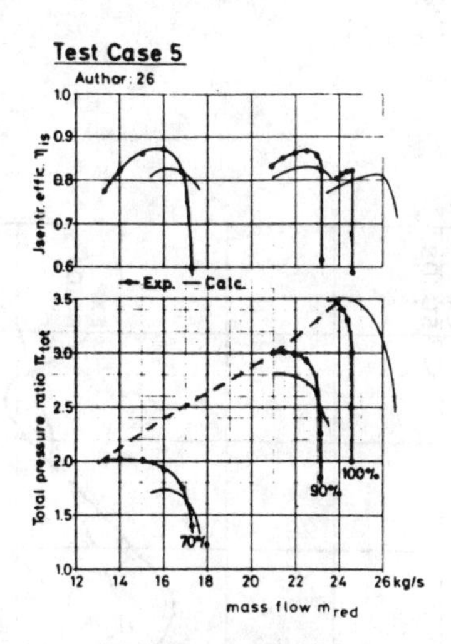

Fig. 4 Overall performance of a 4 stage transonic compressor compared to performance predictions (from (18)).

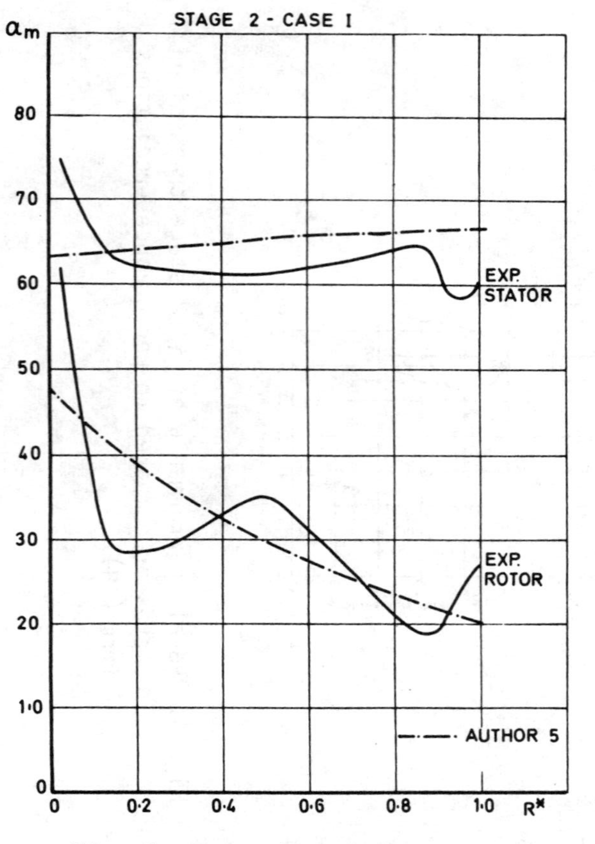

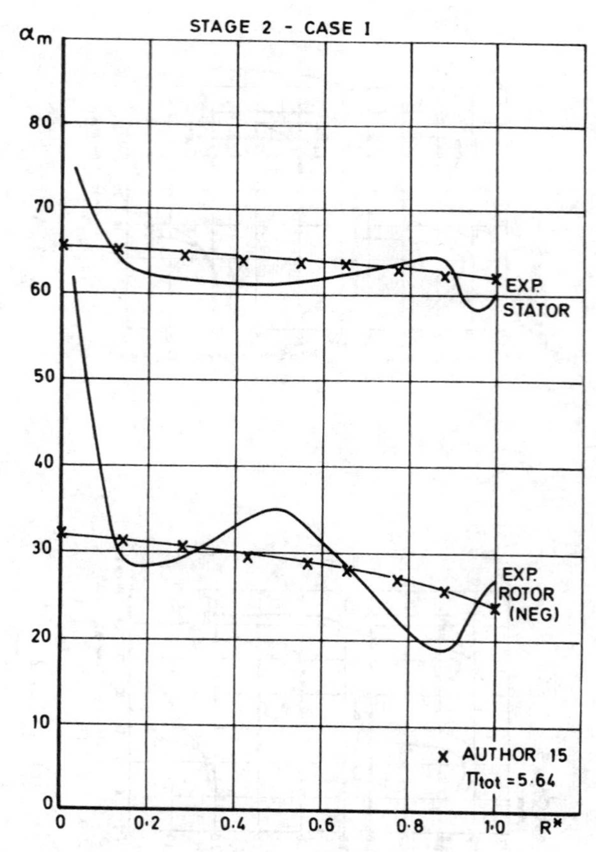

Fig. 5 Comparison between predicted and measured radial distribution of absolute outlet angle from a two stage gas turbine, from (18).

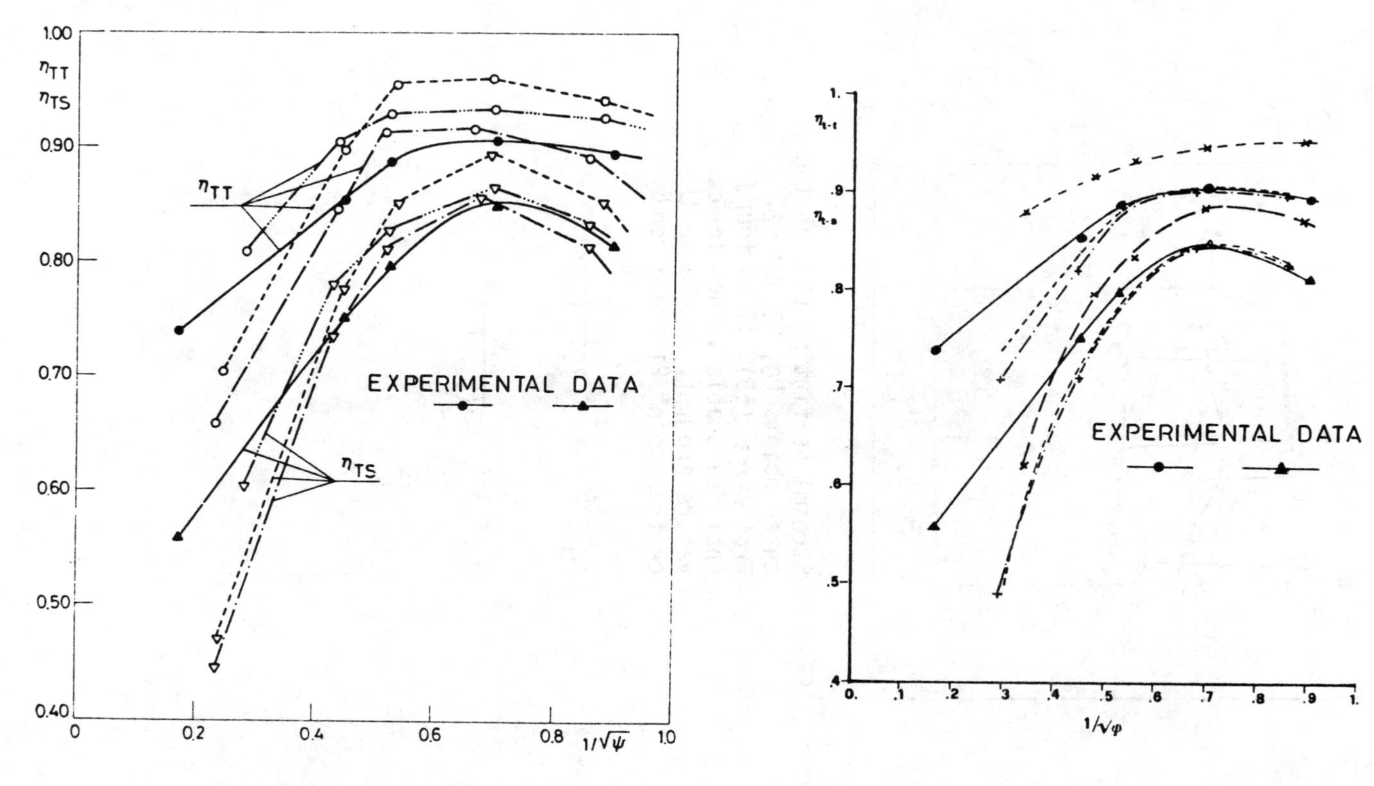

Fig. 6 Overall performance of a single stage axial flow turbine as predicted by various through flow methods (from (20)).

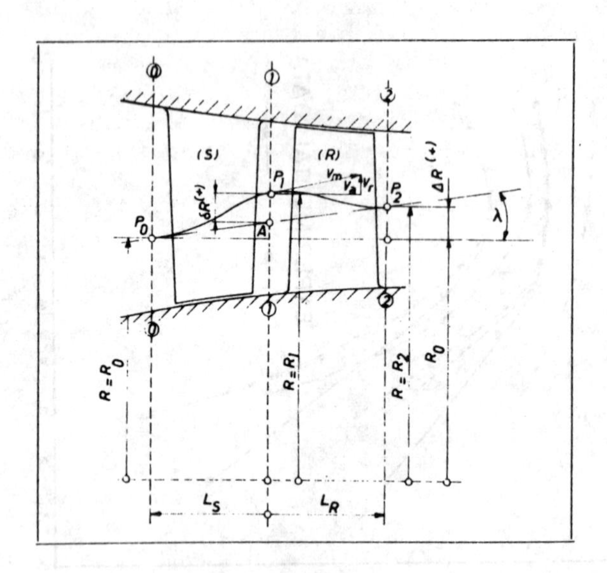

Fig. 7 Streamline fitting through the three points P_0, P_1 and P_2. From Vavra (15), it is found that reasonable values for curvature and hade angle in the central point P_1 are given by:

$$K_m = 5 \cos^{-3} \lambda \frac{4\,\delta R}{(L_S + L_R)^2}$$

$$\lambda = \tan^{-1} \left(\frac{\Delta R}{(L_S + L_R)}\right)$$

For through flow methods, a different situation takes place, since the computing stations are in a larger number and the streamline shape inside the blade can be checked. A variety of interpolating routines is proposed, among which polinomial functions, splines (i.e. a series of cubic with equal first and second derivative in the common points),etc.. Care should be taken to avoid numerical instabilities in the solution, as it could happen with high degree polynomials.

4.2 Loss Prediction

4.2.1 Duct flow methods. As already pointed out, the duct flow methods assume frictionless flow at stations where the equation of motion is solved; however, they must account for losses occurring between computing stations to calculate the entropy gradients. These losses, which are produced by the blade rows placed within the stations, are generally evaluated by making use of loss prediction methods such as those reviewed by Sieverding (23) for axial flow turbines, or by several authors in (24) for axial flow compressors.All these correlations have been derived as mean line methods, i.e. they predict loss coefficients which are averaged along the blade height, while the duct flow calculation requires loss prediction on each individual streamline. Various approaches have been proposed to radially distribute losses:

1. The loss coefficient is calculated at mean radius and assumed constant for all streamlines.
2. The loss coefficients are calculated at each individual streamline, as a function of local geometry and velocity triangles.
3. The losses are divided in various groups (profile + secondary + tip leakage + others) and treated in different ways: for instance, profile losses are individually computed for each streamline,while other losses are calculated for mean line characteristics and then radially distributed with various methods.

The opinions on this point are largely divergent. It should be pointed out that all procedures are highly arbitrary and cannot take account of the real complex physics occurring in a blade row, nor can be validated by experience, due to lack of well documented test cases. In the author's opinion some points can be made:

- some loss correlation methods, empirically derived,give surprisingly good results in the prediction of on-design overall performance of axial flow turbines and compressors; these correlations require the knowledge of main geometric characteristics of bladings (for turbines see for instance tab. 3, from (23)). These informa-

PROFILE LOSSES

Correlation	α_1	α_2	g/c	*Profile*	te	M_2	Re	*Off-design*
Ainley-Mathieson	X	X	X	t/c	X	X	X	X
Traupel	X	X	X		X	X	X	X
Baljé-Binsley	X	X	X				X	
Dejc-Trojanovski	X	X	X					
Craig-Cox	X	X	g/b	CR	X	X	X	X

SECONDARY LOSSES

Correlation	*Turning angle*	*Velocity or Angle ratio*	*Aspect ratio*	g/c	*Profile*	*Inlet B.L.*	M_2	Re
Ainley-Mathieson (Dunham-Came)	X	A	id/od	X				
Traupel	X	V	g/h	X		negl.	ind	
Baljé		A	h/c			X		
Dejc-Trojanovski	X	A	h/c	X	X		X	X
Craig-Cox	X	V	h/b				ind	

Tab. 3 Necessary informations to calculate blade loss coefficients for axial flow turbines, according to various correlations (from (19)).

tions are available in prediction programs, but not in design programs, which must rely on simpler but less accurate correlations. For this reason, often prediction programs are used as design programs as well;

- the accuracy of predictions of overall performance at off-design is quite lower, showing that loss correlations are not reliable at incidence angles far from nominal;
- the radial loss distribution is a very difficult task. Although it seems obvious to concentrate the secondary losses at blade extremities, attention should be paid to generate too high entropy radial gradients, which yield velocity distributions far from reality, as demonstrated in (11). An outlook to detailed loss measurements in actual machines (see for instance fig. 8 , from (25), makes it clear that actual phoenomena are much more complex than the one described under axisymmetry hypothesis. In particular, it

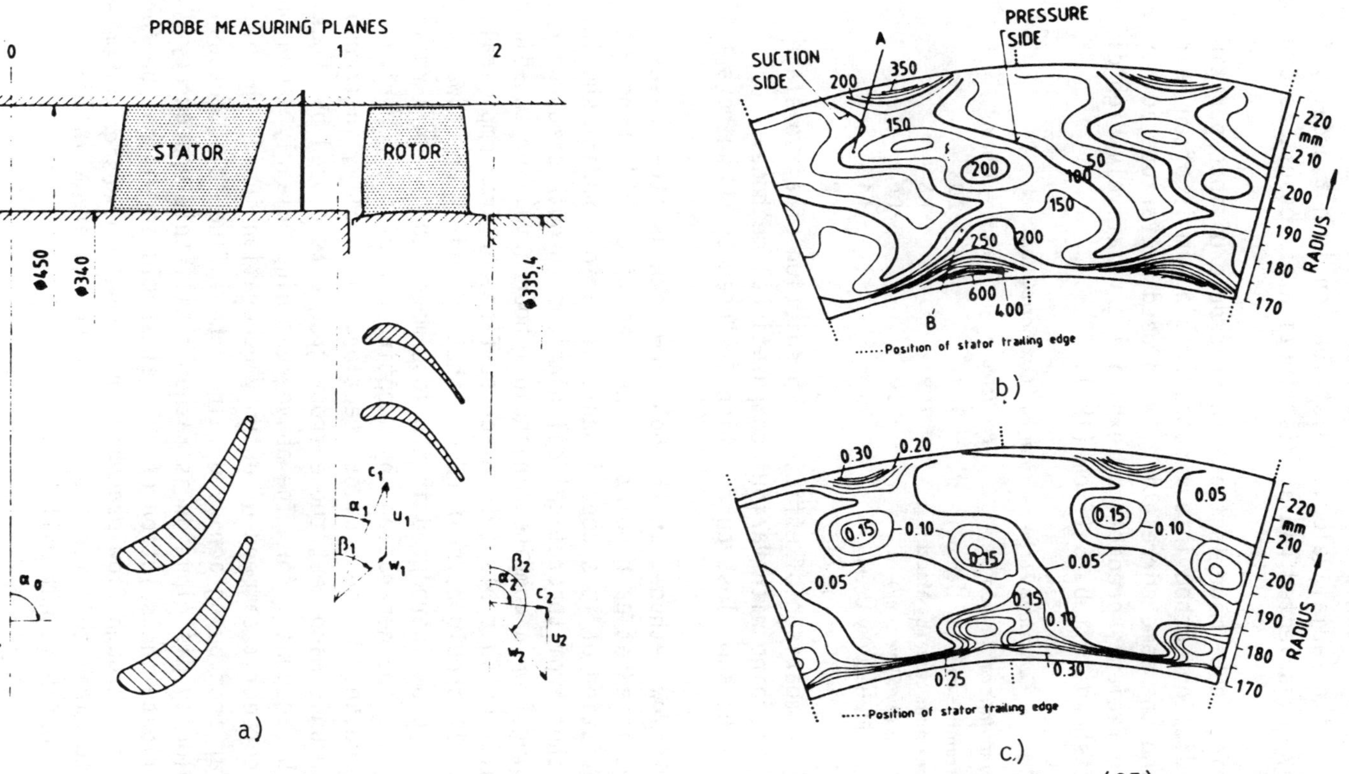

Fig. 8 Detailed flow measurements on the exit of a turbine stator, from (25).

a) Turbine geometry.
b) Contours of time integrated mean square of random velocity fluctuation (m^2/s^2).
c) Contours of total pressure loss coefficients.

should be remembered that the assumptions made don't allow for mixing processes, nor for losses migrations, which occur in reality as shown for instance in fig. 9, from (20);

- fortunately, the radial velocity distribution is not greatly influenced by the selected radial distribution of losses, as argued in (19);
- a final point to be made is that the procedure of incorporating a loss prediction method into a computer program is often rather complex, and the same correlation, when used by various authors, can yield efficiency predictions exhibiting relatively high differences, as shown in tab. 5 (from (18)).

Cambridge Turbine – Craig and Cox Correlations

Computational Method	*Predicted η*	*Pred. η – Test η*
Streamline curv. (by Denton)	91.47	−2.33
Streamline curv. (by Macchi)	91.9	−1.8
NISRE (by Sieverding)	92.2	−1.6
Finite Element (by Uçer)	90.4	−3.4

Tab. 5 Comparison of efficiency prediction of a two stage turbine with various computational methods.
The same loss correlation was used by all authors (19).

4.2.2 Through flow methods. In through flow methods, the losses should be specified at each point of the computing grid. Hence, the chordwise variation of loss coefficients at various radii should be known, either from theoretical calculations (growth of boundary layer resulting from a combined inviscid + boundary layer two-dimensional calculation in a blade-to-blade surface) or from empirical correlations. It should be pointed out that secondary or endwall losses must also be introduced in the loss model to obtain a realistic overall entropy generation, by adopting one of the empirical approches discussed in the previous section. Eventually, in transonic and supersonic machines, the entropy gradients caused by shocks should also be accounted for. The above reasonings clearly demonstrate that the correct treatment of entropy generation inside the blade rows is far from being achieved. Since the influence of losses on radial velocity distribution is rather small, this unaccuracy doesn't invalidate the solution of REE. It should anyway not be forgotten that the through flow programs, when used for design optimization aim to minimize a function (the losses generate inside of the machine), which is not well known.

4.3 Flow Angle Calculation

The correct flow angle prediction is probably the most critical item in duct (or through) flow calculations. In fact, it is a common experience that very accurate flow predictions can be obtained, if correct flow angles are specified. The results of fig. 9 (from (26)) demonstrate this point, which was theoretically justified by Denton in (19), by comparing the order of magnitude of the various terms present in REE and showing that small changes in the spanwise gradient of exit angle have a large effect upon the velocity profile.

As it was discussed in the previous point for losses,the through flow calculations require much more information than duct flow ones, since the flow angles must be known at each computing station, while available deviation correlations provide the flow angles only after the blade rows. The use of inviscid two-dimensional blade-to-blade calculations provides the required passage averaged flow angle, and their use rather than simple empirical correlation is recommended. In (19) it is pointed out that simply going from a linear to a parabolic variation of angular momentum in function of axial distance inside the blade rows can lead in some cases to variations as high as 10% on the predicted overall pressure ratio.

For two-dimensional flows, the flow angle at the exit of a blade row can be computed in various manners:

- by experimental data on cascades; this way is commonly utilized in axial flow compressor calculations, which have blade sections of well established geometry (27);
- by empirical or theoretical correlations; for turbines, the simple so-called "sine rule"or other correlations, reviewed for instance in (11),are often utilized; for supersonic flow at blade row exit, care should be taken to account for flow after expansion, to obtain both more reasonable results, and, most of all, the convergence in numerical solution;
- by blade-to-blade calculations; for inviscid calculations, the obtained exit flow angle depends upon the applied closure condition. A better accuracy is generally obtained by adding to inviscid also boundary layer computations and calculating the downstream flow angle resulting after the mixing process by conservation equations (28).

The last method is the most attractive, since it can automatically incorporate the influence of effects which are important in through flow calculations, like incidence, stream surface divergence, radius

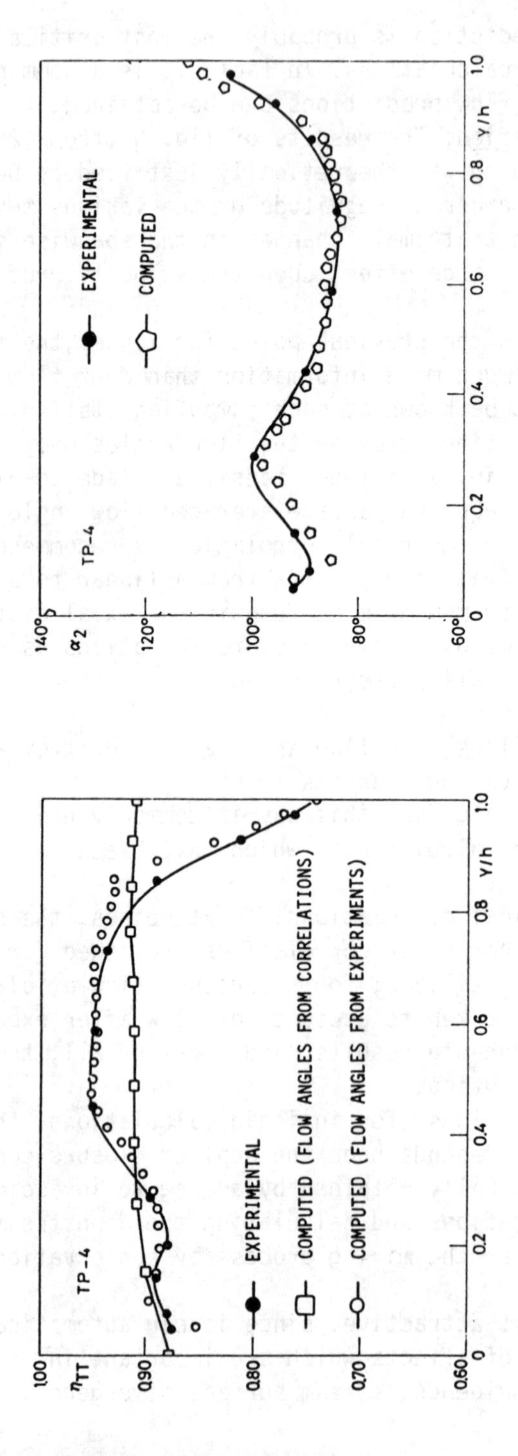

Fig. 9 Comparison of calculations performed by duct-flow method with flow angles as input data from experiment.

change, etc. The required computing time for these calculations is nowadays acceptable.

While two-dimensional exit flow angles can be predicted with reasonable accuracy, a different situation unfortunately takes place for secondary deviations. In (19), the results of an attempt of using the Bardon (29) correlation for secondary deviations are presented (fig. 10-12): it can be seen that they are rather disappointing, since neither the magnitude, nor the extent of the secondary deviation can be predicted. The lack of reliable theories for predicting the secondary deviation is at the moment the major drawback of through flow calculations, since it precludes the possibility of obtaining accurate velocity profiles.

4.4 Annulus wall boundary-layer

Calculations of annulus wall boundary layer are commonly incorporated in through flow calculations of axial-flow compressors. A review of existing theories can be found in (18). It is argued that they can provide:

- a blockage factor which includes secondary flow effects;
- the "force defect", i.e. the variation of the force exerted by the blade on the flow inside the boundary layer. The force defect is the mechanism through which a continuous build-up of boundary layer is prevented in multi-stage compressors after few stages;
- radial variations of incidence and turning angles in the wall regions.

These methods can be easily incorporated in through or duct flow calculation methods and have been found to provide good results, both as 1) blockage values, which influence the continuity equation, and therefore establish the flow rate-pressure drop characteristic, and 2) losses, which contribute to the overall efficiency evaluation of the machine. These correlations require however the adoption of some arbitrary coefficients; with values of these coefficients suggested by Mellor (30), the author obtained the results shown in fig. 13 and 14; a good agreement with the experimental values is found near design, while large errors appear at low flow rates. Similar procedures are not warranted for axial flow turbines, where an annulus boundary layer does not really exist within the blade rows: in fact, a complex process, not yet fully understood, takes place, during which the boundary layer entering a blade row is swept off the end walls and ends up in a passage vortex located at some radial distance (about half chord (18)) from the annulus walls. As already pointed out, reliable methods to predict the radial di-

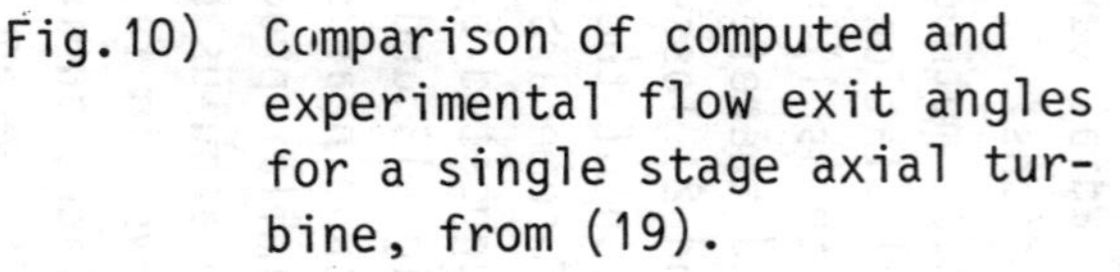

Fig.10) Comparison of computed and experimental flow exit angles for a single stage axial turbine, from (19).

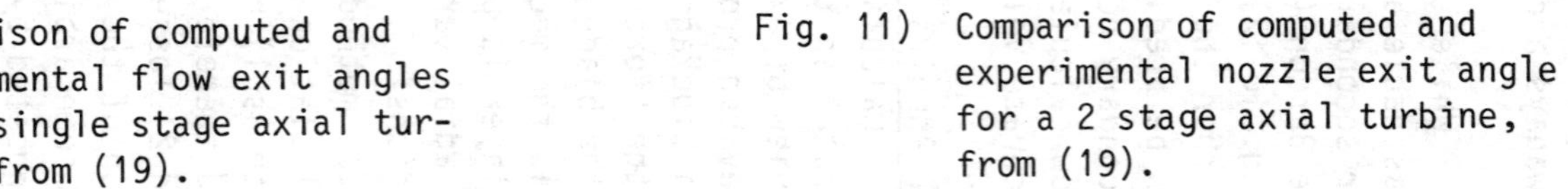

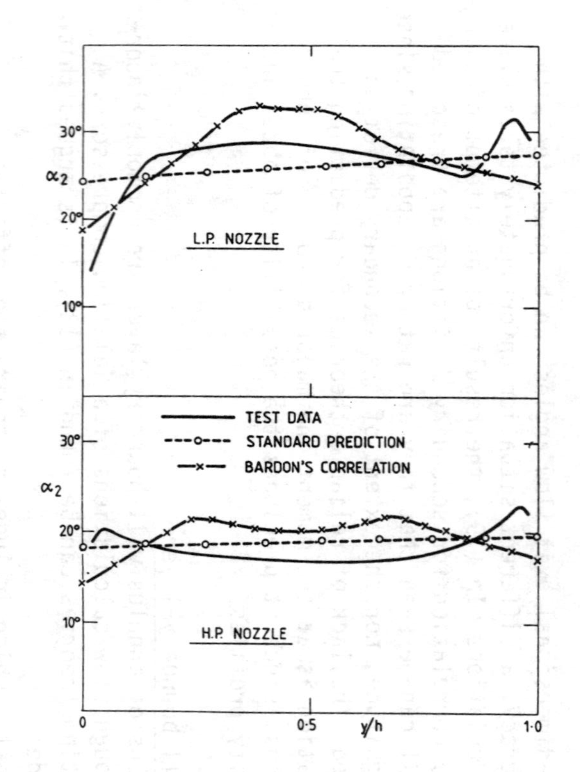

Fig. 11) Comparison of computed and experimental nozzle exit angle for a 2 stage axial turbine, from (19).

stribution of the high losses and flow deviations resulting from this process are not available.

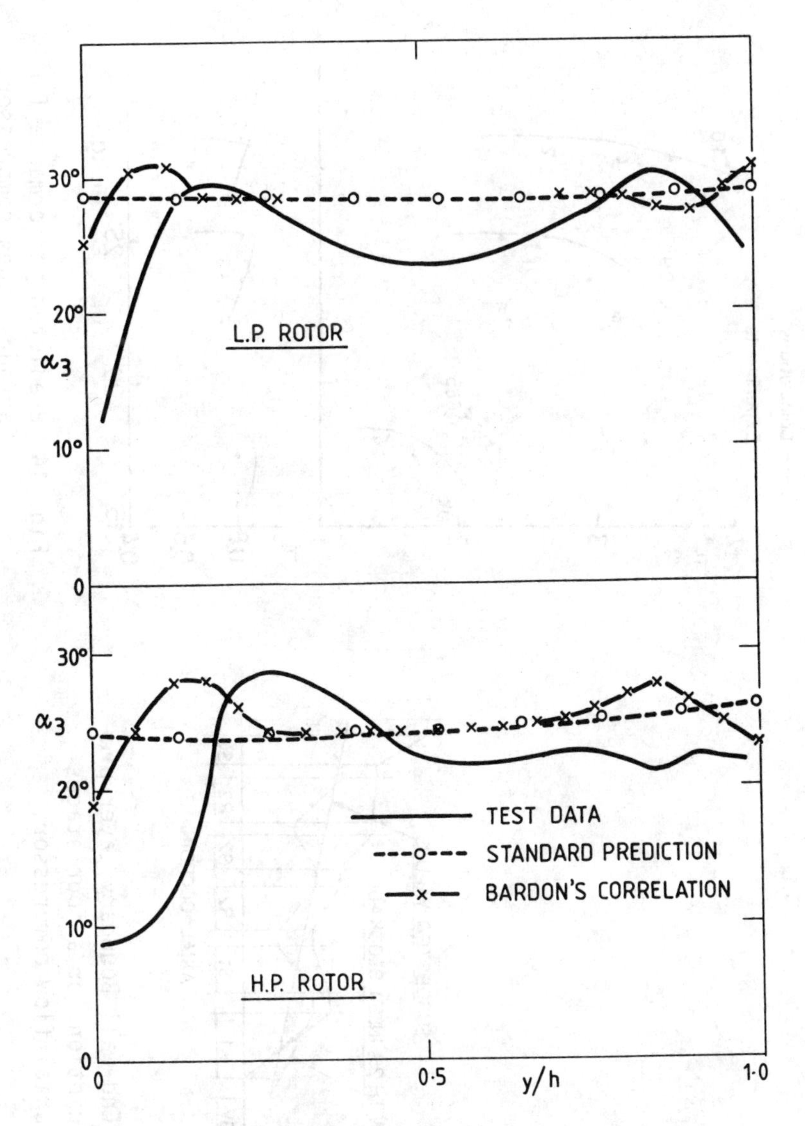

Fig. 12 Comparison of computed and experimental rotor exit angles from a 2 stage gas turbine, from (19).

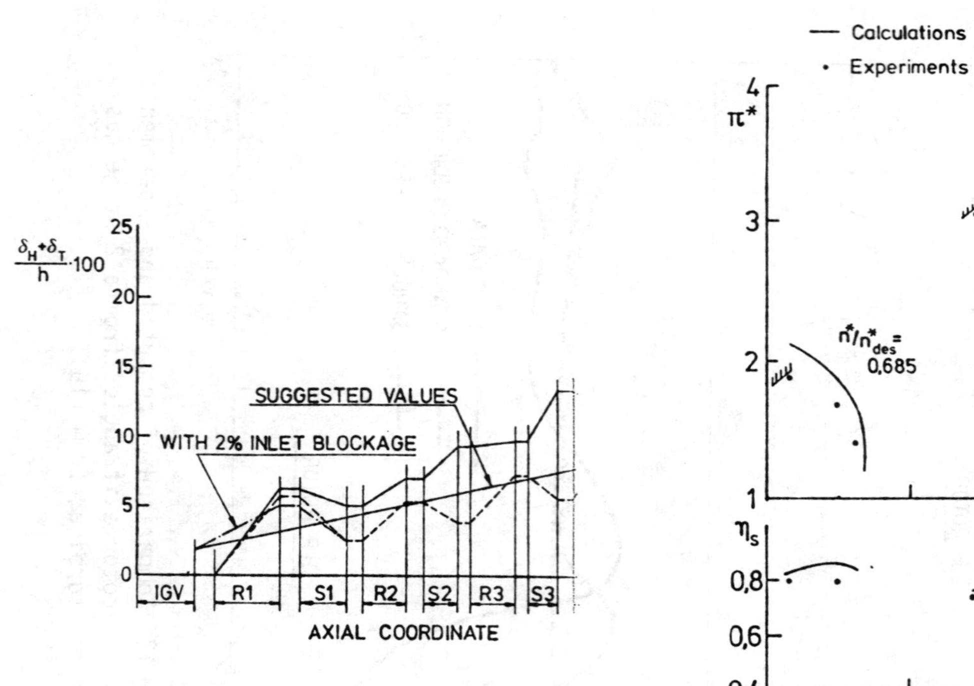

Fig. 13 End-wall Boundary Layer prediction for a four stages axial flow compressor.

Fig. 14 Predicted performance for a four stages compressor.

REFERENCES

1. Wu, C.H.. A General Theory of Three-Dimensional Flow in Subsonic and Supersonic Turbomachines of Axial-, Radial-, and Mixed-Flow Types. NACA TN 2604, 1952.
2. Jennions, I.K. and Stow, P.. A Quasi-Three-Dimensional Turbomachinery Blade Design System: Part 1 - Throughflow Analysis. ASME paper 84-GT-26, London,1984.
3. Arts, T.. Calculation of the Three-Dimensional, Steady, Inviscid Flow in a Transonic Axial Turbine Stage. ASME paper 84-GT-76, London, 1984.
4. Holmes, D.G. and Tong, S.S.. A Three-Dimensional Euler Solver for Turbomachinery Blade Rows. ASME paper 84-GT-79, London, 1984.
5. Quinghnan, W. and others. Quasi-Three-Dimensional and Full Three-Dimensional Rotational Flow Calculations in Turbomachines. ASME paper 84-GT-185, London,1984.
6. Povinelli, L.A.. Assessment of Three-Dimensional Inviscid Codes and Loss Calculations for Turbine Aerodynamic Computations. ASME paper 84-GT-187, London, 1984.
7. Xiao-lu, Z. and others. A Simple Method for Solving Three-Dimensional Inverse Problems of Turbomachine Flow and the Annular Constraint Condition. ASME paper 84-GT-198, London, 1984.
8. Pierzga, M.J. and Wood, J.R.. Investigation of the Three-Dimensional Flow Field within a Transonic Fan Rotor: Experiment and Analysis. ASME paper 84-GT-200, London, 1984.
9. Serovy, G.. Axial-Flow Turbomachine Through-Flow Calculation Methods, in Through Flow Calculations in Axial Turbomachines. AGARD-AR-175, pp. 285-299.
10. Chauvin, J. and Weyer, H.. Turbomachinery Through-Flow Calculation Methods - Technical Evaluation Report in AGARD-CP-195 pp. VII-IX.
11. Vavra, M.H.. Axial Flow Turbines,in Lecture Series 15, Von Karman Institute for Fluid Dynamics, Bruxelles, 1969.
12. Macchi, E.. Computer Program for Prediction of Axial Flow Turbine Performance, U.S. Naval Postgraduate School, NPS- 57Ma 70081a.
13. Jennions, J.K. and Stow, P.. A Quasi Three-Dimensional Turbomachinery Blade Design System: Part II - Computerized System.ASME paper 84-GT-27, London, 1984.
14. Horlock, J.H.. On Entropy Production in Adiabatic Flow in Turbomachines, Journal of Basic Engineering, Dec. 1971, pp. 587-593.
15. Vavra, M.H.. Aerothermodynamic and Flow in Turbomachines, Wiley and Sons, N.Y., 1960, pp. 439-470.
16. Silvester, M.E. and Hetherington, R.. Three-Dimensional Compressi

ble Flow Through Axial Flow Turbomachines, Numerical Analysis. An Introduction, edited by J. Walsh, Ch. II, Pt. III, 1966, pp. 182-189.

17. Frost, D.H.. A Streamline Curvature Through-Flow Computer Program for Analysing the Flow Through Axial-Flow Turbomachines, Aeronautical Research Council, R. & M. 3687, 1972.
18. AGARD Propulsion and Energetics Panel. Through-Flow Calculations in Axial Turbomachinery. AGARD-CP-195, 1976.
19. AGARD Propulsion and Energetics Panel, WG 12. Through Flow Calculations in Axial Turbomachines. AGARD-AR-175, 1981.
20. Macchi, E.. Methods of Calculation on Fluid-dynamics of Turbomachines, ed. CLUP, 1982 (in Italian).
21. Macchi, E.. Test-Cases on Fluid-dynamics of Turbomachines. Ed. CLUP, 1983 (in Italian).
22. Macchi, E. and Angelino, G.. Computation of Thermodynamic Properties of Carbon Dioxide in the Range 0-150 deg. C. ASME Paper 69-GT-118, Cleveland, 1969.
23. Sieverding, K.. Review of Performance Prediction Methods, in AGARD-AR-175, 1981, pp. 13-27.
24. Hirsch, C.. Review of Loss and Deviation Prediction Methods, in AGARD-AR-175, 1981, pp. 111-211.
25. Binder, A. et al.. An Experimental Investigation Into the Effect of Wakes on the Unsteady Turbine Rotor Flow. ASME paper 84-GT-178, London, 1984.
26. Lazzati, P. and Macchi, E.. Comparison between Theoretical and Experimental Results of a Single Stage Axial Turbine, in (20), pp. 139-156.
27. Centemeri, L. and Macchi, E.. Fluiddynamic Analysis of Multistage Axial Flow Compressors. LA TERMOTECNICA, Vol. 28, N. 1, 1974, pp. 34-54 (in Italian).
28. Macchi, E. et al.. Theoretical Prediction of Deflection Angles in Axial Flow Compressor Cascades. LA TERMOTECNICA, Vol. 31, N.8, 1977, pp. 408-417.
29. Bardon, M.F. et al.. Secondary Flow Effects on Gas Exit Angles in Rectilinear Cascades. ASME Journal of Engineering for Power, January 1975.
30. Balsa, T.F. and Mellor, G.L.. The Simulation of Axial Compressor Performance Using an Annulus Wall Boundary Layer Theory, ASME Journal of Engineering for Power, 1975, pp. 305-318.

LIST OF SYMBOLS

B = blade blockage term (see eq(31))
C_p = specific heat at constant pressure, J/kg K
$\vec{f}$ = viscous force per unit mass, N/kg
F_B = specific blade force, N/kg
F_τ = specific dissipative body force, in streamwise direction, N/kg
H = enthalpy, J/kg
m = direction along a meridional streamline, m
$\dot{m}$ = mass flow rate, kg/s
N = number of blades
p = static pressure, Pa
P = perturbation term, see eq(30), N/kg
q = heat per unit mass added to fluid from external sources, J/kg
R = radius, m
R_G = gas constant, J/kg K
S = entropy, J/kg K
S^* = non dimensional entropy, see eq(17)
t = time, s
T = static temperature, K
U = peripheral speed, m/s
V = absolute velocity, m/s
W = relative velocity, m/s
x = axial coordinate, m
X = non dimensional radial coordinate, see eq(15)
Y = non dimensional meridional velocity component, see eq(16)
α = absolute whirl angle
β = relative whirl angle
δ = radial blade lean
λ = hade angle
Θ = anomaly

ω = angular velocity, r/s

ρ = density, kg/m^3

Superscripts

$-$ = passage average

$-$ = referred to mean streamline

$\sim$ = density weighted average

$'$ = perturbation about the passage average

$''$ = perturbation about a density weighted average

Subscripts

A = axial component

E = equivalent conditions (see eq(10))

i = inner radius

m = meridional component

o = outer radius

p = pressure surface

r = radial component

R = total relative conditions

s = suction surface

u = tangential component

t = total conditions

Θ = circumferential component

Ψ = along a streamline

THE APPLICATION OF SINGULARITY METHODS TO BLADE TO BLADE CALCULATIONS

R. VAN DEN BRAEMBUSSCHE

von Karman Institute for Fluid Dynamics

1. INTRODUCTION

The use of singularities, vortices and sources, to calculate the potential flow in cascades or around single airfoils is based on the superposition principle. A sum of potential functions is also a potential function and the velocity components corresponding to the new potential functions equals the sum of the velocity components corresponding to the original potential functions.

The complex potential function corresponding to the flow around arbitrary airfoils can therefore be obtained by superposition of simple, well defined potential functions, corresponding to a parallel flow, vortices and sources. The conditions used to define this combination are the boundary conditions upstream and downstream and the kinematic conditions on the profiles. The main advantage of this procedure is the fact that one can calculate the whole flow field in an exact way by describing the flow only on the blade contour.

A first group of methods, where the center of vortices and sinks are located inside the blade contour, has the disadvantage that even for an increasing number of singularities the solution is generally not converging to the exact one [1,2,3].

A second group of methods, uses only vortices, placed on the blade contour. As will be shown later, this method converges to an analytically correct solution for an increasing number of singularities. These methods have been proposed for the direct problem (analysis of a given cascade) by Isay [4], Martensen [5] and Wilkinson [6] for incompressible flows and will be discussed first.

The use of these methods for blade design (inverse methods) and the extension to compressible flows, separated flows, flows in tandem cascades and flows on non cylindrical stream surfaces will be discussed in subsequent sections.

The incompressible potential flow in two dimensional cascades is defined by the condition of irrotationality

$$\frac{\partial u}{\partial y} - \frac{\partial v}{\partial x} = 0 \tag{1}$$

and the continuity equation

$$\frac{\partial u}{\partial x} + \frac{\partial v}{\partial y} = 0 \tag{2}$$

where u and v are the velocity components in x and y direction. The solution of previous set of equations is equivalent to solving the Laplace equation

$$\frac{\partial^2 \phi}{\partial x^2} + \frac{\partial^2 \phi}{\partial y^2} = 0 \tag{3}$$

2. INCOMPRESSIBLE POTENTIAL FLOW IN TWO DIMENSIONAL CASCADES

2.1 Basic potential flows

In the method discussed here, the flow in a cascade is defined by the sum of the potential functions corresponding to a uniform flow and a continuous distribution of vortices on the blade contour.

The velocity components of a uniform flow are defined by figure 1.

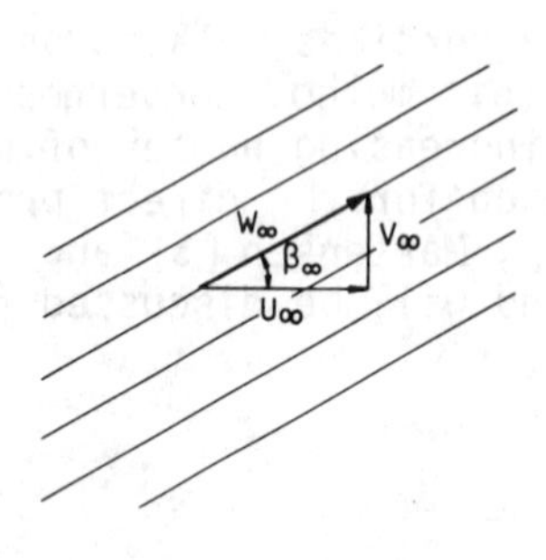

FIG. 1 - UNIFORM FLOW COMPONENTS

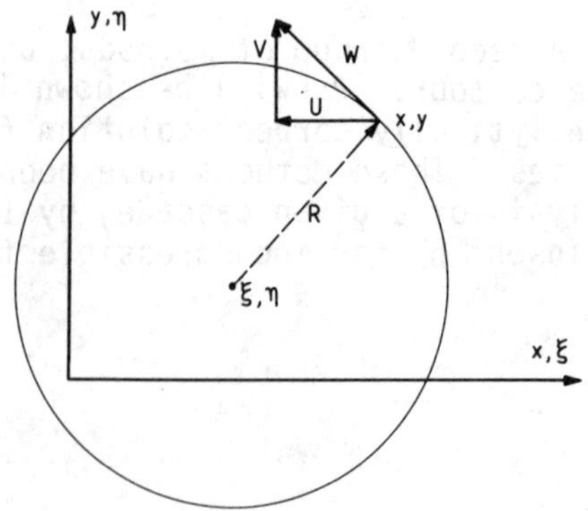

FIG. 2 - FREE VORTEX FLOW

$$u_\infty = W_\infty \cos\beta_\infty$$
$$v_\infty = W_\infty \sin\beta_\infty \qquad (4)$$

where W and β are constant. Such a velocity field satisfies automatically the condition of irrotationality (1) and continuity (2).

A free vortex flow is one where the streamlines are concentric circles and the velocity W along the circle is defined by figure 2.

$$W = \frac{\gamma}{2\pi R}$$

where γ = strength of the vortex located at the center
R = radial distance from the center
The velocity in radial direction is zero.

One can prove that (a) the circulation $\oint W_S . ds$ around any arbitrary closed contour equals the sum of the vortices inside the contour, (b) the total mass flux $\oint W_n . ds$ through any arbitrary closed contour is zero.

This means that the velocity field outside the center of the singularities is irrotational and satisfies the equation of continuity. This can also be proven by substituting the following velocity components into (1) and (2).

$$u = -\frac{\gamma}{2\pi}\ \frac{y-\eta}{(x-\xi)^2+(y-\eta)^2} \qquad (5)$$

$$v = \frac{\gamma}{2\pi}\ \frac{x-\xi}{(x-\xi)^2+(y-\eta)^2} \qquad (6)$$

where x,y are the coordinates of the points where the velocity is calculated.
ξ,η are the coordinates of the point where the vortex is located (γ is positive in the anti-clockwise direction).
Remark the singular behaviour of the expressions (5) and (6) when $x = \xi$ and $y = \eta$.

2.2 Periodic velocity field created by a vortex distribution and uniform flow

The conjugate velocity $\overline{W} = u-iv$ in a point x,y due to a single vortex $\gamma(\xi,\eta)$ is given by

$$\overline{W}(x,y) = \frac{-i}{2\pi}\ \frac{\gamma(\xi,\eta)}{(x-\xi)+i(y-\eta)}$$

In case of a cascade, the flow must be periodic, which means that all singularities must be repeated along the y-axis at an interval t equal to the blade pitch (Fig. 3). The velocity components in a point x,y are function of all the vortices and are defined by the following summation.

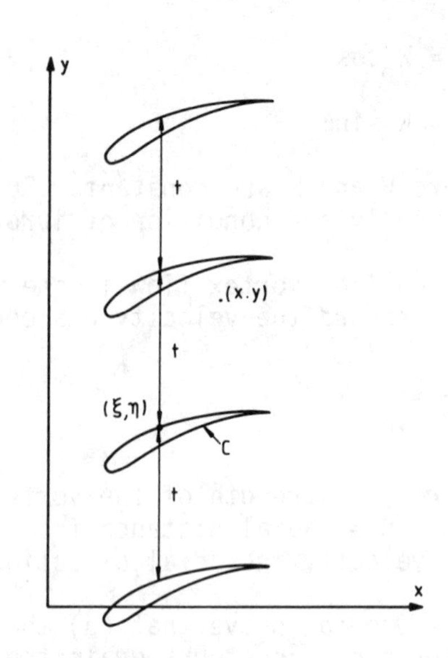

FIG. 3 - DEFINITION OF PERIODIC GEOMETRY

$$\overline{W}(x,y) = \frac{-i\gamma(\xi,\eta)}{2t} \sum_{n=-\infty}^{+\infty} \frac{t}{\pi[(x-\xi)+i(y-\eta-nt)]}$$

The Σ term corresponds to the series development of cotgh and considering a vortex distribution on each blade contour C, one must make the contour integral.

$$\overline{W}(x,y) = -\frac{i}{2t} \oint_C \gamma(\xi,\eta) \mathrm{cotgh} \frac{\pi}{t} \left[(x-\xi)+i(y-\eta)\right] ds(\xi,\eta)$$

This equation can be split into its real part u and imaginary part -v, and after addition of the uniform flow components we obtain the following expressions for the velocity components in (x,y)

$$U(x,y) = u_\infty - \frac{1}{2t} \oint_C \gamma(\zeta,\eta) \frac{\sin Y}{\cosh X - \cos Y} ds(\xi,\eta) \quad (7)$$

$$V(x,y) = v_\infty + \frac{1}{2t} \oint_C \gamma(\xi,\eta) \frac{\sinh X}{\cosh X - \cos Y} ds(\xi,\eta) \quad (8)$$

where $X = \frac{2\pi}{t}(x-\xi)$ $\qquad Y = \frac{2\pi}{t}(y-\eta)$

In these expressions the unknowns are u_∞, v_∞ and the vortex distribution $\gamma(\xi,n)$.

2.3 Singular points

Before defining the unknowns one will first examine in a general way, the variation of the velocity tangential and normal to a line vortex distribution.

S of figure 4 is a closed contour enclosing an elementary part of the vortex distribution on the blade contour C. The following equation expresses that the total circulation on S equals the total vorticity inside the contour

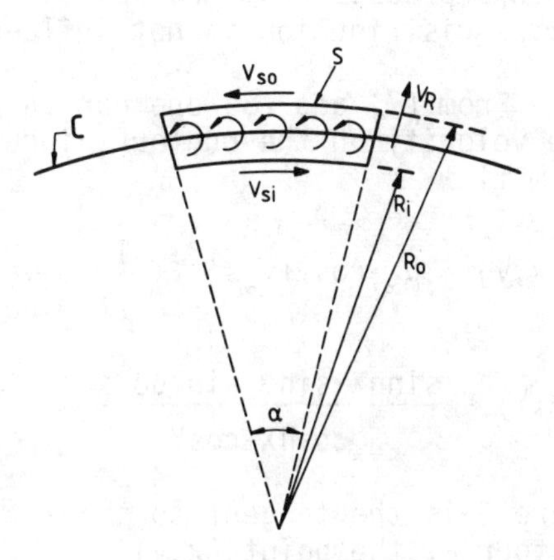

FIG. 4 - VORTEX DISTRIBUTION ON PROFILE CONTOUR

$$\int_0^{\alpha} W_{s_0} R_0 d\alpha + \int_{R_0}^{R_i} W_R \cdot dR - \int_0^{\alpha} W_{si} R_i d\alpha + \int_{R_i}^{R_0} W_R dR = \int_0^{\alpha} \gamma R d\alpha$$

Taking the limit for R_0, $R_i \to R$ one obtains

$$\int_0^{\alpha} W_{s_0} d\alpha - \int_0^{\alpha} W_{si} d\alpha = \int_0^{\alpha} \gamma d\alpha$$

and because this is valid for any value of α, one can conclude that

$$W_{s_0} - W_{si} = \gamma \qquad (9).$$

This means that the difference between the tangential velocities on both sides of a vortex distribution equals the local vortex strength.

Applying also the continuity equation on the S contour

$$\int_0^{\alpha} W_{R_0} \cdot R_0 d\alpha + \int_{R_0}^{R_i} W_s dR - \int_0^{\alpha} W_{Ri} \cdot R_i d\alpha + \int_{R_i}^{R_0} W_s dR = 0$$

Taking again the limit for R_i, $R_0 \to R$, one obtains

$$W_{R_0} = W_{Ri} \tag{10}$$

which expresses that the velocity component perpendicular to a vortex distribution is not influenced by the vortex distribution.

From (7) and (8) one can calculate the tangential component of the velocity on the contour, induced by the γ distribution and uniform flow

$$W_s(x,y) = u_\infty \cdot \cos\beta + v_\infty \sin\beta + \frac{1}{2t} \oint \gamma(\xi,\eta) K(X,Y) ds(\xi,\eta)$$

$$K(X,Y) = \frac{\sinh X \cdot \sin\beta - \sin Y \cos\beta}{\cosh X - \cos Y} \tag{11}$$

where β is the tangent to the contour at the point (x,y)

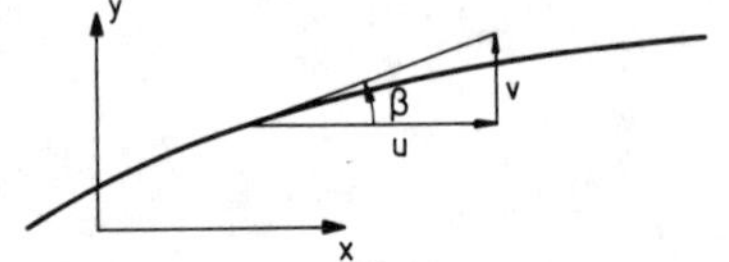

FIG. 5 - BLADE ANGLE DEFINITION

The integrand K(X,Y) is a regular function for every value of X and Y, except for X = Y = 0 where the integrand expresses the contribution to the tangential velocity of the vortex at the point where the velocity is calculated. As already mentioned the tangential velocity is discontinuous on the contour, which explains that the integrand is not determined in a unique way. One will therefore replace the integrand at X = Y = 0 by its mean value, calculated by de l'Hopital's rule

$$K(X{=}Y{=}0) = \frac{t.ds}{2\pi R_c} \tag{12}$$

where R_c is the local curvature radius of the profile contour. The tangential velocity calculated in this way by (11) and (12) will be the mean value of the velocity on outer and inner side of the contour.

$$W_s(x,y) = \frac{1}{2}\left[W\ \ (x,y) + W\ \ (x,y)\right]$$

and together with (9) one obtains

$$W_{s_0}(x,y) = W_s(x,y)+\gamma(x,y)/2 \qquad (13)$$

$$W_{si}(x,y) = W_s(x,y)-\gamma(x,y)/2 \qquad (14).$$

2.4 Boundary and kinematic conditions

The uniform flow field is determined in function of the cascade inlet and outlet condition. Calculating the limit of (7) and (8) for $x = \pm\infty$, one obtains

$$W_1\cos\beta_1 = u(-\infty,y) = u_\infty \qquad (15)$$

$$\begin{aligned} W_1\sin\beta_1 &= v(-\infty,y) = v_\infty - \frac{1}{2t}\oint \gamma(\xi,\eta)ds(\xi,\eta) \\ W_2\cos\beta_2 &= u(+\infty,y) = u_\infty \\ W_2\sin\beta_2 &= v(+\infty,y) = v_\infty + \frac{1}{2t}\oint \gamma(\xi,\eta)ds(\xi,\eta) \end{aligned} \qquad (16)$$

This allows to calculate the uniform velocity components

$$u_\infty = W_1\cos\beta_1 = W_2\cos\beta_2$$

$$v_\infty = (W_1\sin\beta_1+W_2\sin\beta_2)/2 \qquad (17)$$

and to relate the total vorticity per blade to the inlet and outlet conditions

$$\frac{1}{t}\oint \gamma(\xi,\eta)ds(\zeta,\eta) = W_2\sin\beta_2-W_1\sin\beta_1 \qquad (18)$$

The vortex distribution $\gamma(\xi,\eta)$ is defined by imposing zero tangential velocity at the inner side of the profile contour, as a kinematic condition. This is obtained by combining equations (11) and (14) and results in the following Fredholm integral equation, which must be respected in every point on the profile contour.

$$u_\infty\cos\beta +v_\infty\sin\beta+ \frac{1}{2t}\oint\gamma(\xi,\eta)K(X,Y)ds(\xi,\eta)- \frac{1}{2}\gamma(x,y) = 0 \qquad (19)$$

However, this equation is not sufficient to determine $\gamma(\xi,\eta)$ in a unique way, because the transposed of the homogeneous part of the integral is automatically zero.

$$\frac{1}{2t}\oint \gamma(\xi,\eta)\, K(X,Y)ds(x,y)- \frac{\gamma(\xi,\eta)}{2} = 0 \qquad (20)$$

The total amount of circulation around the profile induced by a vortex $\gamma(\xi,\eta)$ equals $\gamma(\xi,\eta)/2$ if the contour is passing through the vortex, which was assumed in the definition of $K(X,Y)$. This problem will be further analyzed in the next chapter.

One can easily prove that a velocity distribution determined by the equations (17) and (19) corresponds to a potential flow in a cascade, and that the local velocity on the profile equals the γ distribution.

- The velocity distribution between the blades is potential, because the components (uniform and vortex flow) are potential outside the singularities. Let ψ be the corresponding stream function.
- The flow is periodic in the y-direction because (7) and (8) contain only periodic functions in Y with period t. The flow is uniform, far upstream and downstream, because (15) and (16) are independent of x and y.
- The calculated velocity is tangential to the contour. Applying Green's theorem on the inner side of the profile contour, one obtains

$$\iint_{C_1} \left[\psi\nabla\psi + \left(\frac{\partial\psi}{\partial x}\right)^2 + \left(\frac{\partial\psi}{\partial y}\right)^2\right] dxdy = \oint_{C_1} \psi \left.\frac{\partial\psi}{\partial n}\right|_{in} ds$$

$\nabla\psi = 0$ because ψ is harmonic outside the singularities

$\left.\frac{\partial\psi}{\partial n}\right|_{in} = W_{si} = 0$ according to the kinematic condition (19)

one thus obtains :

$$\iint \left\{v^2(x,y)+u^2(x,y)\right\} dxdy = 0$$

which can be satisfied only if both u and v are also zero everywhere inside the contour. Therefore the normal velocity component inside the contour is zero and according to (10) also on the outer side.
- The velocity on the outer side of the profile contour equals γ. Combining (9) with the kinematic condition (19) one obtains

$$W_{so} = W_{si} + \gamma = \gamma$$

2.5 Solution of the Fredholm integral equation

The general solution of an integral equation consists of a particular solution of the inhomogenious equation and a general solution of the homogeneous part. From (19) one can conclude that if $\gamma_0(\xi,\eta)$ is a solution of the inhomogeneous equation and $\mu(\xi,\eta)$ is a solution of the homogeneous part, then also $\gamma_0(\xi,\nu)+\Gamma\cdot\mu(\xi,\eta)$

will be a solution for any values of Γ. This means that all possible solutions of (19) have a total vorticity which differs by a value $\Gamma \cdot \oint \mu(\xi,\eta)ds(\xi,\eta)$. An obvious choice for the additional condition to define a unique solution is one in which the total circulation is fixed.

In order to avoid that the integral equations must be solved again, each time the inlet or outlet flow conditions are changed, also the solution of the inhomogeneous equation will be split into two components. The total solution can therefore be written as follows

$$\gamma(\xi,\eta) = u_\infty \cdot \kappa(\xi,\eta) + v_\infty \cdot \lambda(\zeta,\eta) + \Gamma \cdot \mu(\xi,\eta) \tag{21}$$

where κ, λ and μ are defined by

$$\mu(x,y) - \frac{1}{t} \oint \mu(\xi,\eta) K(X,Y) ds(\xi,\eta) = 0 \tag{22}$$

$$\lambda(x,y) - \frac{1}{t} \oint \lambda(\xi,\eta) K(X,Y) ds(\xi,\eta) = 2\sin\beta \tag{23}$$

$$\kappa(x,y) - \frac{1}{t} \oint \kappa(\xi,\eta) K(X,Y) ds(\xi,\eta) = 2\cos\beta \tag{24}$$

Equations (22), (23) and (24) are obtained by substituting (21) into (19) and separating the terms in function of u_∞, v_∞ and those independent of them.

Because only a particular solution of κ and λ is required, one can impose the following additional conditions.

$$\oint \kappa(\xi,\eta) ds(\xi,\eta) = 0 \tag{25}$$

$$\oint \lambda(\xi,\eta) ds(\xi,\eta) = 0 \tag{26}$$

and because there is still the undefined constant Γ in (21), no restriction is imposed by putting

$$\oint \mu(\xi,\eta) ds(\xi,\eta) = 1 \tag{27}$$

As a conclusion one can say that $\kappa(\xi,\eta)$ is the vortex distribution corresponding to an inlet flow parallel to the x-axis without turning (Fig. 6), $\lambda(\xi,\eta)$ corresponds to a flow in the y direction and $\mu(\xi,\eta)$ corresponds to a flow circulating around each blade.

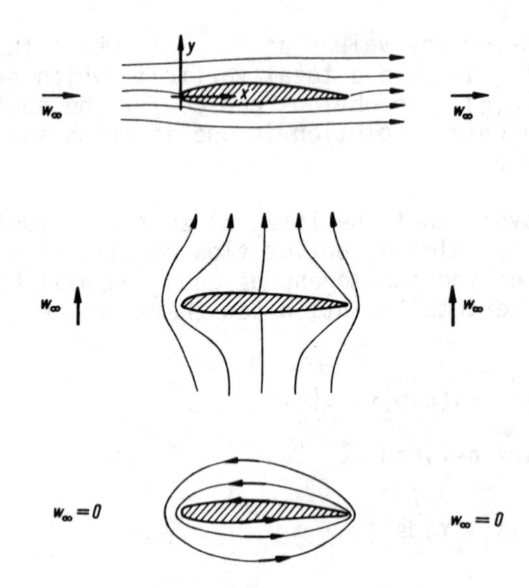

FIG. 6 - BASIC POTENTIAL FLOWS AROUND A BLADE

The total vorticity in function of these three components is given by

$$\oint\gamma(\xi,\eta)ds(\xi,\eta) = u_\infty\oint\kappa(\zeta,\eta)ds(\zeta,\eta)+v_\infty\oint\lambda(\xi,\eta)ds(\xi,\eta)+\Gamma\oint\mu(\zeta,\eta)ds(\xi,\eta)$$

and with (25), (26) and (27) one obtains

$$\oint\gamma(\xi,\eta)ds(\zeta,\eta) = \Gamma \tag{28}$$

Different methods are available to define Γ. In case the inlet and outlet flow conditions are given, the total vorticity can be defined by (18) and

$$\Gamma = t\cdot\left|(W_2\sin\beta_2 - W_1\sin\beta_1)\right| \tag{29}$$

However, when analysing a given cascade, only the inlet flow conditions are known. In this case the outlet flow angle can be defined by the Kutta condition, imposing zero velocity at the trailing edge stagnation point. Because the velocity on the contour equals the local value of the vorticity, one simply has to impose that at the trailing edge (TE)

$$u_\infty\kappa(TE)+v_\infty\lambda(TE)+\Gamma\cdot\mu(TE) = 0$$

Replacing u_∞ and v_∞ by (15) and with (28) one obtains the following expression for Γ.

$$\frac{W_1\cos\beta_1\kappa(TE)+W_1\sin\beta_1\lambda(TE)}{\lambda(TE)/2t-\mu(TE)} = \Gamma \qquad (30)$$

In case of a rounded trailing edge, the location of the stagnation point is not well defined. For this case, an alternative condition has been proposed by Wilkinson [6]. Based on the observation that the real flow separates at the point where the trailing edge round-off starts (TE-ε and TE+ε on Fig. 7) resulting in a separated flow zone of constant pressure, he proposes to impose zero loading on the trailing edge region, or equal velocity in the points TE-ε and TE+ε. Because of the different sense of rotation of the vortices on pressure and suction side, this results in

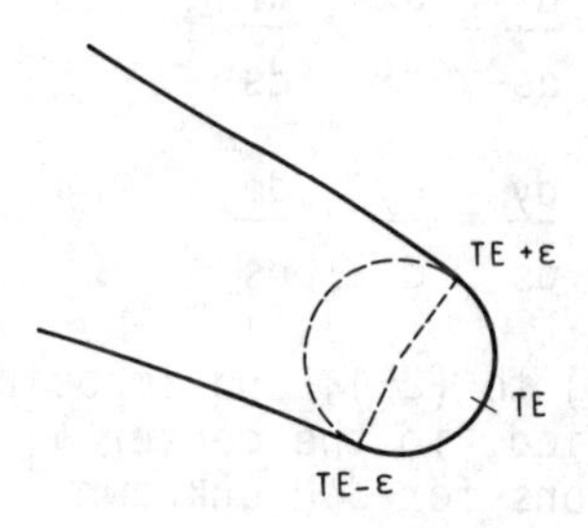

FIG. 7 - DEFINITION OF TRAILING EDGE GEOMETRY

$$u_\infty\kappa(TE-\varepsilon)+v_\infty\lambda(TE-\varepsilon)+\Gamma\cdot\mu(TE-\varepsilon)+u_\infty\kappa(TE+\varepsilon)+v_\infty\lambda(TE+\varepsilon)+\Gamma\cdot\mu(TE+\varepsilon) = 0 \quad (31)$$

2.6 Numerical solution

The equations (15,21,22,23,24,25,26,27,28) together with one of the equations (29,30,31) allow to calculate in an "exact" way the vortex distribution and hence the velocity distribution around the blades in a cascade, with given inlet conditions. However, this requires an analytical solution of the integral equations which is generally not possible.

The contour integrals are therefore replaced by a summation of N elementary integrals over an interval ΔS, in which the vorticity $\gamma(S)\Delta S$ is assumed to be concentrated at the center of the interval.

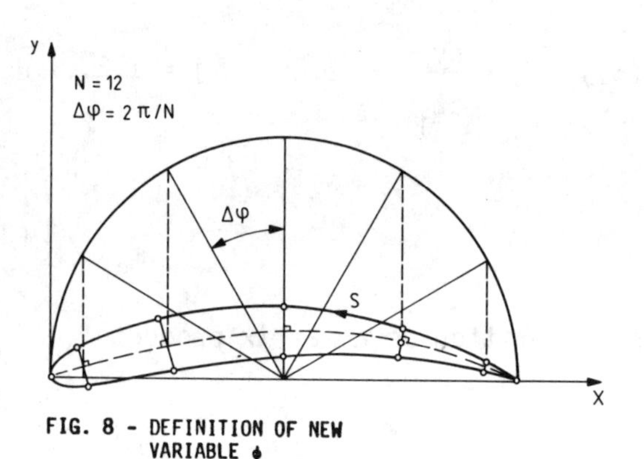

FIG. 8 - DEFINITION OF NEW VARIABLE ϕ

FIG. 8

The change of variables, defined in figure 8, allows to have a finer distribution of intervals close to the leading and trailing edge, for an equidistant distribution of $\Delta\phi = 2\pi/N$.

After substitution of

$$\gamma(s)\Delta s = \gamma'(\phi)\cdot\Delta\phi \quad \text{where} \quad \gamma'(\phi) = \gamma(s)\cdot ds/d\phi$$

$$\cos\beta = \frac{dx}{ds} = \frac{dx}{d\phi}\cdot\frac{d\phi}{ds}$$

$$\sin\beta = \frac{dy}{ds} = \frac{dy}{d\phi}\cdot\frac{d\phi}{ds}$$

in (22) to (27), and imposing that the kinematic conditions must be satisfied, in the center ϕ_j of the N intervals, one obtains 3(N+1) equations for 3.N unknown.

$$\mu'(\phi_j) - \frac{2\pi}{Nt}\sum_{i=1}^{N} K(\phi_j,\phi_i)\mu'(\phi_i) = 0 \qquad j = 1,2,\ldots N$$

$$\frac{2\pi}{N}\sum_{j=1}^{N} \mu'(\phi_j) = 1 \tag{32}$$

$$\kappa'(\phi_i) - \frac{2\pi}{Nt}\sum_{i=1}^{N} K(\phi_j\cdot\phi_i)\kappa'(\phi_i) = 2\cdot\frac{dx}{d\phi_j} \qquad j = 1,2,\ldots N$$

$$\frac{2\pi}{N}\sum_{j=1}^{N} K'(\phi_j) = 0 \tag{33}$$

$$\lambda'(\phi_j) - \frac{2\pi}{Nt}\sum_{i=1}^{N} K(\phi_j,\phi_i)\lambda'(\phi_i) = 2\cdot\frac{dy}{d\phi_j} \qquad j = 1,2,\ldots N$$

$$\frac{2\pi}{N}\sum_{j=1}^{N} \lambda'(\phi_j) = 0 \tag{34}$$

These systems can also be written as a matrix product

$$|K|\cdot|A| = |X| \tag{35}$$

$$|K| = \begin{vmatrix} 1-K(1,1) & -K(1,2) & \dots & -K(1,N) \\ -K(2,1) & 1-K(2,2) & \dots & -K(2,N) \\ -K(3,1) & -K(3,2) & \dots & -K(3,N) \\ \dots & \dots & \dots & \dots \\ -K(N,1) & -K(N,2) & & 1-K(N,N) \\ 2\pi/N & 2\pi/N & & 2\pi/N \end{vmatrix}$$

where $K(i,j) = \dfrac{2\pi}{Nt} K(\phi_i, \phi_j)$

$$A = \begin{vmatrix} \kappa_1' & \lambda_1' & \mu_1' \\ \kappa_2' & \lambda_2' & \mu_2' \\ \kappa_3' & \lambda_3' & \mu_3' \\ \cdot & & \\ \cdot & & \\ \cdot & & \\ \kappa_N' & \lambda_N' & \mu_N' \end{vmatrix} \qquad X = \begin{vmatrix} 2\dot{x}(1) & 2\dot{y}(1) & 0 \\ 2\dot{x}(2) & 2\dot{y}(2) & 0 \\ 2\dot{x}(3) & 2\dot{y}(3) & 0 \\ \cdot & & \\ \cdot & & \\ \cdot & & \\ 2\dot{x}(N) & 2\dot{y}(N) & 0 \\ 0 & 0 & 1 \end{vmatrix}$$

Although there are more equations than unknowns, these systems of equations are not overdetermined. According to (20), there exists a relation between the K(i,j) coefficients of the N equations

$$\sum_{j=1}^{N} K(j,i) = 1 \tag{36}$$

One can therefore eliminate one of the N equations in each system.

The diagonal terms of matrix (K) are function of the local curvature radius, and therefore more difficult to calculate. Martensen [5] therefore proposes to calculate these terms in function of the others, by means of (36). However, this has only a minor influence on the precision.

In this numerical solution, the main error is due to the incorrect approximation of line vortex by a point vortex, in case of thin airfoils.

The difference between the velocity distribution, due to a line vortex, and a point vortex at a distance Δn perpendicular to the vortex distribution Δs, is function of $\Delta n/\Delta s$ (Fig. 9). A first

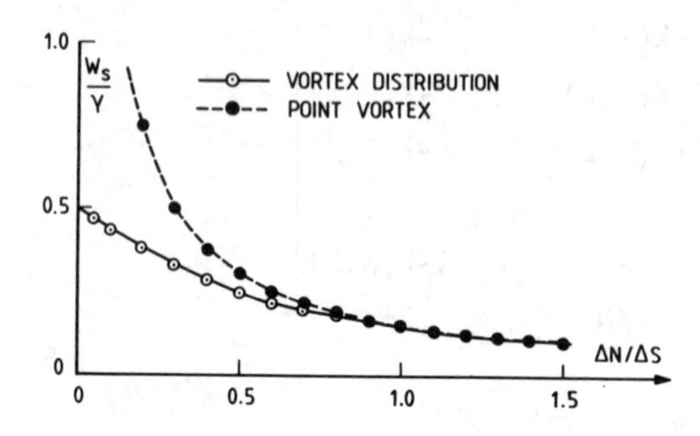

FIG. 9 - VARIATION OF TANGENTIAL VELOCITY

method to decrease the error is by descreasing the interval ΔS (increase of N). However, this has a negative effect on the computing time. A very important improvement is obtained, by calculating the coefficient, K(j,N+2-j), relating the local tangential velocity to the vortex on the opposite side of the profile, in function of other coefficients according to (36).

This results in a very important improvement without an increase in computer time, and certainly justifies the calculation of the blade curvature radius.

The system of equations (35) can be solved by elimination and the velocity in a point i on the profile is defined by

$$W(i) = C(i) \cdot \left[W_1 \cos\beta_1 \kappa'(i) + W_1 \sin\beta_1 \lambda'(i) + \Gamma\left(\mu'(i) - \frac{\lambda'(i)}{2t}\right)\right]$$

$$C(i) = \left[\left(\frac{dx(i)}{d\phi}\right)^2 + \left(\frac{dy(i)}{d\phi}\right)^2\right]^{-1/2}$$

2.7 Calculated results

The main advantages of previous methods are the short calculation time and the applicability to thin and high cambered blades of arbitrary shape. The precision is very good and comparable to exact analytical methods. The possibility to concentrate the calculation points in the regions where the velocity changes rapidly (leading edge and trailing edge region) allows to predict the influence of local changes in blade geometry.

Figure 10 compares the analytical results obtained by conformal mapping and the predictions by previous method for an eliptical blade with a thickness to chord ratio of .05.

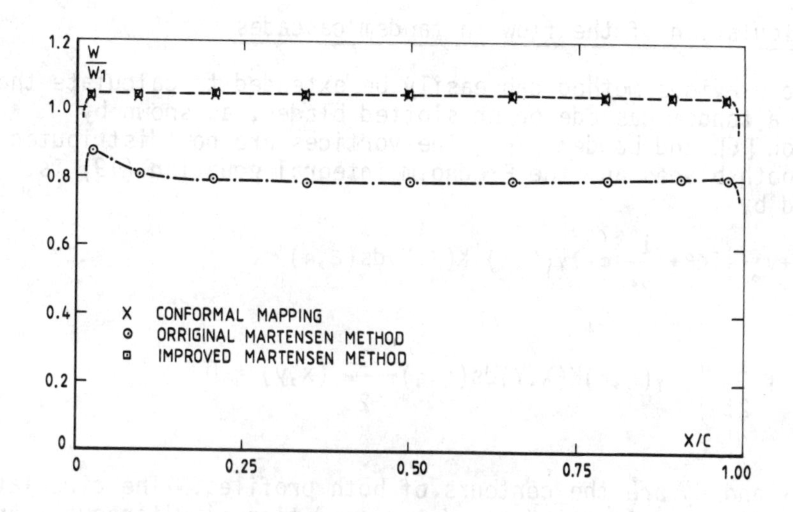

FIG. 10 - COMPARISON BETWEEN CALCULATED AND EXACT SOLUTION

The calculations are made with only ten vortices on each side of the blade. The results obtained with the original method (indicated by circles) show a large discrepancy when compared to the analytical solution (indicated by x). However, the improved method using relation (36) to calculate the opposite points, shows a remarkable agreement (indicated by squares).

Figure 11 illustrates the possibility to predict the flow also in extreme cases.

The highly cambered blade is from an analytically derived cascade, described in reference 8, and is operated at 70° negative incidence. The only discrepancy which can be observed is due to a small error in blade coordinates on the rear part of the suction side.

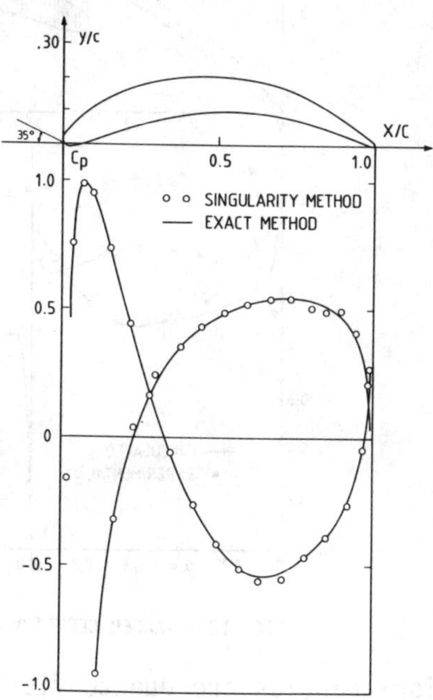

FIG. 11 - CALCULATED RESULTS AT HIGH INCIDENCE

3. EXTENSIONS OF THE SINGULARITY METHOD

3.1 Calculation of the flow in tandem cascades

The previous method can easily be extended to calculate the flow in a tandem cascade or on slotted blades, as shown by Wilkinson [6] and Loudet [7]. The vortices are now distributed around both blades and the Fredholm integral equation (19) is replaced by

$$u_\infty \cdot \cos\beta + v_\infty \cdot \sin\beta + \frac{1}{2t} \oint_{C_1} \gamma(\xi,\eta)\, K(X,Y) ds(\xi,\eta)$$

$$+ \frac{1}{2t} \oint_{C_2} \gamma(\xi,\eta) K(X,Y) ds(\xi,\eta) - \frac{1}{2}\gamma(x,y) = 0 \tag{37}$$

where C_1 and C_2 are the contours of both profiles. The circulation can be defined by imposing the Kutta condition simultaneously on both blades.

A comparison of experimental and theoretical results, obtained by Loudet [7] is shown in figure 12.

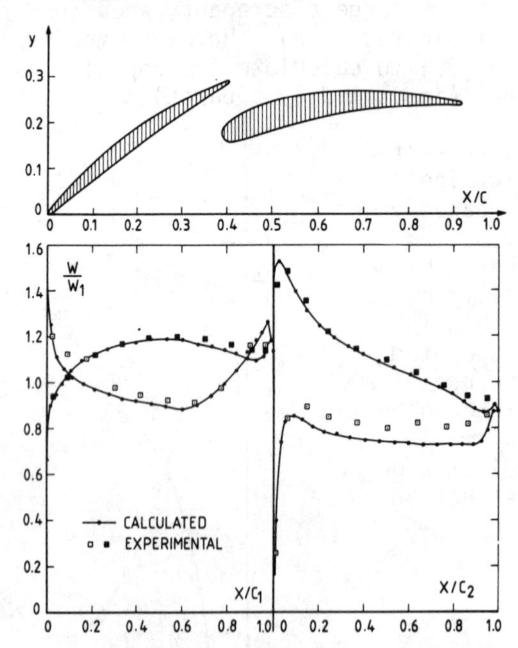

FIG. 12 - TANDEM GEOMETRY AND RESULTS

The discrepancies are due to viscous effects.

One of the advantages of singularity methods is the possibility to calculate streamlines by defining the coordinates for which the stream function has a constant value

$$\psi(x,y) = cst + u_\infty y - v_\infty x + \frac{1}{2\pi} \oint_{C_1+C_2} F(X,Y)\cdot\gamma(\xi,\eta)ds(\xi,\eta)$$

where

$$F(X,Y) = -\frac{1}{2} \ln (\cosh X - \cos Y)$$

This also allows to calculate the slope and the velocity on the stagnation streamline starting at the first blade trailing edge. This was a valuable input in [7] to calculate the development of the first blade wake under the pressure gradient, imposed by the second blade.

3.2 Inverse problem

The use of a singularity in an inverse approach was demonstrated by Wilkinson [6] and Murugesan & Railly [9]. In the latter method, a vortex distribution $\gamma_C(\xi,\eta)$ is placed on the preliminary blade contour, of strength equal to the difference between the desired and actual velocity on the contour.

$$\gamma_C = W_D - W_A$$

This vortex distribution creates a velocity component normal to the contour.

$$W_n(x,y) = \frac{1}{2t} \oint \frac{\sinh X\cdot\cos\beta + \sin Y\cdot\sin\beta}{\cosh X - \cos Y} \gamma_C(\xi,\eta)\, ds(\xi,\eta)$$

The local change in slope of a new contour will be obtained by taking the ratio of this normal velocity to the total imposed tangential velocity

$$dn/ds = W_n/W_A$$

Integration of this new slope gives rise to a new contour. The latter one is analyzed and the resulting velocity is again compared to the desired one. This process normally converges within eight to ten iterations.

The blade shape variation and corresponding suction side velocity variation is shown in figure 13 for a typical case.

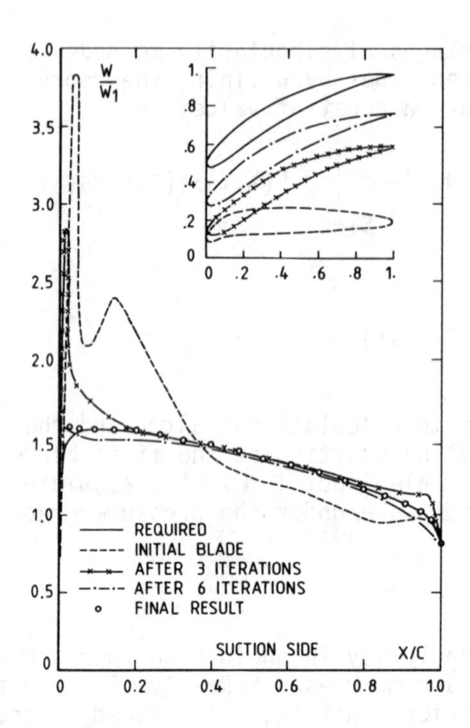

FIG. 13 - VARIATION OF BLADE GEOMETRY AND VELOCITY DISTRIBUTION DURING ITERATIVE PROCEDURE

Although the method is iterative, calculation time is very short, because the basic singularity method is very fast. A further advantage of the method is its automatic convergence to, and control of, the desired solution.

3.3 Calculation of separated flows

Singularity methods have been used to calculate separated flows by Jacob [10,11] for single airfoils and by Geller [12] and Goujon-Dubois [13] for cascades.

Separated flows are characterized by the presence of a dead water zone at constant static pressure, downstream of the separation point. Such flows can easily be modelized by means of additional singularities on the blade contour. One normally uses a source distribution on the contour because this allows to create the blockage corresponding to the separated boundary layer and does not influence the circulation. The source distribution $q(\xi,\eta)$ is defined in such a way to obtain a constant velocity over the separated flow zone. The influence of the source distribution on the tangential velocity distribution is compensated by an additional vortex distribution defined by

$$\delta(x,y) - \frac{1}{t}\oint \delta(\xi,\eta)K(Y,Y)ds(\xi,\eta) =$$

$$\frac{1}{t}\int q(\xi,\eta)\,\frac{\sinh X\,\cos\beta - \sin Y \sin\beta}{\cosh X - \cos Y}\,ds(\xi,\eta)$$

$$\oint \delta(\xi,\eta)ds(\xi,\eta) = 0$$

where $q(\xi,\eta)$ is the source strength in the separated region.

The flow field created by this $q(\xi,\eta)$ and $\delta(\xi,\eta)$ distribution does not create circulation and is of the type shown in figure 14.

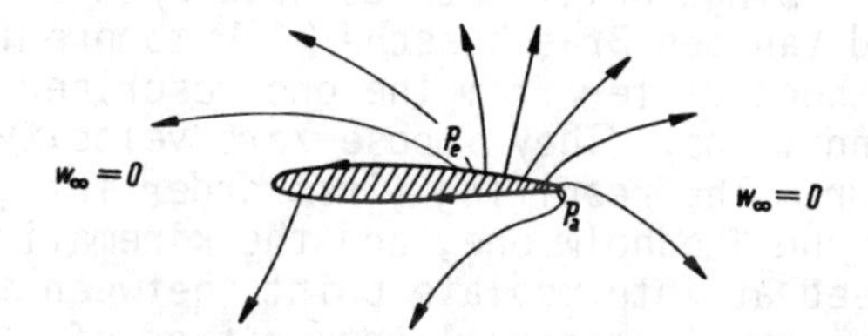

FIG. 14 - BASIC FLOW COMPONENT FOR SEPARATED FLOWS

An extra term accounting for $\delta(\xi,\eta)$ must be added to the definition of the velocity distribution (21)

$$\gamma(\xi,\eta) = u_\infty \cdot \kappa(\xi,\eta) + v_\infty \lambda(\xi,\eta) + \Gamma\mu(\xi,\eta) + \delta(\xi,\eta)$$

The total circulation Γ can no longer be defined by the Kutta condition. However, a good prediction of the outlet air angle is possible by imposing equal velocity (zero loading) on both sides of the wake just downstream the trailing edge.

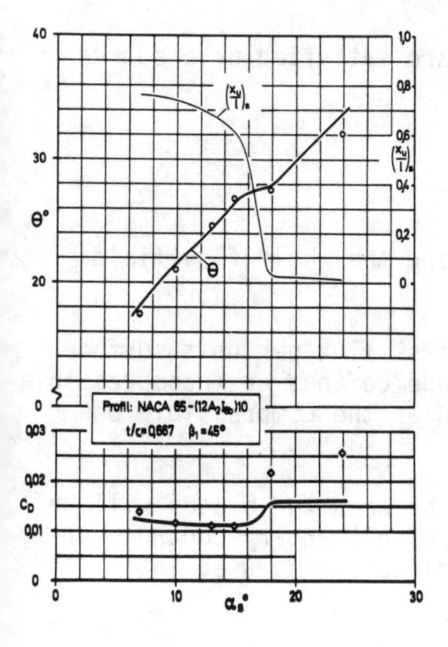

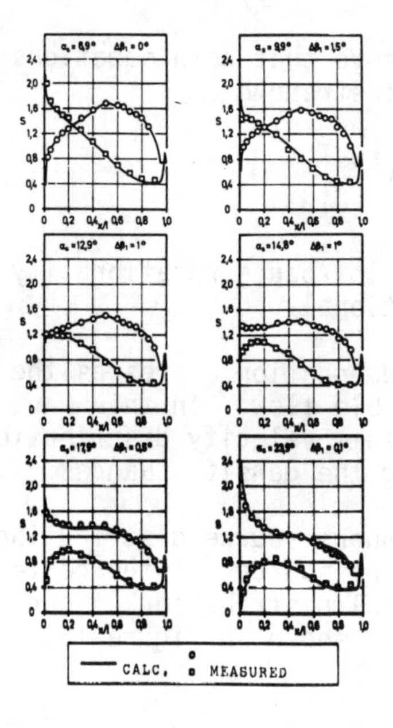

FIG. 15 - RESULTS OF SEPARATED FLOW CALCULATION

Figure 15 shows some typical results for small separated flow zones obtained by Geller [12] .

The calculated pressure distributions on the blades and turning angle θ agree generally well with measured data, while the corresponding drag coefficients show a larger discrepancy.

3.4 Extension to compressible flows

Singularity methods have been used by Isay [14] , Imbach [15] and Van den Braembussche [16] to predict compressible flow. Their methods differ from the one described previously by the kinematic conditions. They impose zero velocity normal to the profile contour. The resulting first order integral equation is not as stable as the Fredholm one, and the kinematic conditions must also be imposed at intermediate points between the vortices. This method also requires a real integration of $\gamma(\xi,\eta)$ $K(X,Y)$ distribution over an interval ΔS, to obtain a sufficient accuracy [17] .

The extension to compressible flow is obtained by means of a supplementary potential function ϕ_K. This function is added to the ones defined for incompressible flow, and must satisfy

$$\frac{\partial u}{\partial y} - \frac{\partial v}{\partial x} = 0$$

$$\frac{\partial u}{\partial x} + \frac{\partial v}{\partial y} = \frac{1}{\rho}\left(u\frac{\partial \rho}{\partial x} + v\frac{\partial \rho}{\partial y}\right)$$

One can prove that both equations are satisfied by a source distribution of strength

$$q = \frac{1}{\rho}\left(u\frac{\partial \rho}{\partial x} + v\frac{\partial \rho}{\partial y}\right)$$

Sources do not introduce rotationality and the whole flow field remains irrotational.

The second equation expresses the mass flux per unit surface for incompressible flow. In this way, one defines an incompressible flow with the same velocity distribution as the compressible one by compensating the density changes.

The continuous source distribution between the blades will be replaced by point sources at the center of elementary control surfaces. A typical distribution of control surfaces, used in reference 16, is shown in figure 16.

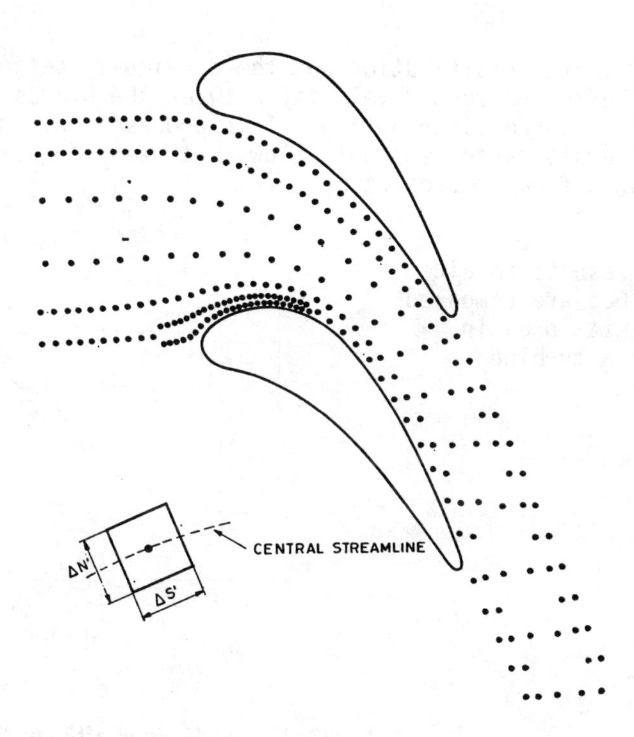

FIG. 16 - DISTRIBUTION OF CONTROL VOLUMES AND SOURCES

They are defined by streamlines and equipotential lines for incompressible flow, and the source strength can be approximated by

$$q = \frac{\Delta N'}{3} (W_0 - W_i)(M_0^2 + M_0 M_i + M_i^2)$$

where M is the local Mach number at inlet (i) and outlet (o) of a particular control surface.

An additional term must be included in (7) and (8) to account for the influence of the N' sources on the total velocity

$$U(x,y) = u_\infty - \frac{1}{2t} \oint \gamma(\xi,\eta) \frac{\sin Y}{\cosh X - \cos Y} ds(\xi,\eta) + \frac{1}{2t} \sum_{n=1}^{N'} q(n) \frac{\sinh X}{\cosh X - \cos Y}$$

$$V(x,y) = v_\infty + \frac{1}{2t} \oint \gamma(\xi,\eta) \frac{\sinh X}{\cosh X - \cos Y} ds(\xi,\eta) - \frac{1}{2t} \sum_{n=1}^{N'} q(n) \frac{\sin Y}{\cosh X - \cos Y}$$

Both velocity components and the local tangent to the blade contour, allow to express the condition of zero normal velocity, and thus to define the corrected vortex distribution on the blades.

This new vortex distribution and the previously defined sources, allow to define the corrected velocity between the blades and to calculate a new source distribution. This process converges in about three iterations for subsonic flow, but never converges if local supersonic flow is present.

Typical results obtained with this method are compared with experimental ones in figure 17 for a turbine cascade

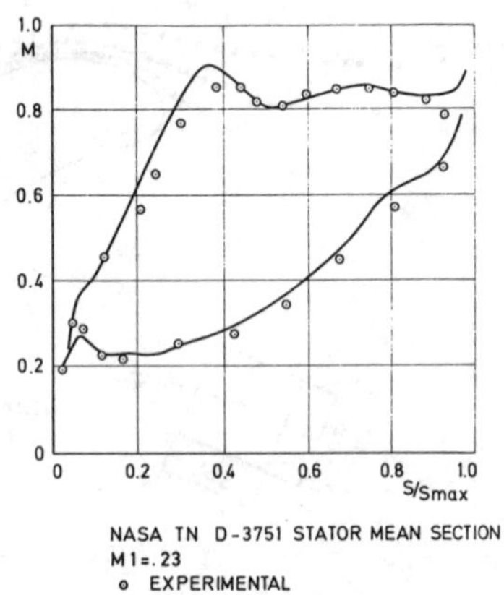

FIG. 17 - COMPARISON BETWEEN EXPERIMENTAL AND THEORETICAL RESULTS

The method allows to predict correctly the local acceleration and deceleration at the leading edge pressure side, and the correct location of the stagnation point. This is very important from the point of view of heat transfer. The correct description of the trailing edge geometry, and velocity together with the Kutta or Wilkinson condition, allows to obtain a good outlet air angle prediction.

Cascades with varying blade height (h) satisfy the following continuity condition.

$$\frac{\partial u}{\partial x} + \frac{\partial v}{\partial y} = \frac{1}{\rho h}\left(u\,\frac{\partial \rho h}{\partial x} + v\,\frac{\partial \rho h}{\partial y}\right)$$

The solution of this problem is similar to the two dimensional compressible flows.

3.5 Flow on non cylindrical stream surface

The use of singularities to calculate the flow on an axisymmetric stream surface with varying radius is discussed in [15][16] and [18]. These methods normally start with a conformal transformation of the stream surface into a straight cascade. In the case of a rotating blade row ($\omega \neq 0$) with varying radius, the relative velocity will be rotational. This can be accounted for by introducing a distributed vorticity over the stream surface [18]. Results obtained with this singularity method show again an almost exact agreement with analytical solutions.

Another possibility is to calculate the absolute irrotational rotor flow (as described by Vavra [19]) where the kinematic conditions have to be satisfied by the corresponding relative velocity.

These combinations of a conformal transformation and a singularity method is more complicated and not easy to extend to compressible flow. Some of the advantages of singularity methods are lost.

The direct use of singularity methods to calculate the three dimensional flows in mixed flow rotors is described in reference 15. This method is very attractive in the limiting case of a radial impeller where straight blades are fixed between two parallel walls perpendicular to the axis of rotation.

The velocity components W_R and W_θ in any point of the flow field are given by :

$$W_R(R,\theta) = -\oint \frac{Z\gamma(R,\theta)}{4\pi R} \cdot \frac{\sin Z(\theta-\theta_s)}{\cosh Z\ell n(R_s/R)-\cos Z(\theta-\theta_s)}\, ds + \frac{Q}{2\pi R}$$

$$W_\theta(R,\theta) = \oint \frac{Z\gamma(R,\theta)}{4\pi R} \quad \frac{\sinh Z\ell n(R_s/R)}{\cosh Z\ell n(R_s/R)-\cos Z(\theta-\theta_s)}\, ds + \frac{\Gamma}{2\pi R} - \omega . R$$

where R_S and θ_S are the coordinates of the vortex

R and θ are the coordinates where the velocity is calculated

Z is the number of blades

The mass flow through the rotor is simulated by Q and the prerotation Γ, both placed at the center of rotation. These equations are equivalent to (7) and (8). Further development is similar to the one of chapter I.

REFERENCES

1. VON KARMAN, T.: Berechnung der druckverteilung an luftschiffkörpern. Abh a.d. aerodyn., Institut Aachen, Heft 6,
2. PRAGER, W.: Die druckverteilung an Körpern in ebener Potential strömung. Phys. Z., Vol. 29, 1928, pp 865-869.
3. SCHLICHTING, H.: Berechnung der reibungslosen inkompressiblen strömung für ein vorgegebenes ebenes schaufelgitter. VDI Forschungheft 447, 1955.

4. ISAY, W.H.: Beitrag zur Potentialströmung durch axiale Schaufelgitter. ZAMM, Band 33, Heft 12, Dec. 1953, p 28.
5. MARTENSEN, E.: Berechnung der Druckverteilung an Gitterprofilen in ebener Potentialströmung mit einer Fredholmschen Integralgleichung. Archive for Rotational Mechanics & Analysis, Vol. 3, No. 3, 1959, pp 235-270.
6. WILKINSON, D.H.: A numerical solution of the analysis and design problems for the flow past one or more aerofoils or cascades. ARC R&M 3545, 1967.
7. LOUDET, C.: Contribution théorique et expérimentale a l'étude des grilles d'aubes en tandem à forte déflection et à forte charge. Ph.D. Thesis, U. Libre de Bruxelles, 1971.
8. GOSTELOW, J.P.: Potential flow through cascades of highly cambered blades. in: Proceedings of the Seminar on Advanced Problems in Turbomachines, VKI LS March 1965.
9. MURUGESAN, K. & RAILLY, J.W.: Pure design mehtod for aerofoils in cascades. J. Mech. Eng. Science, Vol. 11, No. 5, 1969.
10. JACOB, Kl.: Weiterentwicklung eines Verfahrens zur Berechnung der abgelösten Profilströmung mit besonderer Berüksichtigung des Profilwiderstandes. DLR FB 76-36, 1976.
11. JACOB, Kl.: Computation of the flow around wings with rear separation. DFVLR FB 82-22, 1982.
12. GELLER, W.: Berechnung der Druckverteilung an Gitterprofilen in ebener inkompressibler Strömung mit Grenzschichtablösung im Bereich der Profilenden. DLR FB 72-62, 1972.
13. GOUJON-DU BOIS, A.: Calcul d'un écoulement incompressible décollé dans une grille d'aubes. U-VKI 1972-72.
14. ISAY, W.H.: Zur behandlung der Kompressiblen Unterschallströmung durch Axiale und Radiale Schaufelgitter. ZAMM, Band 35, Heft 1/2, 1955.
15. IMBACH, H.E.: Die berechnung der kompressiblen reibungsfreien Unterschallströmung durch räumliche Gitter aus Schaufeln grosser Dicke und Starker Wölbung. Ph.D. Thesis, ETH Zürich, 1964
16. VAN DEN BRAEMBUSSCHE, R.: Calculation of compressible subsonic flow in cascades with varying blade height. in ASME Trans., Series A, - J. Engrg. Power, Vol. 95, No. 4, Oct 1973, pp 345-351
17. GOSIAU, S.: Berekening van de niet samendrukbare losgehechte stroming rond vleugelprofielen bij middel van de singulariteitenmethode. U-VKI 1983-03.
18. LEWIS, R.I.; FISHER, E.H. & SAVIOLAKIS, A.: Analysis of mixed-flow rotor cascades. ARC R&M 3703, 1972.
19. VAVRA, M.: Aero thermodynamics and flow in turbomachines. New York, R.E. Krieger Publ. Co. Inc., 1974.

LIST OF SYMBOLS

h	streamtube height
i	index corresponding to (ξ,η)
j	index corresponding to (x,y)
K	integrand (Eqn. 11)
M	Mach number
N	total number of point vortices
q	source strength
R	radial distance
R_c	curvature radius
s	distance along the contour
t	pitch
TE	trailing edge position
u	velocity component in x direction
v	velocity component in y direction
W	velocity
X,Y	defined in equations 7 and 8
x,y	coordinates of the point where the velocity is calculated
Z	number of blades
β	flow direction measured from axial
Γ	total circulation (Eqn. 28)
γ	vortex strength
δ	vortex strength
ε	distance from trailing edge (Fig. 7)
ϕ	angular coordinate (Fig. 8)
Φ	potential function
κ	vortex strength
λ	vortex strength
μ	vortex strength
ρ	density
θ	circumferential coordinate
Ψ	stream function
ξ,η	coordinate of the point where the vortex is located

Subscripts

A	actual value
c	value of the correction
D	desired value
i	inner side of the contour
n	normal to the surface
O	outer side of the contour
s	along the surface
R	radial component
θ	circumferential component
1	inlet condition
2	outlet condition
∞	uniform flow component

THE APPLICATION OF STREAMLINE CURVATURE TO BLADE-TO-BLADE CALCULATIONS

R. Van den Braembussche

von Karman Institute for Fluid Dynamics

1. INTRODUCTION

The use of streamline curvature methods to calculate the flow in turbomachinery blade rows is well documented. A first group of of methods uses streamlines and normals to the flow [1,2,3]. This results in simpler equations but requires to recalculate the normals each time the streamlines change position. Furthermore, problems related to the periodicity of the flow are not so easy to solve because the conditions apply to different normals.

A second group of methods uses streamlines and quasi-orthogonals [4,5,6,7]. The equations are more complex because the coordinate system is not orthogonal, but at least one coordinate is unchanged during the calculation procedure.

The main problems when using the methods are related to :
- stability and convergence of the solution and the definition of the underrelaxation factor;
- precision and instabilities related to the calculation of the streamline curvature radius;
- numerical integration of the mass flow along a quasi-orthogonal to satisfy continuity.

These problems are discussed in previous references, and have resulted in a calculation method with the following advantages :
- the streamline curvature method is very fast when compared to other methods;
- The precision is sufficiently good for subsonic or pure supersonic flows in channels with a large length over width ratio, and a smooth variation of wall curvature.
The main disadvantages are related to convergence problems.

Some specific problems are encountered in blade-to-blade calculations, such as :
- necessity to define the stagnation streamlines with sufficient accuracy;
- modelization of leading edge and trailing edge shape to avoid discontinuous variations of the curvature radius;
- prediction of the outlet flow angle.

Assuming that the basics of the method are sufficiently known from meridional flow calculations, one will concentrate in the following on the problems which are specific for blade-to-blade calculations.

2. DEFINITION OF STAGNATION STREAMLINES

The flow in the blade-to-blade plane is not calculated on the whole flow field but only in one channel, defined by the suction side of one blade, pressure side of the next blade, and the upstream and downstream stagnation streamlines for both blades (Fig. 1).

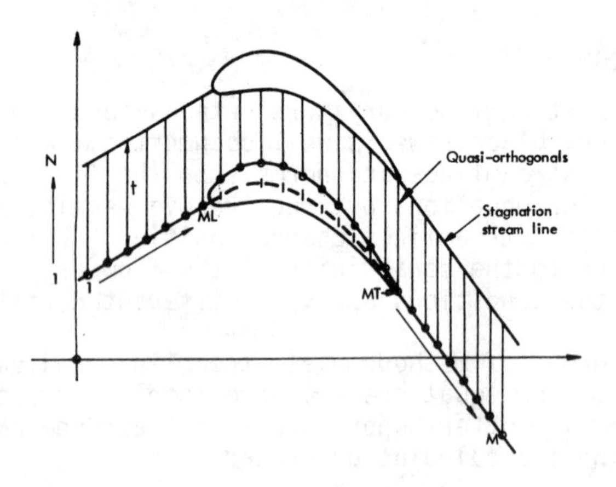

FIG. 1 - PARAMETERS FOR STAGNATION STREAMLINE DEFINITION

The stagnation streamlines are considered as impermeable walls and the velocity is influenced by its location and curvature. In case of large stagger angles or high curvature an important part of the flow channel is formed by the blade suction side and a stagnation streamline. It is obvious that in these regions small errors in the stagnation streamline shape will have an effect on the Mach number distribution. Wilkinson [7] calculated that an area reduction of 4% is sufficient to raise the main Mach number from .8 to 1.0 resulting in an erroneous prediction of choking mass flow.

The following procedure to correct the stagnation streamline position, fully documented in [7], applies to quasi-orthogonals which are tangent to the cascade (Fig. 1). Two successive stagnation streamlines are at one pitch distance from each other

$y(i,N)-y(i,1) = t$

and any modification of one streamline must be applied also on the other streamline.

Periodicity requires to have the same velocity on both streamlines and a change in streamline position will therefore be based on the velocity difference at the intersection of a quasi-ortogonal with the stagnation streamlines.

Assuming that this velocity difference can be distributed uniformly over the quasi orthogonal, one needs a change of local velocity gradient equal to :

$$\Delta \frac{\partial w(i)}{\partial y} = \frac{w(i,N)-w(i,1)}{y(i,N)-y(i,1)} \tag{1}$$

In the case of a two dimensional, inviscid flow in a cascade with uniform inlet conditions, the velocity gradient along a quasi-orthogonal is given by :

$$\frac{\partial w}{\partial y} = \left[\frac{\partial w_s}{\partial s} \cdot \frac{\partial y}{\partial s} + \frac{w}{R_c}\right] \cos\beta \tag{2}$$

A shift of the streamlines will have a direct influence on the streamline slope $\frac{\partial y}{\partial s}$, $\cos\beta$ and curvature radius R_c. The variation of $\frac{\partial w}{\partial y}$ at location i, due to a streamline shift ε at location j, can therefore be expressed as follows :

$$\frac{\partial}{\partial \varepsilon(j)} \left(\frac{w(i)}{\partial y}\right) = F_1 \frac{\partial}{\partial \varepsilon(j)} \left(\cos\beta(i)\right) + F_2 \frac{\partial}{\partial \varepsilon(j)} \left(\frac{y(i)}{\partial s}\right) + F_3 \frac{\partial}{\partial \varepsilon(j)} (R_c) \tag{3}$$

where F_1, F_2 and F_3 can be evaluated by differentiation of (2) and the derivatives can be evaluated from the numerical schemes used to calculate $\cos\beta(i)$, $\partial y(i)/dm$ and R_c.

After summation of (3) for $j = 1 \rightarrow ML$ or $j = MT \rightarrow M$, one obtains the total change $\Delta \frac{\partial w(i)}{\partial y}$ due to all the values of $\varepsilon(j)$ which must be equal to the value defined in (1).

$$\sum_j \varepsilon(j) \frac{\partial}{\partial \varepsilon(j)} \left(\frac{w(i)}{\partial y}\right) = - \frac{w(i,N)-w(i,1)}{y(i,N)-y(i,1)}$$

Satisfying this equation for all i values one obtains a system of equations with the same number of unknowns to define $\varepsilon(j)$.

This procedure is repeated each time the velocity distribution is calculated in function of the curvature radius until convergence is achieved.

3. MODELIZATION OF LEADING EDGE AND TRAILING EDGE

One of the problems of streamline curvature methods is related to the accurate description of the flow in the leading and trailing edge region. Some errors in estimating the curvatures and hence velocity gradients are inevitable in this region because of the discontinuous change in flow direction between stagnation streamline and blade surface. The introduction of a small cusp (Fig. 2) at leading and trailing edge helps in stabilizing the procedure

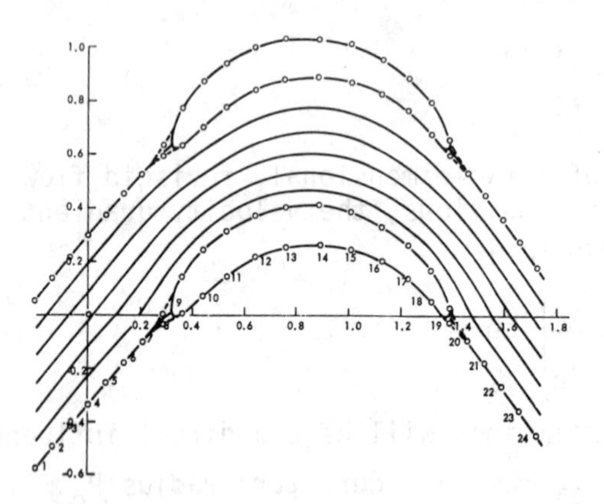

FIG. 2 - DEFINITION OF LEADING EDGE AND TRAILING EDGE CUSP

The initial shape of the cusp is not very important because it will be modified during the calculation procedure; when the stagnation streamlines are changed to satisfy the periodicity condition. The periodicity condition also assures zero loading on the cusp because of equal velocity on both sides. The main influence of the cusp is therefore a smoothing of the velocity distribution in both regions. This is a disadvantage at the leading edge, because one cannot define the stagnation point. However, at the trailing edge, this smoothing has the same effect as boundary layers and trailing edge separation.

4. PREDICTION OF OUTLET FLOW ANGLE

Due to the finite thickness of the trailing edge, it is very unlikely that the flow will follow the blade shape exactly up to the stagnation point. Experiments have shown that the flow normally separates on both sides at the corner of a blunt trailing edge or at the points where the round-off starts. This results in a dead water zone of constant static pressure. This is accounted for in the calculation by the trailing edge cusp and the periodicity conditions imposing equal velocity on both sides.

Wilkinson has shown that the trailing edge shape has only a minor influence on the predicted outlet flow angle. An accuracy of ± 1° can be expected in case of turbine blades. In a compressor cascade, the viscous effects are more important and the trailing boundary layer conditions are very asymmetric. The outlet flow angle prediction is therefore less accurate than in a turbine cascade.

5. RESULTS

A comparison between calculated and epxerimental results is shown in figure 3.

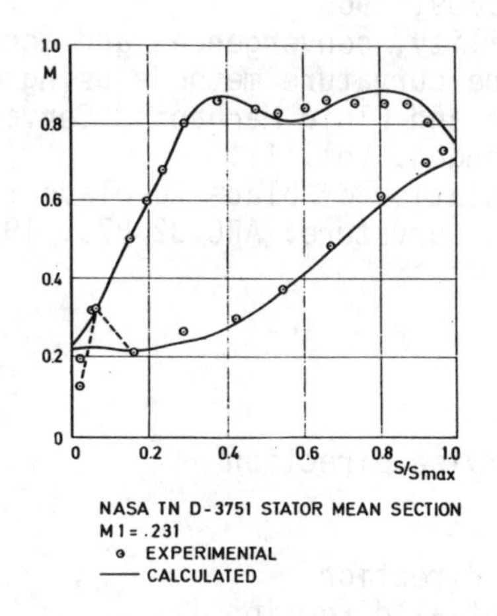

FIG. 3 - COMPARISON BETWEEN CALCULATED AND EXPERIMENTAL RESULTS

The agreement is very good, except at the leading edge, as could be expected. The acceleration and deceleration, close to the pressure side leading edge, are not predicted.

The method is also restricted to subsonic flows because the flow is imposed to be irrotational. Downstream of a shock the method predicts a velocity gradient which is function of the wall curvature. In reality, the velocity gradient just downstream a shock is function of the Mach number distribution just upstream the shock. The method can only be used as an approximation to the so called shock-free flows.

The method always allows, at each section, a subsonic and supersonic solution. An ambiguity exists always in the transonic problem because there is no criterium to choose one of these solutions.

REFERENCES

1. HAMRICK, J.; GINSBURG, A.; OSBORN, W.: Method of analysis for compressible flow through mixed-flow centrifugal impellers of arbitrary design. NACA R 1082, 1952.
2. STOCKMAN, N. & KRAMER, J.: Method for design of pump impellers using a high speed digital computer. NASA TN D 1562, 1963.
3. BINDON, J. & CARMICHAEL, A.: Streamline curvature analysis of compressible and high Mach number cascade flows. Proc. Inst. Mech. Eng.
4. KATSANIS, T.: Use of arbitrary quasi-orthogonals for calculating flow distribution in the meridional plane of a turbomachine. NASA TN D 2546, 1964.
5. KATSANIS, T.: Use of arbitrary quasi-orthogonals for calculating flow distribution on a blade-to-blade surface in a turbomachine. NASA TN D 2809, 1965.
6. WILKINSON, D.: Stability, convergence, and accuracy of two dimensional streamline curvature methods using quasi-orthogonals. Thermodynamics and Fluid Mechanics Convention '70. Int. Mech. Engrs., London, Vol. II.
7. WILKINSON, D.: Calculation of blade-to-blade flow in a turbomachine by streamline curvature. ARC 32 878, 1970.

LIST OF SYMBOLS

M Mach number
R_c curvature radius
s coordinate in streamwise direction
t pitch
w velocity vector
x coordinate in axial direction
y coordinate in tangential direction
β flow direction measured against axial direction
ε streamline shift in y-direction

SUBSONIC AND TRANSONIC FLOW SOLUTIONS USING TIME DEPENDENT FULL POTENTIAL EQUATION

Ahmet Ş. Üçer

Department of Mechanical Engineering
Middle East Technical University
Ankara, Turkey

1. INTRODUCTION

It is well known that flow through turbomachine blade rows is viscous, fully three dimensional and unsteady. The flow may also have weak or strong shocks. In the recent years either the solution of Euler equations are attempted in order to investigate 3-D and transonic flow effects [1,2,3,4], or the effect of viscosity on the flow is investigated by solving 3-D viscous flow equations [5]. The solution of 2 or 3-D steady potential flow is still of some value since it is possible to extend these solutions to transonic flows with shocks [6,7,8,9,23]. It is also possible to introduce time into the governing equation thus making it possible to extend the solution to fully 3-D unsteady potential flows. Due to the mixed elliptic-hyperbolic nature of the transonic flows it has been always necessary to take special precautions for obtaining accurate converged solutions.

For transonic flows three main catagories of mathematical models exist: Euler equation solvers, full potential equation solvers, and transonic small disturbance equation solvers [10,11,12,13,14]. Euler equations are the most complex among the three above and involve less approximations. Although steady solutions are aimed, generally unsteady form is used since the equations are hyperbolic both of subsonic and supersonic regions. Utilizing the assumption of irrotationality a velocity potential may be introduced and a full potential equation may be obtained. It is inherent in this assumption that the entropy generation of the shocks is neglected. Transonic small disturbance equations are preferred in some unsteady computations due to their simplicity.

Several discretization methods are used in the solution of the selected mathematical model. Each of the finite difference [15-20], finite element [6,8,9,21] or finite volume [1,3,4,22] methods have their advantages and disadvantages. Finite difference method although used very frequently needs global coordinate transformations in most of the cases in order to fit the grid system to irregular boundaries. In this method the introduction of Neumann boundary conditions along curvilinier boundaries create problems. In the finite element method higher order discretization using quadratic or biquadratic elements is possible. Locally applied isoparametric mapping and the use of curved sided elements are attractive for the treatment of irregular domains.

The governing equation, for the unsteady motion of a compressible inviscid gas is highly non-linear and hyperbolic. One way is to use the hyperbolic nature of the governing equation and integrate it in the time domain until reaching steady state. This technique is called time marching. Although it is popular it suffers from long computation times. The other branch is the relaxation methods. A lot of work has been performed for obtaining successful differencing techniques at subsonic and supersonic regions [10,11, 15]. In the later years instead of using switching in the differencing scheme an upwind bias in the differencing scheme has been applied at supersonic points by introducing artificial viscosity. It was shown that added artificial viscosity terms introduce the third order derivative of the potential function and looked like the viscous terms of the Navier-Stokes equation. Relaxation methods construct an artificial time-dependent equation and embed the steady flow equation in it to form a pseudo-time integration. Resulting equation may be solved using various techniques. Successive line overrelaxation, alternating direction implicit, approximate factorization,etc.

Another approach which is the explicit implementation of artificial viscosity is artificial compressibility technique [17]. This technique employs a slight modification to the density in the supersonic region, thus the transonic mixed type equation is handled as if it were elliptic. This method introduced new perspectives especially for finite element technique. The elliptic character of Galerkin's weighted residual approach delayed transonic finite element solutions considerably. However, artificial compressibility has provided a unified approach since Galerkin's method can be used all over the solution domain. The artificial viscosity coefficient appearing in the calculation of density in supersonic regions determines the amount of artificial damping and is critical for the success of the method.

In the proceeding subsections, the modelling of compressible unsteady flow is described. In the present formulation and solution technique it is intended to solve time dependent full potential equation using finite element discretization technique both in space and time. Artifical compressibility technique is used for the transonic flow solution.

2. FORMULATION

The following assumptions are made in the formulation of the problem. The flow is assumed inviscid, irrotational and that it takes place on cylindrical surfaces of Figure 1. By ignoring the gravitational effects the equation of motion for inviscid flow becomes.

$$\frac{\partial \vec{V}}{\partial t} + \vec{V}.\nabla\vec{V} = - \frac{\nabla p}{\rho} \tag{1}$$

The above equation is transformed into the following form assuming thermodynamic equilibrium of the fluid elements.

$$\frac{\partial \vec{V}}{\partial t} + \nabla H = \vec{V} \times (\nabla \times \vec{V}) + T\ \nabla s \tag{2}$$

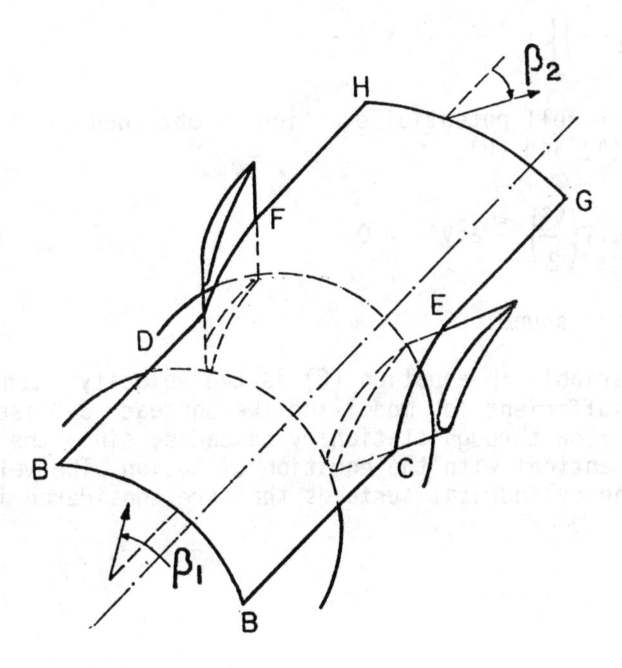

Figure 1. Solution Domain

where H is the total entalpy and s is the entropy. The other variables have their usual meanings. Assuming that the flow at the entry to the blade row is irrotational and that the flow is frictionless adiabatic throughout, equation (2) becomes.

$$\frac{\partial \vec{V}}{\partial t} + \nabla H = 0 \qquad (3)$$

It must be noted here that the flow is assumed to stay isentropic and irrotational through the weak shocks which may occur in transonic flows. In such a flow a velocity potential ϕ my be defined which transforms equation (3) into.

$$\frac{\partial}{\partial t}(\nabla\phi) + \nabla h + \nabla\left(\frac{V^2}{2}\right) = 0 \qquad (4)$$

where h is the enthalpy.

In order to obtain the governing equation for unsteady compressible flow, conservation of mass should be introduced into the equation of motion. The differential form of conservation of mass can be put into the following form by introducing velocity potintial and using the isentropic change of state of the perfect gas.

$$-\frac{1}{a^2}\left[\frac{\partial^2\phi}{\partial t^2} + \frac{\partial}{\partial t}\left(\frac{V^2}{2}\right)\right] + \frac{\nabla\phi.\nabla h}{a^2} + \nabla^2\phi = 0 \qquad (5)$$

The governing unsteady full potential equation is obtained by combining equations (4) and (5)

$$\frac{\partial^2\phi}{\partial t^2} + \frac{\partial}{\partial t}V^2 + \nabla\phi.\nabla\left(\frac{V^2}{2}\right) - a^2\nabla^2\phi = 0 \qquad (6)$$

where a is the speed of sound.

The dependent variable in equation (6) is the velocity potential and the equation is sufficient for modelling the unsteady 3-D isentropic, irrotational flow through stationary cascades, since the energy equation is identical with the equation of motion. The velocity components on the cylindrical surfaces that are considered in this work are

$$V_\theta = \frac{1}{r}\frac{\partial\phi}{\partial\theta} \quad \text{and} \quad V_z = \frac{\partial\phi}{\partial z} \qquad (7)$$

Equation (6) may be written in the following form for a cylindrical surface

$$\frac{\partial^2\phi}{\partial t^2} + \frac{\partial V^2}{\partial t} + \frac{1}{r^2}\left(V_\theta{}^2 - a^2\right)\frac{\partial^2\phi}{\partial\theta^2} + \left(V_z{}^2 - a^2\right)\frac{\partial^2\phi}{\partial z^2}$$

$$+ \frac{2}{r} V_\theta V_z \frac{\partial^2\phi}{\partial\theta\,\partial z} = 0 \tag{8}$$

The equation is in second order quasi-linear form since the coefficients include first order space derivatives of ϕ and speed of sound a,which is also a function of ϕ. Since the quasi-linear form does not allow correct shock capturing and the discretization is troublesome due to the existance of the last term which containes the cross derivative, equation (8) is put into conservative form by multiplying it by $(-\rho/a^2)$

$$-\frac{\rho}{a^2}\frac{\partial^2\phi}{\partial t^2} - \frac{\rho}{a^2}\frac{\partial V^2}{\partial t} + \frac{1}{r}\frac{\partial}{\partial\theta}\left(\rho\,\frac{1}{r}\frac{\partial\phi}{\partial\theta}\right) + \frac{\partial}{\partial z}\left(\rho\,\frac{\partial\phi}{\partial z}\right) = 0 \tag{9}$$

where

$$\rho = \rho_0\left(1 - \frac{V^2}{2c_pT_0}\right)^{\frac{1}{\gamma-1}} \tag{10}$$

$$a^2 = kRT_0\left(1 - \frac{V^2}{2c_pT_0}\right) \tag{11}$$

2.1 Pseudo-Time Dependent and Time Dependent Solutions

Equation (4) may be integrated to give,

$$\frac{\partial\phi}{\partial t} + h + \frac{V^2}{2} = C \tag{12}$$

The constant C has the dimension of enthalpy and expressed as $C = c_pT_1 = a_1{}^2/(\gamma-1)$ where a_1 is the speed of sound which is constant even for a flow which is non-steady. The following relation exists between stagnation speed of sound a_0 and reference speed of sound a_1

$$a_0{}^2 - \frac{\gamma-1}{2}V^2 = a_1{}^2 - (\gamma-1)\left[\frac{V^2}{2} + \frac{\partial\phi}{\partial t}\right] \tag{13}$$

In the case when steady solution is seaked it is assumed that $\partial\phi/\partial t = 0$ thus a_0 and a_1 become identical. It is also possible to define a new variable K which is constant for pseudo-time dependent solutions but variable in unsteady flows.

$$K^2 = \frac{2}{\gamma-1}\frac{a_0{}^2}{a_1{}^2} = \frac{2}{\gamma-1}\left(\frac{a}{a_1}\right)^2 + \frac{V^2}{a_1{}^2} = \frac{2}{\gamma-1} + \tilde{V}_1{}^2 \tag{14}$$

In the case of pseudo-time dependent solutions. The value of K is calculated from the above equation using upstream reference conditions once and for all. $\tilde{V}$ in equation (14) is the non dimensional velocity defined as $\tilde{V} = V/a_1$.

Equations (9), (10) and (11) can be put into non-dimensional form by introducing the following non dimensional quantities

$$\bar{\rho} = \frac{\rho}{\rho_1}\ ,\quad \bar{a} = \frac{a}{a_1}\ ,\quad \bar{t} = \frac{a_1 t}{c_{ax}}\ ,\quad \bar{\phi} = \frac{\phi}{a_1\ c_{ax}}\ ,$$

$$\bar{z} = \frac{z}{c_{ax}}\ ,\quad \bar{r} = \frac{r}{c_{ax}} \tag{15}$$

where c_{ax} is the axial cord of the cascade. After substituting the above relations into equation (9), (10) and (11) and dropping the overbars,

$$\frac{\rho}{a^2}\frac{\partial^2\phi}{\partial t^2} + 2\frac{\rho}{a^2}\left(\frac{\partial\phi}{\partial z}\frac{\partial^2\phi}{\partial z\partial t} + \frac{1}{r^2}\frac{\partial\phi}{\partial\theta}\frac{\partial^2\phi}{\partial\theta\partial t}\right)$$

$$- \frac{\partial}{\partial z}\left(\rho\frac{\partial\phi}{\partial z}\right) - \frac{1}{r}\frac{\partial}{\partial\theta}\left(\rho\frac{1}{r}\frac{\partial\phi}{\partial\theta}\right) = 0 \tag{16}$$

$$\rho = \left(\frac{\gamma-1}{2}\right)^{\frac{1}{\gamma-1}}(K^2-\tilde{V}^2)^{\frac{1}{\gamma-1}} \tag{17}$$

$$a^2 = \frac{\gamma-1}{2}(K^2-\tilde{V}^2) \tag{18}$$

Substituting equation (17) and (18) into equation (16) gives the principal equation as

$$\frac{2}{\gamma-1}(K^2-\tilde{V}^2)^{\frac{2-\gamma}{\gamma-1}}\frac{\partial^2\phi}{\partial t^2}+\frac{4}{\gamma-1}(K^2-\tilde{V}^2)^{\frac{2-\gamma}{\gamma-1}}\left(\frac{\partial\phi}{\partial z}\frac{\partial^2\phi}{\partial z\partial t}+\frac{1}{r^2}\frac{\partial\phi}{\partial\theta}\frac{\partial^2\phi}{\partial\theta\partial t}\right)$$

$$-\frac{\partial}{\partial z}\left[(K^2-\tilde{V}^2)^{\frac{1}{\gamma-1}}\frac{\partial\phi}{\partial z}\right]-\frac{1}{r}\frac{\partial}{\partial\theta}\left[(K^2-\tilde{V}^2)^{\frac{1}{\gamma-1}}\frac{1}{r}\frac{\partial\phi}{\partial\theta}\right]=0 \quad (19)$$

2.2 Boundary and Initial Conditions

The solution domain is shown in Figure 1. The blade passage extends upstream and downstream to ensure uniform flow at the far upstream and downstream boundaries. At the inlet, non-dimensional velocity and flow angle are specified. At the exit, it is sufficient to impose either the flow angle or the non-dimensional velocity if steady flow solution is aimed. The blade surfaces are impermeable walls across which there is no mass flux. The lines BD, AC, FH, and EG are the periodic boundaries on which the periodicity of the flow is imposed. The following boundary conditions are used at the indicated boundaries of the solution domain for steady flow solutions.

On A-B, essential boundary condition is applied

$$\phi_1 = \tilde{V}_1\,(z\cos\beta_1 + r\,\theta\sin\beta_1) \quad (20)$$

On H-G, the mass flux normal to the exit boundary is specified.

$$f = (K^2-\tilde{V}^2)^{\frac{1}{\gamma-1}}\;\tilde{V}_2\,\cos\beta_2 \quad (21)$$

On D-F and C-E since blade walls are impermeable following boundary condition is specified.

$$f = (K^2-\tilde{V}^2)^{\frac{1}{\gamma-1}}\;\frac{\partial\phi}{\partial n} = 0 \quad (22)$$

n being the direction perpendicular to the boundary.

On A-C and B-D periodicity of the flow is imposed so that

$$\Delta\phi = h\,\tilde{V}_1\,\sin\beta_1 \quad (23)$$

where h is the pitch and $\Delta\phi$ denotes the potential function difference between two points in θ direction at a constant z.

On E-G and F-H, the flow is again periodic therefore

$$\Delta\phi = h\,\tilde{V}_2\,\sin\beta_2 \qquad (24)$$

At the inlet, stagnation properties, T_0 and ρ_0, inlet Mach number M_1 and the flow angle β_1 are given. In the case of steady isentropic flow T_0 and ρ_0 are constant throughout the flow field. If an unsteady solution is aimed the values of T_0 and ρ_0 are considered to be the constant upstream conditions and the stagnation properties at the solution points change according to equation (13).

All the properties at the exit can be found,if either exit flow angle β_2 or exit Mach number M_2 is specified by using the continuity equation between inlet and exit and the isentropic flow conditions. In the case when steady solution is seeked the steady flow continity equation is sufficient for setting up the exit boundary condition. However, if unsteady flow solution is aimed, different ways of handling the boundary conditions should be worked out depending on the way the unsteadiness is imposed on the flow.

To be able to start the solution two initial conditions are required since the principal equation containes a second orter time derivative. Because, we generally aim either to a steady solution or to a periodic unsteady solution the initial conditions can be specified arbitrarily. In the solution,the first initial condition is obtained via an incompressible solution and a reasonable potential function distribution is obtained. The second initial condition is found by slightly changing the potential function values by the same amount throughout the flow field.

3. METHOD OF SOLUTION

The problem may be reformulated in an integral form by using finite element method both in space and in time. Galerkin's weighted residual method is utilized in the space for obtaining the semi-discretized differential equation system. Timewise finite element formulation provides a step-by-step recurrence scheme for finding the solution at each time step. When steady state boundary conditions are applied, the solution converges to its steady value after some time. In the case when steady transonic solutions are aimed artificial compressibility method is utilized and shock capturing technique is addopted. The method is a type of time marching method in which the time steps are artificial if a steady solution is of interest. In the case of time dependent solutions a numerically undamped stable timewise integration scheme should be used.

3.1 Spatial Finite Element Formulation

The principal equation (19) is to be solved together with the boundary conditions $\phi-\phi^* = 0$ on S_1 type boundary, and

$$f-(K^2-\tilde{V}^2)^{\frac{1}{\gamma-1}} \frac{\partial\phi}{\partial n} = 0 \tag{25}$$

on S_2 type boundary. Where ϕ^* and f are the known values on the boundary $S_t = S_1+S_2$ and n is the normal to S_2. An approximate solution is substituted in equations (19) and (25), the resulting residuals are multiplied by the weight function N, integrated over the solution domain and forced to be equal to zero. The integration of the steady terms of the resulting equation gives

$$\frac{2}{\gamma-1}\iint_\Omega N(K^2-\tilde{V}^2)^{\frac{2-\gamma}{\gamma-1}} d\Omega \frac{\partial^2\phi}{\partial t^2} + \frac{4}{\gamma-1}\iint_\Omega N(K^2-\tilde{V}^2)^{\frac{2-\gamma}{\gamma-1}}$$

$$\left[\frac{\partial\phi}{\partial z}\frac{\partial^2\phi}{\partial z\,\partial t} + \frac{1}{r^2}\frac{\partial\phi}{\partial\theta}\frac{\partial^2\phi}{\partial\theta\,\partial t}\right] d\Omega + \iint_\Omega (K^2-\tilde{V}^2)^{\frac{1}{\gamma-1}}\left[\frac{\partial N}{\partial z}\frac{\partial\phi}{\partial z} + \frac{1}{r^2}\frac{\partial N}{\partial\theta}\frac{\partial\phi}{\partial\theta}\right] d\Omega$$

$$= \int_{S_2} N(K^2-\tilde{V}^2)^{\frac{1}{\gamma-1}} \frac{\partial\phi}{\partial n}\, ds \tag{26}$$

This integral equation is now converted into an ordinary differential equation system in the smaller solution subdomains, called finite elements.The field variables ϕ, $\partial\phi/\partial t$ and $\partial^2\phi/\partial t^2$ are approximated by the interpolation functions (weight functions) and the nodal values of the field variables. The space derivatives are also defined in a similar way. The field variables and the spatial derivatives of the field variables are:

$$\phi(\theta,z,t) = [N]\,\{\phi(t)\}$$

$$\frac{\partial\phi(\theta,z,t)}{\partial t} = [N]\left\{\frac{\partial\phi(t)}{\partial t}\right\} \tag{27}$$

$$\frac{\partial^2\phi(\theta,z,t)}{\partial t^2} = [N]\left\{\frac{\partial^2\phi(t)}{\partial t^2}\right\}$$

$$\frac{\partial\phi}{\partial\theta} = \left[\frac{\partial N}{\partial\theta}\right]\{\phi\} \quad , \quad \frac{\partial^2\phi}{\partial\theta\,\partial t} = \left[\frac{\partial N}{\partial\theta}\right]\left\{\frac{\partial\phi}{\partial t}\right\}$$

$$\frac{\partial\phi}{\partial z} = \left[\frac{\partial N}{\partial z}\right]\{\phi\} \quad , \quad \frac{\partial^2\phi}{\partial z\,\partial t} = \left[\frac{\partial N}{\partial z}\right]\left\{\frac{\partial\phi}{\partial t}\right\} \tag{28}$$

Equation (26) may now be written in matrix form for an element as

$$\left[\frac{2}{\gamma-1}\iint_{\Omega^e}(K^2-\tilde{V}^2)^{\frac{2-\gamma}{\gamma-1}}[N]^T[N]\,d\Omega\right]\left\{\frac{\partial^2\phi}{\partial t^2}\right\}$$

$$+\left[\frac{4}{\gamma-1}\iint_{\Omega^e}(K^2-\tilde{V}^2)^{\frac{2-\gamma}{\gamma-1}}[N]^T[\phi]^T\left(\left[\frac{\partial N}{\partial z}\right]^T\left[\frac{\partial N}{\partial z}\right]+\frac{1}{r^2}\left[\frac{\partial N}{\partial \theta}\right]^T\left[\frac{\partial N}{\partial \theta}\right]\right)d\Omega\right]\left\{\frac{\partial\phi}{\partial t}\right\}$$

$$+\left[\iint_{\Omega^e}(K^2-\tilde{V}^2)^{\frac{1}{\gamma-1}}\left(\left[\frac{\partial N}{\partial z}\right]^T\left[\frac{\partial N}{\partial z}\right]+\frac{1}{r^2}\left[\frac{\partial N}{\partial \theta}\right]^T\left[\frac{\partial N}{\partial \theta}\right]\right)d\Omega\right]\{\phi\}$$

$$=\int_{S_2}(K^2-\tilde{V}_2^2)^{\frac{1}{\gamma-1}}[N]^T\,\tilde{V}_2\cos\beta_2\,ds \tag{29}$$

The term on the right hand side of the above equation carries the influence of natural boundary condition at the exit of the blade row only. Equation (29) may be written in the following form

$$[M]^e\left\{\frac{\partial^2\phi}{\partial t^2}\right\}^e+[C]^e\left\{\frac{\partial\phi}{\partial t}\right\}^e+[S]^e\{\phi\}^e=\{F\}^e \tag{30}$$

$[M]^e$, $[C]^e$ and $[S]^e$ are the mass, damping and stiffness matrices respectively, and they correspond to the integral expressions in equation (29).

3.2 Spatial Finite Elements

Two types of finite elements are used in the discretization of the solution domain. Initially isoparametric quadratic, quadrilateral elements are utilized. Quadratic elements have eight nodes and they simulate curved boundaries in a better way. Later bilinear elements with four nodes and linear sides are used for checking their suitability to transonic calculations. Computations are performed on a parent element. The results are then transformed to the real elements. Coordinate transformation is accomplished by the use of interpolation functions. The integrations required to obtain $[M]^e$, $[C]^e$ and $[S]^e$ are performed by two-point Gaussion quadrature technique. After establishing mass, damping, and stiffness matrices, eight ordinary differential equations are obtained for each element. The $\{F\}^e$ vector has non zero values only at the exit nodes where natural boundary condition is applied. Since the boundary is a straight line in the tangential direction and the properties are uniformly distributed.

$$\{F\}^e = (K^2-\tilde{V}_2{}^2)^{\frac{1}{\gamma-1}} \tilde{V}_2 \cos\beta_2 \int_{S_2} [N]^T ds \qquad (31)$$

The global matrix for the whole solution domain is obtained by the aid of common nodes between the elements and the following equation is assembled for the whole domain.

$$[M]\left\{\frac{\partial^2\phi}{\partial t^2}\right\} + [C]\left\{\frac{\partial\phi}{\partial t}\right\} + [S]\{\phi\} = \{F\} \qquad (32)$$

The global matrices are banded and this property is taken into account while storing them. Instead of the whole matrices, only the elements in the banded portion are stored. Due to the non-symmetry of matrix [C] it is required to store the whole band instead of the half band width.

3.3 Timewise Finite Element Formulation

A discretization in time domain is applied to the semi-descretized system in equation (32) using weighted residual approach. One-dimensional, 3 point elements with parabolic shape functions are used. A linear system of algebraic equations are obtained as a result. The field variable at each node of the spatial solution domain is written in terms of timewise shape functions and nodal values. The timewise shape functions $N_i(t)$ are taken to be equal for each spatial node. These shape functions have to be at least second order in time as there are second order time derivatives in the system of equations (32). A minimum of three sets of ϕ_i are needed to describe approximately the variation of shape functions in the timewise element. Thus, the field variable is written as,

$$\phi(t) = \Sigma\, N_i(t)\, \phi_i$$

where i= n-1, n, n+1 . n denotes the timewise calculation step. The shape functions are normalized to an interval of $2\Delta t$ by defining the normalized time as $\delta=t/\Delta t$. In the interval $-1<\delta<1$ the shape functions of the second order element may be written as:

$$N_{n+1} = \frac{1}{2}\,\delta(1+\delta), \quad N_n = (1-\delta)(1+\delta), \quad N_{n-1} = -\frac{1}{2}\,\delta(1-\delta) \qquad (33)$$

In the solution procedure the potential function values ϕ_n and ϕ_{n-1} are assumed to be known as initial conditions and ϕ_{n+1} values remain to be determined. A weighted residual equation of the form below may be written excluding the term coming from the outlet boundary condition if the boundary is time independent

$$\int_{-1}^{1} W_j\left\{[M]\left(\{\phi_{n-1}\}\frac{\partial^2 N_{n-1}}{\partial t^2} + \{\phi_n\}\frac{\partial^2 N_n}{\partial t^2} + \{\phi_{n+1}\}\frac{\partial^2 N_{n+1}}{\partial t^2}\right)\right.$$

$$+ [C]\left(\{\phi_{n-1}\}\frac{\partial N_{n-1}}{\partial t} + \{\phi_n\}\frac{\partial N_n}{\partial t} + \{\phi_{n+1}\}\frac{\partial N_{n+1}}{\partial t}\right)$$

$$\left. + [S]\left(\{\phi_{n-1}\} N_{n-1} + \{\phi_n\} N_n + \{\phi_{n+1}\} N_{n+1}\right)\right\} d\delta = 0, \ (j=1) \quad (34)$$

If we now assume that matrices [M], [C] and [S] remain constant at each time step and have the values calculated at the n'th step, equation (34) may be integrated to give the expression below.

$$([M] + \alpha\Delta t\,[C] + \beta\Delta t^2\,[S])\,\{\phi_{n+1}\} =$$

$$-(-2[M] + (1-2\alpha)\Delta t[C] + (\tfrac{1}{2} - 2\beta+\alpha)\Delta t^2[S])\{\phi_n\}$$

$$-([M]-(2-\alpha)\Delta t[C] + (\tfrac{1}{2}+\beta-\alpha)\Delta t^2[S])\{\phi_{n-1}\} \quad (35)$$

$$\text{where } \alpha = \int_{-1}^{1} W_j\,(\delta+\tfrac{1}{2})\,d\delta \Big/ \int_{-1}^{1} W_j\,d\delta$$

$$\beta = \int_{-1}^{1} W_j\,\tfrac{1}{2}\,(1+\delta)\,d\delta \Big/ \int_{-1}^{1} W_j\,d\delta \quad (36)$$

This three point recurrence scheme for second order equation in hand is the direct integration method known as Newmark [24] method. If the forcing function, in this case the boundary conditions are time dependent same interpolation may be used for {f} giving

$$\{f\} = \{f\}_{n+1}\beta + \{f\}_n\,(\tfrac{1}{2} - 2\beta+\alpha) + \{f\}_{n-1}\,(\tfrac{1}{2} +\beta-\alpha) \quad (37)$$

The use of different weighting functions in the interval correspond to a large range of integration schemes [25]. Equation (34) can now be written in the following form.

$$[S_n]\,\{\phi_{n+1}\} = \{F_{n,n-1}\} \quad (38)$$

The essential and periodic boundary conditions are now applied to the above equation. The natural boundary condition to be imposed on the load vector, is either dealt with in the timewise integration scheme if the boundary is time dependent or imposed at this stage if it is constant. It should be noted that the second term in equation (32) makes matrix $[S_n]$ non-symmetrical. Therefore some non-standard operations are needed in the application of periodic boundary condition. The value of $\{\phi_{n+1}\}$ is obtained by solving equation (38) using Gauss elimination method suitable for banded, unsymmetrical matrices.

3.4 Stability and Convergence

The stability and convergence of the solution depends on the values of α and β. Since the values of α and β depend on the selection of weighting function which governs the type of integration scheme, a large variety of methods may be used and best possible alternative for a fast, stable convergence can be selected. A stability analysis for similar linear problems has been performed [25]. A rigorous stability search for the solution technique which has been developed was made. It was established that the stability behavior of the present method is similar to the one obtained for linear problems. Unconditional stability exists if,

$$\beta \geqslant 0.25\ (0.5 + \alpha)^2$$

$$\alpha \geqslant 0.5$$

$$0.5 + \alpha + \beta \geqslant 0 \tag{39}$$

For values $\alpha = 0.5$ and $\beta = 0.25$ no numerical damping is applied to the solution and a stable oscillating solution is obtained. For $\alpha = 1.5$ and $\beta = 1.0$ the highest numerical damping is applied to the solution. The integration scheme in this case is the backward differencing scheme, which always gives a stable solution. Numerical damping does not effect the accuracy of final steady result, although it has an effect on the results of intermediate time steps. If steady results are of interest the values of α and β can be selected as 1.5 and 1.0 respectively. The rate of convergence of this scheme is found to be the highest. Although the effect of time step is found to be irrelevant in unconditionally stable schemes non-dimensional time is kept in the order of 1.0 in most of the calculations. For the convergence criteria the maximum relative change of ϕ at consecutive time steps (n and n+1) is used.

$$\text{Max.} \left| \frac{\phi_i^{n+1} - \phi_i^{n}}{\phi_i^{n}} \right| \leqslant \varepsilon \tag{40}$$

Solution is continued until equation (40) is satisfied.

3.5 Shock Capturing

For subsonic solution, density is updated in each time step from the known kinematics and the stagnation properties. The following equation is used for the density calculation

$$\rho_{n+1} = \rho_0 \left\{ 1 - \frac{1}{2c_p T_0} \left[\left(\frac{1}{r} \frac{\partial \phi}{\partial \theta} \right)_n^2 + \left(\frac{\partial \phi}{\partial z} \right)_n^2 \right] \right\}^{\frac{1}{\gamma - 1}} \tag{41}$$

At supersonic regions artificial compressibility technique is utilized [17]. Before the density which is calculated, by equation (41) and fed into the solution of the next time step it is modified by the equation below

$$\rho_{n+1} = \rho_n - \mu_n \left(\frac{\partial \rho}{\partial s} \right)_n \Delta s \tag{42}$$

where n and (n+1) denote successive time steps. μ is the artificial viscosity coefficient, Δs is the element size in the streamline direction. The density derivative is evaluated by a backward differencing formula. Since μ is zero in the subsonic regions no modification of density calculated from equation (41) is applied at those nodes. However, in supersonic regions densities at the following time step are calculated by modifying the values of ρ of the previous time step using equation (42). If grid lines are more or less alined with streamlines equation (42) becomes simply

$$\rho_{n+1} \simeq (1-\mu_n)\, \rho_n + \mu_n\, \rho_{un} \tag{43}$$

where ρ_{un} is the density of the second nearest upstream node. With the application of artificial compressibility technique, infinite gradients are put into finite form and possible shocks are captured by calculating the supersonic regions. In this technique shocks evolve without any special treatment. Since the principal equation is solved in its conservative form, it is quaranteed to calculate correct shock jump condition.

It is also possible to substitute the modified density in the governing equation and apply finite element discretization to the resulting equation. This will result in having additional terms due to artificial viscosity in the matrix equation. It is established that the form given by equation (43) is easier to implement and satisfactory.

Artificial viscosity coefficient (μ) is a sensitive parameter which determines the amount of artificial damping applied to supersonic points. Several alternatives exist in the literature for computing (μ) [18,19,26,27]. The following are used in the computations

$$\mu = c_m \max\left(0, 1 - \frac{1}{M^2}\right) \quad c_m = \text{constant} \tag{44}$$

$$\mu = c_m \max\left(0, 1 - \frac{1}{M^2}\right) \quad c_m = \text{constant} \times M^2 \tag{45}$$

where c_m is a coefficient of magnification to remove preshock overshoots. Utilization of artificial damping as given by equations (44) or (45) is found to be completely satisfactory for steady flows where time steps are un-physical. However, it may require modification for real time-dependent flows.

4. RESULTS

4.1 Subsonic Flow Solutions

The method of solution is tested using a number of test cases both at low subsonic and high subsonic speeds. A typical finite element grid used for the solution of VKI LS 82-1 turbine cascade is shown in Figure 2. As it is seen, a non-orthogonal grid is used

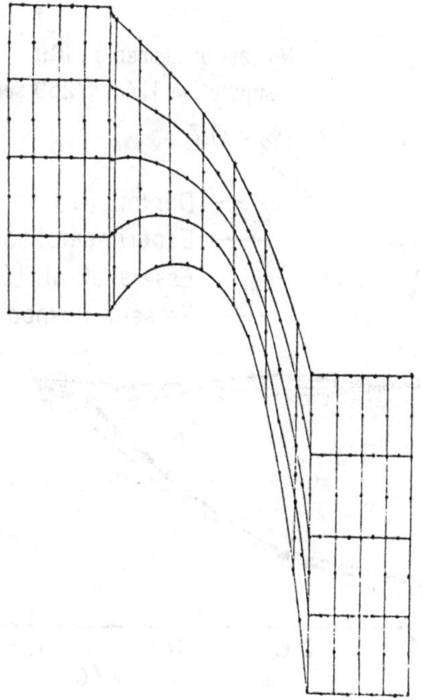

Figure 2. Finite Element Mesh for VKI LS 82-1 Turbine Cascade

at the leading and trailing edges of the blades. Although grid refinement increases the accuracy of the solution at leading and trailing edges, when non-orthogonal grid is used there is always a slight discrepancy between the analytical and numerical solutions [23].

Figure 3 shows the comparisons made for VKI LS 82-1 cascade. The cascade geometry and the experimental blade surface Mach number distributions for various isentropic outlet Mach number values are given in reference [30]. Several numerical solutions are also compared for this cascade in the same reference. Two of these numerical calculation methods which show good agreement with experimental Mach number distribution are compared with the present calculation. Calculations are performed on a finite element mesh shown in Figure 2. For this test case, the inlet and outlet flow angles are $\beta_1 = 0^0$ and $\beta_2 = -79.5^0$ respectively. The outlet Mach number is 0.70. The calculation took 259 seconds CPU time on Burroughs 6900 computer using 104 elements in the calculation domain. Figure 3 shows the blade surface Mach number distributions obtained from present method together with the results of Denton [1] and Essers [29]. The computing time which Denton used on an IBM 370/165 computer is in the same order as that is used by the present calculation. However, calculation performed by using quasi-natural time dependent method [29] takes approximately five times more computing time on VAX 11/780 computer.

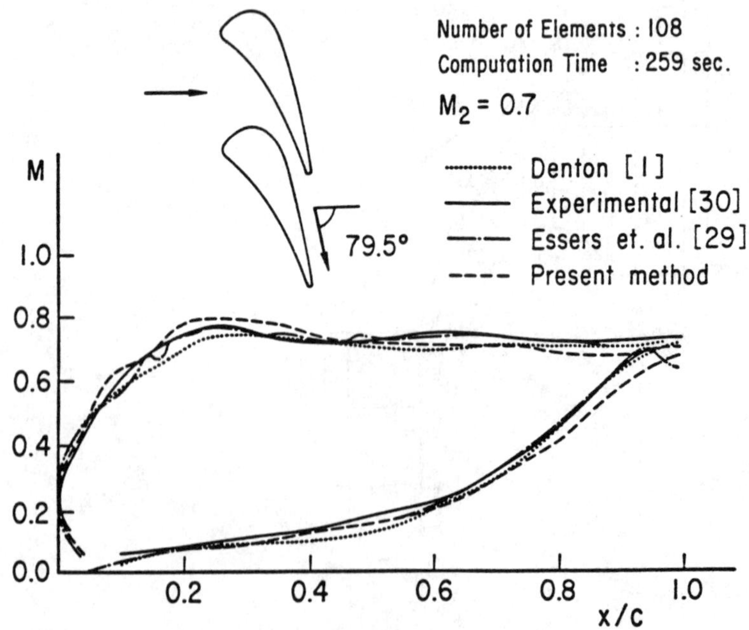

Figure 3. VKI LS 82-1 Turbine Cascade

VKI LS 59-2 blade is the mean rotor blade section designed for high subsonic outlet Mach numbers. The cascade geometry and detailed experimental results are available in reference [28]. The present code is tested at two different outlet Mach Numbers. These Mach Numbers are 0.62 and 0.75. The inlet and outlet flow angles are $\beta_1 = 30^0$ and $\beta_2 = -65.8^0$ respectively. The computation for $M_2 = 0.62$ takes 179 seconds. The computing time increases to 243 seconds for $M_2 = 0.75$. 247 nodes are used in these calculations. Figure 4 and 5 show the blade surface Mach number distributions.

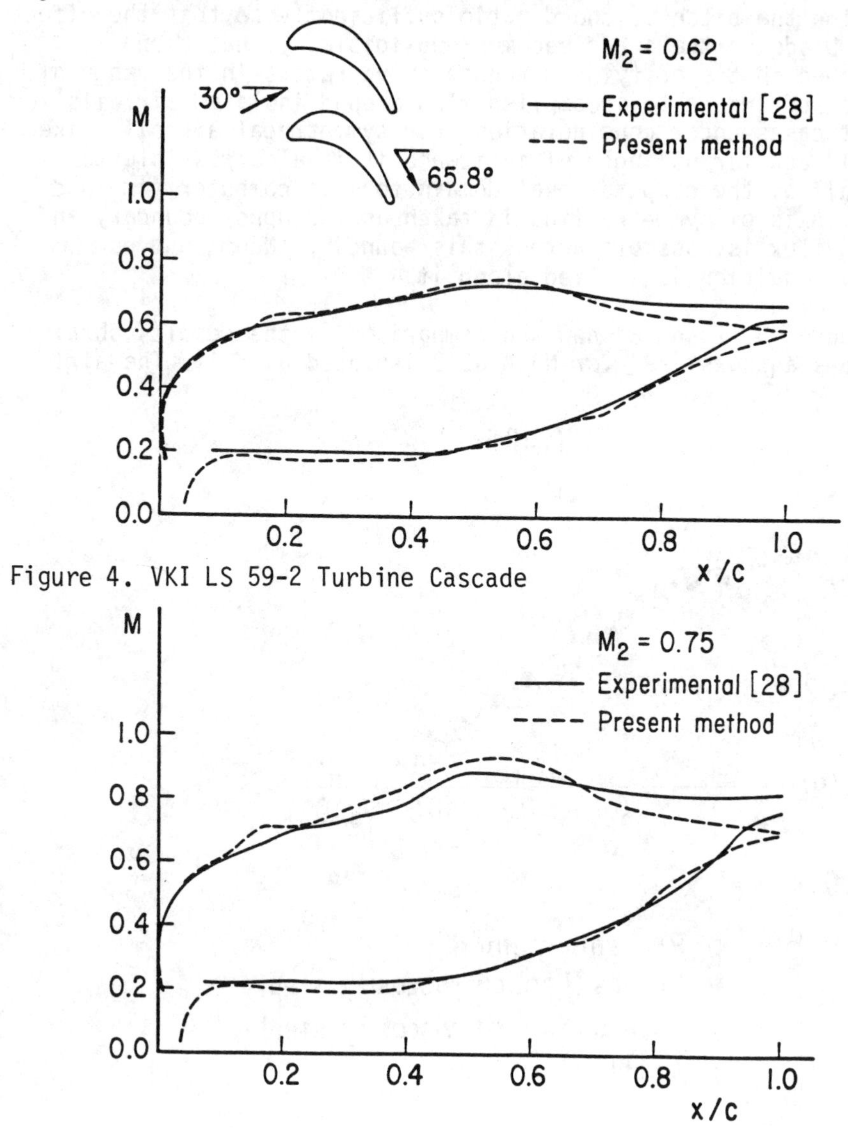

Figure 4. VKI LS 59-2 Turbine Cascade

Figure 5. VKI LS 59-2 Turbine Cascade

The number of iterations for a convergent solution to an accuracy of $\varepsilon < 0.001$ took 4 to 6 iterations depending on the exit Mach number.

As the first step to the transonic flow calculations with shocks NACA 0012 airfoil is tested in a flow with no incidence and at low upstream Mach numbers. The existing code is developed for solving flow through cascades. However, it can as well be used for the solution of isolated airfoil problems. This is accomplished by increasing the pitch to chord ratio sufficiently so that the effect of each blade on the other becomes negligible. It has been established that specifying pitch to chord ratios in the range of 10-13 is sufficient to accomplish flow around isolated airfoils for the test cases under consideration. For symmetrical airfoils like NACA 0012 and for no angle of incidence it is also possible to solve half of the computational domain to save computer time and storage. Axis of symmetry line is taken as the upper boundary and since no flux is possible across this boundary. Neumann type boundary condition is applied along it.

Figures 6, 7 and 8 show the comparison of the results obtained by various authors [29] for NACA 0012 isolated airfoil. The airfoil

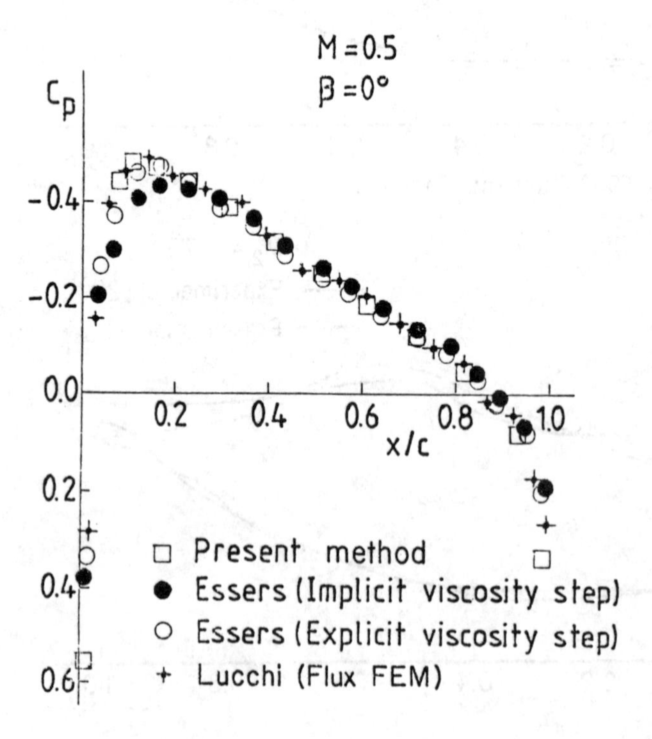

Figure 6. NACA 0012 Airfoil $M_1 = 0.5$

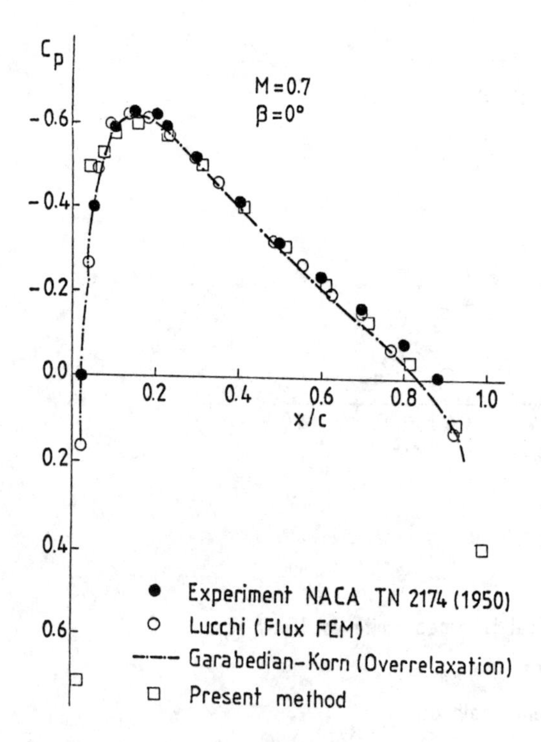

Figure 7. NACA 0012 Airfoil $M_1 = 0.7$

is tested at free stream Mach numbers of 0.5 and 0.7. A relatively fine mesh is used in both tangential and axial directions. The computation takes 433 seconds for $M_\infty = 0.5$ and 774 seconds for $M_\infty = 0.7$ on Burroughs 6900 computer for getting an accuracy of less than 10^{-4} for the maximum relative change of potential function.

4.2 Transonic Flow Solution

The basic full potential equation solver uses 8-noded isoparametric elements and calculates the derived properties at the nodes. Aritmetic averaging is used for finding the property at the inter elemental nodes. Two more versions of the code are developed. Both of them assume that the field variable derivatives and associated properties are to be constant in each element during the calculation of elemental matrices. One version uses 8 noded isoparametric elements while the other uses 4 noded linear elements. Several tests has been performed for establishing the accuracy of these three different versions. For shock capturing, It has been concluded that the original version of the program gives best results.

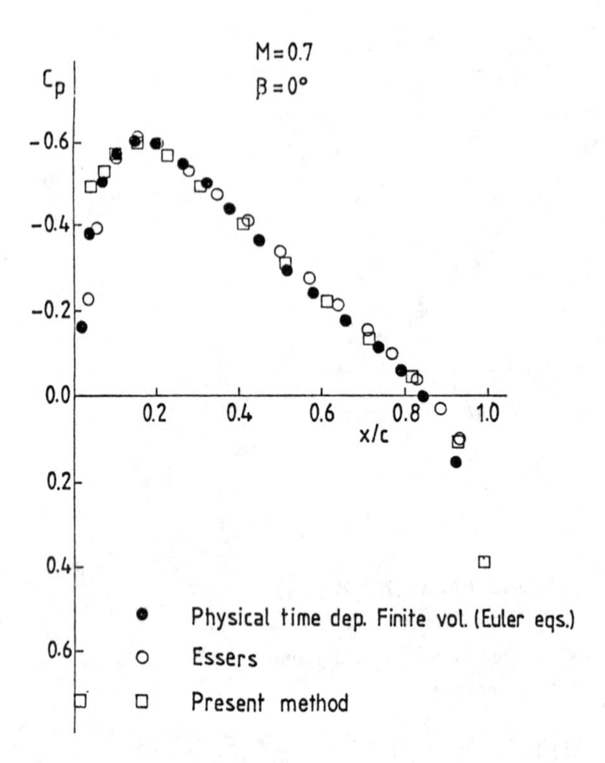

Figure 8. NACA 0012 Airfoil M_1 = 0.7

However, if sufficiently fine mesh is not used some non-smooth property variations are expected at the locations where properties change fast. Therefore in the regions where rapid changes of properties occur such as at the leading and trailing edges, and at shock regions it is necessary to apply mesh refinement. It can also be said that when properties are calculated at the centroids as recommended by the various authors [26,31], although smoother property variations are obtained some changes in shock strength and in the location at the shock is expected. The effect of calculating the properties at the centroids or plotting the Gaussion-point properties for surface data give similar results.

The coefficient of magnification in equation (44) or the constant in equation (45) has a typical value of 2.0 . The shock strength was found to be dependent on the amount of artificial viscosity. High values of artificial viscosity coefficient at supersonic regions have a decreasing effect on the shock strength. It was established that the pre-shock overshoots can be removed by the use of equation (45). The type of grid and the amount of refinement slightly effects the value of magnification coefficient.

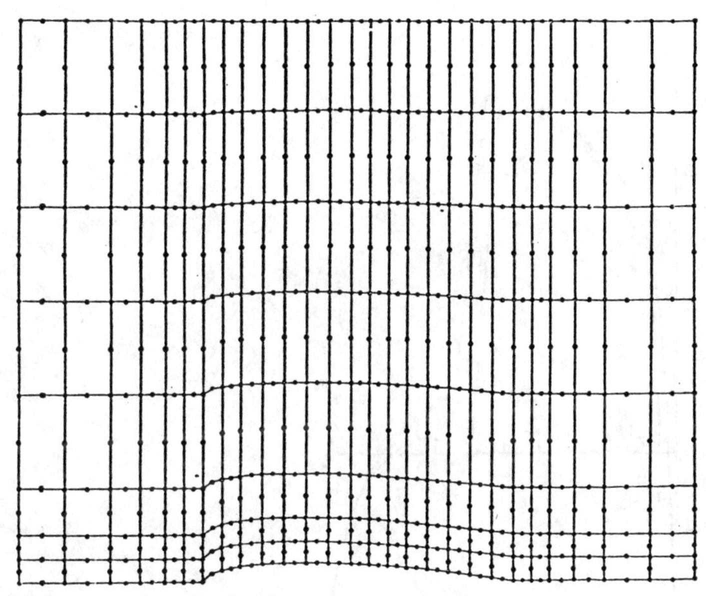

Figure 9. Computational Grid With 8-Noded Elements (Test Case 1)

Figure 9 shows the finite element grid for the calculation of NACA 0012 airfoil of an inlet Mach number 0.8 and pitch to chord ratio of 3.6 (test case 1). In Figure 10 the result of the present code is compared with shock capturing calculation of reference [31]. The result with least artificial viscosity is used for the comparison. The re-expantion after the shock predicted by the present method is greater than that of the compared result. The difference between the shock locations is 5% of the chord length. Although no experimental evidence has been found as to the location of the shock Ecer and Akay had shown that shock fitting shifts the shock downstream by 10% of the chord length.

NACA 0012 airfoil is then tested at a free stream Mach Number of 0.816 (test case 2). Figure 11 shows the comparison of the present result with the result of Essers [22] and an experimental result. Essers uses finite volume technique in his calculations. A stronger shock is predicted by the present solution method when compared to the measurements. This may be due to the viscous effects.

Figures 12 and 13 illustrate the comparison of computational predictions made by different authors for NACA 0012 airfoil at free stream Mach number of 0.8 (test case 3). Deconinck and Hirsch [6] in their multigrid method show that the shock location shifts down stream as the solution converges. Results in Figure 13 are taken

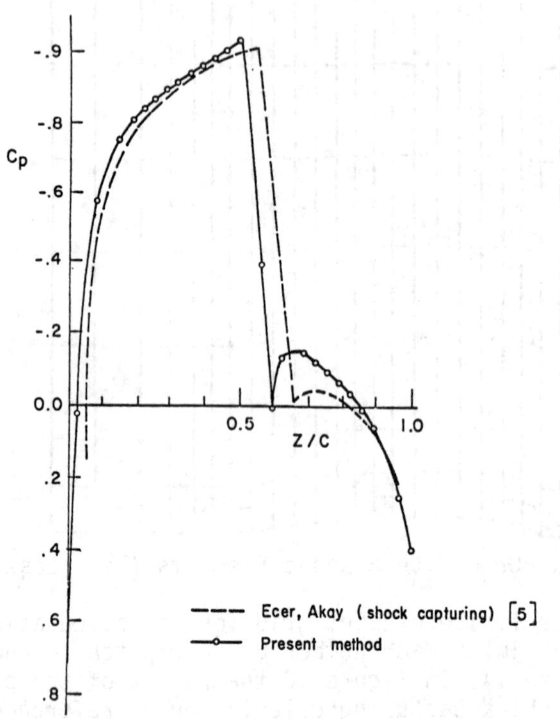

Figure 10. NACA 0012 Cascade $M_1 = 0.8$, Pitch/Chord = 3.6

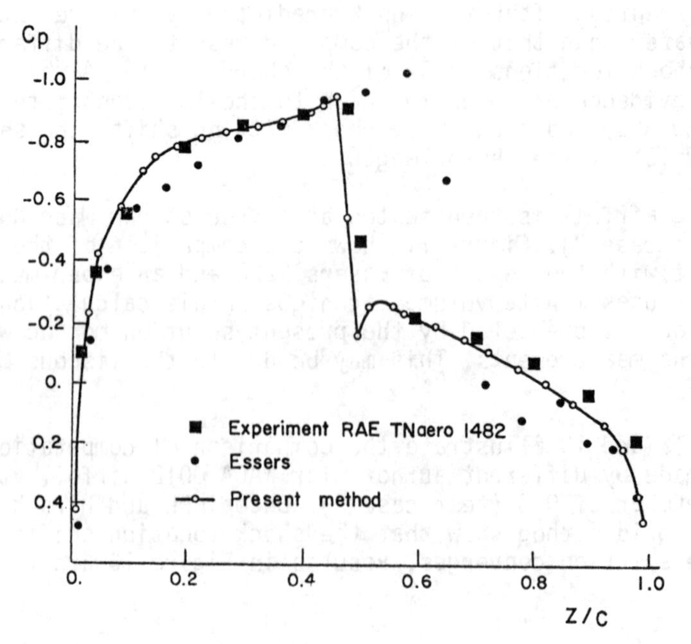

Figure 11. NACA 0012 Airfoil $M_1 = 0.816$

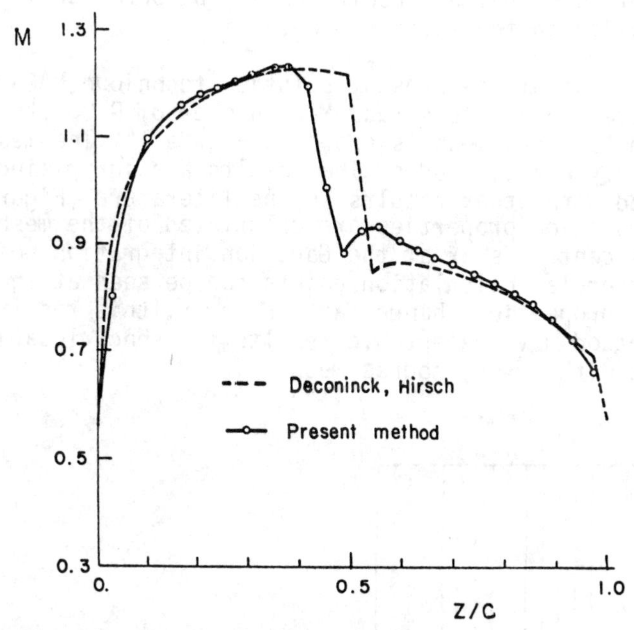

Figure 12. NACA 0012 Airfoil M_1 = 0.80

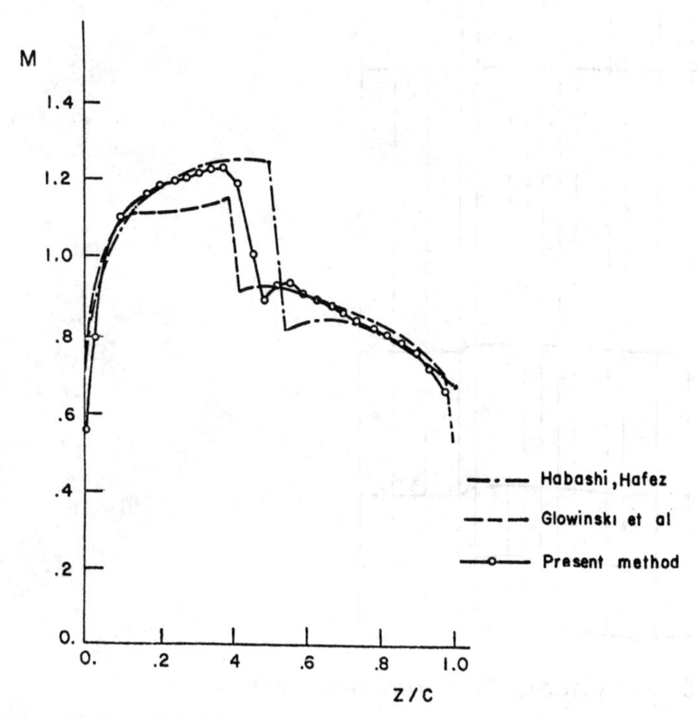

Figure 13. NACA 0012 Airfoil M_1 = 0.80

from reference [26]. The resolution of the shock in the present calculation is not good. Better resolution may be obtained if mesh refinement is applied to the shock region.

As a further test of the present solution technique NACA 0012 airfoil is subjected to a free stream Mach number of 0.85 (test case 4). The computational mesh is shown in Figure 14. The mesh is very course and only has 249 nodes. The results are surprisingly good when compared with other results in the literature (Figure 15, 16). In this calculation properties are calculated of the mesh points not at the centroids or at the Gaussion integration points. Therefore some irregular calculation points can be seen at the regions where the properties change fast. However, this result shows that the method give exceptable results for shock location and strength even with a very course mesh.

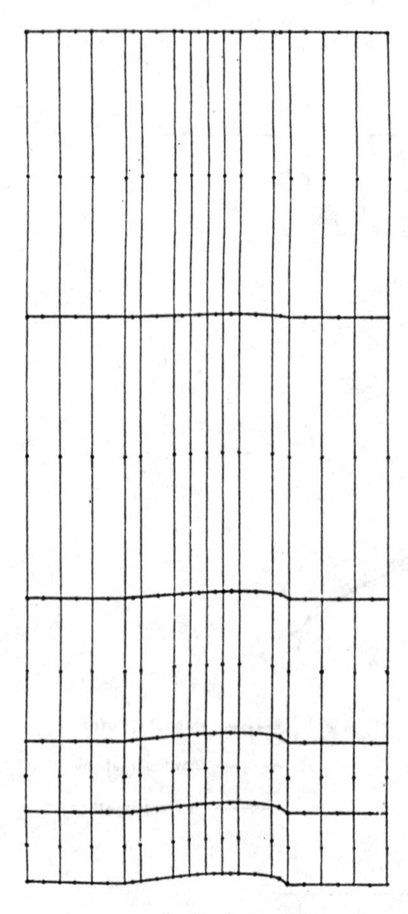

Figure 14. Computational Grid for Test Case 4

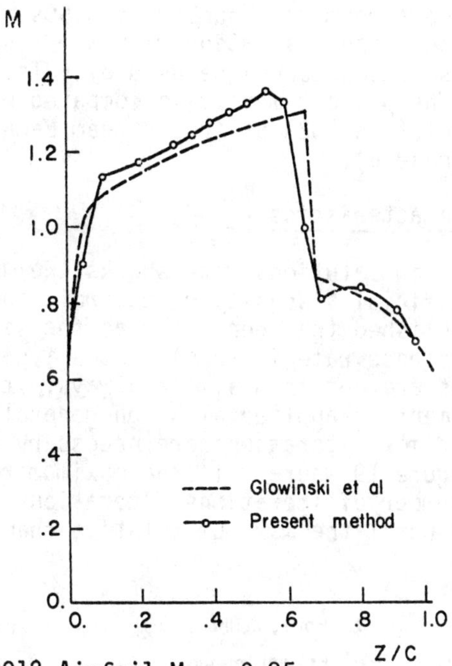

Figure 15. NACA 0012 Airfoil M_1 = 0.85

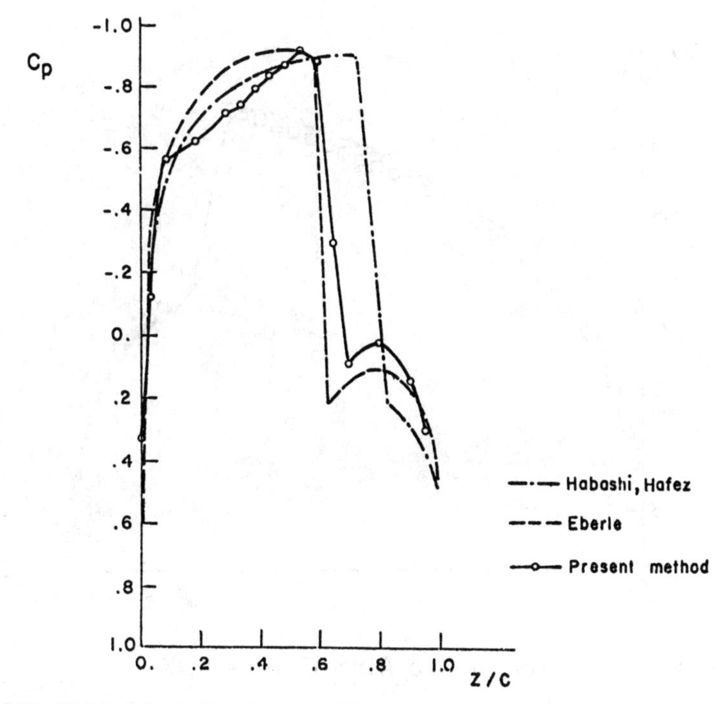

Figure 16. NACA 0012 Airfoil M_1 = 0.85

Two more tests are performed one on a 10% circular-arc airfoil at free stream Mach number of 0.84, and the other on 6% circular arc airfoil at a free stream Mach number of 0.909. Figure 17 shows the comparison of the present solution with the result of Holst and Ballhaus [18]. The solution technique used by Holst is a finite difference scheme. The present method is compared with experimental and a theoretical solution [32] at free stream Mach number of 0.909 in Figure 18 (test case 6).

4.3 Convergence Characteristics

For transonic flow solutions with shocks the magnification coefficient for artificial viscosity c_m controls the convergence history. It is established that generally as the value of c_m increases the convergence rate increases. At a typical value of 2, convergent solutions are obtained almost always. It is observed that if grid refinement is applied, although generally more accurate results are obtained more iterations are necessary for convergence. This is shown in Figure 19 where R is the maximum relative change in ϕ and N is the number of iterations. Iterations in the time domain are extended until the maximum relative change in potential

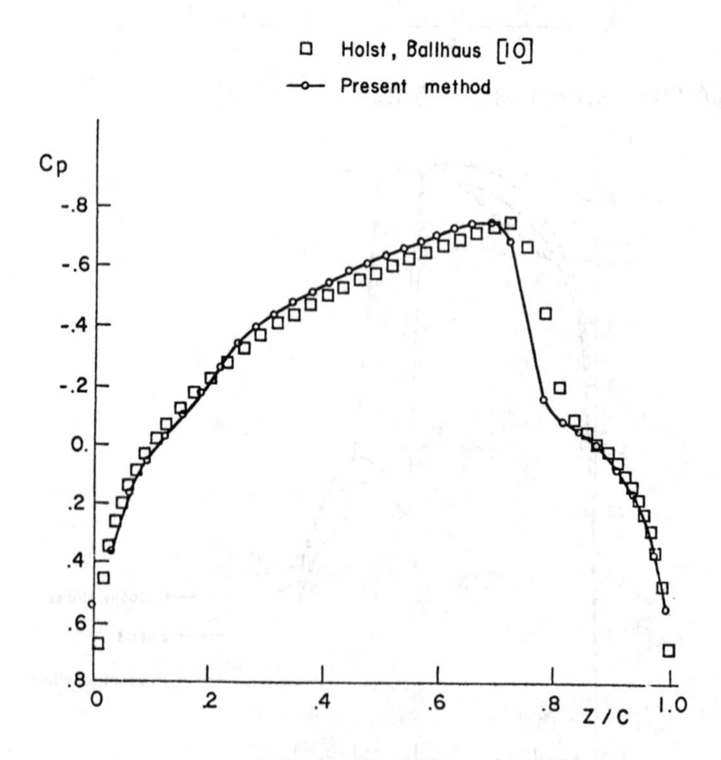

Figure 17. 10% Circular Arc Airfoil M_1 = 0.84

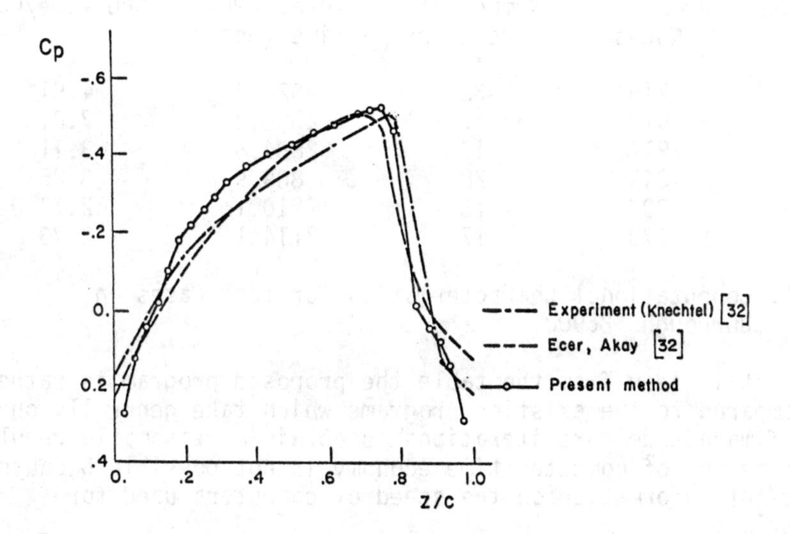

Figure 18. 6% Circular Arc Airfoil M_1 = 0.909

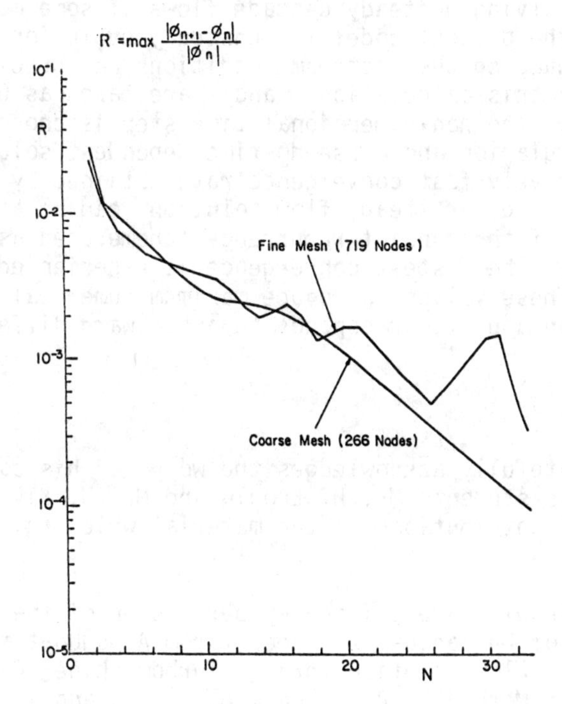

Figure 19. Convergence Characteristics

function drops to a value of 10^{-4} in almost all computations. Table 1 shows a sample of calculations made for various test cases and their convergence characteristics.

Test Case	Number of Nodes	Number of Time Steps	Total CPU Time (Sec.)	CPU Time/Node
1	719	32	3529.1	4.91
2	918	17	2580.2	2.81
3	917	17	2854.8	3.11
4	249	28	809.9	3.25
5	797	15	2210.1	2.77
6	773	17	2114.1	2.73

Table 1. Computational Characteristics for Test Cases on Burroughs B6900.

As it is seen from the table the proposed program is rather fast compared to the existing programs which take generally one order of magnitude more iterations to obtain a reasonable result. The comparison of computer time economy is not possible because of the lack of information on the speed of computers used for calculations.

As it is stated before the proposed scheme has an inherent characteristic of solving unsteady cascade flows if some modifications are made to the present code. The unsteady evolution of the flow by suddenly changing the upstream conditions was predicted satisfactorily. For this calculation and are taken as 0.6 and 0.3025 respectively. The non-dimensional time step is chosen as 0.0002 in this calculation and a pseudo-time dependent solution is obtained. The relatively fast convergence rate obtained by the program when it is used for steady flow solutions can be attributed to the utilization of three point recurrence scheme used as an iterative procedure. The fastest convergence is experienced when $\alpha=1.5$ and $\beta=1.0$. These values introduce maximum numerical damping during the iteration and it corresponds to a backward differencing scheme in time.

Acknowledgements

The author gratefully acknowledges the work of his colleque Dr. I. Yegen and his students Mr. H. Eroglu and Mr. U. Kıvrak who made very valuable contributions to the material which this lecture covers.

The subsonic results are printed by permission of the Council of the Institution of Mechanical Engineers, from "A psudo-time-dependent Potential Flow Solution Through Turbomachine Cascades Using Finite Element Method" by A.Ş. Üçer, H. Eroglu and I. Yegen.

REFERENCES

1. Denton, J.D. A Time Marching Method for Two and Three Dimensional Blade to Blade Flows. Aeronautical Research Council R and M, 377, 1975.
2. Gopalakrishnan, S. and Bozzola, R.A. A Numerical Technique for Calculation of Transonic Flow in Turbomachinery Cascades. ASME Paper No 71-GT-42, 1971.
3. Jameson, A., Schmidt, W., and Turkel, E. Numerical Solution of Euler Equations by Finite Volume Methods Using Runge-Kutta Time Stepping Schemes. AIAA 14 Fluid and Plasma Dynamics Conference, Palo Alto, California, 1981.
4. Casier, F., Deconninck, H., Hirsch, C. A Class of Centeral Bidiagonal Schemes with Implicit Boundary Conditions for the Solution of Euler Equations. AIAA 21st Aerospace Sciences Meeting, Jan 10-13, 1983.
5. Moore, J. and Moore, J.G. Three-Dimensional, Viscous Flow Calculations for Assessing the Thermodynamic Performance of Centrifugal Compressors-Study of the Eckard Compressor. AGARD-CP-282, 1980.
6. Deconinck, H., Hirsch, Ch. Finite Element Solution Methods for Transonic Blade-To-Blade Calculation in Turbomachines. ASME Journal of Eng. for Power, Vol.103, No.4, 1981.
7. Veuillot, J.P., Viviand, H. A Pseudo-Unsteady Method for the Computation of Transonic Potential Flows. AIAA paper 79-007, 1979.
8. Habashi, W.G., Kotiuga, P.L. Finite Element Solution of Subsonic and Transonic Cascade Flows. Proc. of Second Int. Conf. on Numerical Method in Laminar and Turbulent Flows, Venice, July 1981.
9. Akay, H.U., Ecer, A. Transonic Flow Computations in Cascades Using Finite Element Method. ASME J. of Eng. for Power, Vol.103, 1981.
10. Murman, E.M., Cole, J.D. Calculation of Plane Steady Transonic Flow. AIAA Journal, Vol.9, No.1, 1971.
11. Murman, E.M. Analysis of Embedded Shock Waves Calculated by Relaxation Methods. AIAA Journal, Vol.12, No.5, 1974.
12. Martin, E.D. A Fast Semi-Direct Method for Computing Transonic Aerodynamic Flows, AIAA Journal, Vol.14, No.7, 1976.
13. Hafez, M.M., Cheng, H.K. Shock Fitting Applied to Relaxation Solutions of Transonic Small Disturbance Equations. AIAA Journal, Vol.15, No.6, 1977.
14. Hafez, M.M., Cheng, H.K. Shock Fitting in Transonic Flow Computation. From Transonic Flow Problems in Turbomachinery. Ed. by Adamson, T.C., Platzer, M.F., Hemisphere Publishing Corp., Washington, 1976.
15. Yoshihara, H. Finite Difference Procedure for Unsteady Transonic Flows. A review, from Transonic Flow Problems in Turbomachinery Ed. by Adamson, T.C., Platzer, M.F., Hemisphere Publishing Corp., Washington 1976.

16. Deleney, R.A. Time Marching Analysis of Steady Transonic Flow in Turbomachinery Cascades Using the Hopscotch Method. ASME Paper 82-GT. 152, 1982.
17. Hafez, M., South, J., Murman, E. Artificial Compresibility Methods for Numerical Solutions of Transonic Full Potential Equation. AIAA Journal, Vol.17, No.8, 1979.
18. Holst, T.L., Ballhaus, W.F. Fast Conservative Schemes for the Full Potential Equation Applied to Transonic Flows. AIAA Journal Vol.17, No.2, 1979.
19. Doria, M.L., South, J.C. Transonic Potential Flow and Coordinate Generation for Bodies in a Wind Tunnel. AIAA paper 82-0223, 20th Aerospace Meeting, 1982.
20. Martin, E.D. A Fast Semi-Direct Method for Computing Transonic Aerodynamic Flows. AIAA Journal, Vol.14, No.7, 1976.
21. Üçer, A.Ş., Yeğen, I., Çetinkaya, T. A Finite Element Solution of Compressible Flow Through Cascades of Turbomachines Proc. of Second Int. Con. on Num. Meth. in Laminar and Turbulent Flows, Venice, 1981.
22. Essers, J.A., Kafyeke, A., New Fast Artificial Evolution Method for Transonic Flows in Turbine Cascades. VKI Tech. Not 136, 1980.
23. Üçer, A.Ş., Eroğlu, H., Yeğen, I. A Pseudo-Time-Dependent Potential Flow Solution Through Turbomachine Cascades Using Finite Element Method. Proceedings of I Mech. E C59/84, 1984.
24. Newmark, N.M. A Method of Computation for Structural Dynamics. ASCE Journal of Engineering Mechanics Division. Vol 85, 1959.
25. Zienkiewicz, O.C. The Finite Element Method. THird Edition. McGraw Hill Book Company, 1977.
26. Habashi, W.G., Hafez, M.M. Finite Element Method for Transonic Cascade Flows. AIAA Paper 81-1472, 17th Joint Propulsion Conference, 1981.
27. Hirsch, Ch., Deconinck, H. Transonic Flow Calculations with Finite Elements from Numerical Methods for The Computation of Inviscid Transonic Flows with Shock Waves, a GaMM Workshop.
28. Sieverding, C. Base Pressure Measurements in Transonic Turbine Cascades. VKI LS84, 1976.
29. Essers, J.A. A Quasi Natural Time Dependent Method for Two-Dimensional Cascade Flows, in Numerical Methods for Flows in Turbomachinery Bladings. VKI LS 1982-05, Vol.1, 1982.
30. Sieverding, C., Test Case 1: 2-D Transonic Turbine Nozzle Blade, in Numerical Methods for Flows in Turbomachinery Bladings. VKI LS 1982-05, 1982.
31. Ecer, A., Akay, H.U. Finite Element Analysis of Transonic Plows in Cascades: Importance of Computational Grids in Improving Accuracy and Convergence. NASA Contractor Report 3446, 1981.
32. Ecer, A., Akay, H.U. Treatment of Shocks in the Computation of Transonic Flows Using Finite Elements Third International Conference on Finite Elements in Flow Problems, Canada, 1980.

NOTATION

a	Speed of sound
c	Chord length
$[C]$	Damping matrix
c_p	Specific heat, pressure coefficient
f	Flux term proportional to mass outflux perpendicular to outlet boundary
$\{F\}$	Load vector
g	Gravitational acceleration
h	Enthalpy
H	Total enthalpy
M	Mach Number
$\tilde{V}$	Dimensionless speed, v/a
$[M]$	Mass matrix
N	Shape function
p	Pressure
r	Radius of cylindrical surface
R	Gas constant
$[S]$	Stiffness matrix
S_1	Essential type boundary
S_2	Natural type boundary
t	Time
T	Temperature
V	Velocity vector
W	Weighting function for timewise residual equation
z	Axial coordinate
α	A parameter used in the recurrence scheme
β	Flow angle, A parameter used in the recurrence scheme
ε	Convergence criterion
γ	Specific heat ratio
δ	$t/\Delta t$
Δt	Time step
θ	Tangential coordinate

ρ Density
ϕ Potential function
Ω Solution domain

Subscripts

ax Axial
n Time step
z Axial direction
θ Tangential direction
1 At the inlet of the cascade
2 At the outlet of the cascade

Superscripts

e In element level
T Transpose of a matrix

FINITE ELEMENT SOLUTIONS OF STEADY INVISCID FLOWS IN TURBOMACHINERY

H. U. Akay and A. Ecer

Computational Fluid Dynamics Laboratory
School of Engineering and Technology
Purdue University at Indianapolis
Indianapolis, Indiana - USA

1 INTRODUCTION

The numerical solutions of inviscid, compressible flow equations through stationary and rotating curved passages are of practical interest in the design of turbomachinery components. Often, the viscous effects may either be neglected or added as corrections to the inviscid solutions in the form of boundary layer type calculations. The obtained inviscid solutions can also be coupled with simple loss models to provide fully three-dimensional flow characteristics.

Customarily, the inviscid flow equations are treated in two distinct categories:

i) Potential flow equations in which the rotationality of the flow is neglected.
ii) Euler equations which take the rotationality of the flow into account.

In the former case only the conservation of mass equation need be solved which, in three-dimensions, can conveniently be expressed in terms of a single scalar variable: ϕ, the velocity potential function. On the other hand, the solutions to Euler equations are usually obtained by solving the momentum equations in addition to the conservation of mass. In this case, the equations are usually cast in terms of so called primitive variables, $\underline{u}$ and ρ where $\underline{u}$ is the velocity vector and ρ is the mass density. With isentropic and isoenergetic inflow conditions in fully subsonic flows, the two approaches yield, at least in theory, the same solution. However,

Euler equations are inherently more difficult to solve and they usually require an order of magnitude more computational effort than the potential flow equation in obtaining steady state solutions. Furthermore, traditionally, relaxation schemes are more popular for analyzing steady potential flows while unsteady Euler equations are employed even for analyzing steady rotational flows.

Mathematically, steady potential equation and unsteady Euler equations belong to two separate classes. When expressed in terms of the velocity potential function, ϕ, the potential flow equation is a second-order elliptic (for subsonic flows) or mixed elliptic-hyperbolic equation (for transonic flows). Whereas the unsteady Euler equations with primitive variables constitute a system of first-order hyperbolic equations. Hence the numerical techniques employed for the solution of these two sets of equations are also quite different. As mentioned above, the steady state solutions to potential flow equations are commonly obtained through relaxation-type techniques. Very efficient algorithms are available for these, both in subsonic as well as transonic regimes with shocks. On the other hand, Euler equations are usually solved using time marching techniques even when steady state solutions are sought. Comparisons of Euler and potential solutions for transonic flows, have been attempted only recently and for fairly simple flow configurations, e.g., [1-4].

The limitations of potential flow equations over Euler equations in analyzing flows through passages may be cited as follows:

i) Non-uniform inflow conditions, possibly due to upstream viscous effects in internal flow calculations, cannot be properly treated. Consequently, the motion of resulting vorticities and the secondary velocity components cannot be predicted.

ii) Since the normal component of momentum across a shock wave is not satisfied, the entropy changes across the shocks cannot be accounted for.

One has to employ Euler equations rather than potential equations in order to include the above effects. Consequently, the solution of Euler equations requires a higher-level accuracy in terms of modeling more complicated details of such flows.

In the following lecture notes, the use of a relaxation-type solution scheme is presented for steady Euler equations which contains the potential equation with velocity potential function as a subset of the Euler equations. This formulation is accomplished through a transformation of variables from the primitive to a so-called Clebsch form [5]. The transformation is a direct consequence of a generalized form of the well-known Bateman's variational principle [6]. Following this formulation, a relaxation-type finite element algorithm is developed, for the solution of both steady Euler and potential equations in a unified form. Typical solutions are presented for analyzing flows through stationary and rotating passages, illustrating the important aspects of different flow parameters.

2 EULER EQUATIONS

The general problem of interest is the fully three-dimensional analysis of steady, inviscid and rotational flows in a rotating passage. For axisymmetric rotors, a cylindrical frame of reference (r,θ,z) rotating with a constant angular velocity Ω about the z-axis (figure 1) is often employed and a single passage is considered. The z-axis coincides with the axis of rotor. The angle θ is measured relative to the rotating frame of reference and increases in the direction of rotation Ω. The radius r is measured from the center of rotation. The governing flow equations for such systems are [7]:

a) Conservation of mass

$$\underline{\nabla}\cdot(\rho\underline{u}) = 0 \tag{1}$$

b) Momentum

$$\underline{u}\cdot\underline{\nabla}\ \underline{u} = -\frac{1}{\rho}\nabla p - 2\underline{\Omega}\times\underline{u} - \underline{\Omega}\times(\underline{\Omega}\times\underline{r}) \tag{2}$$

c) Energy

$$\rho\underline{u}\cdot\underline{\nabla}I = 0 \tag{3}$$

where $\underline{u}$ is the relative velocity vector, ρ the mass density, p the static pressure, I the total (stagnation) relative enthalpy called rothalpy. The variables ρ, $\underline{u}$ and I are the primitive variables. In equations (2), the second and third terms on the right-hand-side respectively denote Coriolis and centripedal accelerations due to rotation of the passage.

The vector differential operator $\underline{\nabla}$ in cylindrical coordinates is

$$\underline{\nabla} = \underline{i}_r\frac{\partial}{\partial r} + \underline{i}_\theta\frac{1}{r}\frac{\partial}{\partial\theta} + \underline{i}_z\frac{\partial}{\partial z} \tag{4}$$

where $\underline{i}_r$, $\underline{i}_\theta$ and $\underline{i}_z$ are the unit vectors respectively in radial, tangential and axial directions.

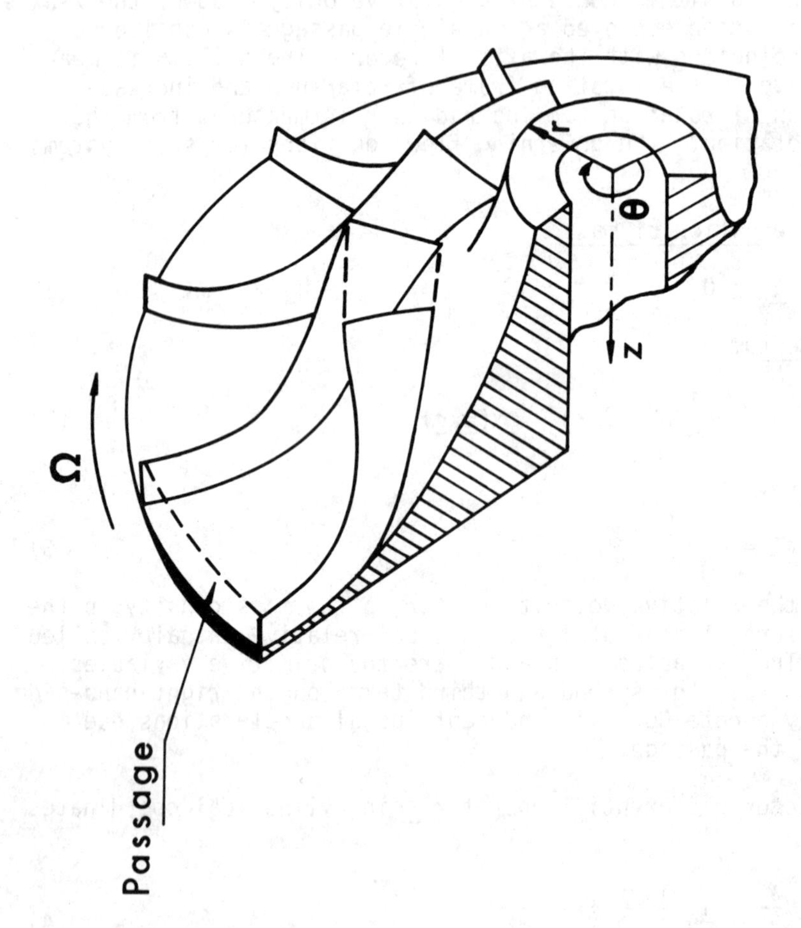

Figure 1 A typical blade passage of a compressor wheel.

For the solution of equations (1)-(3), the following auxiliary variables and relations are also needed:

a) Equation of state

$$p = \frac{\gamma-1}{\gamma}\,\rho h = \kappa\rho^{\gamma} e^{(\gamma-1)S/R} \tag{5}$$

where h is the specific enthalpy, γ is the ratio of specific heats, S is the entropy, R is the gas constant and κ is a constant,

b) Absolute velocity vector

$$\underline{v} = \underline{u} + \underline{i}_{\theta}\Omega r \;, \tag{6}$$

where $\underline{v}$ is the absolute velocity vector,

c) Enthalpy

$$\begin{aligned} h &= I - \tfrac{1}{2}\underline{v}\cdot\underline{v} + \underline{v}\cdot\underline{i}_{\theta}\Omega r \\ &= I + \tfrac{1}{2}\underline{v}\cdot\underline{v} - \underline{u}\cdot\underline{v} \\ &= I - \tfrac{1}{2}\underline{u}\cdot\underline{u} + \frac{\Omega^2 r^2}{2} \;, \end{aligned} \tag{7}$$

d) Total (stagnation) enthalpy

$$H = h + \tfrac{1}{2}\underline{v}\cdot\underline{v} \;. \tag{8}$$

The governing equations for fixed passages can be retrieved simply by setting $\Omega=0$. In that case, $\underline{u}=\underline{v}$ and H=I, and the use of an absolute coordinate system suffices.

For defining steady flow conditions in rotating passages, the tangential variation of all flow properties are assumed to be zero and the following inlet conditions are specified:

$$\underline{v} = \underline{v}\,(\xi) \tag{9}$$

$$S = S\,(\xi) \tag{10}$$

$$I = \text{constant} \tag{11}$$

$$H = \text{constant} \tag{12}$$

where, for example, $\xi = r$ for a compressor wheel with an axial inlet, and $\xi = z$ for a turbine wheel with a radial inlet. In non-rotating passages, such as curved ducts, $\xi = (x,y)$ for which, the inlet cross-section coincides with the stationary (x,y) plane.

Under the above conditions, equation (3) is identically satisfied and the flow is said to be iso-energetic. Only such flows will be treated for the remainder of these notes. A further simplification may be obtained if $\underline{v}$ is assumed uniform at the inlet. In this case, the entropy S also becomes uniform. Consequently the flow is irrotational everywhere until shock waves interfere. Barring the appearance of shocks or ignoring the entropy changes at shocks, one can apply the condition of irrotationality. In that case, only the conservation of mass equation need be solved. Hence, the potential flow equation

$$\underline{\nabla}\cdot(\rho\underline{\nabla}\phi-\underline{i}_\theta\rho\Omega r) = 0 \tag{13}$$

with

$$\underline{v} = \underline{\nabla}\phi \tag{14}$$

is obtained.

3 CLEBSCH FORMULATION OF EULER EQUATIONS

3.1 Introduction

For the solution of the potential flow equation in the form of (13), the finite element applications involving the use of Bateman's variational principle [6] are well known. The same principle is not directly applicable to the primitive variable form of the Euler equations (1)-(2). However, it can conveniently be extended to the solution of Euler equations using a transformation of variables for the velocity vector as follows:

$$\underline{v} = \underline{\nabla}\phi + S\underline{\nabla}\eta \tag{15}$$

where S denotes entropy, ϕ and η are the Lagrangian multipliers of the related variational principle. This transformation is known as a Clebsch transformation and ϕ,S and η are called the Clebsch variables. When the flow is irrotational, the transformation reduces to the familiar form given by equation (14) with the velocity potential function ϕ as the only primary variable. An extensive treatment of this principle including various possibilities for the choice of Clebsch variables has been given by Seliger and Whitham [8].

Finite element algorithms involving the transformation in equation (15) have previously been developed by the present authors [3-4, 9-12]. The applicability of the method has been demonstrated for both subsonic and transonic steady flow problems. Finite element formulations involving Clebsch variables with different solution strategies have also been reported by Lacor and Hirsch [13] and Detyna [14]. Finite difference applications are due to Grossman [15], Chang and Adamczyk [16] and Roberts [17].

3.2 An Eulerian Variational Principle

It can be shown that an equivalent form of the Euler equations defined by equations (1)-(2) can be obtained through the stationary values of the following constrained variational functional:

$$\Pi = \int_{\overline{V}} \{\frac{1}{2}\rho\underline{v}\cdot\underline{v} - \rho E(\rho,S) + \rho I + \phi[\underline{\nabla}\cdot(\rho\underline{u})]$$

$$+ \eta[\underline{\nabla}\cdot(\rho\underline{u}S)]\}dV \qquad (16)$$

where $\overline{V}$ is the flow domain and dV is the differential volume.

In the above functional, the first term denotes the kinetic energy, second and third terms denote the internal energy and rothalpy. Through the use of Lagrangian multipliers ϕ and η, the conservation of mass and entropy equations are respectively added to the functional as constraints, where E is the internal energy,

$$E = \frac{1}{\gamma-1}\frac{P}{\rho} = \frac{h}{\gamma} . \qquad (17)$$

The variables of the functional in equation (16) are $\underline{u},\rho,S$, ϕ and η, and momentarily no variations of these are allowed on the boundaries. The variation of equation (16) with respect to the velocity vector produces a Clebsch transformation in the following form:

$$\delta\underline{u} = \delta\underline{v}: \quad \underline{v} = \underline{\nabla}\phi + S\underline{\nabla}\eta . \qquad (18)$$

The variations with respect to the remaining variables ρ,S,ϕ and η respectively provide the following relations:

$$\delta\rho: \quad h = I + \frac{1}{2}\underline{v}\cdot\underline{v} - \underline{u}\cdot\underline{v} \qquad (19)$$

$$\delta S: \quad \rho\underline{u}\cdot\underline{\nabla}\eta = - p/R \qquad (20)$$

$$\delta\phi: \quad \underline{\nabla}\cdot(\rho\underline{u}) = 0 \qquad (21)$$

$$\delta\eta: \quad \underline{\nabla}\cdot(\rho\underline{u}S) = 0 \ . \tag{22}$$

If equations (18) and (19) are assumed to hold a priori, the variables ϕ,η and S remain as the only primary unknowns. Then, the flow equations corresponding to these variables become

$$\underline{\nabla}\cdot(\rho\underline{\nabla}\phi) + \underline{\nabla}\cdot(\rho S\underline{\nabla}\eta) - \underline{\nabla}\cdot(\rho\underline{i}_\theta\Omega r) = 0 \tag{23}$$

$$\rho\underline{u}\cdot\underline{\nabla}\eta = - p/R \tag{24}$$

$$\rho\underline{u}\cdot\underline{\nabla}S = 0 \ , \tag{25}$$

where

$$\underline{u} = \underline{v} - \underline{i}_\theta\Omega r \tag{26}$$

$$\underline{v} = \underline{\nabla}\phi + S\underline{\nabla}\eta \tag{27}$$

$$\rho = c\bar{\theta}h^{(\bar{\theta}-1)}\,e^{-S/R} \tag{28}$$

$$p = ch^{\bar{\theta}}\,e^{-S/R} \tag{29}$$

$$\bar{\theta} = \frac{\gamma}{\gamma-1} \ , \quad c = \kappa(\kappa\bar{\theta})^{-\bar{\theta}} \ . \tag{30}$$

The constant rothalpy can be calculated from the given conditions at station 1 as:

$$I = I_1 = H_1 - \underline{v}_1\cdot(\underline{\Omega}x\underline{r}_1) = H - \Omega(rv_\theta)_1 \tag{31}$$

where $(rv_\theta)_1 = \lambda$ = constant is the prerotation or swirl at the entrance of the passage.

The vorticity vector in this formulation can be calculated from its kinematic definition as

$$\underline{\zeta} = \underline{\nabla}x\underline{v} = \underline{\nabla}Sx\underline{\nabla}\eta \ . \tag{32}$$

By using this form of the vorticity vector, in conjunction with the first law of thermodynamics, one can write the following expression:

$$\underline{u} \times \underline{\zeta} = \underline{\nabla}I - S\underline{\nabla}h \ . \tag{33}$$

With the above relationships, it is easy to verify that for iso-energetic flows, equations (23)-(25) are precisely equivalent to

Euler equations (1)-(2).

If variations of the variables are allowed on the boundaries, the natural boundary conditions for the first two equations (23) and (24) are respectively obtained as

$$\delta\phi: \quad f_\phi = \rho \underline{u} \cdot \underline{n} \tag{34a}$$

$$\delta S: \quad f_S = S\rho \underline{u} \cdot \underline{n} \ , \tag{34b}$$

where $\underline{n}$ is the unit outward vector on the boundary Γ.

Finally, the representation of the velocity vector given by equation (18) is not unique. Depending upon the characteristics of the flow one may select alternate Clebsch variables in obtaining the corresponding Euler equations (see references 9-12 for other choices).

3.3 Equivalence with Bateman's Principle

Applying integration by parts to the terms involving the Lagrangian multipliers ϕ and η in the functional equation (16), it can be shown that the integrand of the function becomes

$$\frac{\gamma}{\gamma-1} p - \rho E = p \tag{35}$$

Therefore, the variational principle simply reduces to finding the stationary values of the functional

$$\Pi = \int_{\overline{V}} p dV \tag{36}$$

which is the Bateman's functional. The same principle applies to either rotational or irrotational flows. The transformation in equation (18) is used for rotational flows, while equation (14) is used in the irrotational case. Incompressible flow equations can be obtained directly by setting the mass density to constant in either case.

3.4 Transformation to Second-Order

As it is noted, the conservation of mass equation (23) is a second-order quasi-linear partial differential equation, whereas equations (24) and (25) are first-order convective equations. Because the first-order equations are inherently more difficult to compute without the use of an artificial dissipation mechanism, we choose to convert them to equivalent second-order forms. This, we do by pre-multiplying both equations with the convective operator. Then, equations (23)-(25) take the following final form:

$$\underline{\nabla}\cdot(\rho\underline{\nabla}\phi) + \underline{\nabla}\cdot(\rho S\underline{\nabla}\eta) = \underline{\nabla}\cdot(\rho\underline{i}_{\theta}\Omega r) \tag{37}$$

$$\rho\underline{u}\cdot\underline{\nabla}(\rho\underline{u}\cdot\underline{\nabla}\eta) = -\rho\underline{u}\cdot\underline{\nabla}(p/R) \tag{38}$$

$$\rho\underline{u}\cdot\underline{\nabla}(\rho\underline{u}\cdot\underline{\nabla}S) = 0\ . \tag{39}$$

3.5 Pseudo - Unsteady Form

We attempt to solve these steady equations with a relaxation scheme which can be considered a pseudo-time integration scheme. The set of nonlinear equations (37)-(39) are cast into a suitable pseudo-unsteady form so that a step-by-step time integration scheme can be developed in obtaining steady-state solutions. For this we augment equations (37)-(39) by introducing additional terms with time derivatives on the primary variables ϕ,S and η as follows:

$$\frac{\Delta t}{\omega}\underline{\nabla}\cdot(\rho\underline{\nabla}\phi,_t + \rho S\underline{\nabla}\eta,_t + \rho S,_t\underline{\nabla}\eta) + \underline{\nabla}\cdot(\rho\underline{\nabla}\phi + \rho S\underline{\nabla}\eta)$$

$$-\ \underline{\nabla}\cdot(\rho\underline{i}_{\theta}\Omega r) = 0 \tag{40}$$

$$\frac{\Delta t}{\omega}\,\rho\underline{u}\cdot\underline{\nabla}(\rho\underline{u}\cdot\underline{\nabla}\eta,_t) + \rho\underline{u}\cdot\underline{\nabla}(\rho\underline{u}\cdot\underline{\nabla}\eta) + \rho\underline{u}\cdot\underline{\nabla}p/R = 0 \tag{41}$$

$$\frac{\Delta t}{\omega}\,\rho\underline{u}\cdot\underline{\nabla}(\rho\underline{u}\cdot\underline{\nabla}S,_t) + \rho\underline{u}\cdot\underline{\nabla}(\rho\underline{u}\cdot\underline{\nabla}S) = 0 \tag{42}$$

where

$$\phi,_t = \partial\phi/\partial t,\ \text{etc.} \tag{43}$$

In equations (40)-(42), Δt denotes a pseudo-time step which is used to make the equations dimensionally correct, and ω is a relaxation factor to control the amount of damping introduced by the pseudo-unsteady terms. The above equations now form a set of damped equations. The time asymptotic solutions of these equations yield the original steady Euler equations (37)-(39) as unsteady terms approach zero.

3.6 Weak Variational Form

In order to obtain the finite element approximations for equations (40)-(42), we first express them in the following integral form:

$$\delta\Pi = \int_{\overline{V}} \{[\text{equation (40)}]\delta\phi + [\text{equation (41)}]\delta S$$

$$+ [\text{equation (42)}]\delta\eta\}dV = 0 \qquad (44)$$

where $\delta\phi$, $\delta\eta$ and δS are the spatial test functions chosen respectively for equations (40)-(42). After integrating equation (44) by parts, the weak form becomes

$$\delta\Pi = - \int_{\overline{V}} \{[\frac{\Delta t}{\omega}(\rho\underline{\nabla}\phi,_t + \rho S\underline{\nabla}\eta,_t + \rho S,_t\underline{\nabla}\eta)$$

$$+ \underline{\nabla}\phi + \rho S\underline{\nabla}\eta + \underline{i}_\theta\rho\Omega r]\cdot\underline{\nabla}\delta\phi$$

$$+ (\frac{\Delta t}{\omega}\rho\underline{u}\cdot\underline{\nabla}\eta,_t + \rho\underline{u}\cdot\underline{\nabla}\eta + \frac{P}{R})(\rho\underline{u}\cdot\underline{\nabla}\delta S)$$

$$+ (\frac{\Delta t}{\omega}\rho\underline{u}\cdot\underline{\nabla}S,_t + \rho\underline{u}\cdot\underline{\nabla}S)(\rho\underline{u}\cdot\underline{\nabla}\delta\eta)\}dV$$

$$+ \int_{\Gamma_{2\phi}} f_\phi\delta\phi ds + \int_{\Gamma_{2\eta}} f_\eta\delta\eta ds$$

$$+ \int_{\Gamma_{2S}} f_S\delta S ds = 0 , \qquad (45)$$

where

$$f\phi = \rho\underline{u}\cdot\underline{n} \qquad (46)$$

$$f_S = (\rho\underline{u}\cdot\underline{\nabla} + p/R)\rho\underline{u}\cdot\underline{n} \qquad (47)$$

$$f_\eta = (\rho\underline{u}\cdot\underline{\nabla}S)\rho\underline{u}\cdot\underline{n} \qquad (48)$$

are the natural boundary conditions. In view of equations (24) and (25) respectively, both f_S and f_η values vanish.

Thus, the problem is defined by a single non-homogenous natural boundary condition, namely the normal mass flux specified at farfield inlet and exit stations on the flow domain. In addition to the mass flux at the inlet, both S and η values must be specified on each incoming streamline. For a known velocity distribution, stagnation temperature and static pressure values at the inlet, the entropy distribution can be determined from the equation of state. Values of η may be calculated from equation

(32) based on the distribution of the normal component of vorticity vector. Since for rotating blade passages in an axisymmetric rotor this component of vorticity is zero, η = constant value is specified at the inlet plane. Specification of exit and periodic boundary conditions for a rotating passage are discussed in section 4.3.

4 FINITE ELEMENT FORMULATION

4.1 Semidiscrete Approximations

The variational statement in equation (45) can now be employed directly in obtaining finite element discretizations. For semi-discrete approximations of the pseudo-time dependent equations, we use the interpolation functions N_i (x,y,z) for all variables such that their distributions within an element V^e are given by

$$\phi^e(x,y,z) = N_i(x,y,z)\phi_i^e(t) = \underline{N}^T\underline{\phi}^e \tag{49}$$

$$\eta^e(x,y,z) = N_i(x,y,z)\eta_i^e(t) = \underline{N}^T\underline{\eta}^e \tag{50}$$

$$S^e(x,y,z) = N_i(x,y,z)S_i^e(t) = \underline{N}^T\underline{S}^e \tag{51}$$

where $\phi_i^e, \eta_i^e, S_i^e$ denote the nodal point values of the variables in each element. Using equations (49)-(51), we express the variational equation (45) for an element V^e and sum the contributions of all elements in the domain of interest so that

$$\delta\Pi = \textstyle\sum_e \delta\Pi_e = 0 \tag{52}$$

This procedure produces the following nonlinear system of ordinary differential equations:

$$\begin{aligned} &\frac{\Delta t}{\omega}(\underline{K}_{\phi\phi}\underline{\phi}_{,t} + \underline{K}_{\phi\eta}\underline{\eta}_{,t} + \underline{K}_{\phi S}\underline{S}_{,t}) \\ &= \underline{f}_\phi - \underline{K}_{\phi\phi}\underline{\phi} - \underline{K}_{\phi\eta}\underline{\eta} = \underline{R}_\phi \end{aligned} \tag{53}$$

$$\frac{\Delta t}{\omega}\,\underline{K}_{S\eta}\underline{\eta}_{,t} = \underline{f}_S - \underline{K}_{S\eta}\underline{\eta} = \underline{R}_S \tag{54}$$

$$\frac{\Delta t}{\omega}\,\underline{K}_{\eta S}\underline{S}_{,t} = \underline{f}_\eta - \underline{K}_{\eta S}\underline{S} = \underline{R}_\eta \tag{55}$$

where

$$\underline{K}_{\phi\phi} = \sum_e \int_{V^e} \rho \underline{N},_i \underline{N},_i^T dV \tag{56}$$

$$\underline{K}_{\phi\eta} = \sum_e \int_{V^e} \rho S \underline{N},_i \underline{N}_i^T dV \tag{57}$$

$$\underline{K}_{\phi S} = \sum_e \int_{V^e} \rho \eta,_i \underline{N},_i \underline{N}^T dV \tag{58}$$

$$\underline{K}_{S\eta} = \underline{K}_{\eta S} = \sum_e \int_{V^e} (\rho u_i \underline{N},_i)(\rho u_j \underline{N},_j^T) dV \tag{59}$$

$$\underline{f}_{\phi} = \sum_e \int_{V^e} \rho \Omega \underline{N},_{\theta} dV + \sum_e \int_{\Gamma_{2\phi}^e} f_{\phi} \underline{N} ds \tag{60}$$

$$\underline{f}_S = - \sum_e \int_{V^e} \frac{P}{R} \rho u_i \underline{N},_i dV \tag{61}$$

$$\underline{f}_{\eta} = \underline{0} \,. \tag{62}$$

$\underline{R}_{\phi}$, $\underline{R}_{\eta}$, $\underline{R}_S$ are residual vectors which vanish as the steady-state solution is reached.

One can observe that when the entropy gradients are ignored, the only equation to be solved is the conservation of mass equation which reduces to

$$\frac{\Delta t}{\omega} \underline{K}_{\phi\phi} \underline{\phi},_t = \underline{f}_{\phi} - \underline{K}_{\phi\phi} \underline{\phi} = \underline{R}_{\phi} \tag{63}$$

Therefore, for potential flows equation (63) is the only equation to be solved, which is merely a subset of Euler equations (53)-(55).

4.2 Pseudo-time Integrations

In order to solve equations (53)-(55) by a step-by-step integration procedure, we employ forward differencing in time and advance the solution from $t = n\Delta t$ to $(n+1)\Delta t$ until a steady-state solution is reached. By substituting the forward-difference approximations in time,

$$\underline{\phi}_{,t} = (\underline{\phi}^{n+1} - \underline{\phi}^{n}) / \Delta t, \text{ etc.},$$

into equations (53)-(55), we obtain the following algorithm which has the same features of a relaxation scheme:

$$\dot{\underline{\eta}}^{n} = (\underline{K}^{n}_{S\eta})^{-1} \underline{R}^{n}_{S} \tag{64}$$

$$\dot{\underline{S}}^{n} = (\underline{K}^{n}_{\eta S})^{-1} \underline{R}^{n}_{\eta} \tag{65}$$

$$\dot{\underline{\phi}}^{n} = (\underline{K}^{n}_{\phi\phi})^{-1} \underline{R}^{n}_{\phi} - \underline{K}^{n}_{\phi S} \dot{\underline{S}}^{n} - \underline{K}^{n}_{\phi\eta} \dot{\underline{\eta}}^{n} \tag{66}$$

Integrations are continued using

$$\underline{\eta}^{n+1} = \omega \dot{\underline{\eta}}^{n} + \underline{\eta}^{n} \text{ , etc.}, \tag{67}$$

until the residual vectors $\underline{R}^{n}$ diminish.

The computational advantages offered by the above relaxation scheme are significant since the variables need not be solved simultaneously. Also, as a consequence of the employed transformation to a second-order form, the coefficient matrices in all of the equations (64)-(66) are fully symmetric. The inverses of the coefficient matrices are obtained by using a frontal Gaussian-elimination solver which takes advantage of symmetric and banded structure of these matrices. Moreover, since the coefficient matrices in equations (64) and (65) are the same, one forward-elimination is performed for both systems at the same time. This provides additional savings in the computations. It has been observed that the efficiency of the scheme can further be improved by replacing the coefficient matrix to be inverted in equation (66) with a constant matrix $\underline{K}^{o}_{\phi\phi}$ such that equation (66) becomes

$$\dot{\underline{\phi}}^{n} = (\underline{K}^{o}_{\phi\phi})^{-1} \underline{R}^{n}_{\phi} - \underline{K}_{\phi S} \dot{\underline{S}}^{n} - \underline{K}_{\phi\eta} \dot{\underline{\eta}}^{n} \tag{68}$$

Since the forward-elimination of $\underline{K}^{o}_{\phi\phi}$ is performed only for the first time step, the subsequent steps can be calculated by relatively inexpensive forward- and backward-substitutions. The constant coefficient matrix in this case may be computed using the farfield density ρ^{o}. This simplification does not appreciably affect the convergence rate of equation (66) to the steady-state solution. However, because the coefficient matrix in equations (64) and (65) are purely convective in nature, it is essential that they be updated at every step.

4.3 Application of Boundary Conditions

The inflow boundary conditions are easy to implement as discussed in section 3.6. For the outflow, the mass flux distribution is unknown and it is influenced by the rotationality developed in the approaching flow. In the scheme developed here, the outflow mass fluxes are applied iteratively at each time step from their most recently calculated values in the elements closest to the boundary. The overall mass conservation is assured by appropriately scaling the fluxes so that

$$\left|\sum_e \int_{\Gamma^e_{2\phi(\text{inflow})}} f_\phi \underline{N} ds\right| = \left|\sum_e \int_{\Gamma^e_{2\phi(\text{outflow})}} f_\phi \underline{N} ds\right| \tag{69}$$

is satisfied at each time step. On solid walls, the flow tangency condition is satisfied by setting $f_\phi=0$. Also, since the application of only Neumann-type boundary conditions does not dictate a unique ϕ distribution, the value of ϕ is set to an arbitrary constant at one node in the flow domain.

For rotating blade passages, periodicity must be imposed along the streamwise boundaries of the extensions provided to the blade passage (figure 2). Axisymmetry dictates that all flow variables be equal along periodic sides. This is achieved by setting

$$\phi_{AC} = \phi_{BD} + \lambda_1 \tag{70}$$

$$S_{AC} = S_{BD} \tag{71}$$

$$\eta_{AC} = \eta_{BD} \tag{72}$$

at the inlet extension, where λ_1 is the circulation calculated on surface AB. Similarly at the exit station we have

$$\phi_{FH} = \phi_{EG} + \lambda_2 \tag{73}$$

$$S_{FH} = S_{EG} \tag{74}$$

$$\eta_{FH} = \eta_{EG} + \lambda_3 \tag{75}$$

where

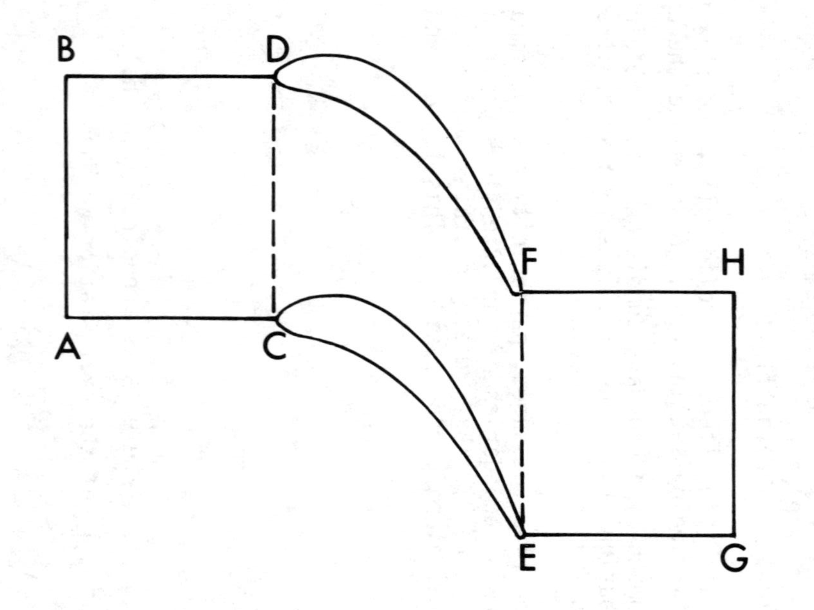

Figure 2 Periodic boundaries of a blade-to-blade surface.

$$\lambda_2 = \phi_F - \phi_E \tag{76}$$

$$\lambda_3 = \eta_F - \eta_E \tag{77}$$

as calculated from a previous time step. This equalizes the pressure at the trailing edges (E and F) of the blades in subsonic rotational flows. The situation is more complicated in the transonic case since entropy values at the trailing edges E and F will not necessarily be the same.

4.4 Transonic Flows

It can be shown that equation (37) is elliptic for subsonic relative flows. On the other hand, when the relative Mach number $|\underline{u}|/a$ is greater than one, equation (37) becomes hyperbolic and the integration scheme defined by equation (66) is no more stable. To account for the hyperbolicity of the equations at supersonic relative flows, the mass density in an element e is modified using an upwinding technique as follows

$$\tilde{\rho}_e^n = \rho_e^n - \alpha_e \Delta s_e \rho_{e,s}^n \tag{78}$$

where $\tilde{\rho}_e^n$ is the modified mass density at time step n, Δs_e the element size in the flow direction s, α_e the coefficient of artificial viscosity. For sufficiently streamlined and uniform distribution of elements, equation (78) simplifies to

$$\tilde{\rho}_e^n = \alpha_e \rho_{e-1}^n + (1-\alpha_e)\rho_e^n \tag{79}$$

where ρ_{e-1} is the mass density of the nearest element located directly at the upstream side of element e. This has a stabilizing effect on the numerical integrations of the equation. The shocks are captured during the course of integrations as sharp transitions from supersonic to subsonic regions.

A one-dimensional stability analysis [18] yields the following limits on the value of the artificial viscosity coefficient α_e and the relaxation factor ω:

$$\alpha_e = \mu_e(1-1/M_e^2) \quad , \quad \omega < 1/(\alpha_e M_e^2) \tag{80}$$

where μ_e is a multiplier (usually 1 to 3), M_e is the relative Mach number in element e. Higher values of μ_e provide faster convergence but accuracy suffers due to excessive artificial viscosity.

The above shock capturing technique can be combined [4] with a shock fitting algorithm so that the entropy changes across the shocks are calculated from the Rankine-Hugoniot relation:

$$\Delta S = \frac{R}{(\gamma-1)}\{\log_e \left[1 + \frac{2\gamma}{\gamma+1}(M_n^2 - 1)\right] - \gamma \log_e \frac{(\gamma+1)M_n^2}{(\gamma-1)M_n^2 + 2}\} \qquad (81)$$

where ΔS is the increase in entropy, M_n is the normal component of the upstream Mach number at the shock. This change in entropy is neglected in potential flows.

5 NUMERICAL RESULTS

5.1 Analysis of Potential Flows in Rotating Blade Passages

In this section, we illustrate the applications of a three-dimensional computer code, TFLOW, specifically developed for analyzing potential flows through rotating blade passages [19]. Results include subsonic as well as transonic flows through compressor and turbine wheels. Due to the potential flow assumption, only the discrete form of conservation of mass equation as defined by equation (68) is solved in this case ($\dot{\underline{S}}^n = \dot{\underline{\eta}}^n = \underline{0}$).

The program TFLOW has a built-in finite element grid generation scheme for describing blade-passage geometries conveniently. This grid generation scheme provides the necessary flexibility in choosing an appropriate grid for a blade-passage configuration with minimum input of information from the user. The three-dimensional grid to be used for the analysis consists of a total of (NRxNTxNZ) elements as shown in figure 3, where NR, NT and NZ respectively denote the number of elements in the radial, peripheral and meridional directions. All elements are 8-noded brick elements with trilinear approximation functions.

Considering a cylindrical relative coordinate system (r,θ,z) at selected stations along the meridional direction, the user specifies the (r,θ,z) coordinates of the blade centerline at the hub and the shroud as well as the corresponding blade thicknesses. The program determines the geometry of the second blade of the channel using the given information on the number of total blades in the rotor. Between two blades, the flow region is divided into NRxNT subdivisions with equal spacings. In the meridional direction, the positions of (NZ+1) number of stations are supplied by the user depending on the requirements of the geometry and the flow

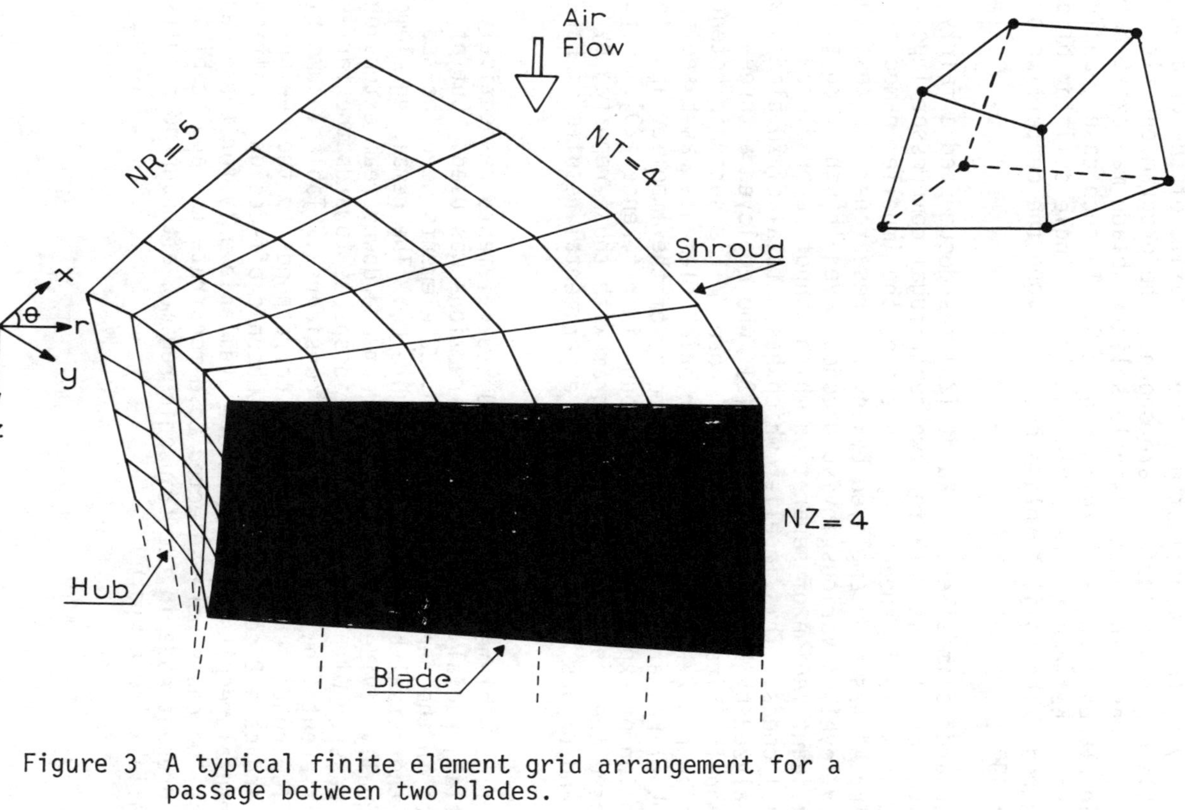

Figure 3 A typical finite element grid arrangement for a passage between two blades.

pattern for accuracy. All intermediate coordinates are generated by the computer using a linear interpolation. A sample grid for a typical passage with (5x6x13) elements is shown in figure 4, where for clarity in the visualization, the boundaries of the grid only on the blade and hub surfaces are shown. The program can also analyze the cases in which a single splitter blade may exist between two main blades such as shown in figure 5. In such cases, analysis of a passage between two main blades with a single splitter blade is required. The splitter blade extends along the centerline of the passage.

5.1.1 Eckardt's impeller. Eckardt [20] has documented a fairly extensive set of test data for two centrifugal compressor impellers running at subsonic speeds. Among these, the impeller named "impeller A" has been employed by several researchers to test the accuracy level of various inviscid codes, even though it contains a significant region of separation which cannot be predicted by potential codes. This impeller which has a total of 20 blades was also analyzed by Prince and Bryans [21] who employed a three-dimensional finite element potential code. This subsonic potential code which has 27-noded higher-order brick elements was based on the work of Laskaris [22]. The geometry of the impeller and the description of experiments can be found in reference [20]. Here we compare the results of TFLOW program with those reported by Prince and Bryans [21] and with the experimental results of Eckardt [20].

A finite element mesh of 9 x 10 x 32 elements respectively in tangential, radial and meridional directions was used. Out of 30 elements in the meridional direction, 22 elements were located within the actual blade passage (figure 6). The remaining elements were equally divided along the upstream and downstream extensions of the passage with stretching of grid spaces towards the farfield. Four different flow conditions were considered. Table 1 summarizes these flow conditions, where case 1 corresponds to the design condition and case 2 corresponds to a subsonic off-design condition. Experimental results by Eckardt are available only for these cases. The last two cases were selected to demonstrate the applicability of the present code in the transonic regime, although no results are available to compare.

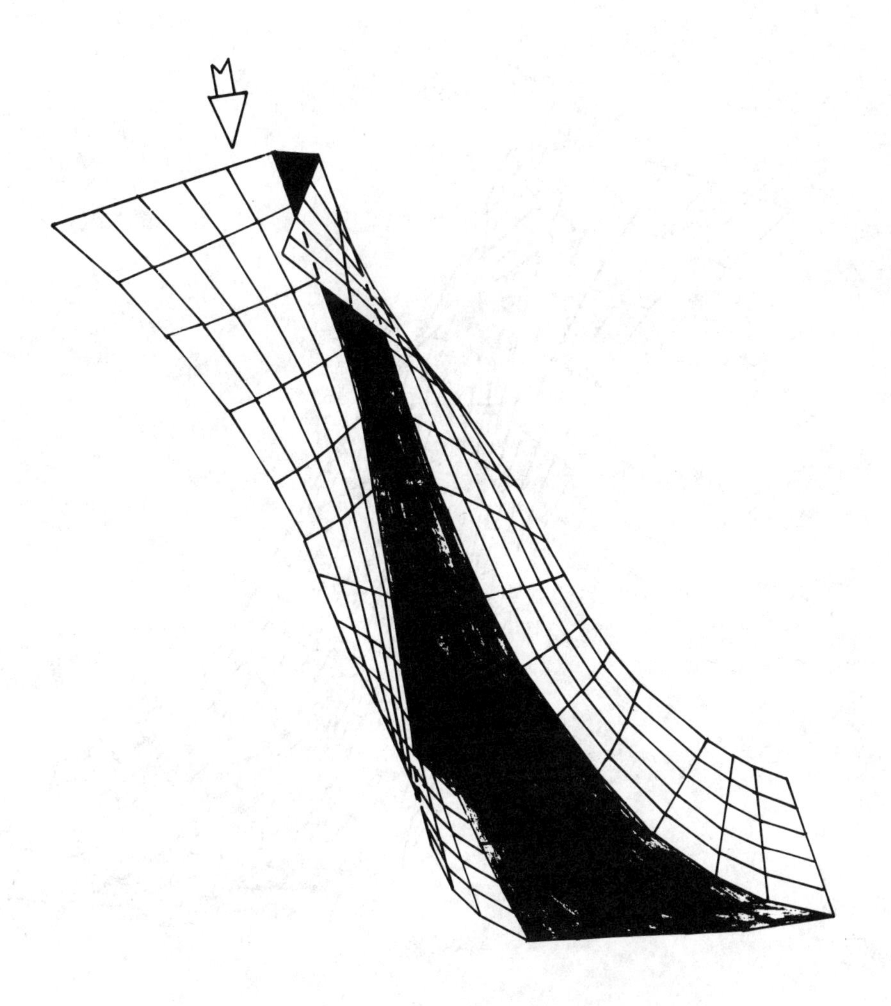

Figure 4 The boundaries of a typical grid on the blade and hub surfaces of a passage.

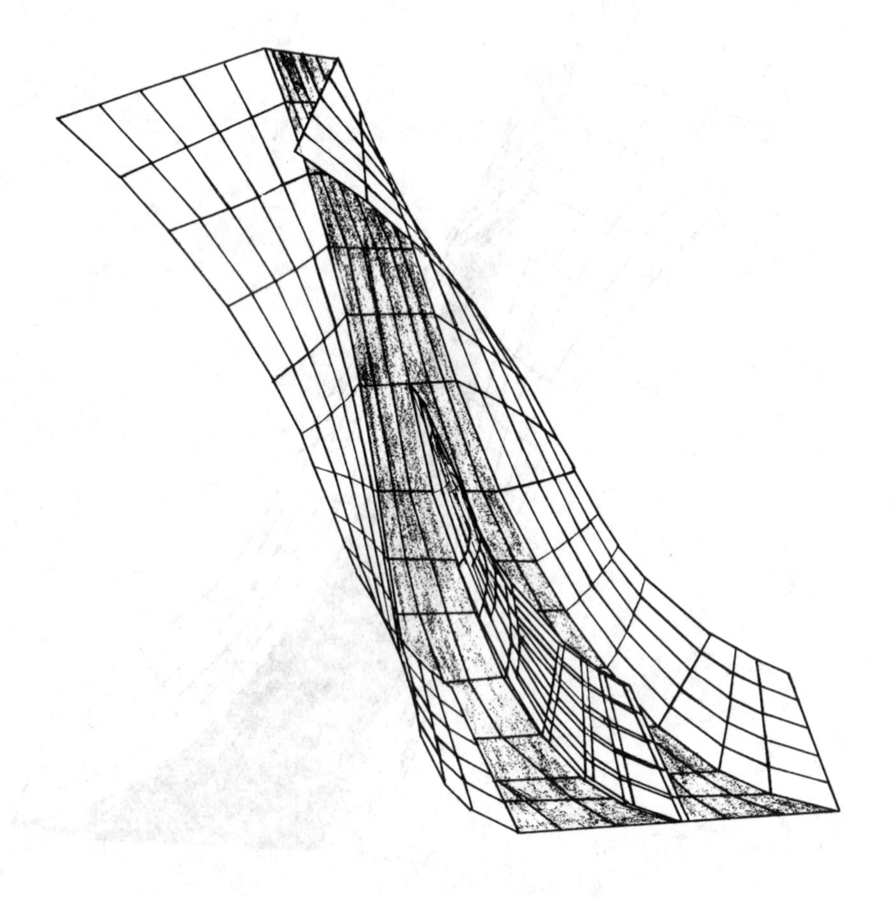

Figure 5 A typical grid arrangement for a passage with a splitter blade.

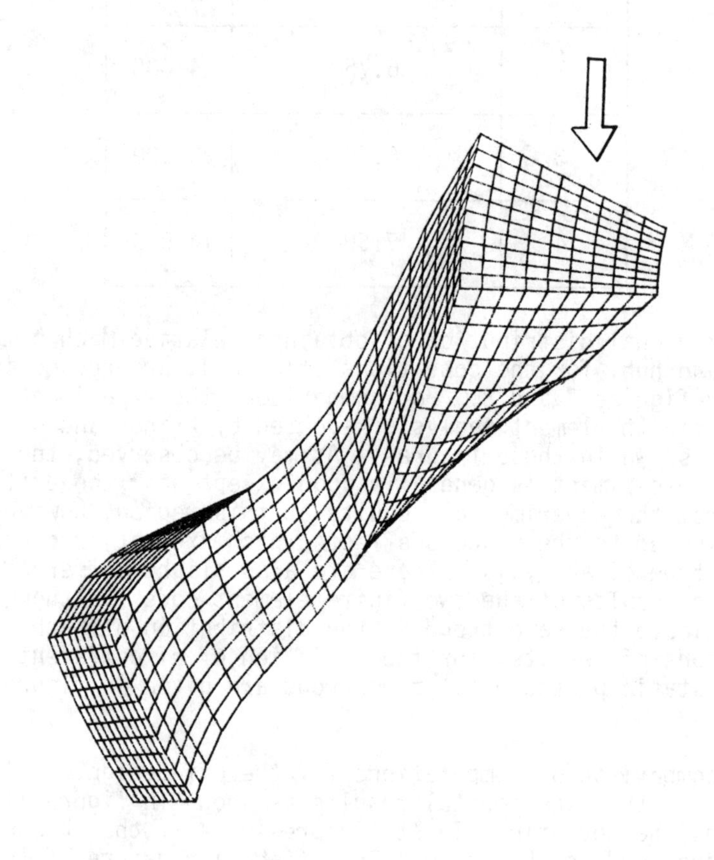

Figure 6 The finite element grid used for a blade passage of Eckardt's impeller A. (9x10x22 trilinear elements)

Table 1 Flow conditions for Eckardt's compressor impeller

Case	Mass Flow (kg/s)	Ω (rpm)
1	4.54	14 000
2	6.75	14 000
3	6.75	25 000
4	7.50	14 000

Meridional distribution of obtained relative Mach numbers at shroud and hub, for the conditions of case 1, are respectively given in figures 7 and 8. For comparison, the experimental results and the finite element analysis reported by Prince and Bryan [21] are also shown in these figures. As may be observed, the agreement with the experiment is generally good except near the exit. Deviations from the experimental results at this region, however, may be attributed to the viscous effects which are not accounted for by this type of analysis. There are also slight differences between the results of the two finite element codes, although they both indicate the same trend in the distribution of Mach numbers. Comparisons of results, for the variation of circumferential average static pressures at the shroud are given in figures 9 and 10.

A comparison of computations for the off-design condition of case 2 with the experimental results is shown in figure 11. As expected, the deviations in static pressures at the shroud are more pronounced in this case. The off-design nature of the flow is clearly depicted by the appearance of unloading around the leading edges. As may be observed from the curves for the relative Mach number at the shroud shown in figure 12, the flow is still subcritical with the relative Mach number slightly below 1.

The shroud Mach number distributions of cases 3 and 4 are respectively given in figures 13 and 14. The flow becomes transonic in both of these cases but is not yet choked. The shock remains close to the leading edges in case 3, but moves towards the trailing edges in case 4. Local mesh refinements in the neighborhood of shocks do produce sharper shocks as indicated in figure 13. The changes were insignificant in the subsonic cases.

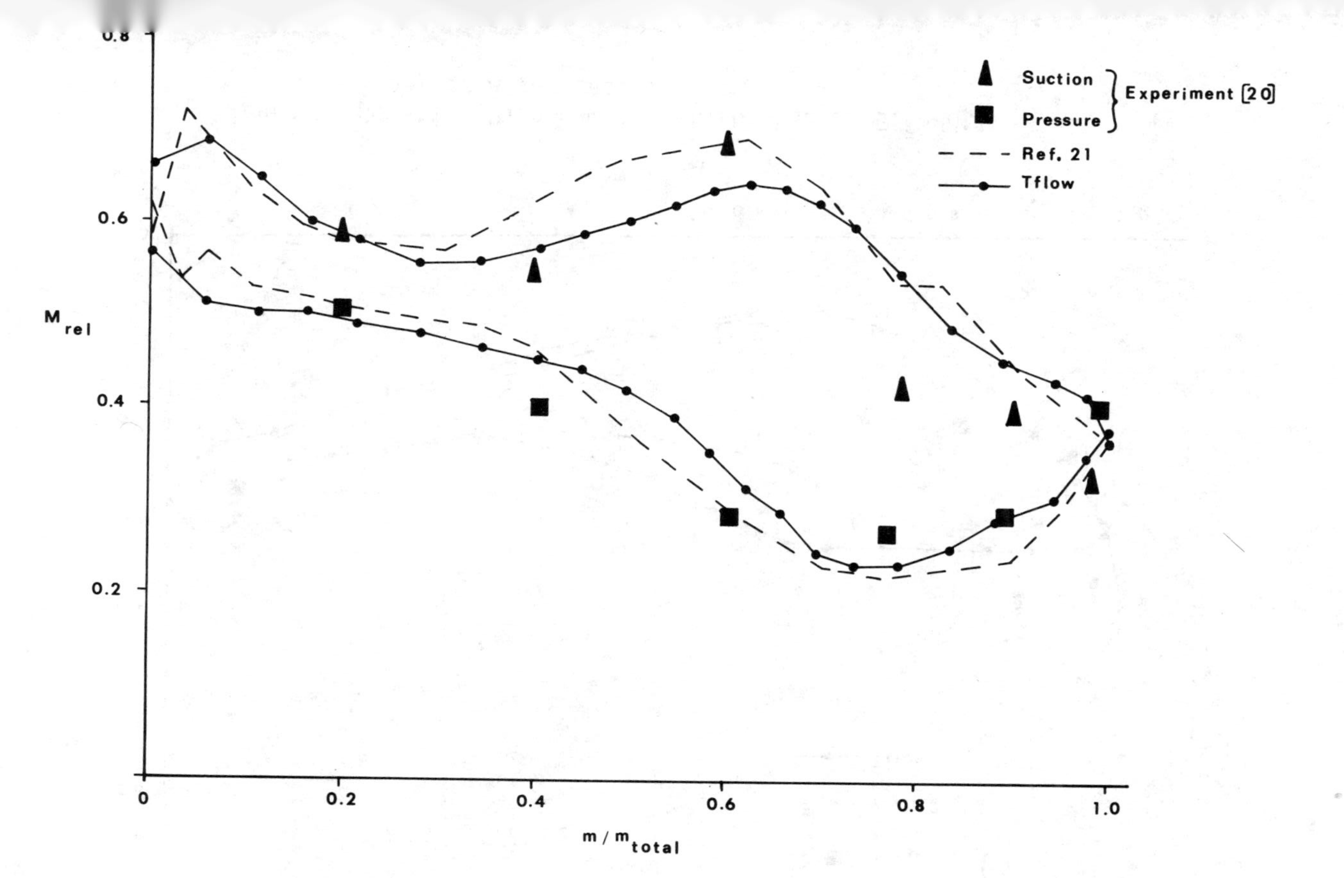

Figure 7 Relative Mach numbers at the shroud of Eckardt's impeller A for case 1.

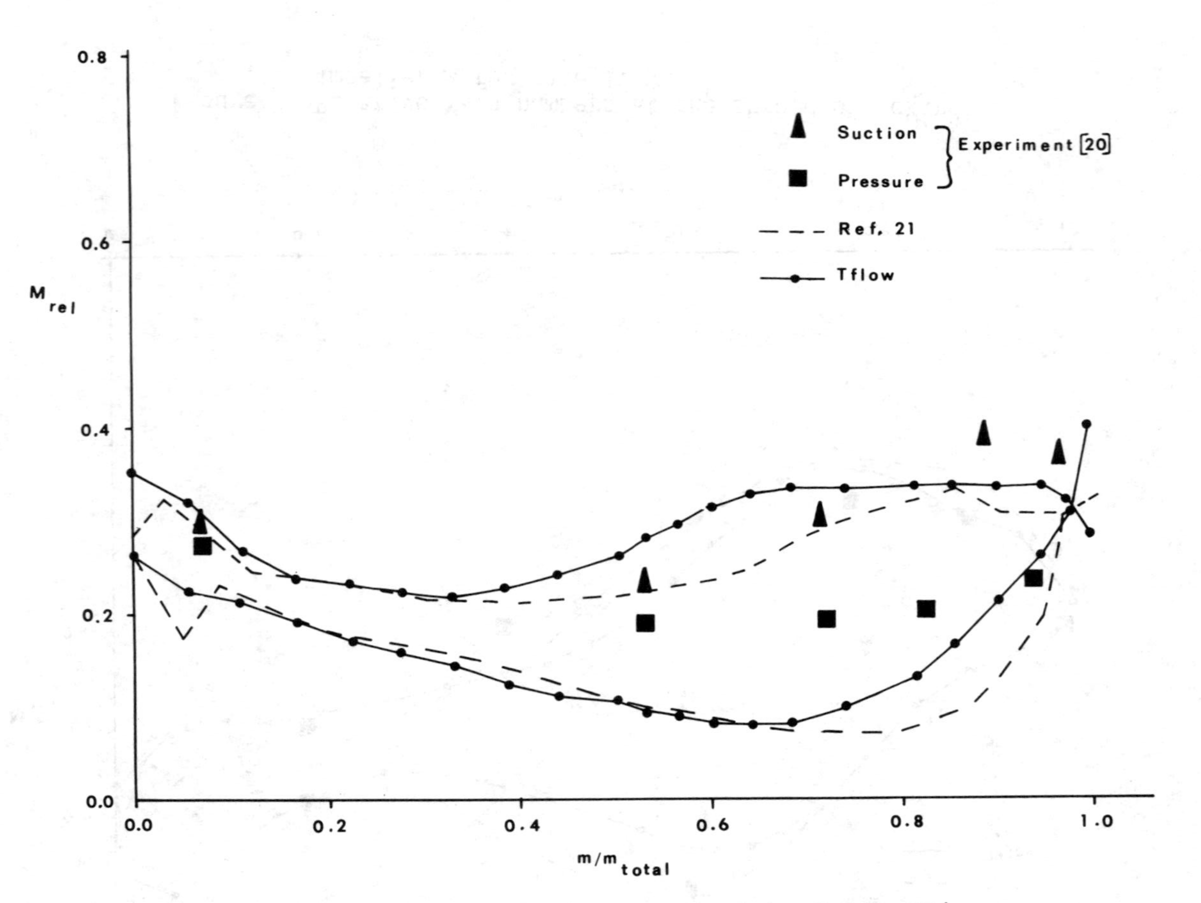

Figure 8 Relative Mach numbers at the hub of Eckardt's impeller A for case 1.

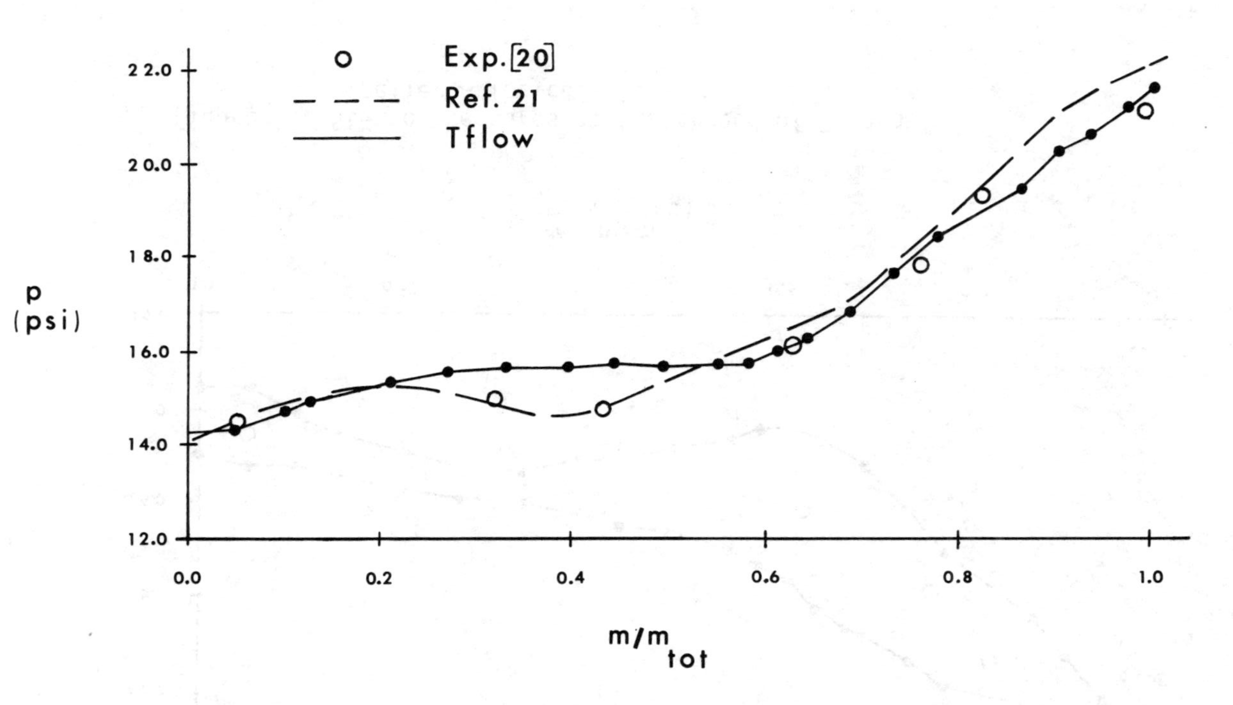

Figure 9 Average static pressures at the shroud of Eckardt's impeller A for case 1.

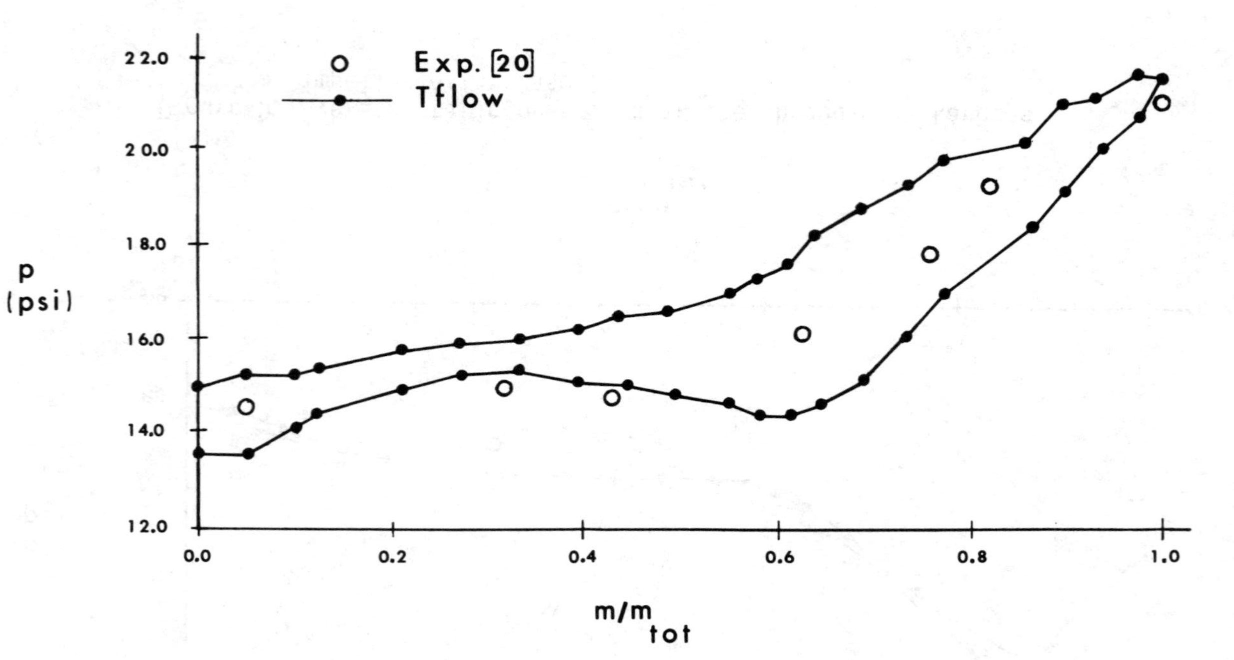

Figure 10 Static pressures at the shroud of Eckardt's impeller A for case 1.

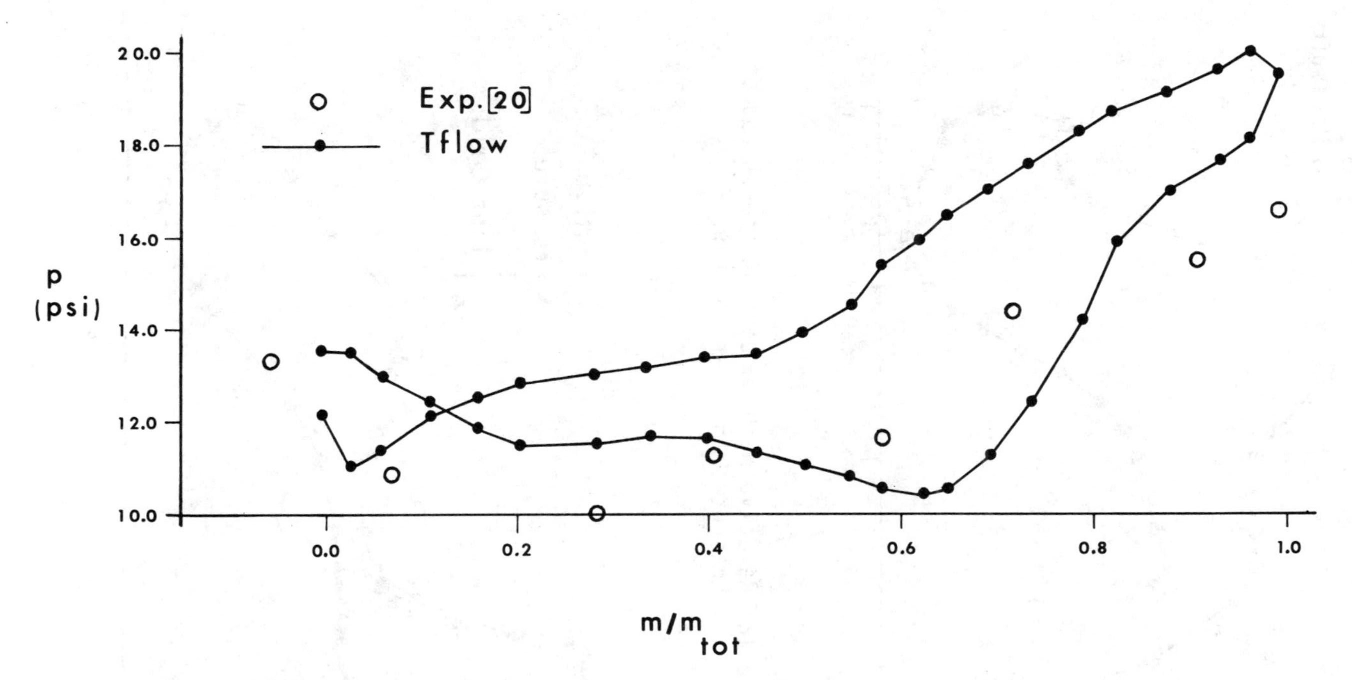

Figure 11 Static pressures at the shroud of Eckardt's impeller for case 2.

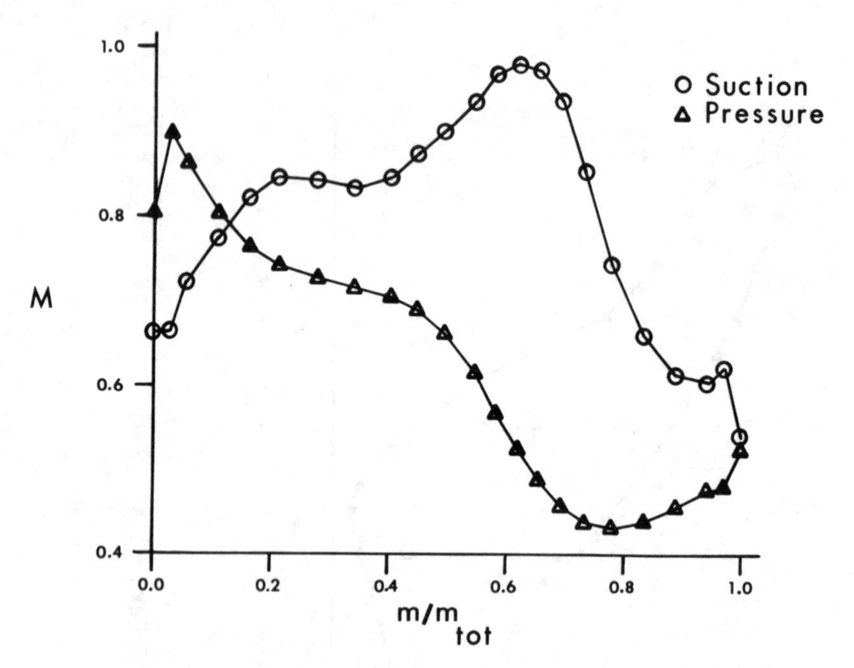

Figure 12 Relative Mach numbers at the shroud of Eckardt's impeller for case 2.

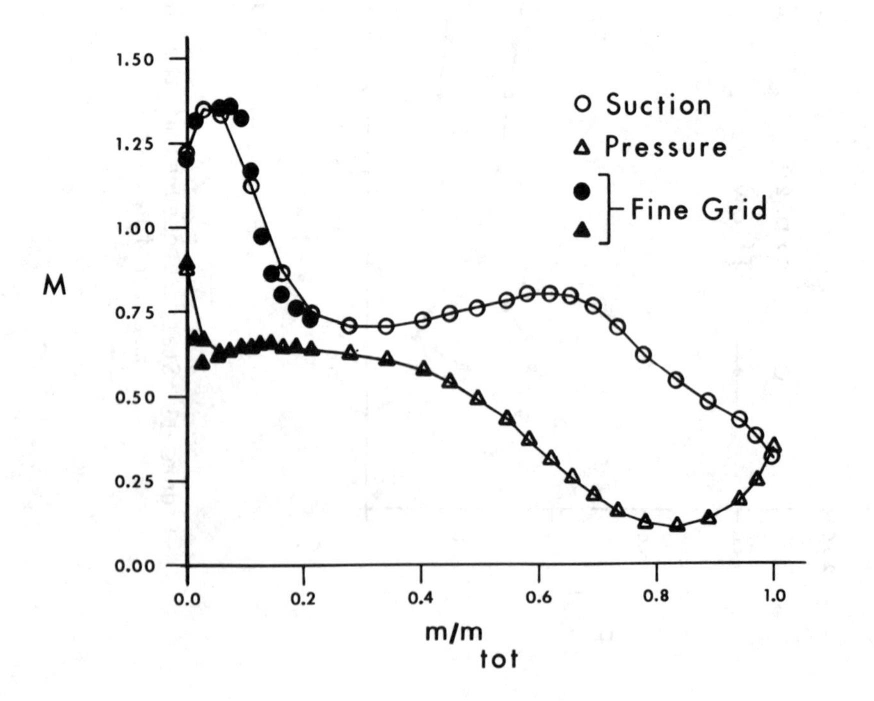

Figure 13 Relative Mach numbers at the shroud of Eckardt's impeller for case 3.

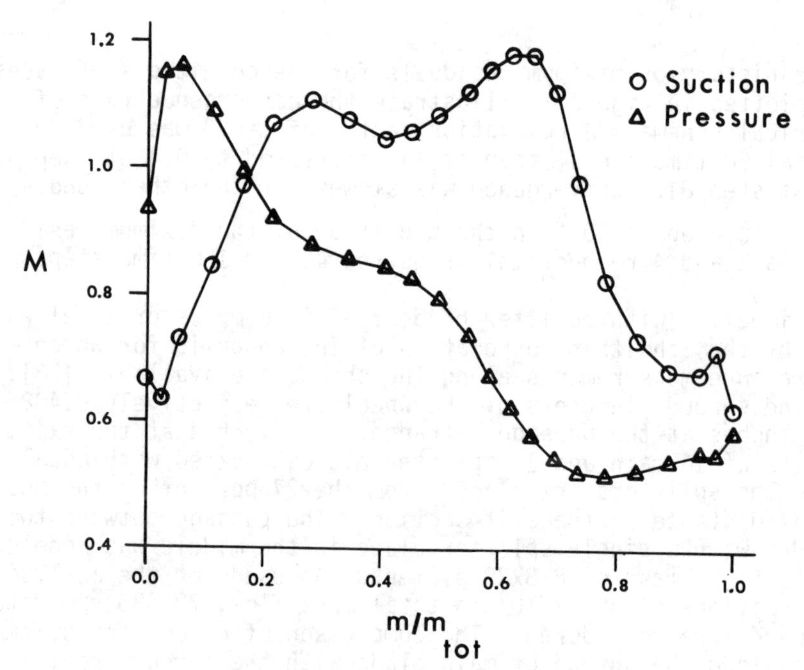

Figure 14 Relative Mach numbers at the shroud of Eckardt's impeller A for case 4.

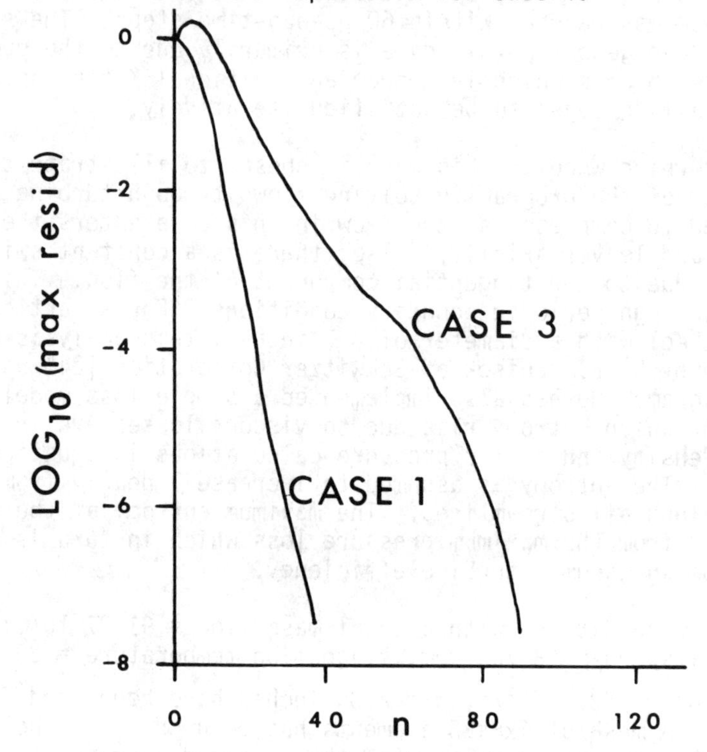

Figure 15 Residual histories for cases 1 and 3 of Eckardt's impeller A.

Time history of maximum residuals for the conditions of cases 1 and 3 plotted in figure 15 illustrate the convergence rate of the numerical scheme. A relaxation factor of ω=0.2 was used during the initial 50 time steps, then it was increased to 0.4 at step 51, and 0.6 at step 81. Convergence was slower for the other conditions. For a drop of 10^{-5} in the magnitude of the maximum residuals, cases 2 and 4 respectively required 40 and 100 time steps.

5.1.2 An impeller with splitter blades. This compressor wheel was designed by the Schwitzer Corporation of Indianapolis for which static pressure measurements along the shroud are available [23]. The hub and shroud diameters of the wheel are respectively 0.408 and 1.23 inches at the passage entrance, 1.71 inches at the exit. It consists of 14 main and 14 splitter blades located with equal spacing. The splitters are placed from the 27 percent of the hub meridional distance to the exit section. The passage between two main blades with a single splitter blade in the middle was modeled in this case. A mesh of 8x8x29 elements was used for the analysis. Inflow conditions of 35.67 lb/min total mass flow, 70 680 rpm rotational speed were considered. The comparison of calculated static pressures along the shroud of main blade with the circumferential average values of the experiment is given in figure 16. The agreement is again good with some deviations at the exit. The full convergence was reached within 60 pseudo-time steps. The slower rate of convergence in this case is primarily due to the presence of splitter blades which introduce an additional Kutta constraint at the trailing edges to be satisfied iteratively.

5.1.3 A turbine wheel. This case is chosen to illustrate the capability of the program in solving flows through turbine wheels. As opposed to compressors, the flow in this case enters the passage radially and leaves axially. Also, there is a constant swirl at the inlet due to the tangential component of the flow. This is modeled through periodic boundary conditions. For a particular turbine wheel with a diameter of 4.7 inches, this analysis was performed by E. F. Benisek of Schwitzer Corporation [24] using the TFLOW program. He has also implemented a simple loss model to the program in which entropy rise due to viscous losses are introduced to mass density and static pressure calculations in equations (28) and (29). The entropy is assumed to increase linearly from inlet to exit along all streamlines. The maximum entropy at the exit is calculated from the maximum pressure loss which in turn is determined from an assumed turbine efficiency.

Inflow conditions with a total mass flow = 91.97 lb/min, rotational speed = 48 705 rpm, stagnation temperature = 306.8^{o}F, prerotation = 163.3 ft^2/s, r = 2.35 inches have been used for the analysis. A mesh of 6x8x23 elements has been chosen. The comparison of the obtained results with the measured average static

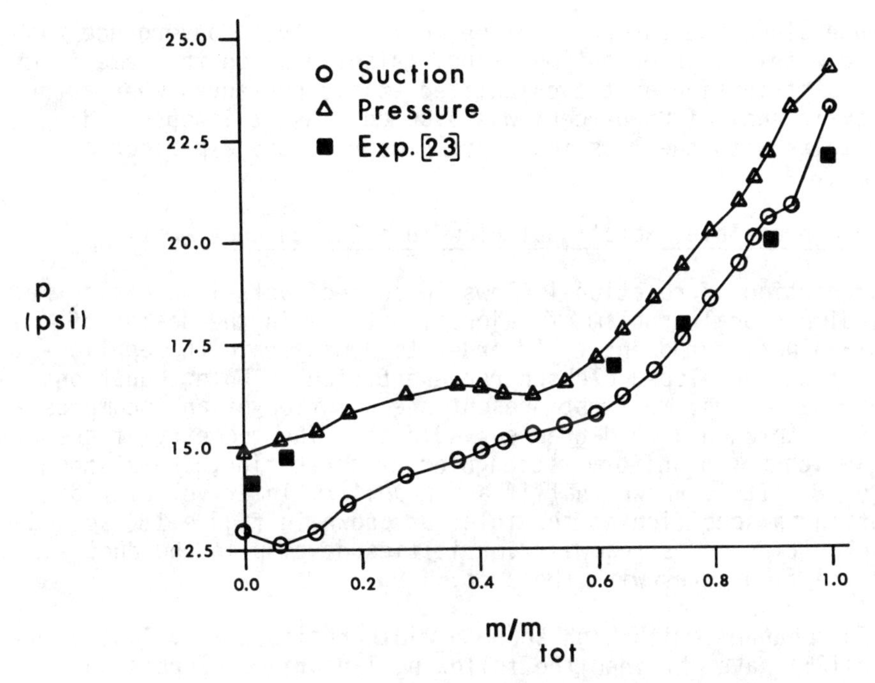

Figure 16 Static pressures at the shroud of a compressor wheel with splitter blades.

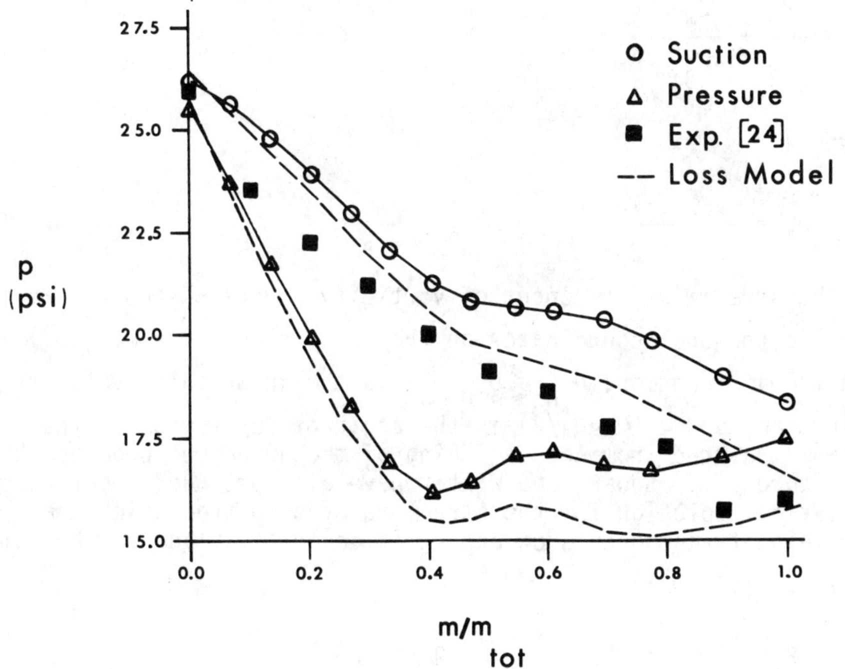

Figure 17 Static pressures at the shroud of a turbine wheel.

pressure along the shroud is given in figure 17. Convergence was reached within 16 pseudo-time steps. Also shown on the same figure is the distribution of the calculated static pressures when a turbine efficiency of 85 percent was assumed. As it is observed, the results with the loss model are closer to the experimental values.

5.2 Incompressible, Rotational Flow in a Curved Duct

The prediction of rotational flows in curved ducts is a basic three-dimensional problem of major importance in the design of turbomachinery components. In order to demonstrate the applications of the developed Clebsch representation of Euler equations to internal flows, we first present the solution of an incompressible flow through a 90 degree curved duct. The geometry of the duct, which has a uniform rectangular cross-section, is as shown in figure 18. It is known that if a non-uniform inlet velocity distribution is specified at the inlet as shown in figure 18, secondary velocities and streamwise vorticities develop in the duct while the flow turns with the duct.

For channels with high span to width ratios (ℓ/b), Squire and Winter [25] have obtained the following linearized expression governing the secondary flow on each cross-plane:

$$\frac{\partial^2 \psi}{\partial \eta_2^2} + \frac{\partial^2 \psi}{\partial \eta_3^2} = - \zeta_1 \tag{82}$$

where

$$\zeta_1 = - 2\varepsilon \frac{\partial U_{in}}{\partial \eta_3} \tag{83}$$

is the streamwise component of vorticity. Here η_2 and η_3 are the local orthogonal coordinates on the cross-plane, $\psi = \psi(\eta_2,\eta_3)$ is the stream function, $U_{in} = U_{in}(\eta_3)$ is the known inlet velocity distribution, and ε $(0<\varepsilon<\pi/2)$ is the angle of deflection of the stream measured from the beginning of the 90 degree bend as shown in figure 18b. Square and Winter have also provided a series type analytical solution for the stream-function ψ from which the secondary velocity components u_2 and u_3 at any cross-plane can be computed as follows:

$$u_2 = \partial\psi/\partial\eta_3, \quad u_3 = - \partial\psi/\partial\eta_2 \tag{84}$$

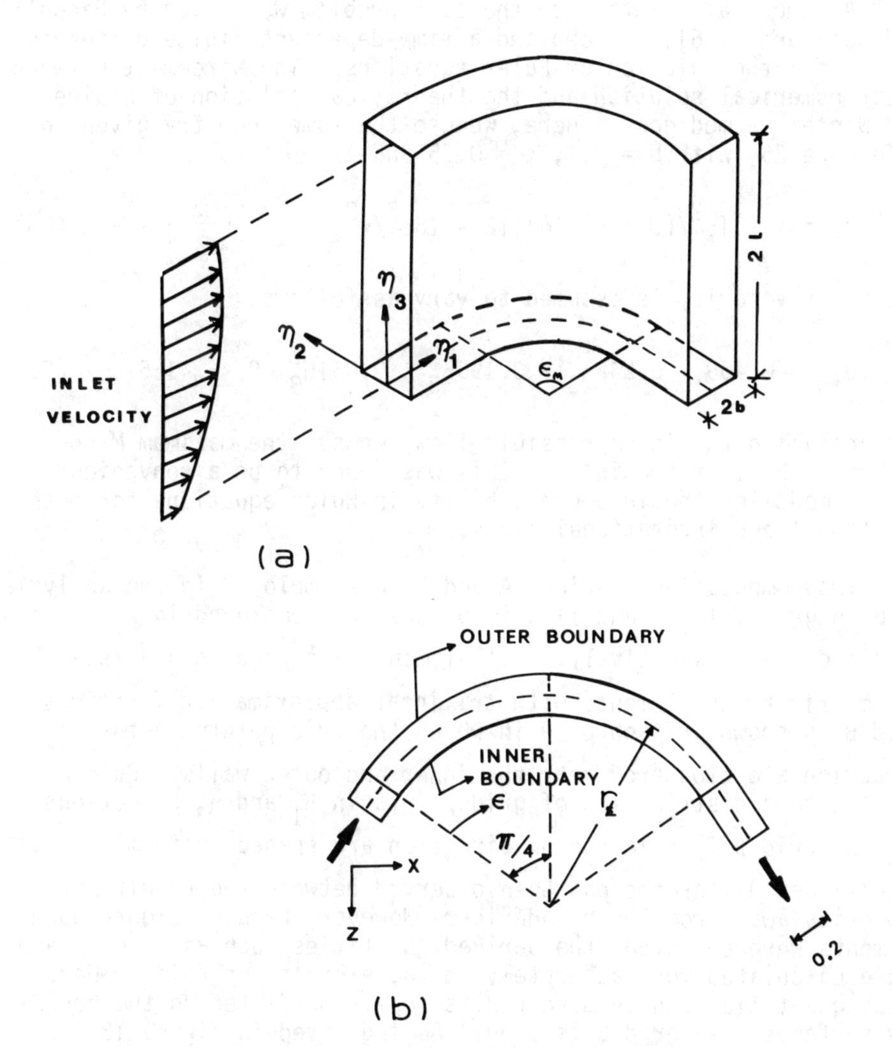

Figure 18 The geometry and the coordinate system of Squire's duct.

They have compared their results with measurements of the secondary flow in a curved duct and showed satisfactory agreement between the theory and the experiment.

A numerical solution to the same problem was given by Pandolfi and Colasurdo [26], who applied a time-dependent finite difference scheme for the solution of Euler equations. The agreement between their numerical solution and the theoretical solution of Squire and Winter seemed good. Here, we use the same geometry given in reference 26, with $b = 0.1$, $\ell = 0.75$ and

$$r_c = \ell - [\varepsilon^2/(8 + \pi^2/16)][2 - 16\varepsilon^2/\pi^2] \tag{85}$$

The inlet velocity is assumed to vary as follows:

$$U_{in} = 0.469 + 0.242\eta_3 - 0.196\eta_3^2 + 0.051\eta_3^3 \quad 0 \leq \eta_3 \leq 1.5 \tag{86}$$

For analyzing the incompressible flow, we set the maximum Mach number to 0.01 at the inlet. This was found to be a convenient way of modeling the incompressibility in Euler equations for both rotational and irrotational flows.

Two computational grids, A and B were employed in the analysis. In both grids, 17, 9 and 11 grid points were employed in η_1, η_2 and η_3 directions respectively. All elements are eight-noded iso-parametric brick elements with trilinear approximation functions. Grid B is shown in figure 19 in which the grid points in the η_2 direction are clustered near the inner and outer walls. Grid A has the same distribution of grid points in η_1 and η_3 directions but the grid points in the η_2 direction are spaced uniformly. No significant difference has been observed between the results of two grids away from the boundaries. However, because eight-noded elements were employed, the derived quantities such as ρ, $\underline{u}$, p and $\underline{\zeta}$ are calculated more accurately at the element centroids. Hence, these quantities can be more realistically predicted on the boundary surfaces when grid B is used. As indicated in figure 18b, straight line extensions are provided at inlet and exit of the channel in order to assure streamwise uniformity of the flow before and after the 90 degree bend.

The plots of secondary velocity vectors at the exit plane ($\varepsilon = \pi/2$), obtained using grids A and B, are compared with the solution of Squire and Winter in figure 20. As may be observed from these plots, the agreement is good in spite of the several simplifying assumptions of the theory by Squire and Winter. The

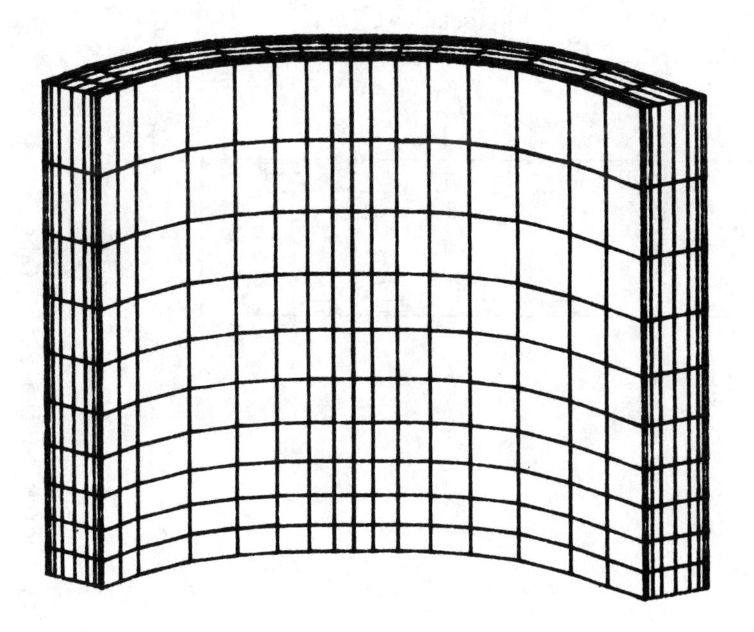

Figure 19 The computational grid B used for Squire's duct. (16x8x10 trilinear elements)

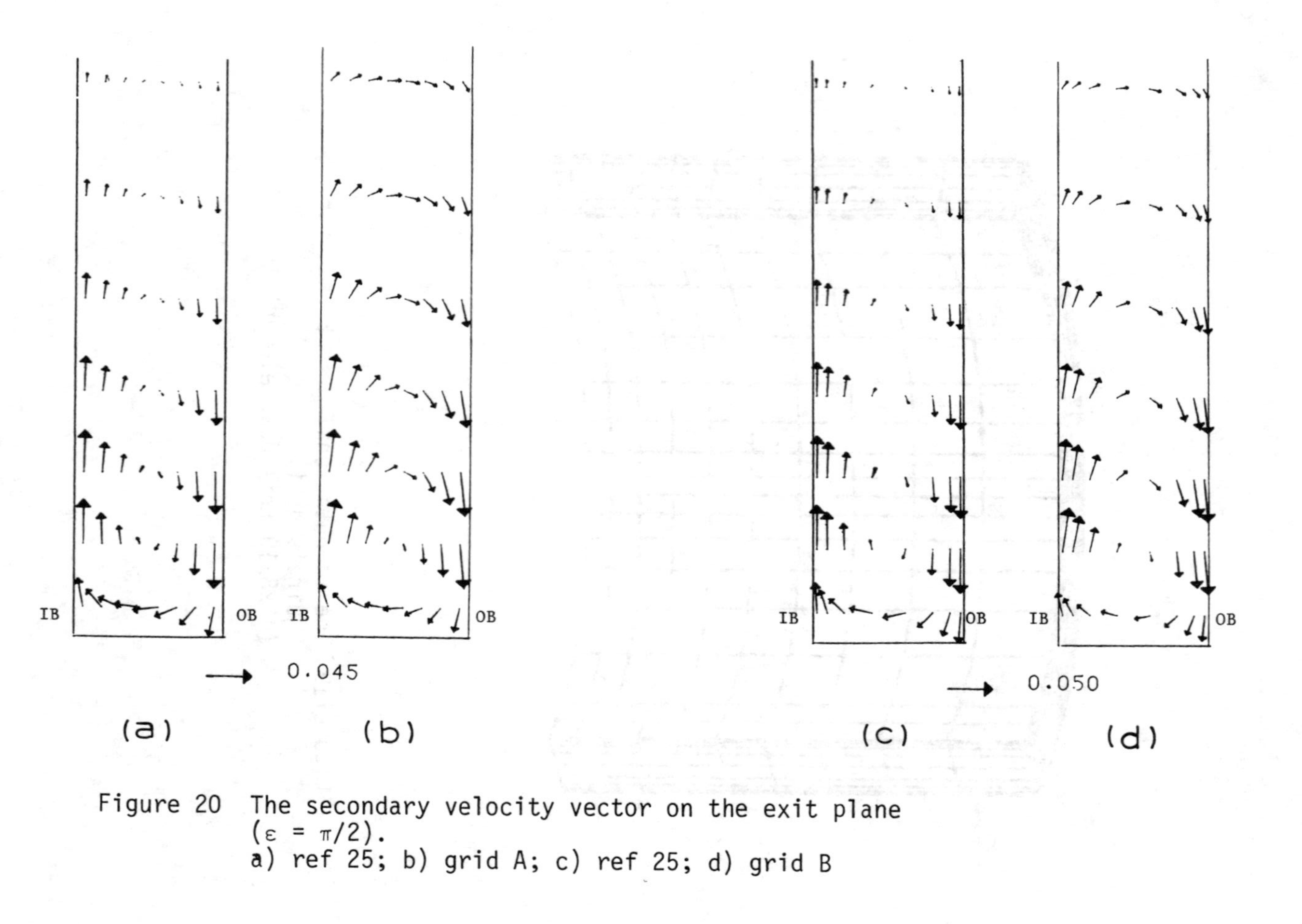

Figure 20 The secondary velocity vector on the exit plane ($\varepsilon = \pi/2$).
a) ref 25; b) grid A; c) ref 25; d) grid B

Euler solution yields slightly larger magnitudes for u_2, however, it is a smaller component of the secondary velocity. The magnitudes of the resultant vectors are very close in the region where the major secondary flow activity is $(0 \leq \eta_3 \leq 0.4)$.

In figure 21, the variation of the spanwise velocity component u_3 at two different elevations of the exit plane are plotted along the channel width and compared with the theory. It can be seen from these results that the agreement between the two solutions is excellent. Both grids essentially predict the same straight line variation of u_3 across the channel width.

In figure 22, the spanwise variation of u_3 at the exit plane is compared with the theoretical solution. While the linearized theory assumes the same magnitude on the inner and outer boundaries, the calculated values on inner and outer boundaries differ slightly with the Euler solution. This is attributed to the presence of the curvature effects in the full three-dimensional analysis.

The differences between the results of theory and Euler equations are more significant in the values of the streamwise vorticities as can be seen in figure 23. From equation (83), the linearized theory assumes that the streamwise vorticity is constant along constant η_3 values on a cross-plane. No such behavior is is true for the Euler equation. Close agreement with the theory was obtained only away from the solid boundaries as it is evident from this figure.

The residual histories showing the convergence of three Clebsch variables ϕ, S and η are shown in figure 24. As may be observed, the convergence rate of conservation of mass equation for ϕ is considerably faster than the convergence of two convective equations for S and η.

5.3 Compressible, Rotational Flow in a Curved Duct

In order to illustrate the applicability of the developed technique to compressible internal flow problems, flow through an elbow with a 90 degree turning is analyzed. This particular duct has been designed by Stanitz [27] to provide an attached boundary layer through the entire bend. Later several experiments have been conducted by Stanitz, et al. [28] on the same elbow to investigate the development of secondary flows due to various inflow conditions. The geometry and the employed finite element grid of the elbow is shown in figure 25. Only half of the passage was modeled by assuming symmetry across the mid-span surface.

The geometry of the duct was originally designed by Stanitz

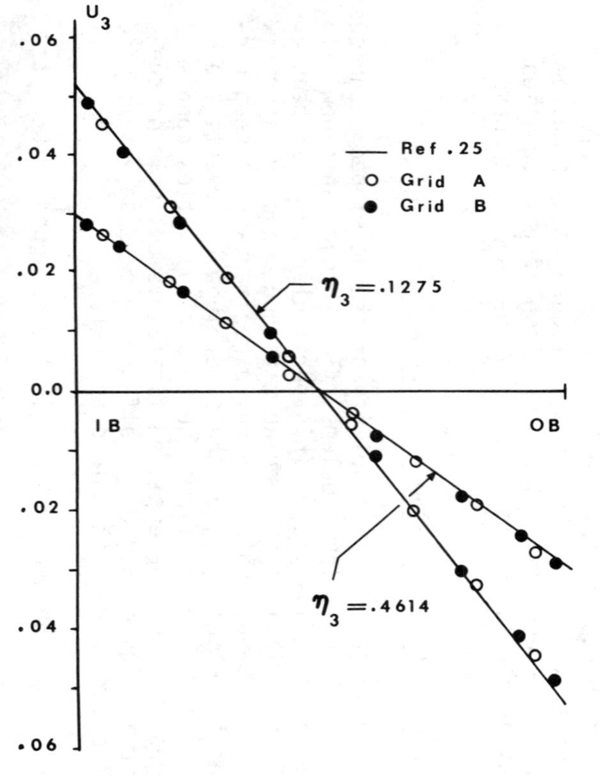

Figure 21 The secondary velocity component u_3 along the width of the exit plane.

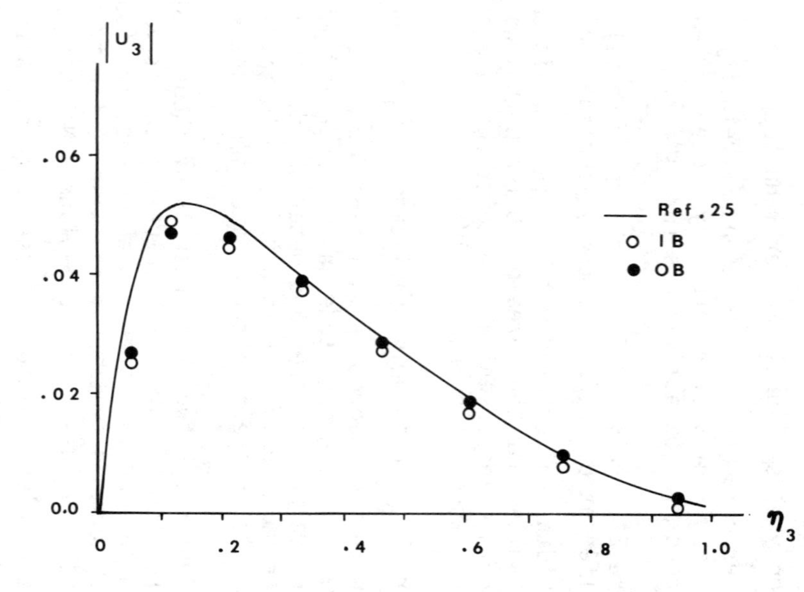

Figure 22 Spanwise variation of the secondary velocity component u_3 on the exit plane.

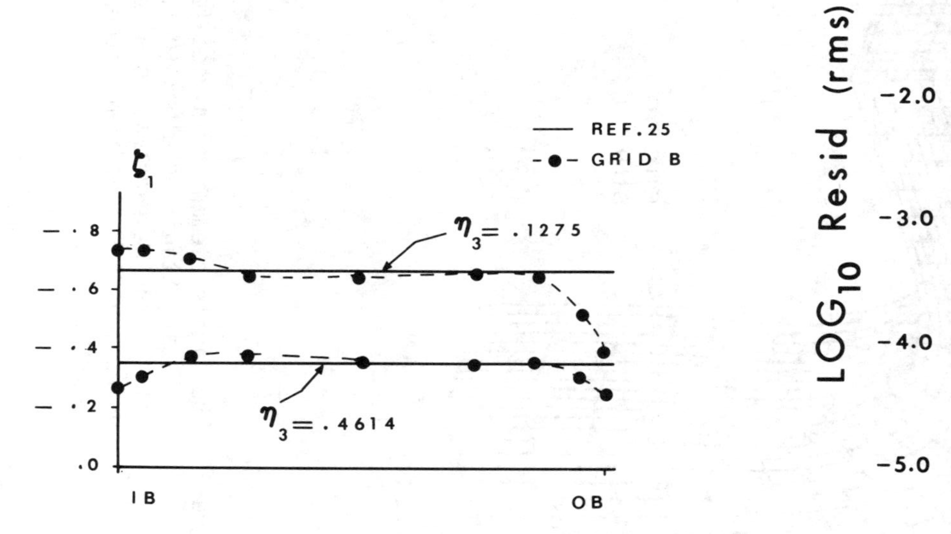

Figure 23 Streamwise vorticity component ζ_1 along the width of the exit plane.

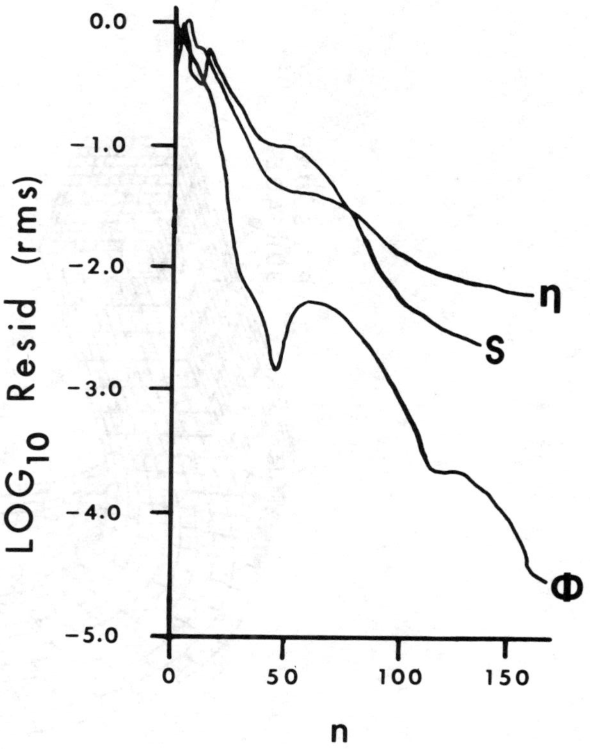

Figure 24 Residual histories of the variables ϕ, S and η for Squire's duct.

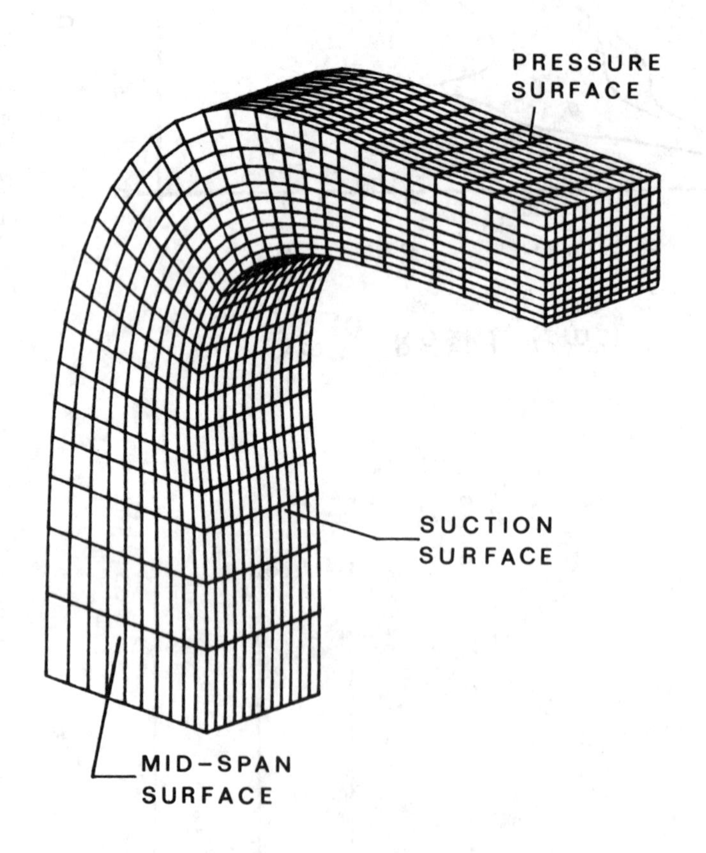

Figure 25 The finite element grid used for analyzing Stanitz's elbow (only half of the passage was modeled using 10x12x29 trilinear elements).

using a two-dimensional potential solution at the mid-span surface shown in figure 26. The dimensions of the duct are given in the figure which has a 6 inch straight extension at the exit. The variation of velocity potentials from -0.75 to 6 are also shown on the same figure. These values of velocity potentials are employed later in this paper to describe the flow at different cross-sections of the duct, e.g., $\phi = 5$ denotes the exit section for the elbow while $\phi = 6$ is the exit section past the 6 inch straight extension [28]. This particular geometry was tested for different inflow conditions by Stanitz, et al. Spoilers of sizes 0, 0.5, 1., 1.5, 2. and 2.5 inches were placed at the entrance section to generate boundary-layer thicknesses on the plane walls. Also, for each spoiler size, two different inlet pressure conditions have been considered producing main stream Mach numbers of 0.26 and 0.40 at the exit. In reference 28, measurements of the secondary flow effects are presented for all of these flow conditions.

Here, the spoiler size of 2.5 inches with 0.40 nominal exit Mach number was considered as a test case. This corresponds to a tank pressure, p_T, of 46 inches of water at the elbow inlet. The measured velocity profile at the inlet of this case is shown in figure 27. Shown in the same figure is the profile used for the present inviscid computations. The remaining inflow conditions are tabulated in Table 2. These values were determined so that the stagnation pressure corresponding to the maximum inlet flow is equal to the total tank pressure p_T. The employed finite element grid, shown in figure 25, has 11 grid points between suction and pressure surfaces, 13 grid points in the spanwise direction and 30 grid points in the streamwise direction. Only half of the flow was modeled in the spanwise direction by assuming symmetry across the mid-span.

Table 2 Inflow conditions used for Stanitz's duct

Static pressure (N/m^2)	108 938
Tank temperature (^{o}C)	25
Maximum velocity (m/s)	79.3
Maximum Mach number	0.230

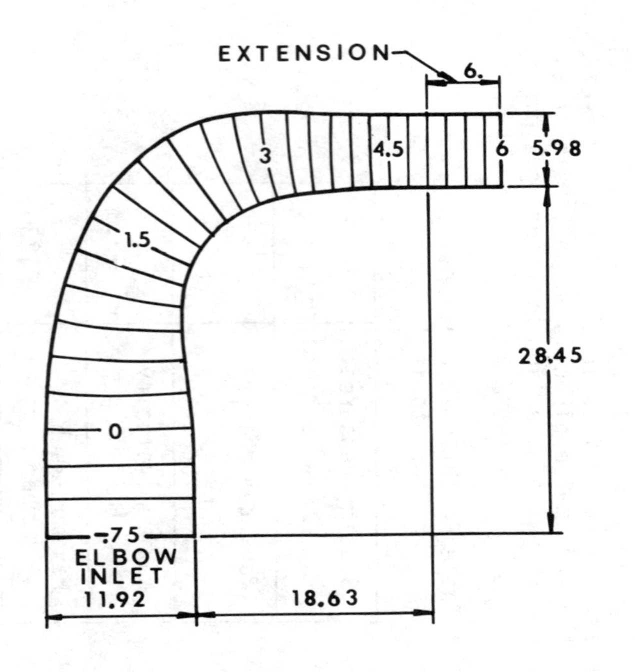

Figure 26 Velocity potentials for the mid-span surface of Stanitz's elbow [28]. (All dimensions are in inches.)

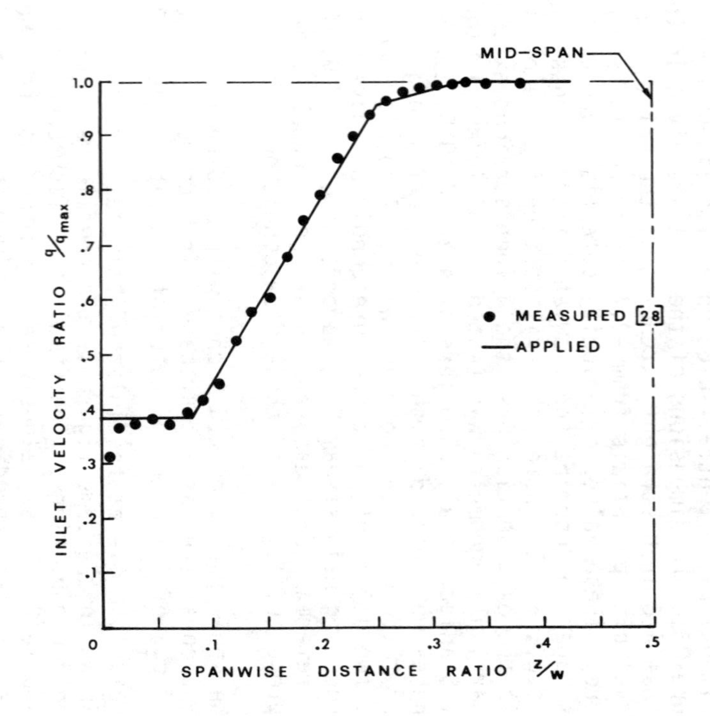

Figure 27 The inlet velocity profile for 2.5 in spoiler of Stanitz's elbow.

Stanitz, et al. [28] defined two types of dimensionless pressures in presenting the experimental results. The first one is the pressure ratio P which is expressed as,

$$P = (p - p_a)/(p_T - p_a) \tag{87}$$

where p is the static pressure, p_a is the atmospheric pressure and p_T is the stagnation pressure of the tank at the inlet. The second dimensionless quantity is the loss in the stagnation pressure p_t in comparison with the tank stagnation pressure p_T at the inlet. This non-dimensional stagnation pressure loss is defined as follows:

$$\Delta P_t = (p_T - p_t)/(p_T - p_a) \tag{88}$$

During the computations, both of these values were calculated for the same problem.

The plots of calculated secondary velocity vectors at various cross-sections of the bend are shown in figure 28. The arrows on these figures show both the direction and relative magnitude of the velocities in the planes described by the equi-potential surfaces of figure 26. The scales for each of the figures are different as indicated. The development of the vortex and the increase in magnitude of secondary velocities through the sections can be visualized from these figures. For several cross-sections, the contours for stagnation pressure losses ΔP_t are shown in figure 29. As can be seen from these figures, the pressure losses as defined by the inlet velocity profile rotate and accumulate towards the suction surface as the flow approaches the exit.

Contours for the stagnation pressure losses (ΔP_t) measured by Stanitz, et al. at the exit section of the duct (i.e., $\phi = 6.0$) are shown in figure 30. When compared with the computed results in figure 29, it is observed that there is a considerable entropy generation in the experimental results at the corner of the suction surface and the mid-span passage which cannot be predicted by the Euler solution.

The obtained numerical results were compared also with another Euler solution. Moore and Moore [29] have solved the same problem, also using an inviscid code and presented the variation of static pressure ratio along the elbow. Figure 31 shows the comparison of calculated pressure ratio distributions on three surfaces of the Stanitz' elbow with those of reference 29. The agreement between the two solutions is satisfactory.

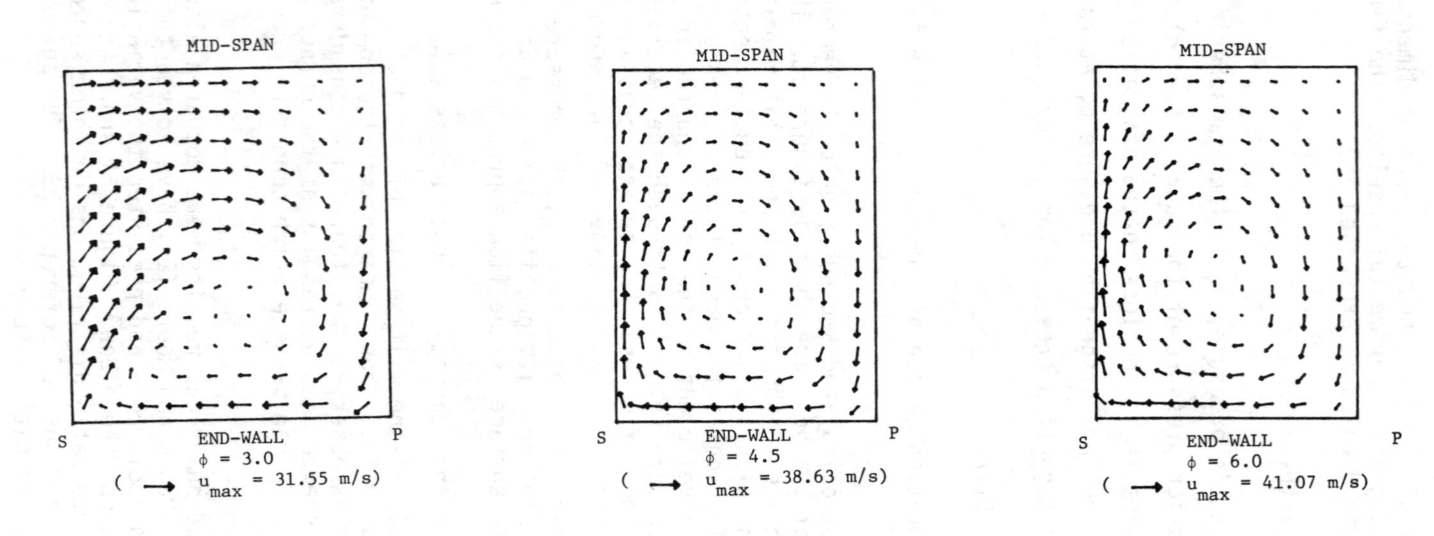

Figure 28 The secondary velocity vectors at different cross-sections of the elbow.

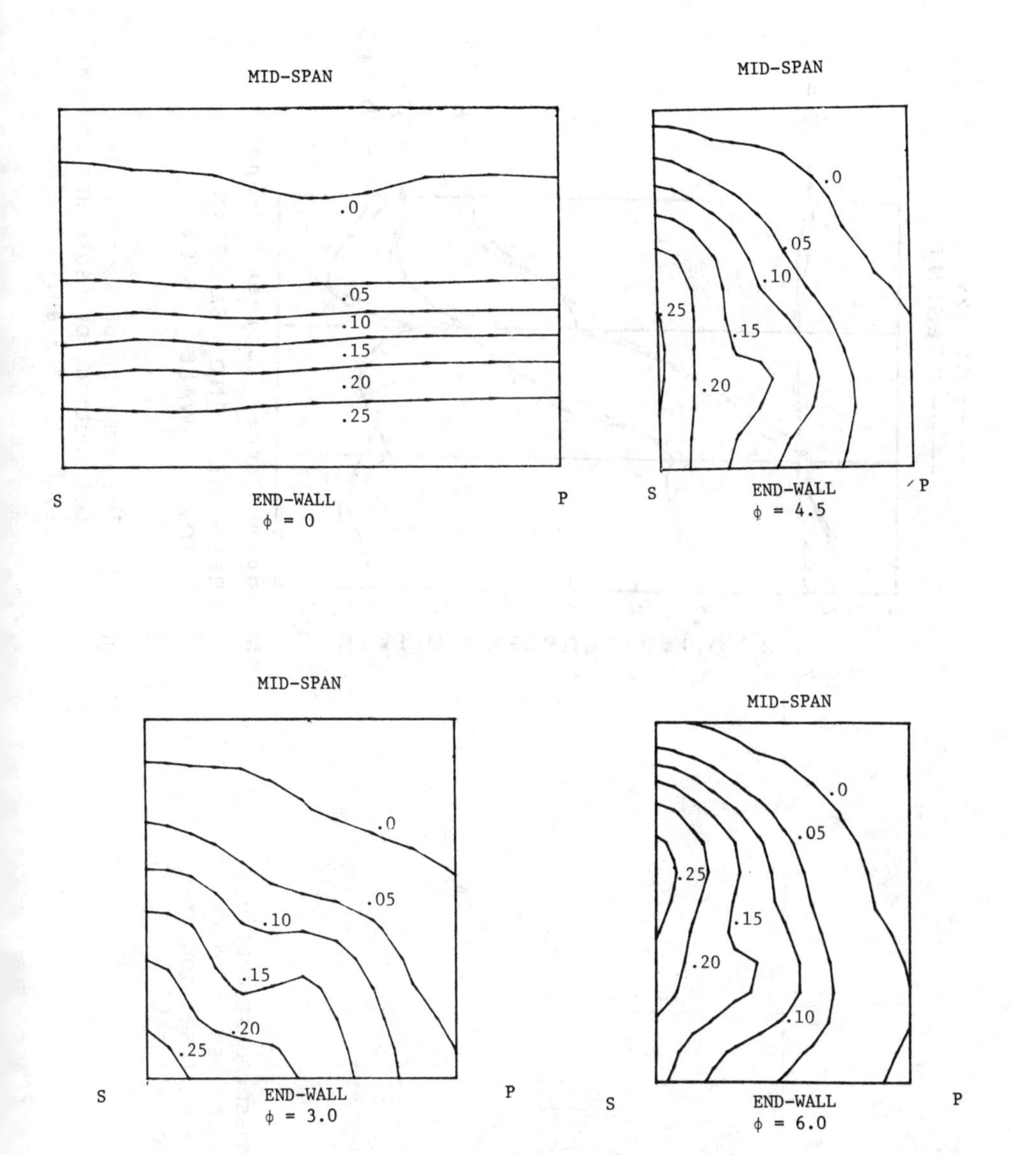

Figure 29 The stagnation pressure loss contours at different cross-sections of the elbow.

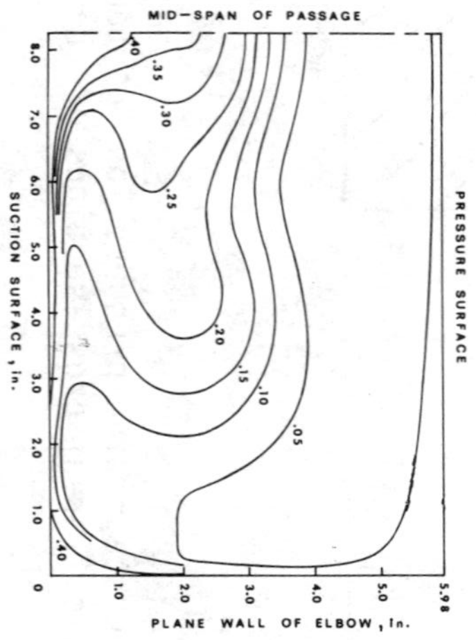

Figure 30 The measured stagnation pressure loss contours at the exit section of the bend (ϕ = 0).

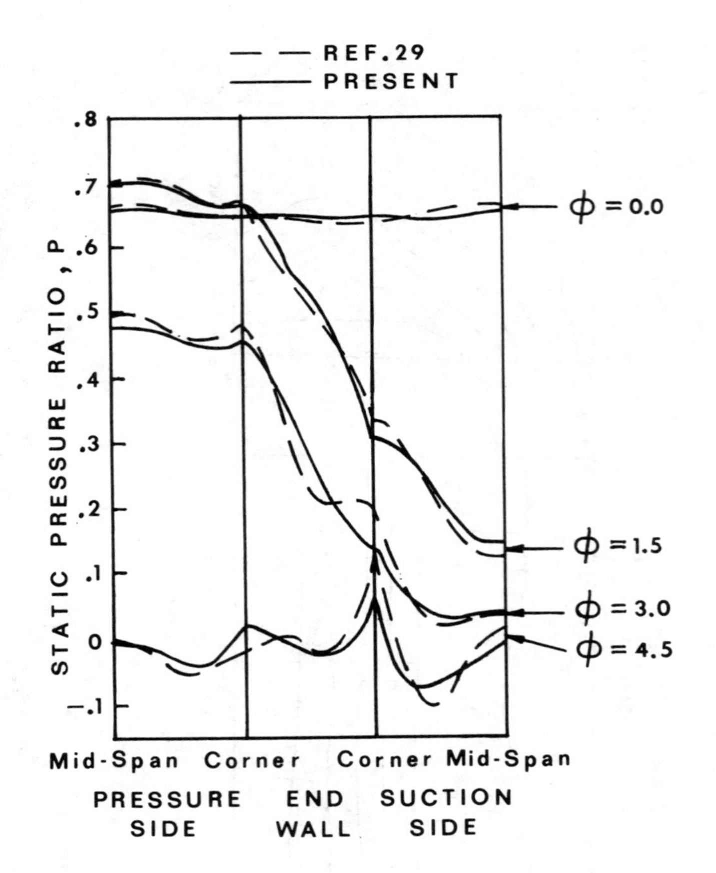

Figure 31 The computed pressure distribution on the suction, pressure and end-wall surfaces of Stanitz's elbow.

ACKNOWLEDGEMENTS

The authors wish to express their appreciation to the management of Schwitzer Corporation for partially supporting the development of the computer program TFLOW, and allowing the publication of their experimental data.

REFERENCES

1. Jameson, A., Schmidt, W. and Turkel, E. Numerical Solutions of Euler Equations by Finite Volume Methods Using Runge-Kutta Time Stepping Schemes. AIAA Paper 81-1259, AIAA 14th Fluid and Plasma Dynamics Conference, Palo Alto, California, 1981.

2. Nixon, D. and Klopfer, G. H. Nonisentropic Potential Formulation for Transonic Flows. AIAA Journal, Vol. 22, No. 6, June 1984, pp. 770-776.

3. Ecer, A. and Akay, H. U. A Finite Element Formulation of Euler Equations for the Solution of Steady Transonic Flows. AIAA Paper 82-0062, AIAA 20th Aerospace Sciences Meeting, January 11-14, 1982, Orlando, Florida. (Also AIAA Journal, Vol. 21, No. 3, 1983, pp. 343-350.)

4. Akay, H. U. and Ecer, A. Application of a Finite Element Algorithm for the Solution of Transonic Euler Equations. AIAA Paper 82-0970, AIAA/ASME 3rd Joint Thermophysics, Fluids, Plasma and Heat Transfer Conference, St. Louis, Missouri, June 7-11, 1982. (Also AIAA Journal, Vol. 21, No. 11, 1983, pp. 1518-1524.)

5. Clebsch, A. Ueber Eine Allgemeine Transformation d. Hydrodynamichen. J. Reine Agnew., Math, No. 56, 1859, pp. 1-6.

6. Bateman, H. Notes on a Differential Equation Which Occurs in the Two-Dimensional Motion of a Compressible Fluid and the Associated Variational Problems. Proc. Roy. Soc., (London), Section A., Vol. 125, No. 700, 1929, pp. 598-618.

7. Vavra, M. H. Aero-Thermodynamics and Flow in Turbomachines. John Wiley and Sons, New York, 1960.

8. Seliger, R. L. and Whitham, G. B. Variational Principles in Continuum Mechanics. Proceedings of Royal Society A., Vol. 305, 1968, pp. 1-25.

9. Ecer, A. and Akay, H. U. Applications of Variational Principles in Computing Rotational Flows. Recent Advances in Numerical Methods in Fluids, Vol. IV, edited by W. G. Habashi, Pineridge Press, Swansea, U. K., 1984.

10. Akay, H. U. and Ecer, A. Finite Element Formulation of Rotational Transonic Flow Problems. Finite Elements in Fluids, Vol. V, edited by R. H. Gallagher, et al., John-Wiley, New York, 1984.

11. Ecer, A., Akay, H. U. and Sener, B. Solution of Three-Dimensional Inviscid Rotational Flows in a Curved Duct. AIAA Paper 84-0032, AIAA 22nd Aerospace Sciences Meeting, Reno, Nevada, January 9-12, 1984.

12. Ecer, A., Akay, H. U. and Sener, B. A Finite Element Solution of Three-Dimensional Inviscid Rotational Flows Through Curved Ducts. ASME Symposium on Computation of Internal Flows, February 11-17, 1984, New Orleans, Louisiana.

13. Lacor, C. and Hirsch, Ch. Rotational Flow Calculations in Three-Dimensional Blade Passages. ASME Paper: 82-GT-316, ASME 27th International Gas Turbine Conference, March 1982, London, England.

14. Detyna, E. Finite Element Method with Clebsch Representation. Numerical Methods for Fluid Dynamics, edited by K. W. Morton and M. J. Baines, Academic Press, 1982, New York.

15. Grossman, B. The Computation of Inviscid Rotational Gasdynamic Flows Using an Alternate Velocity Decomposition. AIAA Paper 83-1900, Proceedings of AIAA 6th Computational Fluid Dynamics Conference, July 13-15, 1983, Danvers, Massachusetts.

16. Chang, S. C. and Adamczyk, J. J. A Semi-Direct Solver for Compressible Three-Dimensional Rotational Flow. AIAA Paper 83-1909, Proceedings of AIAA 6th Computational Fluid Dynamics Conference, July 13-15, 1983, Danvers, Massachusetts.

17. Roberts, A. The Treatment of Shocks in Fast Solving Methods. Numerical Methods in Applied Fluid Mechanics, edited by B. Hunt, Academic Press, 1981, New York.

18. Ecer, A. and Akay, H. U. Finite Element Analysis of Transonic Flow in Cascades. NASA Contractor Report 3446, July 1981.

19. Ecer, A., Akay, H. U. and Mattai, S. Finite Element Analysis of Three-Dimensional Flow Through a Turbocharger Compressor Wheel. ASME Paper: 83-GT-92, 28th International Gas Turbine Conference, March 27-31, 1983, Phoenix, Arizona.

20. Eckardt, D. Flow Field Analysis of Radial and Backsweep Centrifugal Compressor Impellers. Performance Prediction of Centrifugal Pumps and Compressors, edited by Gopalakrishnan, et al., ASME 1980, pp. 77-95.

21. Prince, T. C. and Bryans, A. C. Three-Dimensional Inviscid Computation of an Impeller Flow. ASME Paper: 83-GT-210, 28th International Gas Turbine Conference, March 27-31, 1983, Phoenix, Arizona.

22. Laskaris, T. E. Finite Element Analysis of Three-Dimensional Potential Flow in Turbomachines, AIAA Journal, Vol. 16, No. 7, July 1978, pp. 717-722.

23. Analysis of S-86 Compressor Stage. Internal Report, Schwitzer Corporation, Indianapolis, Indiana, 1984 (in preparation).

24. Benisek, E. F. TFLOWT - A Three-Dimensional Finite Element Computer Program for Designing Flows in Compressors and Turbines. Report No: YS-60, Schwitzer Corporation, Indianapolis, Indiana, April 1984.

25. Squire, H. B. and Winter, K. G. The Secondary Flow in a Cascade of Airfoils in a Nonuniform Stream. Journal of Aeronautical Sciences, Vol. 18, No. 4, April 1951, pp. 271-277.

26. Pandolfi, M. and Colasurdo, G. Three-Dimensional Inviscid, Compressible, Rotational Flows. Flow in Primary, Non-Rotating Passages in Turbomachines, edited by H. J. Herring, et al., The Winter Annual Meeting of the ASME, December 2-7, 1979, New York, New York.

27. Stanitz, J. D. Design of Two-Dimensional Channels with Prescribed Velocity Distributions Along the Channel Walls. I - Relaxation Solutions. NACA TN 2593, January 1952.

28. Stanitz, J. D., Osborn, W. M. and Mizisin, J. An Experimental Investigation of Secondary Flow in an Accelerating Rectangular Elbow with 90° of Turning. NACA TN 3015, October 1953.

29. Moore, J. and Moore, J. G. A Calculation Procedure for Three-Dimensional, Viscous, Compressible Duct Flow. Part II - Stagnation Pressure Losses in a Rectangular Elbow. Flow in Primary, Non-rotating Passages in Turbo-machines, edited by H. J. Herring, et al., The Winter Annual Meeting of the ASME, December 2-7, 1979, New York, New York.

NOMENCLATURE

a = speed of sound
e = a finite element
E = internal energy
f_ϕ, f_S, f_η = boundary fluxes
h = enthalpy
H = total enthalpy
I = rothalpy
$\underline{K}_{ij}$ = coefficient matrices
m = meridional distance
M = relative Mach number
n = pseudo time-step
$\underline{n}$ = normal vector
p = static pressure (absolute)
R = gas constant
$\underline{R}_i$ = residual vectors
(r,θ,z) = cylindrical coordinates
S = entropy
t = pseudo-time direction
$\underline{u}$ = relative velocity vector
$\underline{v}$ = absolute velocity vector
V = volume
(x,y,z) = cartesian coordinates
α = artificial viscosity
γ = ratio of specific heats
Γ_i = boundary surfaces
δ = variational operator
$\underline{\zeta}$ = vorticity vector
η = Lagrangian multiplier
λ = prerotation
λ_i = circulation
μ = artificial viscosity multiplier
Π = variational functional
ρ = mass density (absolute)
ϕ = velocity potential
ω = relaxation factor
Ω = rotational speed
$\underline{\nabla}$ = vector differential operator

SOLUTION OF THE EULER EQUATIONS FOR TURBOMACHINERY FLOWS
PART I BASIC PRINCIPLES AND TWO DIMENSIONAL APPLICATIONS

J.D. Denton

Whittle Laboratory, Cambridge, U.K.

1. INTRODUCTION

Numerical methods of calculating the flow in turbomachinery are well developed and are now almost universally used by designers. Of the many methods proposed and developed over the last 25 years for the calculation of blade to blade flows only a few have survived and are in everyday use for the design of blade rows. The most recent method to come into widespread use is one involving solution of the Euler equations, usually using time dependent techniques. This is the method to be described in this lecture and to understand its place in the designer's 'tool kit' we must compare and contrast it with the other methods he is likely to have available.

The flow of fluid through blade rows is governed by the equations for conservation of mass, energy and momentum. These equations can be rearranged to derive an equation for the conservation of entropy which can then be used insteady of any one of the other equations i.e. in two dimensions we have five conservation equations (mass,two momentum, energy and entropy) only four of which are independent. If we confine ourselves to steady,adiabatic reversible flow, as is often done in turbomachinery, then the energy and entropy equations reduce to statements of conservation of total enthalpy and entropy along streamlines. Hence a simple imposition of conservation of these quantities along streamlines effectively solves two of the four equations. The equations remaining to be solved are the continuity equation and the momentum equation in a direction perpendicular to the streamlines.

Three of the four methods in common use for blade to blade calculations effectively make these assumptions, i.e. that the flow is steady, adiabatic and reversible (frictionless with no shocks) and their application is therefore limited to situations where these assumptions are reasonable. These 3 methods will be outlined below before introducing the methods which are the main subject of this lecture.

1.1 The Streamline Curvature Method

This widely used method (Wilkinson (1)) usually uses the true streamlines as grid lines and imposes conservation of enthalpy and entropy along them. The momentum equation in a direction roughly perpendicular to the streamlines is solved using the curvature of the streamlines obtained from a previous iteration. The constant of integration needed to complete the integration of the momentum equation is obtained iteratively from the need to satisfy continuity of mass flow across the passage. This introduces difficulties in dealing with choked flow although patches of supersonic flow can be handled without great difficulty. The method therefore solves two equations numerically and involves two nested iterations. Nevertheless it is reasonably fast and accurate except in regions of very high curvature such as around the leading edge and is widely used for design purposes.

1.2 The Stream Function Method

This method is similar to the last method in that it makes the same assumptions and effectively solves the same equations. However, the equations are usually solved on a fixed grid and the momentum equation perpendicular to the streamlines is written in terms of the stream function. (Smith and Frost (2)). This brings the advantage that the continuity equation is automatically satisfied by the boundary conditions so only a single level of iteration is needed. The method is therefore fast and accuracy can be obtained by concentrating grid points in regions of rapid change. Its only disadvantage is that it is usually limited to strictly subsonic flow because of the ambiguity in obtaining the velocity from the gradient of stream function near the sonic point. Recently Zhao (3) has shown how this can be overcome and the method used to calculate supersonic flows with shocks. Despite its attractions and speed the stream function method does not appear to be widely used for blade to blade flows in design applications.

1.3 Methods using the Potential Function

The use of a velocity potential automatically implies irrotational flow and hence a relationship between the two velocity components. This effectively removes one of the momentum equations and if used with the assumption of constant enthalpy and entropy leaves only the continuity equation to be solved in finite difference form. This is most usually written as a requirement of zero mass flow into a series of interlocked control volumes, (e.g. Caspar (4)). Hence only a single equation for the potential needs to be solved and very fast numerical techniques are available for this. The method is strictly limited to shock free (but not subsonic) flow but techniques for using it to capture weak (hence nearly isentropic) shock waves are well developed. The velocity potential method is therefore one of the most important tools of the blade deisgner but its use is limited to flows with only weak shock waves.

In contrast to the methods described above the methods to be described in this lecture involve direct solution of the conservation equations (Euler equations) and make no assumption of isentropic flow. Even the assumption of adiabatic flow is not necessary although it is usually made. Hence, in two dimensions, three or possible four conservation equations need to be solved in finite difference form. The methods available are therefore more time consuming than those previously described but since fewer assumptions are made they are not so limited in application. In particular supersonic flows with strong shock waves can be calculated. Also, if solved by a time dependent technique, the shock wave position does not need to be known before the solution is started. If the correct equations are solved and the correct boundary conditions are applied then shock waves must appear in the solution exactly as in the real flow. Unfortunately, the dropping of the isentropic assumption makes much greater demands on the accuracy of the finite differencing techniques used and accurate solutions with shocks are still difficult to obtain. Nevertheless the extreme flexibility of the methods coupled with the decreasing cost of computer power has led to a continued increase in the use of the Euler solvers over the last decade.

2. EQUATIONS

The conservation equations are simple statements of the laws of Nature and are most naturally applied to a control volume. (Fig.5). If V_o is such a volume which is fixed in space then the equations simply say that the rate of flow of the conserved quantity into the volume is equal to the rate of change of the amount of that quantity inside the volume. When interpreted in this way the pressure acting on a surface must be regarded as equivalent to a flow of momentum across it. If $\underline{dA}$ is the inwards normal vector to an element of the surface of the control volume, the equations become:

$$\text{Continuity} \quad \iint \rho\underline{V}.\underline{dA} = V_o \frac{\partial \rho}{\partial t} \quad \ldots \quad (1)$$

$$\text{Momentum} \quad \iint (P\underline{dA} + \rho(\underline{V}.\underline{dA})\underline{V} = V_o . \frac{\partial(\rho\underline{V})}{\partial t} \quad \ldots \quad (2)$$

$$\text{Energy} \quad \iint \rho(\underline{V}\ .\underline{dA})H = V_o . \frac{\partial(\rho E)}{\partial t} \quad \ldots \quad (3)$$

where H is the stagnation enthalpy defined by

$$H = C_p T + 1/2V^2$$

and E is the specific internal energy

$$E = C_v T + 1/2V^2$$

The equations must be closed by an equation of state for the fluid which for a perfect gas would be

$$P = \rho RT$$

where R is the gas constant.

When written in this form the values of fluid properties on the LHS of the equations are values on the faces of the elements whilst those on the RHS are average values within the elements. This form of the equations is usually referred to as the 'Finite Volume' form.

The same equations are often regarded as differential equations relating the spatial gradients of flow properties at a point. In this case they may be written as:

$$\text{Continuity} \quad \nabla.(\rho\underline{V}) = -\frac{\partial \rho}{\partial t} \quad \ldots \quad (4)$$

$$\text{Momentum} \quad \nabla P + \rho(\underline{V}.\nabla)\underline{V} = -\frac{\partial(\rho\underline{V})}{\partial t} \quad \ldots \quad (5)$$

$$\text{Energy} \quad \nabla.(\rho\underline{V}H) = -\frac{\partial(\rho E)}{\partial t} \quad \ldots \quad (6)$$

The equations may be solved in either form. In the finite volume form the flow properties are usually (but not always) stored at the centre of the elements and the values on the cell boundaries which are needed to form the fluxes on the LHS of the equations are found by interpolation. When solved in the finite difference form the gradients at the nodal points must be obtained from finite difference approximations in terms of the values at surrounding point

It is not immediately apparent whether numerical differentiation or numerical interpolation is likely to be most accurate. In fact on a uniform mesh the two approaches are exactly equivalent. Hence there would at first sight seem to be little to choose between them. However, on an irregular mesh the finite volume approach has the decisive advantage that, since all flows (of mass, energy and momentum) leaving one cell enter into neighbouring cells, global conservation of these quantities is ensured. This means that in the steady state the mass flow rate out of the whole domain must exactly equal that entering it and the momentum change must exactly balance the applied force (e.g. lift). Such global conservation is not automatically obtained with the finite difference approach, especially on highly irregular meshes and this can lead to serious errors in calculating internal flows. Hence the finite volume approach has become almost universal for calculation of flows in turbomachinery.

The energy equation, Eqn. 3, is an essential part of the system of equations governing unsteady flow. However, for steady adiabatic flow in a fixed coordinate system the energy equation reduces to the simple result that H is constant along streamlines. Hence, if H is uniform at inlet it is everywhere constant and if we are only interested in steady flow we can replace Eqn. 3 by H = const. This reduces the number of equations to be sovled numerically by one with a proportional saving of computer time, (Denton (5)). This simplification has also become almost universal for turbomachinery flow calculations.

In a rotating coordinate system the constant enthalpy assumption may be replaced by one of constant rothalpy ($H - \Omega r V_\theta$) along streamlines. Provided that the rothalpy is uniform at inlet it remains constant through the whole flow field. This restriction is not serious in 2D or quasi-3D calculations but is not generally acceptable for fully 3D calculations where the inlet flow properties may vary considerably in the spanwise direction.

3. CHOICE OF GRID SYSTEM

Turbomachinery flow calculations are extremely demanding in terms of grid definition compared to, say, isolated aerofoil calculations. This is because of the large changes in flow direction which occur coupled with the need to satisfy a periodicity condition upstream and downstream of a blade row. The periodicity condition requires that all flow properties at corresponding points one blade pitch apart must be identical and this is most easily satisfied if grid points on the upper periodic boundaries are displaced one pitch circumferentially from those on the lower boundaries. The natural choice of grid to achieve this requirement is that shown in Figure 1. where the grid lines are formed by

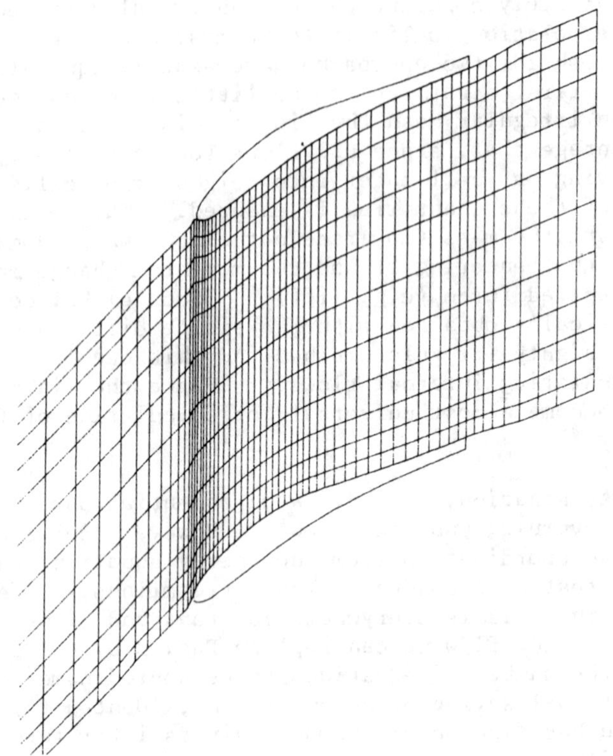

Figure 1 Grid for a Compressor Blade

quasi-streamlines and by pitchwise lines and this type of grid has been by far the most common choice in turbomachinery calculations. However, its use imposes severe limitations on accuracy when the elements become highly skewed as they do in a staggered cascade. The problem is most easily illustrated by considering the finite difference approach although the consequences are exactly the same for the finite volume method.

Consider the 5 grid points shown below:

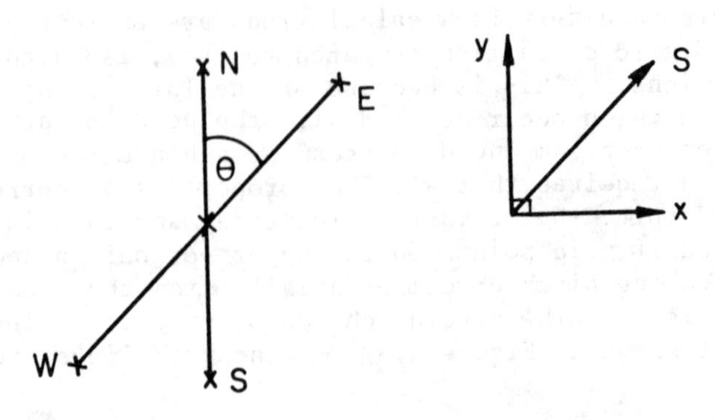

We wish to evaluate $\frac{\partial Q}{\partial x}$ and $\frac{\partial Q}{\partial y}$ for use in equations 4 - 6, where Q is any flow property. From points N,S,E,W we can easily obtain:

$$\frac{\partial Q}{\partial y} = \frac{Q_N - Q_s}{\Delta y}$$

$$\frac{\partial Q}{\partial S} = \frac{Q_E - Q_w}{\Delta S}$$

Now $\frac{\partial}{\partial S} = \frac{\partial}{\partial x} \sin\theta + \frac{\partial}{\partial y} \cos\theta$

$$\frac{\partial Q}{\partial x} = \frac{Q_E - Q_w}{\Delta S} \frac{1}{\sin\theta} - \frac{Q_N - Q_s}{\Delta y} \cot\theta \qquad (7)$$

so when θ becomes small the $\frac{\partial}{\partial x}$ term becomes the difference of two large quantities, both of which are being evaluated approximately by finite difference means, and so the accuracy to which it is determined becomes poor.

Hence there is a strong incentive to use a more nearly orthogonal mesh for highly staggered cascades or whenever θ is small. This is not possible without sacrificing the use of periodic grid points and consequently the introduction of considerable extra complication in satisfying the periodicity condition. This difficulty has dissuaded most workers from developing such grids. The only known exceptions are Delaney (11) and Holmes and Tong (12) who have made use of non-periodic grids. Delaney used an orthogonal mesh, shown in Fig. 2, obtained by solving a set of Laplace equations. Holmes used a much simpler geometric construction shown in Fig. 3. Both of these methods appear to obtain better accuracy per grid point than is obtained from the simple mesh of Fig. 1 but whether the extra complication involved is less undesirable than the use of more mesh points on the simple grid is not yet resolved. The choice of grid for highly staggered cascades therefore remains an area of active development. In three dimensions the use of anything but very simple meshes becomes prohibitive.

Having chosen a grid the problem remains of whether to locate the points where the variables are stored at the centres of the finite volume cells or at the intersections of grid lines. Most methods choose the former location and then have to interpolate to obtain properties on the cell boundaries and extrapolate to find them on the blade surfaces. Ni (6) and Denton (7) choose the latter location which removes the need for interpolation but requires a means of distributing the changes computed for a cell between the surrounding nodes. It is important to recognise that this distribution of the changes only affects the stability of a method.

The accuracy is determined by the accuracy with which the fluxes are evaluated in the final steady state.

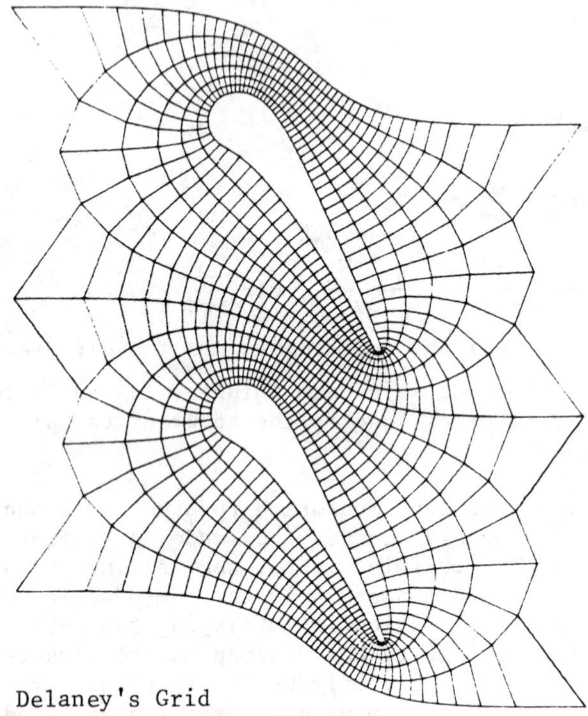

Figure 2. Delaney's Grid

4. NUMERICAL SCHEMES

Integration of equations 1 - 3 through time to reach a steady state solution requires very specialised techniques. A very large number of schemes have been developed and these may be classified broadly as implicit or explicit according to whether a point is updated using the new values or the old values of the variables at the surrounding points. Implicit schemes inevitably involve some form of matrix inversion but allow much larger time steps to be taken. Hence the steady state is reached in fewer but more costly time steps than with an explicit scheme. For inviscid turbomachinery calculations implicit schemes have not proved as successful or as economic as explicit schemes and they will not be considered further.

Explicit schemes are always limited to time steps which satisfy a condition similar to

$$\Delta t < \frac{\Delta x}{C+V}$$

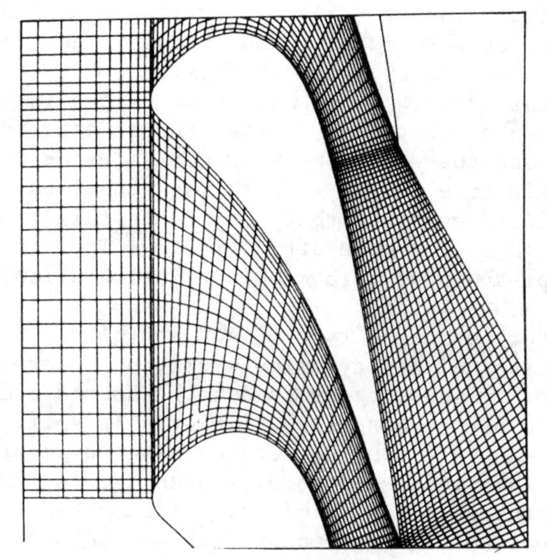

Computational grid, VKI turbine nozzle

Figure 3 Holme's Grid

which essentially says that the fastest moving pressure wave must not move more than one grid spacing per time step. This inevitably means that if there are N grid points from inlet to outlet of a cascade the number of time steps required to reach a steady state from an arbitrary initial guess is of the order 10 N i.e. many hundred time steps will be needed. Hence the comparatively large computer times required by such methods.

The methods used to achieve stability of explicit schemes can be roughly classified as:

(a) Addition of large numerical damping terms to the equations e.g. Lax, McDonald (8).

(b) Two or more step predictor-corrector schemes, e.g. Lax-Wendroff, McCormack, Brailovskaya,Jameson (9).

(c) Deferred correction schemes in which the numerical damping of type (a) schemes is gradually cancelled out e.g. Denton (5), Arts (10).

(d) Single step second order accurate schemes e.g. Delaney(11), Ni (6).

Space does not permit a detailed discussion of the individual schemes but the advantages and disadvantages of each classification will be considered.

The 'Lax' type schemes (a) are very simple and fast but are only first order accurate in time and space. The accuracy in time is not important for steady state calculations but the low accuracy in space means that a very large number of grid points are needed to obtain acceptable solutions. The schemes have advantages of simplicity and robustness and the flow variables need only to be stored at a single time level. With the reduction in cost of computer power it is possible that schemes of this type, with very fine meshes, could become more attractive. They are, however, little used at present for turbomachinery calculations.

The predictor-corrector type of scheme (b) has been the most popular one in external aerodynamics where the McCormack scheme has been widely used. The methods tend to be second order accurate in time and space but because each time step is split into a predictor and a corrector step all flow properties have to be stored at two time levels. This is a serious dissadvantage for 3D calculations where computer storage may be limited. The methods are more complex and appear to be less robust than methods of type (a). They have been used in turbomachinery calculations (e.g. Thompkins (13)) but are not especially popular.

Recently a 4 step Runga-Kutta type of scheme has been developed by Jameson (9). This appears to be very successful for external flow calculations and has been adapted to turbomachinery calculations by Holmes (12). The method permits $2\sqrt{2}$ times longer time steps than most other explicit methods but does require considerable computer storage. Whether or not this scheme can justify its increased complexity remains to be determined.

The deferred correction schemes (c) are very similar to schemes of type (a) but the numerical damping inherent in the use of a first order scheme is progressively cancelled out as the calculation proceeds leaving a solution which is second order accurate in space. This type of scheme is possibly the most widely used in turbomachinery (Denton (5 & 7), Arts (10)), but does not seem to be used in external aerodynamics. It has the advantages of speed, simplicity and low storage requirements of class (a) but the accuracy of class (b).

There are comparatively few schemes of class (d) but Delaney (11) and Ni (6) have developed schemes of this type for turbomachinery calculations. Delaney's scheme is based on the hopscotch appraoch of updating odd and even points on successive time steps. Ni's method uses the second derivative in time to obtain a formula for distributing the changes between the surrounding nodes. Although both these schemes seem to be very successful they do not appear to have been copied by others.

It can be seen that a very large number of numerical schemes are available for solving the Euler equations in turbomachinery cascades. The accuracy of the schemes will be considered later but when assessing the schemes the advantages of simplicity and robustness are considered by the author to be of paramount importance. Computational efficiency is rapidly becoming less important whilst computer storage requirements are now only important for 3D applications.

5. BOUNDARY CONDITIONS

Most of the boundary conditions applied to cascade flow calculations are very straightforward. They may be understood physically as being those applied to a cascade fed from a plenum at constant stagnation pressure and temperature and exhausting to a plenum at constant static pressure. If the downstream pressure is suddenly changed a pressure wave will move upstream through the cascade and, on reaching the inlet boundary, will change the mass flow entering. Further pressure waves will be produced by interaction of the changed mass flow with the cascade and these will eventually escape from the system leaving a new steady flow. Thus the total to static pressure ratio of the cascade is fixed and the mass flow is a product of the calculation.

With subsonic inflow the total temperature, total pressure and flow direction must be fixed at the upstream boundary whilst only the static pressure must be fixed at the downstream boundary. With supersonic relative inflow it is not usually possible to fix the flow direction and a condition of specified tangential velocity at inlet may be used instead. This allows the inlet flow direction to change as the calculation proceeds. If the axial Mach number is greater than unity all flow conditions must be fixed at the upstream boundary since pressure waves can no longer reach it to change the inflow conditions. The static pressure at the upstream boundary must be obtained either by extrapolation from the interior flow field or by use of characteristics type relationships applied to upstream travelling pressure waves.

The periodicity condition on the "free" boundaries upstream and downstream of the blades is easily satisfied by considering points outside the calculation domain to have the same flow properties as points one pitch distant within the domain and then equating all flow properties at corresponding points on the boundaries.

The most difficult boundary condition to satisfy accurately is that on the solid surfaces of the blades. A condition of no flow through the surface is always easily applied but the extra conditions needed depend on the type of grid used. If the cell boundaries lie on the blade surfaces and the nodes lie off them then the only property needed on the surface is the pressure. This may be found either by extrapolation from the interior flow field or by applying the momentum equation perpendicular to the boundary in the form $\partial P/\partial n = \rho V^2/r_c$. If the nodes lie on the boundary then both pressures and velocities are needed on the surface. These must effectively be found by using one-sided differencing at the boundary.

The author's latest scheme (Denton (7)) adopts a novel approach in that no surface boundary conditions (apart from zero flow) seem at first sight to be needed. The boundary points are updated exactly as interior points but the solution then obtained depends on the initial guess. This dependence is gradually removed by a slight smoothing of the variables until finally a "smooth" solution is obtained which is independent of the initial guess but makes no special assumptions about the derivatives at the surface.

A Kutta condition at the trailing edge of the blades is strictly needed to determine the exit flow angle. In practice it seems to be universally found that no explicit condition is needed to obtain good predictions of exit angle. This is thought to be because the numerical viscosity inherent in all methods, together with the periodicity condition applied immediately behind the trailing edge, is sufficient to ensure that the flow leaves the trailing edge smoothly.

6. SOURCES OF NUMERICAL ERROR

As already mentioned solutions of the Euler equations are much more demanding in terms of the accuracy of the finite differencing procedures than are methods which impose entropy conservation. This is because numerical errors at a point in the flow will produce a source of entropy, exactly as viscous effects do in a real flow. This entropy is convected with the flow and so will influence the whole flow field downstream from its source. As with real viscous effects the consequences are more serious for a diffusing flow than for an accelerating one. In contrast, in methods such as streamline curvature, where entropy and enthalpy conservation are imposed, numerical errors (e.g. in estimating the curvature) are localised and do not convect. In cases where the stagnation enthalpy is constant these entropy changes manifest themselves as changes in the stagnation pressure which should be constant in an inviscid flow. As an illustration of this loss Fig. 4 shows the mass averaged stagnation pressure variations computed through a turbine cascade over a range of exit Mach numbers. A loss of about ½% occurs at the leading edge and a further loss, which is Mach number dependent, occurs near the trailing edge. Most of the trailing edge loss is genuine shock loss when the exit flow is supersonic.

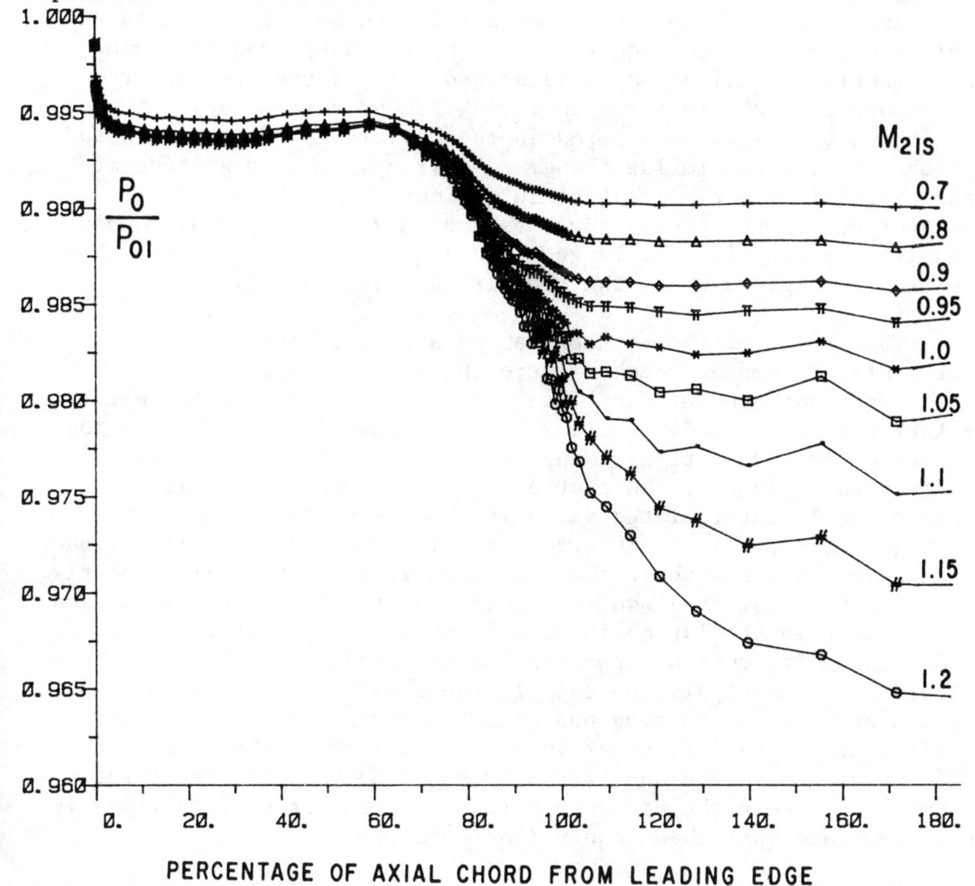

Figure 4 Loss of Stagnation Pressure through a Turbine Cascade

The exact original of this loss has been the source of much debate. It seems clear, however, that it occurs because the finite difference equations which are solved are only approximations to the true governing equations. The accuracy of the approximation should increase and hence the magnitude of the loss should decrease with decreasing good spacing.

In the finite difference form of solution the loss may be seen to arise as a result of the approximations to the differentials. For example in a central difference scheme we have:

$$\left.\frac{\partial Q}{\partial x}\right)_I = \frac{Q_{i+1} - Q_{i-1}}{2\Delta x} - \frac{1}{6}\frac{\partial^3 Q}{\partial x^3}\Delta x^2$$

The last term is an error whose magnitude depends on Δx^2. When such approximations to the differentials are substituted into the solution procedure for equations 4 - 6 it can be seen that the differential equations actually solved are not the true Euler equations but contain terms involving second derivatives which are analogous to the viscous terms in the Navier-Stokes equations and which result in a similar change in entropy. The magnitude of such terms is clearly greatest in regions where the higher derivatives of the flow properties are large and especially when the grid spacing is also large. Hence close grid spacings should be used in regions where the flow is changing rapidly.

The origin of the entropy can be seen even more clearly in the finite volume approach. Here the finite difference approximations are for the fluxes through the faces of the elements rather than for the derivatives of flow properites. Usually the values of the properties on the face are obtained by interpolation between the values at the centre of the elements and it is assumed that a single interpolated value is representative of the whole face. Usually linear interpolation is used and this will produce errors in all cases where the flow properties do not vary linearly. Hence at first sight it would appear that the finite volume method with linear interpolation is only first order accurate. However, it is the difference between the fluxes through two faces of the elements which actually appears in the equations and this will be second order accurate even when the fluxes are only first order. Similar approximations occur in methods such as the author's or Ni's where the flow properties are stored at the corners of the elements. Here the assumption that the properties vary linearly over the face introduces equivalent errors.

An example of the origin of entropy in the finite volume method may be easily obtained by considering the 1D flow in a passage as illustrated below

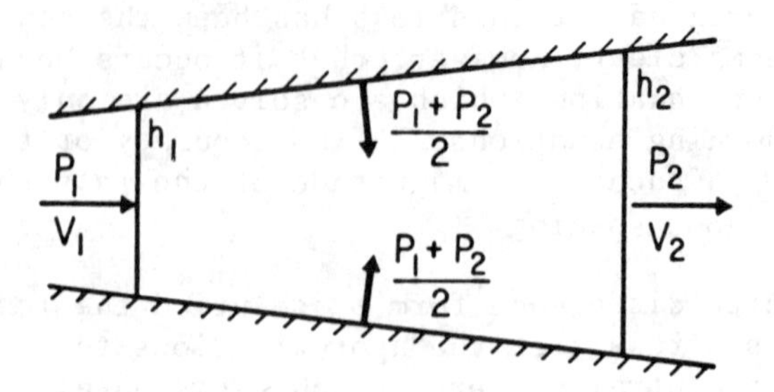

The upper and lower boundaries are solid and we assume that the average pressure acting on them is $\frac{1}{2}(P_1 + P_2)$ and that P_1 and P_2 are uniform over the inflow and outflow boundaries. The momentum equation applied to the control volume gives

$$\dot{m}(V_2 - V_1) = (P_1 - P_2)\ \tfrac{1}{2}\ (h_1 + h_2)$$

If P_1 and P_2 are specified together with the inlet velocity and temperature this equation may be solved iteratively together with the continuity equation to obtain V_2. Because we have assumed a linear pressure variation on the solid walls this solution is not exact but will contain errors due to the term

$\frac{d^2P}{dx^2}$ which has been neglected. From the solution, assuming constant stagnation temperature, we can calculate the change in stagnation pressure. This is plotted in Fig. 5 for a range of inlet and outlet Mach numbers. It can be seen that changes of stagnation pressure of the order of $\frac{1}{2}$% are easily produced when the Mach number changes by more than about 50% between adjacent grid points. This applies even at low Mach numbers and although it is easy to keep the change in Mach number acceptably small over most of the blade surface changes of 100% between adjacent points often occur around the leading edge. It must be remembered that the changes are cumulative along a stream tube so that changes of a few percent can easily be built up.

In two dimensions further errors will also occur due to assuming that the pressure and flow vary linearly over the inflow and outflow boundaries of an element and these are likely to be more serious since the associated pressure forces act nearly in the direction of the flow.

Entropy production must of course occur at shock waves. Since in practice these are calculated as regions of rapid change spread over several grid points rather than as true discontinuities, the mechanism of production must be the same numerical viscosity as produced the unwanted entropy changes elsewhere. Hence some numerical viscosity is essential to enable a method to capture shock waves. If the natural levels are not sufficient then "wiggles" will occur upstream and downstream of the shock and may influence a large part of the flow. In such cases extra numerical viscosity (or smoothing) may be introduced. This should be formulated so that it acts mainly on the high order derivatives occurring in shocks and has little influence elsewhere.

Most schemes smear shocks over 3 or 4 grid points and within the shock the solution is meaningless. However, if momentum, mass flow and energy are conserved across a shock, as they must be in a finite volume method, then the correct overall shock jump

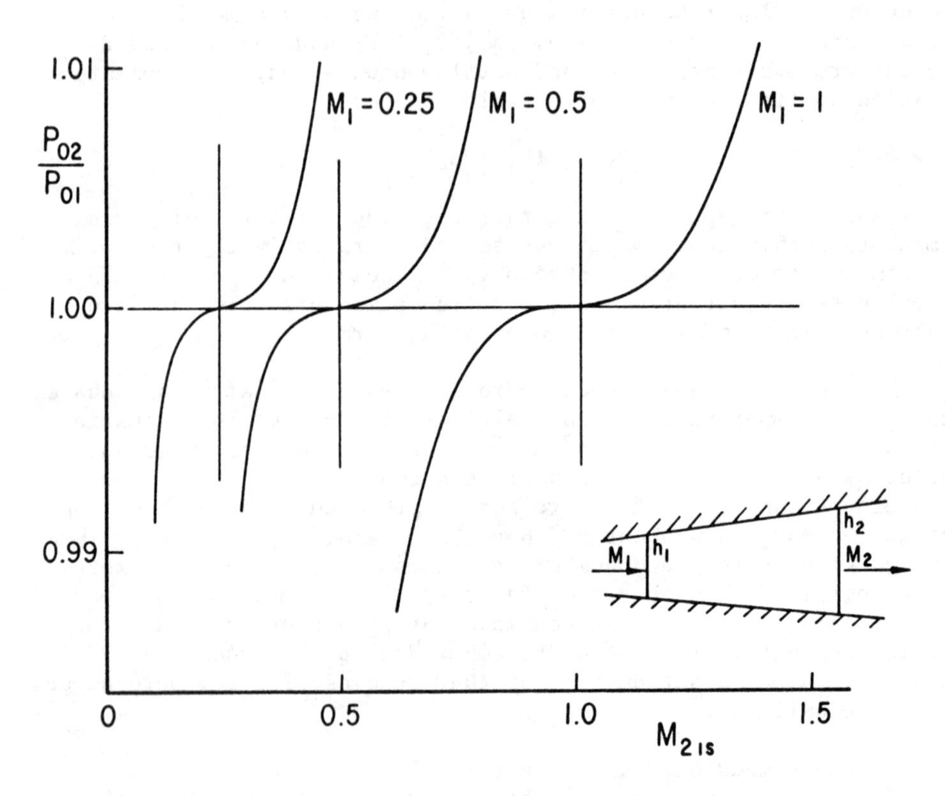

Figure 5 Loss of Stagnation Pressure in ID Flow

conditions must be obtained. When a smeared shock intersects a solid boundary, however, the incorrect pressure within the shock will produce incorrect forces on the boundary and hence an incorrect momentum change. This will in turn influence the downstream flow and as a result accurate predictions of flows with reflected shock waves are still difficult.

In practice it is often found the numerical errors are largest around the leading edge of a cascade because it is here that the higher derivatives of flow properties are largest. The situation is worst for compressor cascades with high leading edge loading (e.g. supercritical blades) where the high derivatives combine with high velocities. In turbine cascades the acceleration on the blade surfaces greatly decreases the effect of any loss at the leading edge and very good predictions can be obtained for subsonic cascades. It is usually found that normal shock waves can be well predicted even on a skewed grid. However, oblique shock waves tend to be more smeared, especially on a skewed grid and as a result their interaction with a solid boundary is often poorly predicted. An important area where this problem occurs is in the trailing edge shock system from a supersonic turbine blade.

7. SOLUTION TECHNIQUES

Many techniques have been developed in order to enhance the speed and/or accuracy of Euler solvers. Some of these are applicable only to one particular program but some are more general, only the latter will be discussed in this section.

One technique which is now used in almost all finite volume methods is the use of spatially varied time steps. Inspection of Eqns. 1 - 3 shows that the requirement for a steady state solution is that the LHS is zero, i.e. that the sum of the fluxes is zero for every element. If this is achieved the correct steady state solution will be reached irrespective of the length of the time step. Hence the maximum stable step can be taken for every element, small steps for small elements and large steps for large elements. This means that there will be little penalty in refining the grid in regions of steep gradients and stretching it towards the inlet and outlet boundaries compared to a solution with a uniform grid with the same number of points. The local time step may be easily varied by a factor of the order 10-100 in this way and considerable ($\sim$ 50%) saving in CPU time obtained. The transient part of the solution loses all physical meaning if this is done but this is of no importance unless unsteady flow is being studied.

The most powerful tool for increasing the rate of convergence, however, is the use of the multigrid technique. This is not yet widely used for the Euler equations and the detailed approach depends to some extent on the numerical scheme. The basic principle is, however, the same in all cases ,that is to permit information to travel by more than a single grid spacing per iteration whilst maintaining much of the simplicity of an explicit scheme. Multigrid was first used in turbomachinery calculations by Ni (6) who was able to use his distribution formula to propagate changes in the variables from one element to a block of 4 elements. A study of Ni's scheme and of other schemes is given by Johnson (14). Several levels of grid can be used and speed increases of the order of 5 can thereby be obtained.

The author has developed a different multigrid method for use with his opposed difference scheme. The method is much simpler to apply than Ni's and larger blocks of elements can be used (e.g. 3 x 3 or 4 x 4). Fig. 6 shows the individual finite volume elements and the blocks of elements which are used. The conservation equations applied to the block will give a rate of

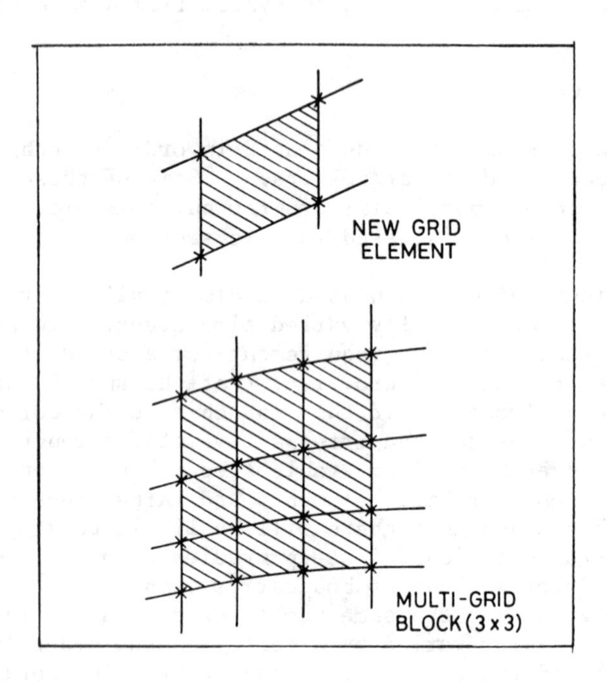

Figure 6 Multigrid Block

change for the block which can be multiplied by a larger stable time step than can be used for the elements. Every node within the block is then updated by a change from its own element and by a larger change from the block. Effectively a course grid solution and a fine grid solution are proceeding in parallel with the rate of convergence being determined largely by the course grid and the accuracy being determined by the fine grid. Several levels of grid can be used to obtain large increases in rate of convergence with very little extra computer work per iteration or extra computer storage. Fig. 7 shows the effect of a single level of multigrid on the mass balance of a turbine cascade.

Another technique which can always be used to increase stability is negative feedback. This effectively varies the local time step in proportion to the local rate of change of the variables. It therefore prevents the growth of a local instability which might otherwise cause failure of the whole calculation. Such local instabilities occur at the start of a calculation due to the very strong transients and would otherwise require the time step to be lowered. If the LHS of equations 1-3 is evaluated before being multiplied by the time step, then its magnitude can be assessed and if it is too large the time step by which it is to be multiplied can be automatically reduced. This has no effect on the steady

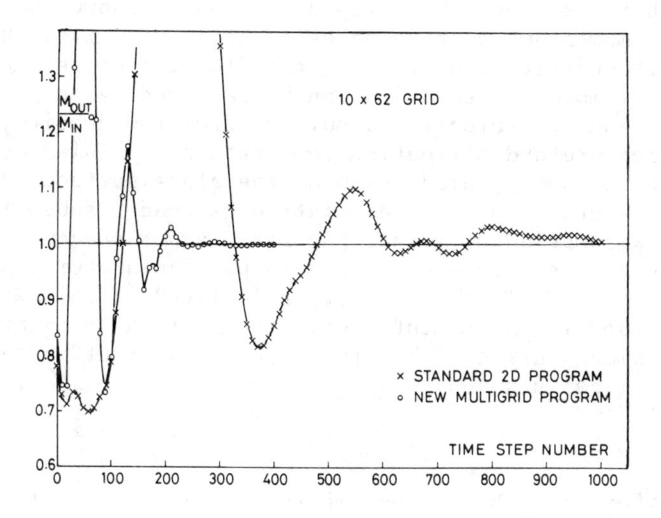

Figure 7 Effect of Multigrid on Convergence

state solution for which all the changes tend to zero. The feedback may have little effect, or even delay, the convergence of a well behaved problem but it contributes enormously to the "robustness" which is an extremely important property of any method.

As has been mentioned previously the leading edge of a blade row causes special difficulties because the gradients are so large. These gradients may be reduced by fitting a cusp to the leading edge so that the flow does not "see" the usual stagnation point. This is only permissible of course if the details of the leading edge flow are not of immediate interest. Such a treatment will reduce entropy production at the leading edge and hence improve the quality of the downstream solution. However the author's experience is that it is not possible to develop a general method of fitting the cusp which will give good results in all cases and unless computer time is severely limited it is better to do without a cusp and to use a very fine mesh around the leading edge.

Cusps may also be used at the trailing edge of a cascade to simulate the wake blockage which occurs in a real flow. Such a trailing edge cusp should not carry any load but should reduce the available flow area as does the real wake. Use of a simple cusp can be dangerous for turbine blades with supersonic trailing edge flow because the blockage will reduce the amount of trailing edge expansion which is an important feature of such flows.

Singh(15) has described a special trailing edge model for use when calculating such supersonic trailing edge flows.

On the blade surfaces the flow direction is known and hence only a single momentum equation needs to be solved, i.e. that along the surface streamlines. If the flow can be assumed to be isentropic this momentum equation can be replaced by the isentropic relationships and the velocity calculated from the density, stagnation pressure and stagnation temperature. This ensures that there is no entropy production on the blade surfaces and so it can be a very useful tool for calculations around a subsonic leading edge. The assumption of isentropic flow cannot of course be used when shock waves are present and in particular for supersonic compressor blades. If the velocity calculated in this way is subsequently used in the momentum equations for the elements bounding the blade surfaces then there is a danger of instability with some numerical schemes.

8. APPLICATIONS

Most authors of publications in this field compare their methods with several test cases which might be either exact solutions or experimental results from cascade tests. In many ways exact solutions are preferable because discrepancies from test data can always be excused by the neglect of viscous effects. In the author's opinion many more test cases than the 2 or 3 usually presented are necessary to prove the generality of a method. Since this lecture is not concerned with verifying any one method the test cases presented will be used to illustrate very difficult flows where problems are likely to occur.

A very simple but challenging test case is the subsonic flow about a circular cylinder or cascade of cylinders. The blunt leading edge produces a very highly distorted grid where numerical errors are likely to be large. These will then be shown up by a lack of symmetry of the computed flow since entropy produced at the leading edge will tend to produce a recirculating flow near the rear of the cylinder. A good method should preserve the symmetry and should compute a near stagnation point at the downstream end of the cylinder. This is only true for subsonic flow since if shock waves occur they will destroy the symmetry and recirculation might occur as a result of the shock loss. Fig. 8 illustrates the results of a calculation on a cascade of cylinders with a peak Mach number of about 0.8.

A very different type of test case can be obtained by considering completely supersonic flow over a cascade of wedges. The familiar shock-deflection relationships enable exact solutions to be obtained comparatively easily in this case. Fig. 9

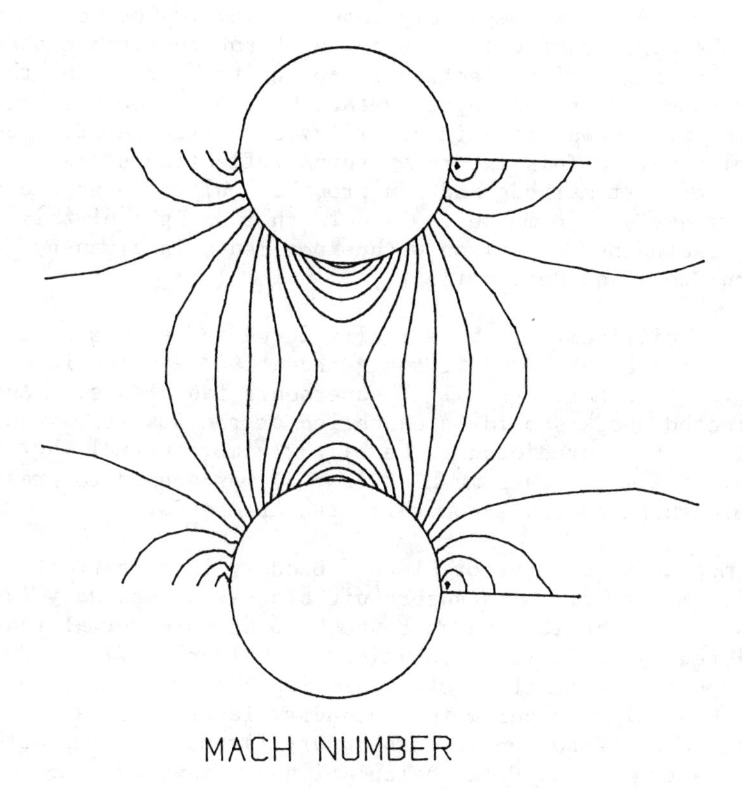

Figure 8 Computer Flow around a Cascade of Cylinders

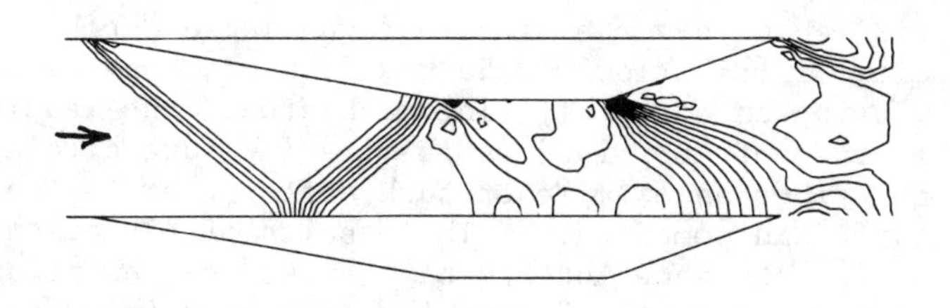

Static pressure contours

Figure 9 Supersonic Flow over an Unstaggered Wedge

illustrates the flow over an unstaggered wedge with an inlet Mach number of 2. The leading edge shock is refelected by the lower surface and should be completely cancelled by the corner when it reaches the upper surface. Because of the inevitable shock smearing the angle of reflection is not quite correct and the computer shock meets the upper surface before the corner. As a result the flow downstream is significantly different from the exact solution. This incorrect shock reflection seems to be a feature of most methods and can probably only be overcome by the use of extremely fine meshes. A further example of this type of flow, including comparison with experiment, is given by Bryanston-Cross and Denton (16).

A more difficult and more realistic wedge case is shown in Fig. 10. Here the wedge is staggered at 60° and the inlet Mach number of 1.6 is more typical of supersonic fan blades. Again the reflected shock should be cancelled at the downstream corner but in fact it is predicted to be slightly more normal that the exact solution. A much finer mesh (29 x 125) was needed to preserve reasonable shock sharpness on this staggered grid.

In most real compressor blades boundary layer effects are extremely important. In supersonic blades the boundary layer blockage causes the leading edge shock to be more normal than would be predicted by an inviscid solution. Calvert (17) has developed a very sophisticated method of predicting the resultant flow by combining an Euler solver with a boundary layer calculation. The method copes with separated boundary layers by using both the boundary layer and the Euler calculations in inverse mode. An example of the resulting computed shock structure is shown in Fig. 11 which shows the leading edge shock to be captured very cleanly.

In turbine blade calculations problems are most likely to be encountered at the leading edge and at the trailing edge. A leading edge at incidence can produce extremely rapid acceleration and deceleration which has the effect of causing a "spike" in the surface velocity distribution. Such a flow can only be computed accurately by using an extremely fine mesh around the leading edge with the order of 10 points on the leading edge circle. Hodson (18) describes an example of such a flow where a "spike" was both measured and computed. The results of his calculation, reproduced in Fig. 12, show that such a "spike" can be predicted by an Euler solver although a stream function or a velocity potential method would probably have been better in this case.

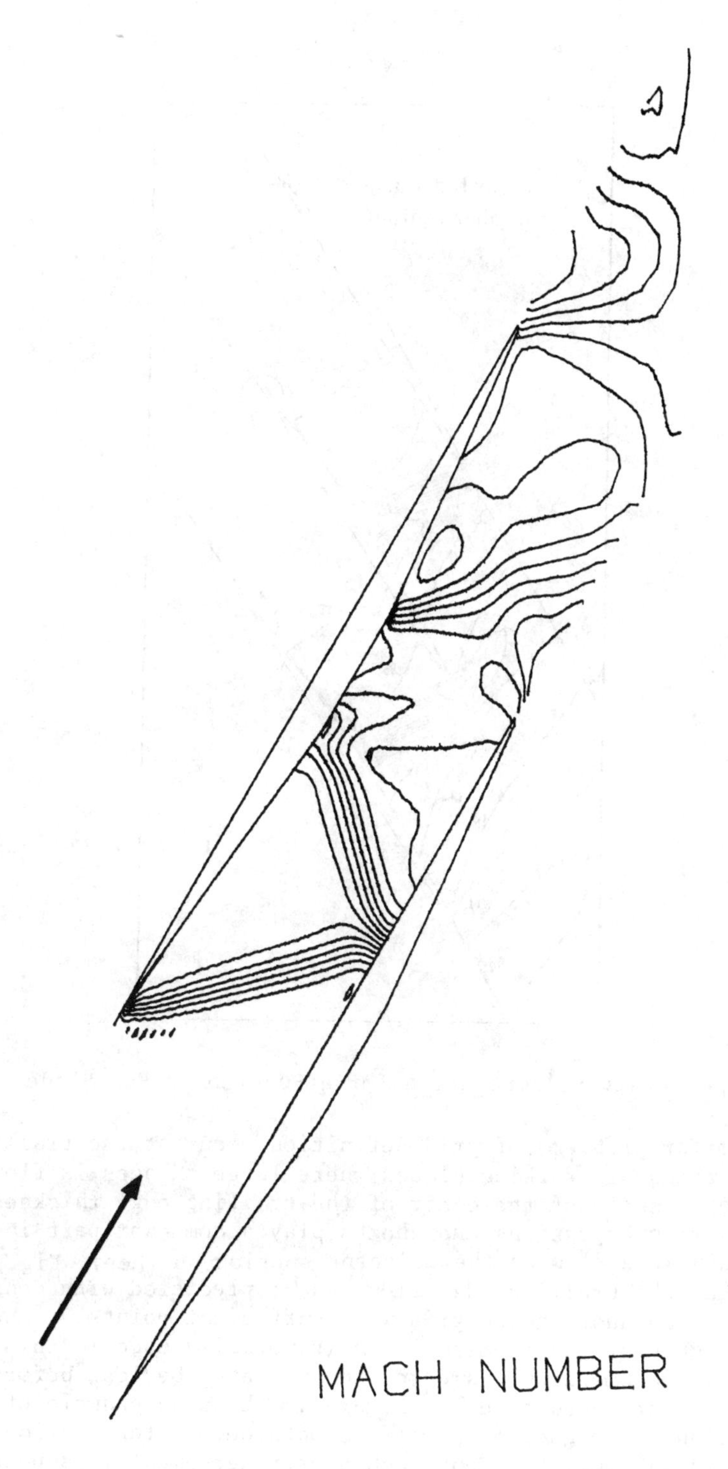

Figure 10. Computed Flow through a Cascade of Staggered Wedges

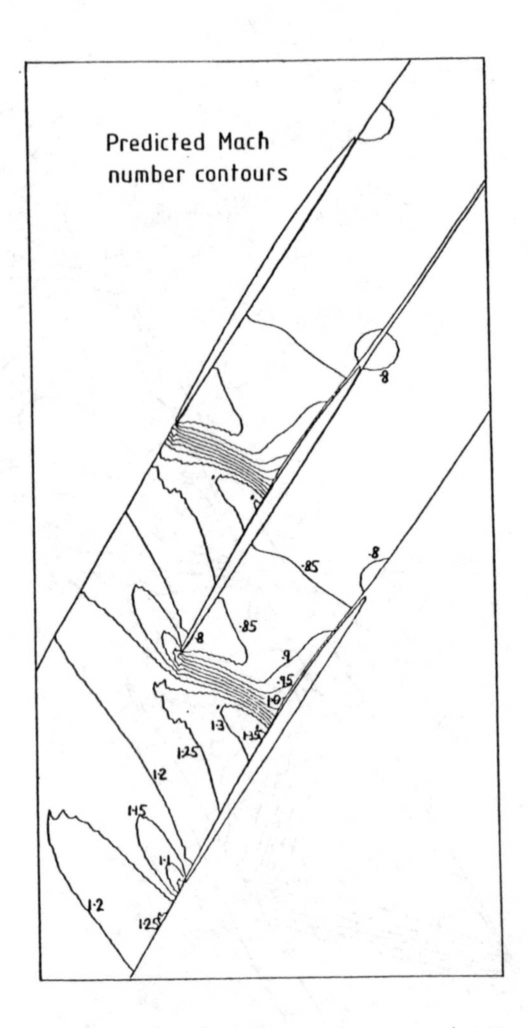

Figure 11 Calvert's Solution for a Supersonic Fan Blade

Similar problems of grid definition occur at the trailing edge of transonic turbine blades where large changes in flow occur in a length scale of the order of the trailing edge thickness. The resulting expansions and shocks play a dominant part in determining the flow on the adjacent suction surface. Fig.13 shows that the trailing edge flow can be predicted with considerable success on an unstaggered grid with sufficient points. However, on the highly staggered mesh which is normally used in this region the shocks become highly smeared and may even be lost before they reach the suction surface. Figure 14 shows an example of such a situation where good agreement is obtained on the suction surface in subsonic flow but much poorer agreement in supersonic flow. Holmes and Tong (12) show that good shock capturing can

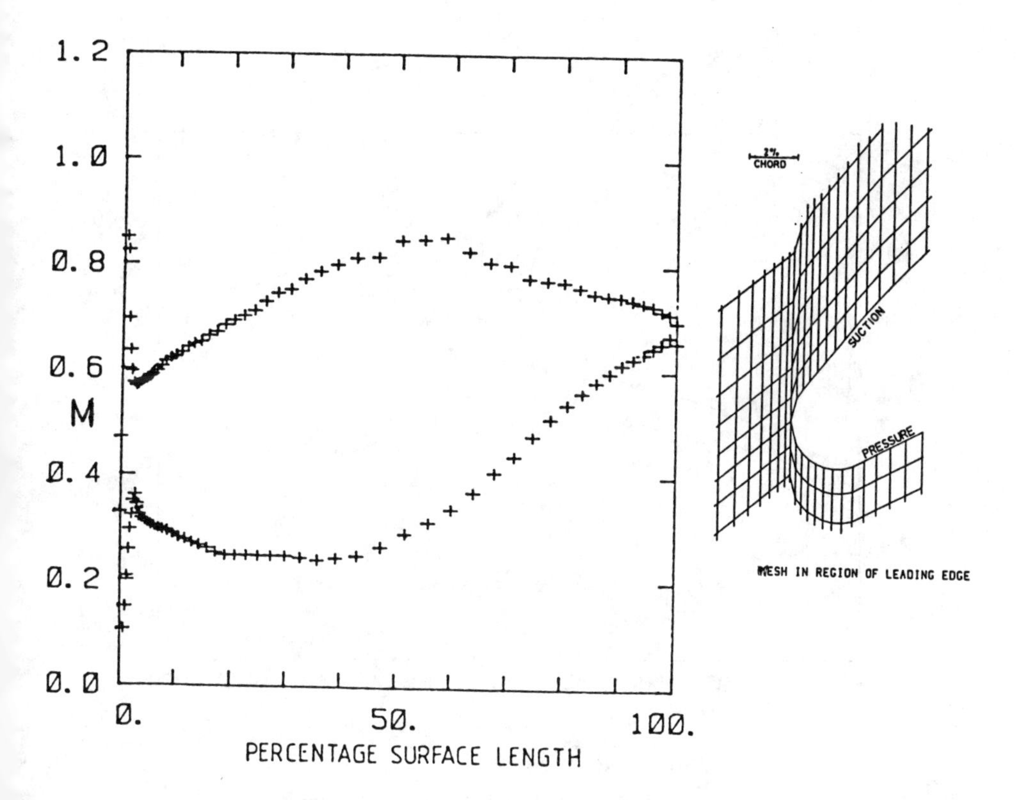

Figure 12 Hodson's Calculation on a Turbine Blade at Incidence

can be obtained at a trailing edge by use of a quasi-orthongal grid with a large number of points.

This section has attempted to emphasize the limitation of Euler solutions when applied to difficult test cases. It must be emphasised, however, that in most practical cases good solutions can be obtained. Were this not the case the methods would not be in use for everyday design applications.

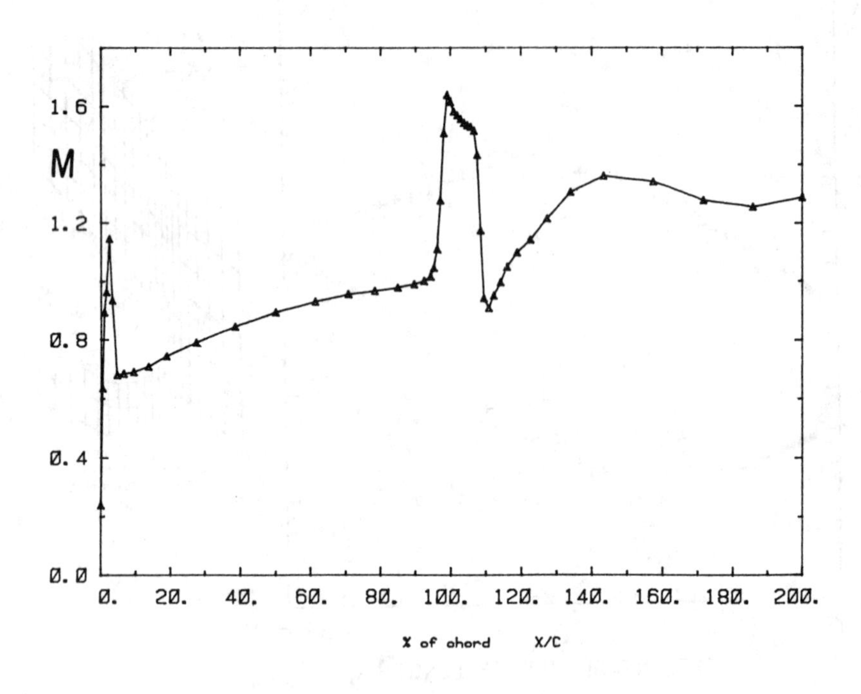

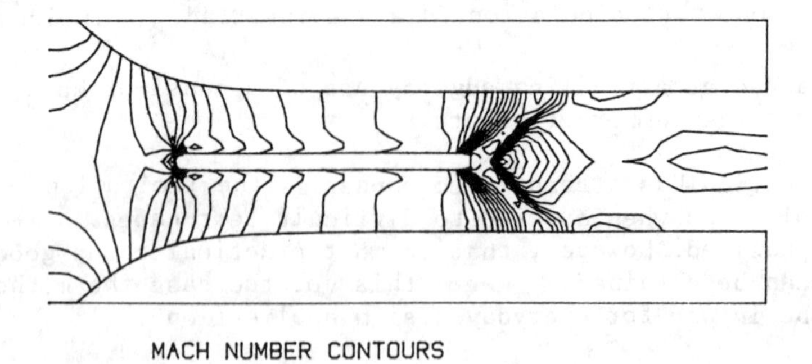

Figure 13. Computed Flow over a Plate in a Wind Tunnel

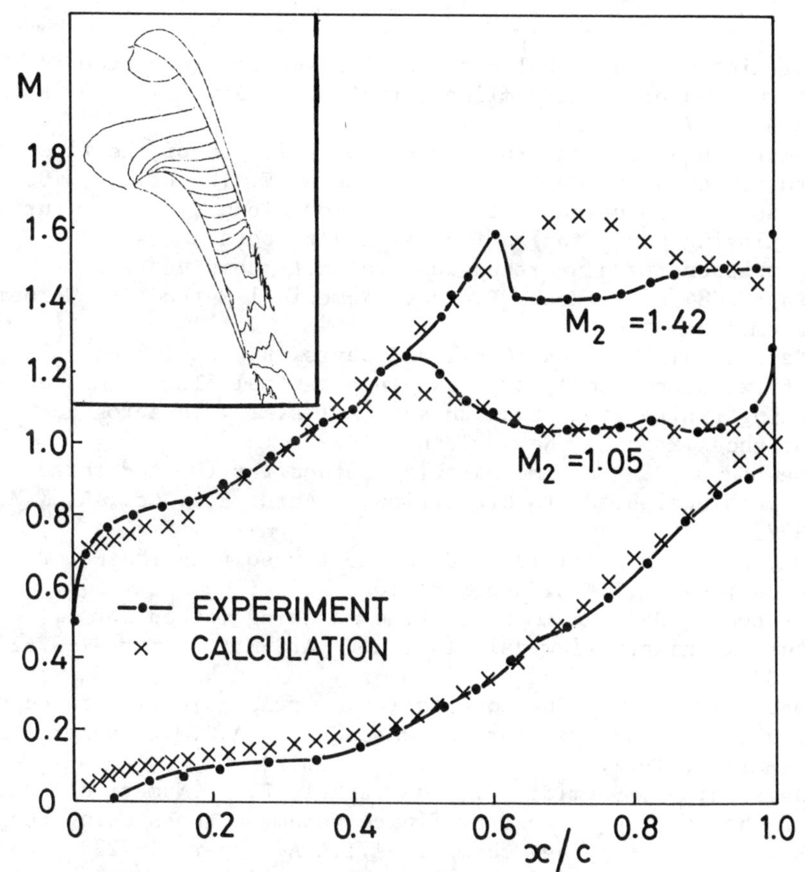

Figure 14 Transonic Turbine Blade with no Attempt to capture the Trailing Edge Shock System.

10. CONCLUSIONS

Euler solvers have been shown to be a very powerful tool for turbomachine blade design. They have been developed to a high level of sophistication over the last decade. However, some problems remain and this lecture has tried to emphasise these rather than the many successes of the methods. The methods developed for two dimensional inviscid flow are especially attractive because they can in principle be extended to deal with both viscous and three dimensional flows. Their extension to the latter is the subject of the next lecture.

REFERENCES

1. Wilkinson, D.H. Calculation of blade to blade flow in a turbomachine by streamline curvature. Aero Res. Co. R. & M. 3704, 1972.
2. Smith, D.J.L., and Frost, D.H. Calculation of the flow past turbomachine blades. Proc. I. Mech. E. 184 Pt. 3G, 1969-70.
3. Zhao, X. Solution of transonic flow along S_1 stream surfaces employing non-orthogonal curvilinear coordinates and corresponding non-orthongonal velocity components. Paper C84/84 I. Mech. E. Conf. Numerical Methods in Turbomachiner Birmingham 1984.
4. Caspar, J.A., Hobbs, D.E., and Davis, R.L. The calculation of two dimensional compressible potential flow in cascades using finite area techniques. A.I.A.A. 17th Aerospace Sciences Meeting, New Orleans, 1979.
5. Denton, J.D. A time marching method for two and three dimensional blade to blade flow. Aero. Res. Co., R. & M., 3775, 1974.
6. Ni, R.H. A multiple grid scheme for solving the Euler equations. A.I.A.A. paper 81-1025, 1981.
7. Denton, J.D. An improved time marching method for turbomachinery flow calculation. A.S.M.E. paper 82-GT-239, 1982.
8. McDonald, P.W. The computation of transonic flow through two dimensional gas turbine cascades. A.S.M.E. paper 71-GT-89, 1971.
9. Jameson, A., Schmidt, W., and Turkel, E. Numerical solutions of the Euler equations by finite volume methods using Runge-Kutta time stepping schemes. A.I.A.A. paper 81-1239, 1981.
10. Arts, T. Cascade flow calculations using a finite volume method. V.K.I. L.S. 1982-05.
11. Delaney, R.A. Time marching analysis of steady transonic flow in turbomachinery using the hopscotch method. A.S.M.E. paper 82-GT-152, 1982.
12. Holmes,D.G., and Tong. S.S. A three dimensional Euler solver for turbomachinery blade rows. A.S.M.E. paper 84-GT-79, 1984.
13. Thompkins, W.T., and Tong, S.S. Inverse or design calculations for non-potential flow in turbomachinery blade passages. A.S.M.E. paper 81-GT-78, 1978.
14. Johnson, G.M. Multiple grid accelerations of Lax-Wendroff algorithms. NASA TM-82843, 1982.
15. Singh, U.K. Computation of transonic flow in turbine cascades with viscous effects. A.S.M.E. paper 84-GT-18, 1984.
16. Bryanston-Cross, P.J., and Denton, J.D. Comparison of interferometric meausrements and computed flow around a wedge profile in the transonic region. A.S.M.E. paper 82-GT-258 1982.

17. Calvert, W.J. Application of an inviscid-viscous interaction method to transonic compressor cascades. AGARD CP351, 1983.

18. Hodson, H.P. Boundary layer separation and transition near the leading edge of a high speed turbine blade. A.S.M.E. paper 84-GT-179, 1984.

NOTATION

dA	Surface area vector
c	Speed of sound
C_p	Specific heat capacity at constant pressure
C_v	Specific heat capacity at constant volume
E	Specific internal energy
H	Stagnation enthalpy
$\dot{m}$	Mass flow rate
M	Mach number
P	Static pressure
P_o	Stagnation pressure
R	Gas constant
r_c	Radius of curvature
V	Velocity
V_o	Volume of element
x	Cartesian coordinate
y	Cartesian coordinate

Subscripts

o	Stagnation conditions
is	Isentropic process
1	Inlet or upstream
2	Outlet or downstream

SOLUTIONS OF THE EULER EQUATIONS FOR TURBOMACHINERY FLOWS
PART 2. THREE DIMENSIONAL FLOWS

J.D. Denton

Whittle Laboratory, Cambridge, U.K.

1. INTRODUCTION

The choice of methods available for flow calculation becomes much more limited when the step from two dimensions to three dimensions is taken. The equations themselves are no more complex but many of the simplifications available in 2D become difficult to apply and the geometrical problems become more complex in 3D flow.

Stream function methods were originally proposed by Wu as being extendable to fully 3D flow by means of iterating between solutions on two families of stream surfaces. However, this approach involves considerable complexity and although it has been developed (Ref.(21)) it has not yet proved popular. Streamline curvature methods also become very difficult to formulate in 3D largely because of the difficulty of tracing streamlines and of satisfying continuity on a whole plane, instead of on a line as in 2D.

Velocity potential methods are fairly readily extendable to 3D as long as the flow remains irrotational. Unfortunately this assumption becomes much more limiting in 3D than in 2D because vorticity may be present at inlet to a blade row and is almost always shed from the trailing edge. The vortex sheet from the trailing edge can be computed by arranging for it to lie on a boundary of the computational domain so that there is no vorticity within the domain but this becomes difficult to achieve for flows with large vorticity. Three dimensional potential methods have been developed, e.g. Laskaris (6), Hirsch (7) and Soulis (8) but they do not appear to have been widely used for design purposes.

Conversely, methods of solving the Euler equations are readily extendable to 3D flow. All that is required in principle is the solution of a third momentum equation, although in practice the energy equation may need to be reintroduced and the geometrical complexity is greatly increased. The first fully 3D solutions were reported about a decade ago, Denton (1975), Thompkins (1976), Bosman and Highton (1979). At first these were too expensive in computer time, too limited in geometry and too temperamental for use in routine design work. However reductions in computer costs and improvements in numerical methods over the last decade have made fully 3D methods useable as design tools and they are now used by most large gas and steam turbine manufacturers. An important factor contributing to the popularity of these methods is their simplicity and ease of physical understanding together with the fact that they are applicable over the whole Mach number range likely to be encountered in turbomachines.

Recent 3D methods have been described by Enselme et al (9), Thompkins (10), Arts (22), Holmes and Tong (23) and Koya and Kotake (24). All of these methods use different numerical schemes which only their authors are competent to describe in detail. Most of them show examples of application to only one type of problem (Refs. 9 and 10 to transonic fans and Refs. 22 and 24 to subsonic turbines). Only Holmes and Tong describe their scheme as being applicable to a range of turbomachine types but they give few examples to demonstrate this.

The author's method started its development in 1973, being at first limited to very simple geometries (Ref. (3)). Its capabilities were gradually extended (Denton (1976), Denton and Singh (1979)), with little change in the numerical method, until it could compute the flow through most types of fixed or rotating blade row. During 1981-82 the scheme was changed to use the new grid and algorithm previously developed in 2D, (Denton (1982)). This change proved even more beneficial in 3D than in 2D and enabled rapid development to include even more complex geometries and boundary conditions. The resulting suite of programs can compute the flow through virtually all types of turbomachine in common use and the solutions obtained range from incompressible flow in centrifugal pumps to flow in low pressure steam turbines with Mach numbers approaching 2. The method has been widely used by other workers and a great deal of user experience is not available, e.g. Refs. 14,15,17,20, 25,26,28,29. This lecture will describe the method in outline, but not in great detail, and will concentrate on giving examples of its use on a wide variety of problems.

2. METHOD

The method is of the finite volume type which has even greater advantages in 3D than in 2D because of the increased complexity of the geometry. The grid is set up and the equations are solved in the physical rather than in the transformed plane because of the greater physical understanding obtained. This ease of understanding is important when dealing with complex geometries. Because all control volumes share common faces, except at the inflow, outflow and solid boundaries, global conservation of all conserved flow properties is automatically obtained no matter how distorted the grid. For turbomachine calculations the natural choice of coordinate system would seem to be a cylindrical r,θ,x system centred on the axis of rotation. However, Holmes and Tong (23) argue that a Cartesian system is still preferable. This simplifies the calculations at the interior mesh points but greatly complicates the application of the boundary conditions. The main complication of using a cylindrical system is that the direction of the radial vector changes around the surfaces of the elements and this must be allowed for when evaluating the radial components of pressure force and momentum flux.

For calculating the flow through rotating blade rows it seems natural to work in a rotating coordinate system and to solve for the relative velocities. There is no special difficulty in doing this although Coriolis and Centrifugal terms need to be introduced into the momentum equation. However, both the author and Holmes and Tong (23) have found it preferable to solve for the absolute flow variables in a rotating coordinate system. The author's reasons for doing this were to avoid Coriolis forces and the use of rothalpy but Holmes also points out the advantages in having a uniform rather than a rotating far field in propeller calculations.

3. EQUATIONS

The equations for conservation of mass, energy and momentum applied to a fixed control volume ΔV for a time interval Δt are simply

$$\sum(\rho V_x \, dA_x + \rho V_\theta \, dA_\theta + \rho \, V_r \, dA_r) = \frac{\Delta V}{\Delta t} \cdot \Delta\rho \tag{1}$$

$$\sum(\rho V_x \, h_o \, dA_x + \rho V_\theta \, h_o \, dA_\theta + \rho V_r \, h_o \, dA_r) = \frac{\Delta V}{\Delta t} \cdot \Delta(\rho E) \tag{2}$$

$$\sum((P + \rho V_x^{\,2}) \, dA_x + \rho V_\theta V_x dA_\theta + \rho V_r \, V_x \, dA_r) = \frac{\Delta V}{\Delta t} \cdot \Delta(\rho V_x) \tag{3}$$

$$\sum(\rho V_x V_r dA_x + \rho V_\theta V_r dA_\theta + (P + \rho V_r^2) dA_r)$$
$$+ \rho\Delta V. \frac{V_\theta^2}{r} + P.\frac{\Delta V}{r} = \frac{\Delta V}{\Delta t} \cdot \Delta(\rho V_r) \quad (4)$$

$$\sum(\rho V_x rV_\theta dA_x + (P + \rho V_\theta^2)r dA_\theta + \rho V_r rV_\theta dA_r)$$
$$= \frac{\Delta V}{\Delta t} \cdot \Delta(\rho rV_\theta) \quad (5)$$

The summations are taken over the faces of the volume and dA_x, dA_r, dA_θ, the projected areas of these faces in the coordinate directions, are taken as positive when directed inwards to the element.

In the above equations V_x, V_θ, V_r are the absolute velocities and x,r,θ are cylindrical coordinates, i.e. the radial and circumferential directions are always defined relative to the point at which the velocity is being evaluated and so vary around the faces of an element. Equation (5) is the moment of momentum equation which is used because it is simpler than the θ momentum equation. Note that in 3D flow it is seldom possible to replace the energy equation (Eqn. 2) by an assumption of constant stagnation enthalpy as is usually done in 2D flow. Hence the present programs always solve the energy equation.

Equations 1 - 5 can all be expressed in the form

$$\sum \text{FLUX} + \text{SOURCE} = \frac{\Delta V}{\Delta t} \cdot \Delta \text{ (property)} \quad (6)$$

where the SOURCE term only occurs in the radial momentum equation and FLUX is the rate of flow of mass, energy, etc., through each face of the element. For use in rotating blade rows the equations could be re-cast in terms of relative velocities, with the inclusion of extra source terms where necessary, and the problem solved in a rotating coordinate system. In practice it has been found simpler to continue to work with absolute velocities and to transform the rate of change calculated from equations 1-5 for a fixed grid to a rotating grid which instantaneously coincides with the fixed grid, using

$$\left.\frac{\partial Y}{\partial t}\right)_R = \left.\frac{\partial Y}{\partial t}\right)_F + \Omega \cdot \frac{\partial Y}{\partial \theta}$$

where Y is ρ, ρE, ρrV_θ or ρV_r depending on which equation is being updated. Integrating this equation over the elemental volume shows that the $\Omega\frac{\partial Y}{\partial \theta}$ term can be expressed simply as an additional flux through the bladewise faces of the elements.

4. GRID

The choice of grid is probably the most critical part of any 3D flow calculation procedure. This is especially true in turbomachinery where extremely complex geometries have to be handled. The use of as simple a grid as possible greatly helps in fitting these complex geometries but a simple grid is unlikely to provide the greatest accuracy possible from a limited number of grid points. In the author's original method the finite volumes consisted of cuboids with a node at the centre. In two dimensions (Ref. (13)) it would found to be simpler and more accurate to use essentially the same cuboids but with a node at each corner. This new type of grid has been extended to 3D where its advantages over the old grid have proved even greater than in 2D. The cuboids are formed by the intersection of 3 families of surfaces as shown in Figures 1 and 2. The streamwise surfaces are are akin to the blade to blade stream surfaces of the Q3D approach. They are surfaces of revolution such that the first surface coincides with the hub of the machine and the last with its shroud or casing. The remainder of the streamwise surfaces are non-uniformly spaced between the hub and the shroud with the actual spacing being chosen by the user.

The bladewise surfaces are such that the first and last of them contain the two blade surfaces forming one blade-blade passage. Upstream and downstream of the blade row these two surfaces are chosen to lie <u>roughly</u> parallel to the relative flow direction and they are one blade pitch apart in the circumferential direction. The remainder of the bladewise surfaces are non-uniformly spaced, as chosen by the user, between the two blade surfaces.

The quasi-orthogonal surfaces are surfaces of revolution lying <u>roughly</u> perpendicular to the meridional flow direction. One of these surfaces contains the leading edge of the blade row and another the trailing edge, the positions of the remainder of the quasi-orthogonal surfaces may be chosen by the user. There are grid nodes, where all flow properties are stored, at each of the 8 corners of the cuboid as shown in Fig. 3. However, except on the boundaries, each corner is shared by 8 cuboids so there is effectively only one grid node per cuboid.

The projected area of every face of the cuboid in the 3 coordinates directions must be evaluated and stored. Since the streamwise and quasi-orthogonal faces lie in the circumferential

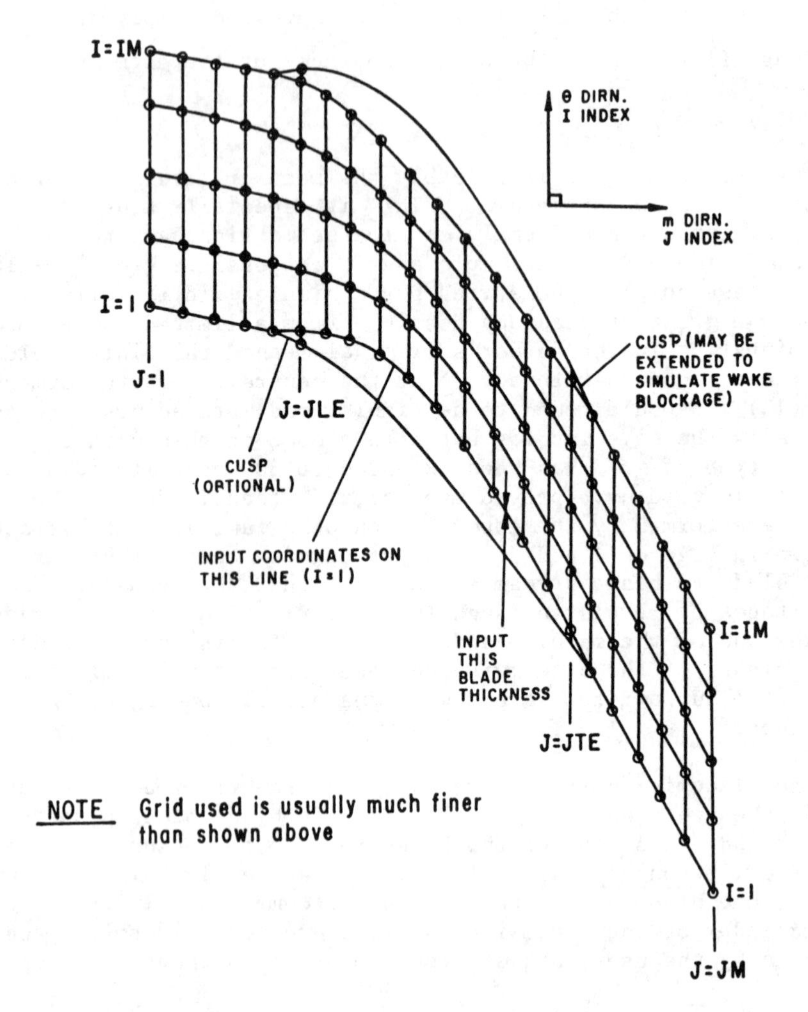

Figure 1 Grid on a Streamwise (1.E m-θ) Surface.

direction this makes a total of 7 areas per element. The areas are projected in coordinate directions defined at the centre of the face concerned.

5. SOLUTION PROCEDURE

The numerical scheme used to solve equations 1 - 5 is that described as scheme 'A' by Denton (1982). The other two schemes (B & C) described in that paper are not stable when the energy equation is included. Hence the scheme used is the exact analog of the opposed difference scheme used in the author's original methods, but is applied to the new grid.

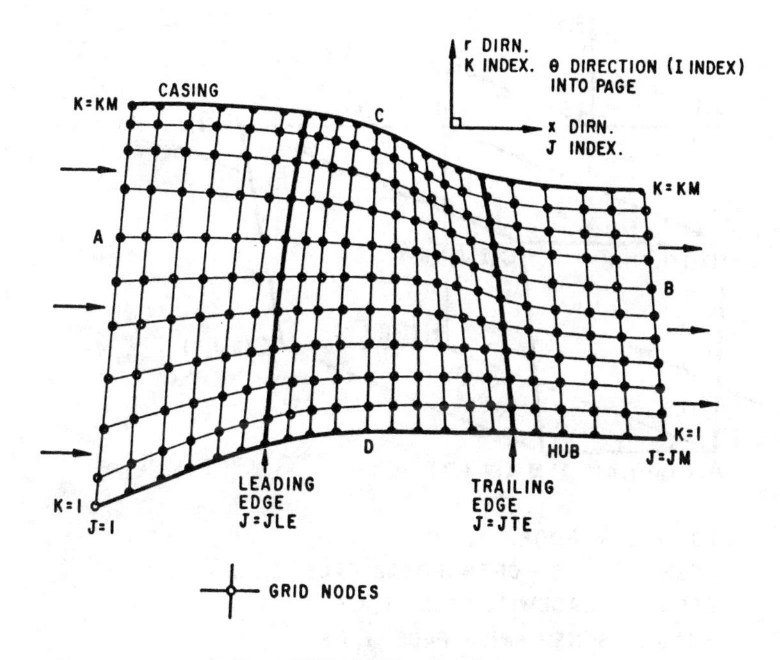

Figure 2 Grid in a Meridional Plane.

In each equation the fluxes though all the faces of the elements are evaluated using averages of the flow properties at the 4 corners of the face concerned. These fluxes are summed to find the change of the conserved property for each element over the time step. One quarter of this change is then added to the values of the property at the 4 downstream corners of the element. Note that the manner of distributing the changes does not effect the final steady state solution, for which the sum of the fluxes must be zero, but it does have a critical influence on the stability of the method.

Stability is enforced by using an effective pressure in the 3 momentum equations rather than the true pressure. This effective pressure is made equal to the current pressure at the next downstream grid point plus a correction i.e.,

$$P_{EFF_J} = P_{J+1} + CF_J$$

The correction CF, is obtained by an interpolation procedure

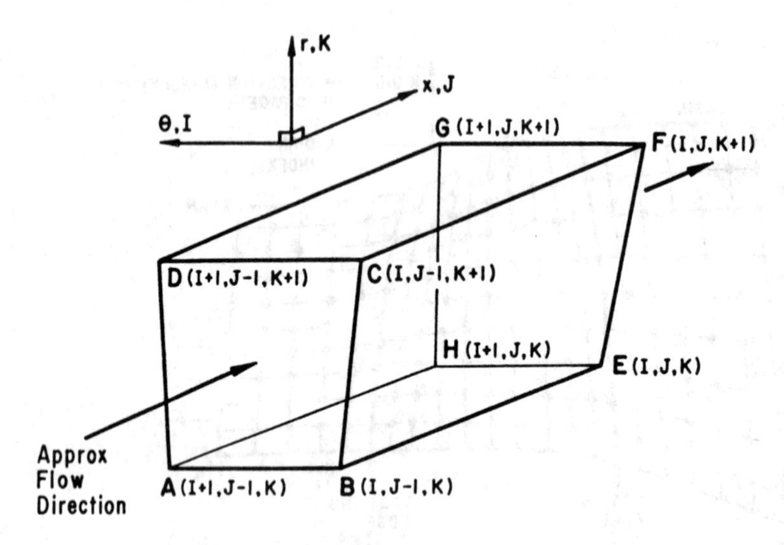

Figure 3 Single Element of Grid.

which must not make use of the true pressure at J. Changes in CF are damped after every time step according to

$$CF_{J,NEW} = (1-RF)* CF_{J,OLD} + RF*CF_{J,INT}$$

where $CF_{J,INT}$ is a new estimated correction factor obtained from the interpolation. RF is a relaxation factor, typically, 0.05. The current scheme uses parabolic interpolation, as illustrated in Fig. 4, and so provides second order accuracy for a continuously changing pressure. Additional artificial viscosity may be provided in shock waves by taking only part of the interpolated correction, the fraction used being a function of the local density gradient so that outside the shock region it is very closely one.

Some small amount of smoothing in the circumferential and quasi-orthogonal directions is necessary to stabilise the scheme. As explained in Ref. (13) this is because there are more nodes than elements and the smoothing is needed to remove this indeterminacy in the problem. The amounts of smoothing needed are very small

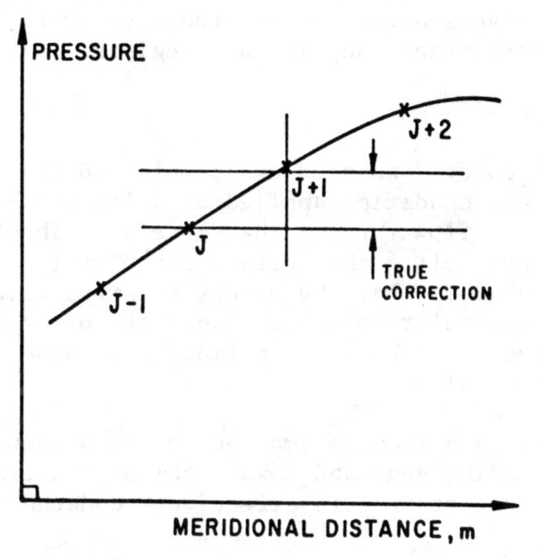

Figure 4 Definition of Correction Factors.

indeed. Typically during every time step all independent variables, such as ρ, are smoothed according to

$$\rho_{NEW} = (1 - SF)*\rho_{OLD} + SF*\rho_{SMOOTHED} \tag{8}$$

where SF is of the order 0.005 and $\rho_{SMOOTHED}$ may be obtained by either linear or non-linear smoothing. The latter is formulated so that it does not change a cubic variation in the variable (e.g. ρ) and so introduces only extremely small amounts of artificial viscosity. Numerical experiments show that the smoothing factor SF may often be increased by about an order of magnitude before it starts to significantly affect the solution.

The scheme as described so far is only stable if the flow velocity is positive in the meridional direction. This is a feature of the opposed difference scheme which was overcome in two dimensions by means of a modified scheme (scheme B of Ref. (13)) applied everywhere in the flow field. The 3D programs overcome the problem by switching to the use of upwinded values of convected velocity (but not mass flux, pressure or enthalpy) wherever the meridional velocity is detected to be negative. This permits large regions of reversed flow to be predicted at the expense of a loss of second order accuracy in such regions.

6. BOUNDARY CONDITIONS

The boundary conditions used are simple and intuitively obvious. The only condition applied on solid boundaries (blade and end-walls) is no flow through the surface. The flow velocity is not resolved parallel to the surface since it is simpler not to do so and not obvious that the errors incurred (i.e. incorrect acceleration perpendicular to the surface) are worse than those caused by resolving (i.e. body force imposed perpendicular to the surface), see Ref. (29).

A periodicity condition is imposed on the bladewise surfaces upstream of the leading edge and downstream of the trailing edge and points on these surfaces are effectively updated exactly as interior nodes.

At the downstream flow boundary only the static pressure needs to be prescribed. This may either be input and held constant at all spanwise positions or only the value on the hub may be fixed and the pressure at other points built up from an assumption of simple radial equilibrium as the calculation proceeds. The latter assumption may not be realistic if the downstream boundary coincides with a region of high meridional curvature.

At the upstream (inflow) boundary the stagnation pressure and stagnation temperature must be specified at each grid node, the values may vary in both the spanwise and circumferential directions. Several options are available for specifying the inlet flow direction. Either the relative or the absolute flow directions in both the streamwise and meridional planes may be specified, or the angle in the streamwise plane may be replaced by a specified absolute whirl velocity. The latter condition can be used when the relative inlet velocity is supersonic and the flow will then automatically satisfy the unique incidence condition.

The final condition needed at the upstream boundary is a means of extrapolating the static pressure from the interior flow field. This may be done by applying a condition of either $dp/ds = 0$ or or $d^2p/ds^2 = 0$ along the quasi-streamlines. The pressure so obtained is used together with the stagnation pressure and temperature and the usual isentropic relationships to obtain the absolute velocity. The flow directions or specified whirl velocity then enable this to be converted into the 3 velocity components.

7. STABILITY AND CONVERGENCE

As with all explicit schemes the maximum stable time step is limited by the CFL condition to

$$\Delta t < \frac{\Delta \ell}{V+c} \tag{9}$$

In complex geometries with highly distorted elements it is difficult to know which values of $\Delta \ell$ and V limit stability. In practice $\Delta \ell$ is taken as min $(\Delta s, \Delta(r\theta), \Delta q)$ for each element and a factor of safety is provided by taking c as the inlet stagnation speed of sound and V as the estimated maximum velocity in the flow field.

Either uniform time steps (which are the same for all elements and equal to the smallest timestep needed by any one element), or non-uniform time steps, may be used. The latter option takes different timesteps for each element, based on the local stability, and usually gives much more rapid convergence. It also enables the timestep to be varied in response to the rate of change of the variable in the element concerned. If the rate of change is found to be large the timestep is automatically reduced so that only a small increment is added to the variable. This 'negative feedback' can be made to exert a powerful stabilising influence and enables acceptable solutions to be obtained over most of the flowfield even when local regions are unstable. It also prevents the timestep having to be reduced in order to overcome initial transients.

The rate of convergence may also be greatly accelerated by the use of a multigrid type of method. This is described in Ref. (13) for 2D flow and carries over directly to 3D. The method consists simply of combining groups of elements into blocks and updating each node by an increment from its adjacent elements and, in the same cycle, by an increment from the adjacent blocks. In the 3D codes two levels of multigrid have been coded and an advantage of the method used is that the second level uses very little extra computer work per cycle. Typically blocks of 3 x 3 x 3 and 9 x 9 x 9 elements are used and the resulting effect on convergence is shown in Fig. 5. In interpreting this graph it should be remembered that when multigrid is used the same change in a variable implies a much smaller (1/3 or 1/9) flux imbalance per element than without multigrid. Hence the factor of 3 increase in convergence rate shown in Fig. 5 is an <u>underestimate</u> of the rate of reduction in flux imbalance.

Convergence is usually taken to occur when the average change in meridional velocity per cycle is less than 5×10^{-5} x an average velocity for the whole flow. The actual number of cycles to

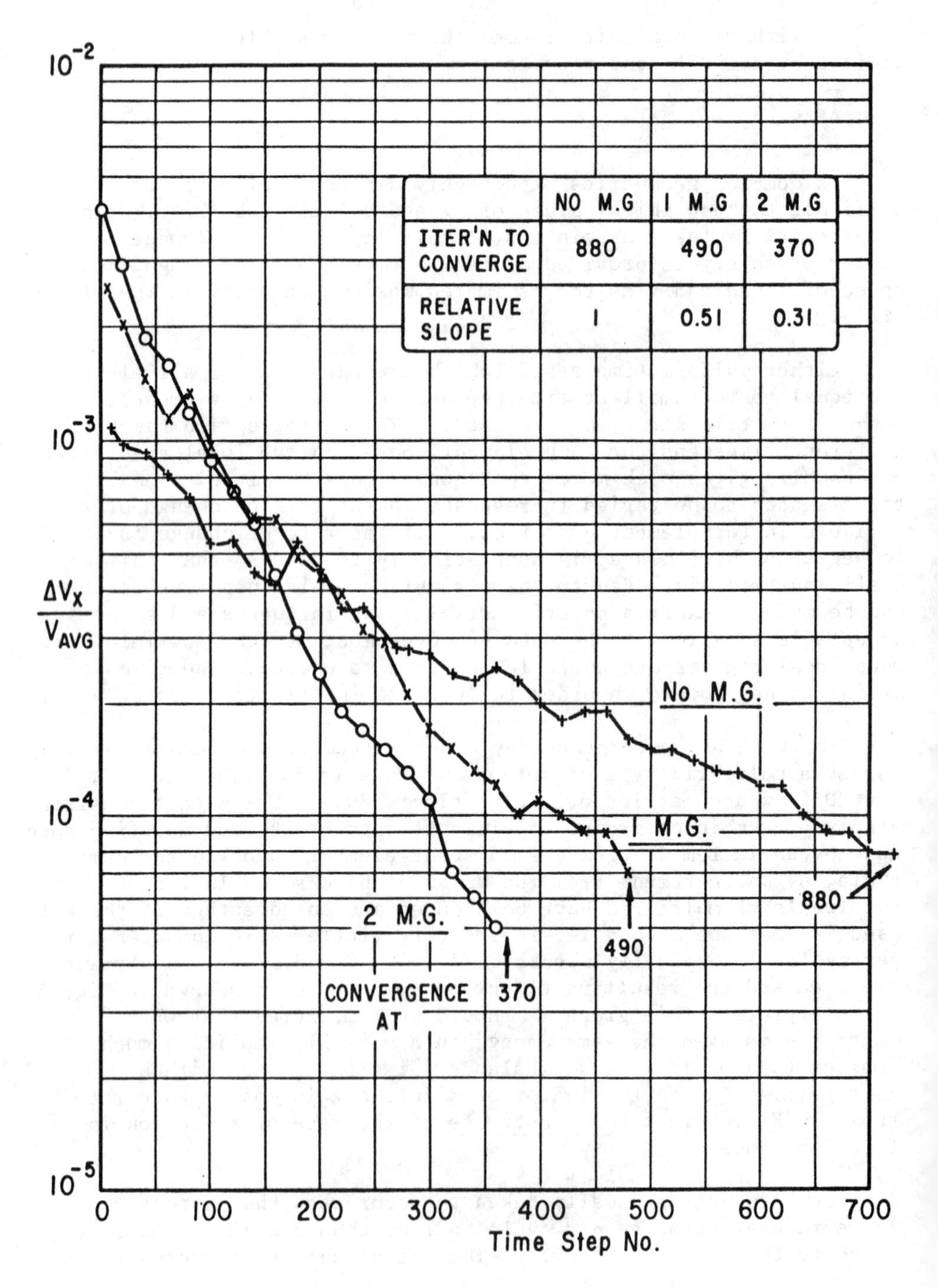

Figure 5 Effect of Multigrid on Convergence.

convergence depends on the number of grid points used. Values for some of the test cases used in this paper are given in Table 1. Convergence in 300-500 cycles is usual for many practical problems.

Computer time per point per cycle clearly depends on the machine used. All the results shown were obtained on a Perkin-Elmer 3230 mini computer for which the time per grid point per cycle is about 1.3×10^{-3} secs. Times on a modern mainframe computer would be about 1/10 of this and for supercomputers such as the CRAY1 times are about 1/100 of the above value.

8. EXAMPLES OF THE USE OF THE BASIC PROGRAM

The basic version of the program can be used for all types of blade row which do not contain splitter blades or shrouds. The only other geometrical restriction is that the hub and casing must be surfaces of revolution. All these limitations have been separately removed in special versions of the code which will be described later.

A comparision of the present method with the author's original method and with test data is shown in Fig. 6. This shows results for the 3D transonic turbine cascade tested by Camus et al (1983). Both methods give similar results over the upstream half of the blade with the new method on the whole being rather better. In particular the methods give almost identical results on the pressure surface at 8.3% span which suggests thatthe discrepancy from experiment in this region may be due to viscous effects. Past experimence has shown that the shock system from the trailing edge of a transonic turbine blade cannot be computed without an extremely fine grid and special treatment of the trailing edge region. No such treatment was included in either method and so neither predicts the flow over the rear of the suction surface with great accuracy. The trends and levels of Mach number are, however, correctly predicted by both programs.

As an illustration of the use of the program to predict shear flows the same test case has been run with a stagnation pressure gradient imposed at the upstream boundary. Accurate prediction of such inviscid secondary flows requires a very fine grid and minimal numerical viscosity. For the present case 19 grid points were used in the spanwise direction with their spacing varied so as to concentrate more points in the simulated boundary layer which was imposed only on the flared end wall. Figure 7 shows contours of stagnation pressure on quasi-orthogonal surfaces through the blade row and illustrates typical features of inviscid secondary flow. Similar results have been obtained by Barber (1981) using the author's original program and further examples of the use of the

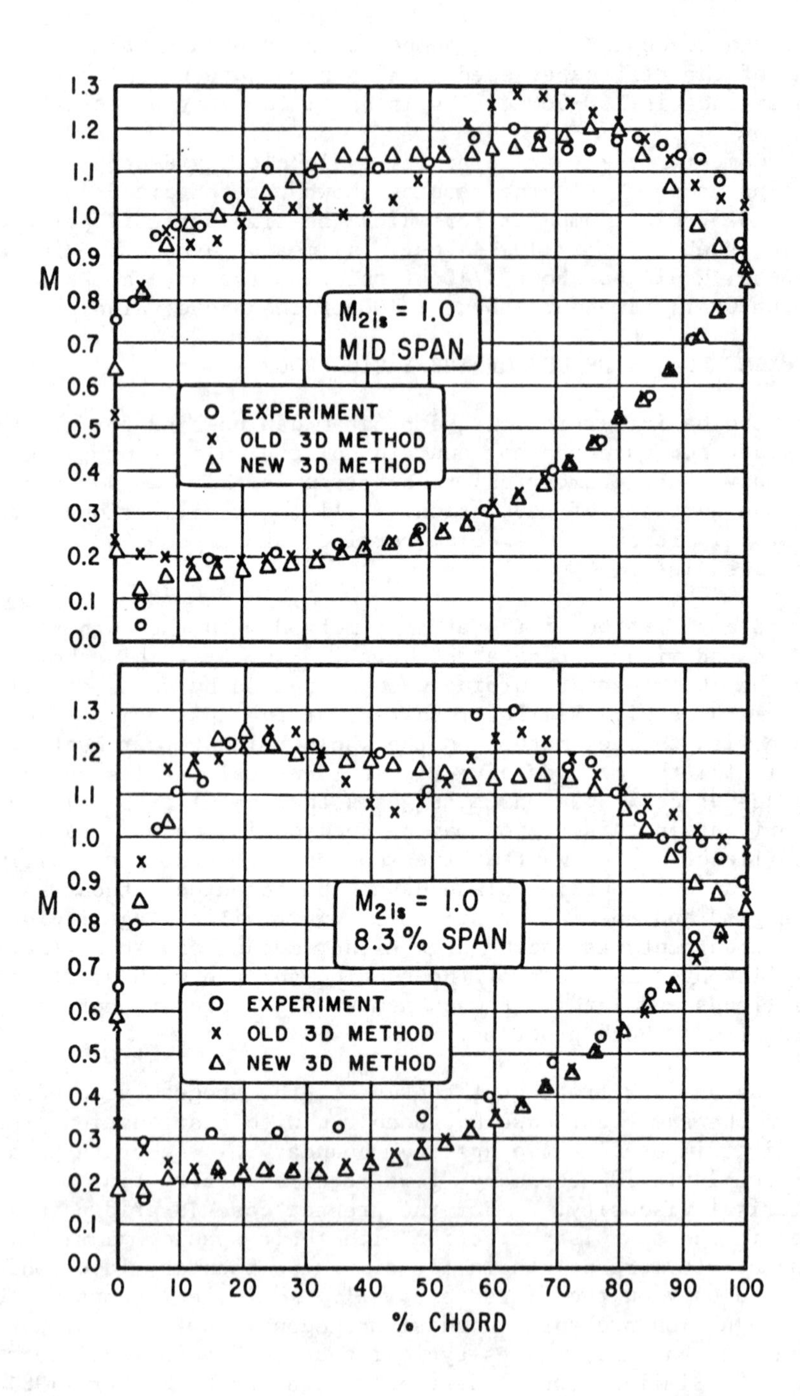

Figure 6 Comparison of Old and New Methods With Experiment on a Transonic Turbine Cascade.

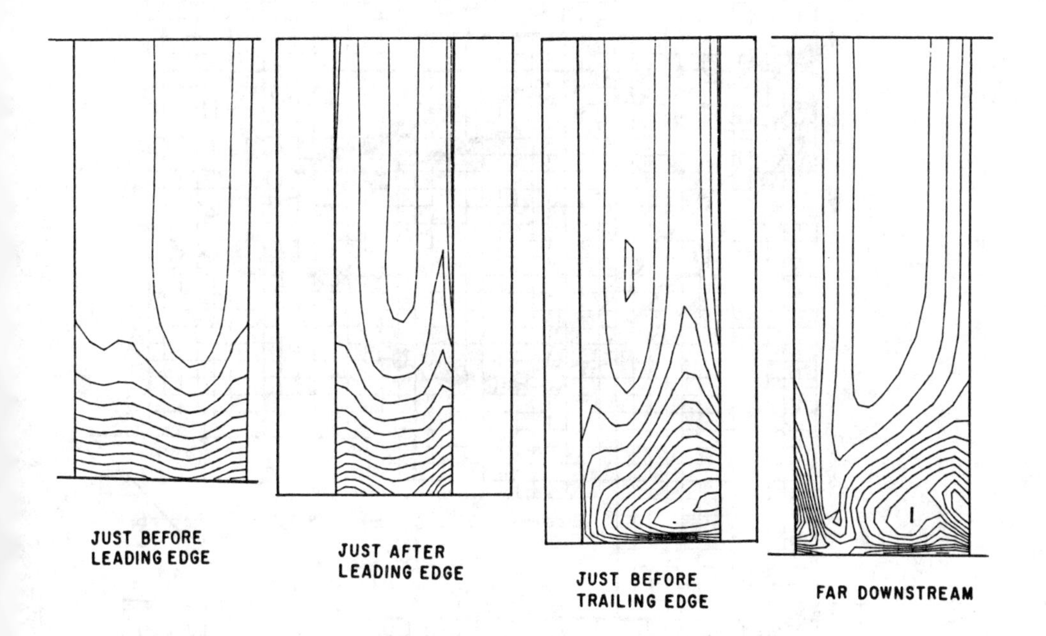

Figure 7 Stagnation Pressure Contours in Transonic Turbine Cascade With Shear Flow.

newer program to predict shear flows are given in Refs (25) and (26).

Swept blade rows provide an extremely simple 3D geometry which nevertheless produces complex 3D flows which are difficult to predict from a quasi-3D approach. Experimental data on such a turbine cascade has been obtained by Gotthardt and Stark and results are available in Ref. (16). Figure 8 compares experimental and computed results for the blade surface pressure distributions close to the end walls. The agreement for this low speed flow can be seen to be very good. Other comparisions of the predictions of the program with test data for simple 3D geometries have been published by Dawes and Squire Ref(17),and by Povinelli Ref. (27).

The use of the program for a rotating blade row of complex geometry is issulstrated in Figure 9. This shows the computed results for a radial inflow turbine with a strong shear flow imposed at inlet. Very strong spanwise flows are set up in this case and the validity of quasi-3D calculations would be dubious. No experimental data are available within the blade row but radial

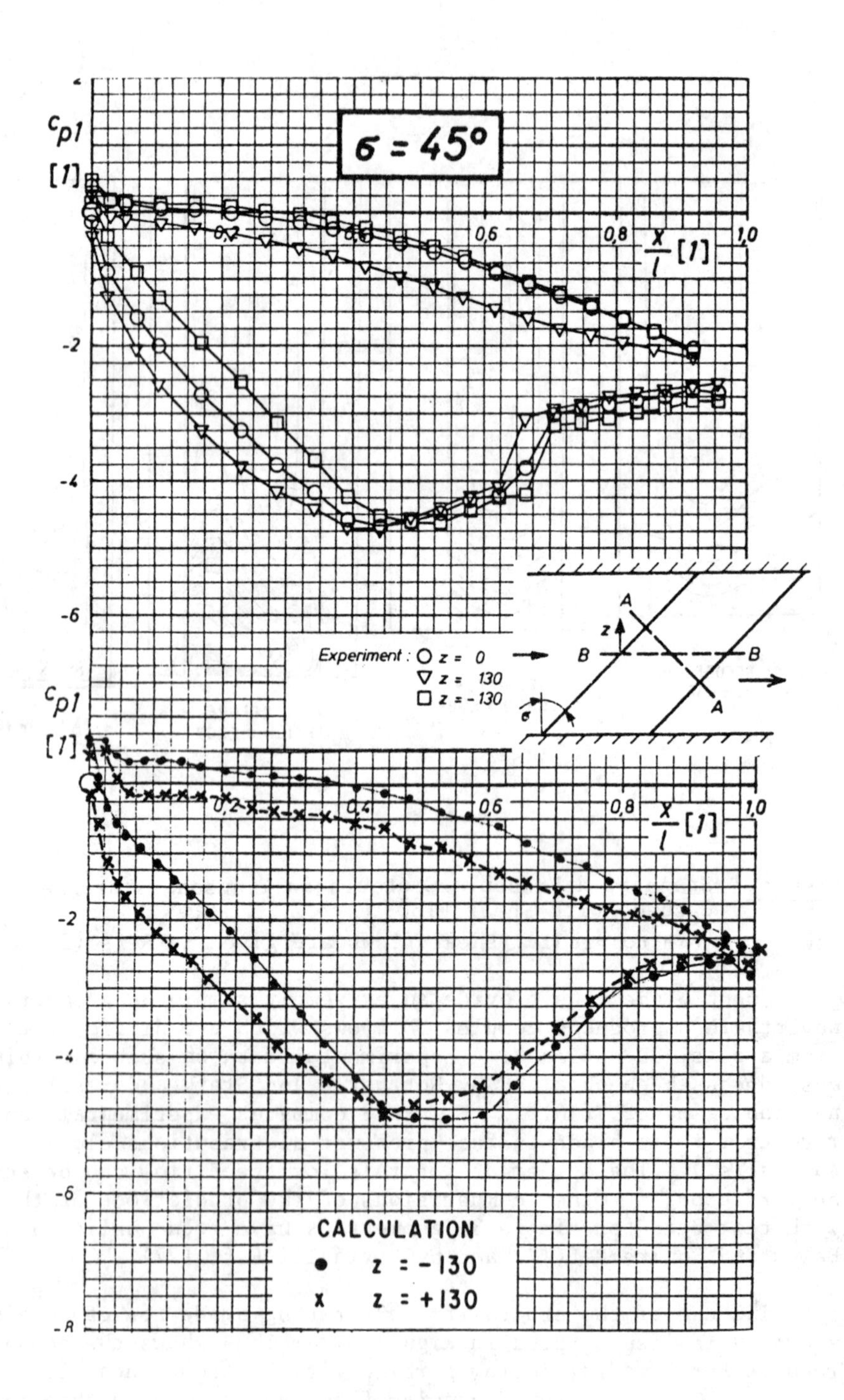

Figure 8 Swept Turbine Cascade.

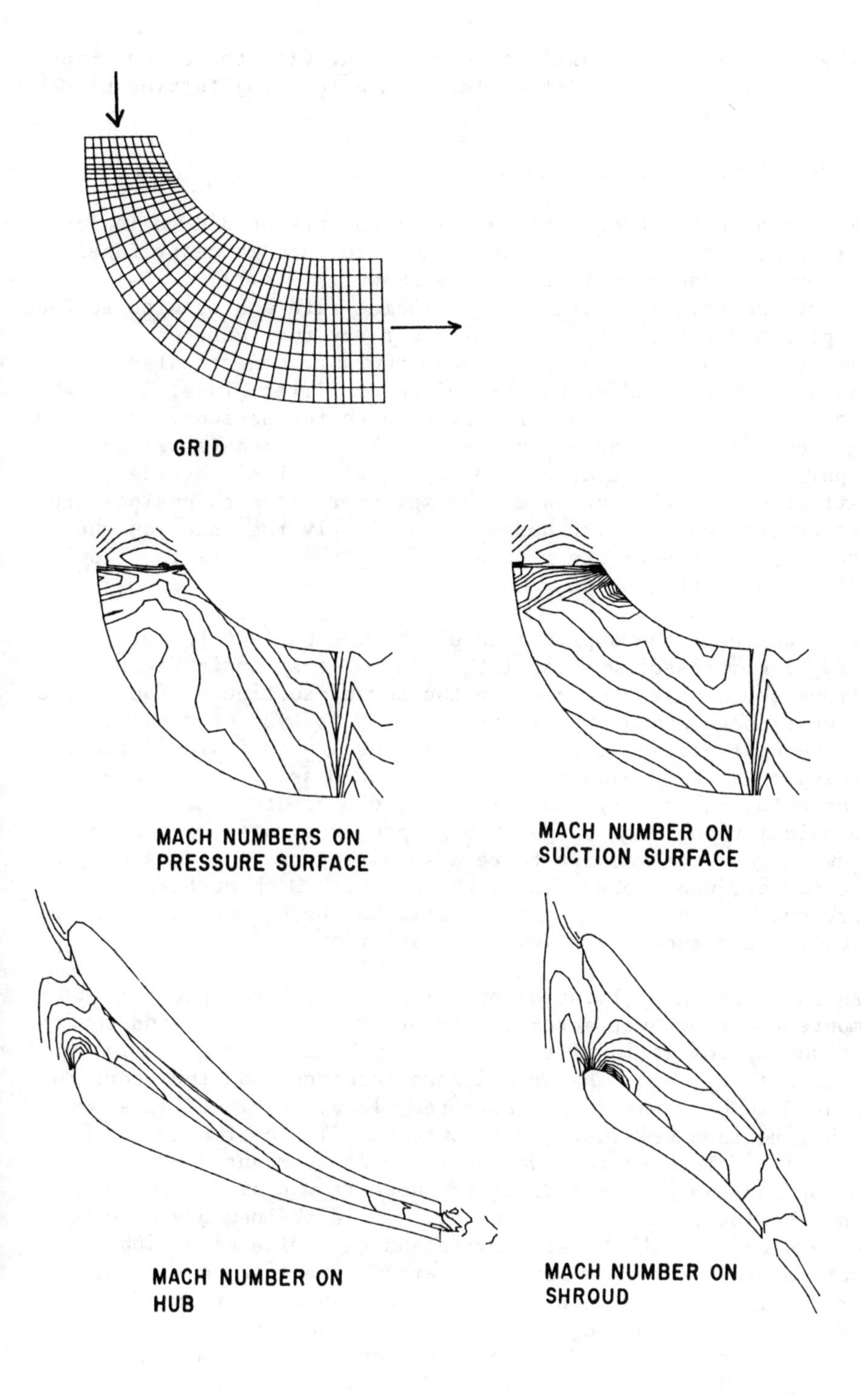

Figure 9 Radial Inflow Turbine

traverses at exit show qualitative agreement with the calculation. Comparisons with test data at outlet from a rotating turbine blade are given in Ref. (25).

9. INCLUSIONS OF SPLITTER BLADES

No fundamental changes to the algorithm are needed to include part-pitch splitter blades of the type often used in centrifugal compressors. The grid is set up as shown in Fig. 10. Each surface of the splitter blade must coincide with a bladewise surface of the grid and off the splitter blades these two surfaces are coincident. The grid nodes on these two surfaces are also coincident upstream and downstream of the splitter blade. These nodes are updated exactly as if they were on the periodic boundaries during each time step and at the end of the step the values at every pair of coincident nodes are set equal to their average. The nett effect is that nodes on the splitter blade extensions are treated exactly as interior nodes. Similarly the nodes on the splitter blade surfaces are treated exactly the same as those on the main blade surfaces.

An example of the application of the method to a transonic centrifugal compressor with splitter blades is shown in Figure 11. The strong shock wave predicted in the shroud-suction surface corner has recently been found experimentally. This type of machine has proved the most difficult to compute because, with inviscid flow, the pressure ratio is almost independent of mass flow. Hence the solution obtained is very sensitive to the pressure ratio imposed on the calculation. A slightly high pressure ratio can cause the flow through the machine to reverse direction, whilst a slightly low one can produce choked flow with very high Mach numbers. However, once the correct pressure ratio has been found no special difficulties are encountered with the solution.

An interesting application of the splitter blade program is to compute the flow over aerofoils or other bodies in a wind tunnel. The top and bottom walls of the tunnel can be treated as 'main' blades with zero camber, zero stagger and zero thickness and the model in the tunnel can be treated as a splitter blade. Figure 12 shows the highly refined grid used to compute the flow over a 2D aerofoil in a tunnel. The aerofoil was asymmetrically mounted by supports from below and the blockage of these supports was simulated by a bulge on the lower wall of the tunnel. Grid refinements of over 50:1 were used in both the streamwise and pitchwise directions in order to provide a very fine grid around the leading edge of the aerofoil. The solution is compared with a hologram of the flow around the leading edge in Fig. 13 and shows remarkably good agreement over a region where the flow accelerates from stagnation to Mach 1.2 in 2.5% of the chord. The solution over the remainder

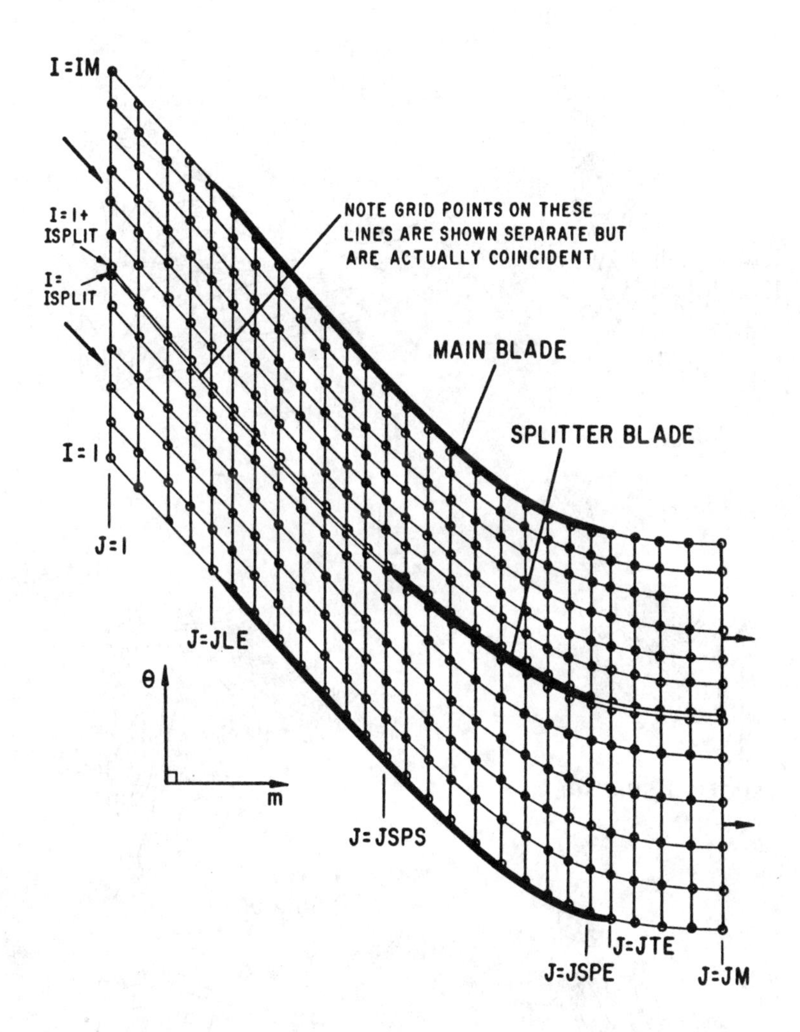

Figure 10 Grid on a Streamwise Surface-With Splitter Blade.

of the surface also compares well with results from both holography and surface static tappings as shown in Fig. 14. This solution could of course have been obtained much more efficiently from a 2D calculation had a suitable program been available. Further details of the calculation are given by Bryanston-Cross and Denton (1984).

10. INCLUSION OF PART SPAN SHROUDS

Exactly the same method as used for part pitch splitter blades can be used to include part span shrouds or part span flow splitters into the calculation. Such shrouds and splitters are commonly used in the 1st stage compressor (fan) of by-pass

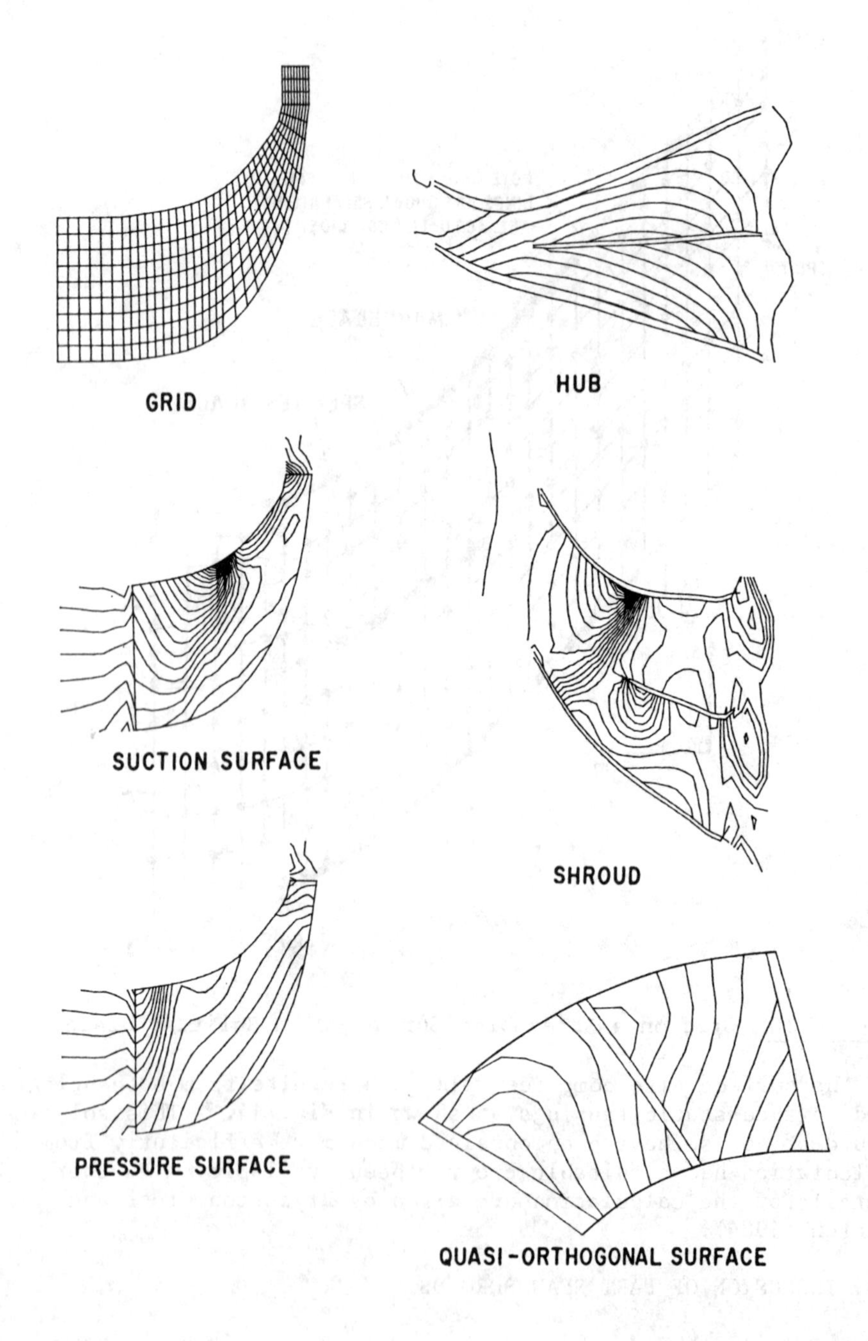

Figure 11 Transonic Sentrifugal Compressor With Splitter Blade-Mach Number Contours.

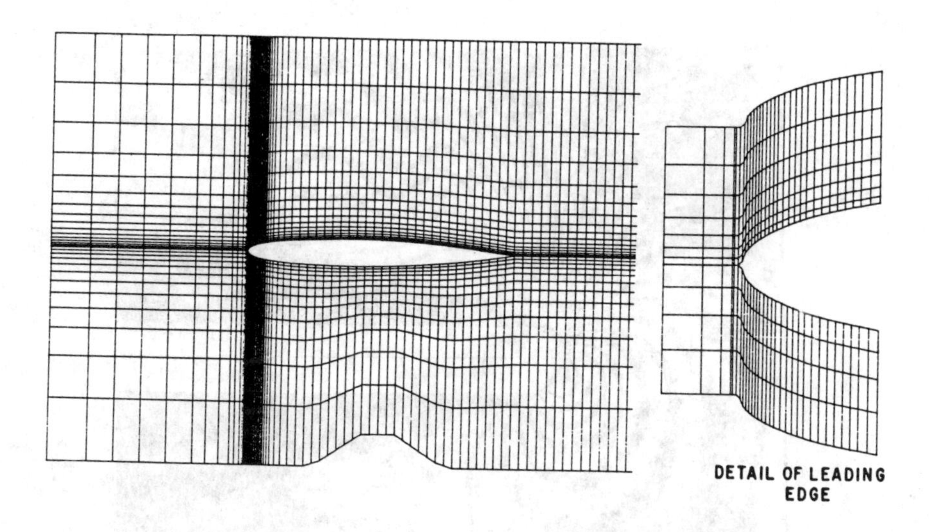

Figure 12 Grid Used for Aerofoil in Wind Tunnel.

engines and the prediction of flow through such transonic fans constitutes one of the main applications of the program. Fig. 15 shows the grid set up for such a case. The two streamwise surfaces which include the upper and lower surfaces of the shroud are coincident upstream and downstream of it. A flow splitter may be treated as a shroud which intersects the downstream boundary and the only additional complication it introduces is that the downstream static pressure must be prescribed independently above and below the splitter.

No experimental data for this type of flow are known to be available so Fig. 16 illustrates the predictions for a hypothetical shroud introduced into the NASA Rotor 33 (Ref. (19)). In this case the shroud has remarkably little effect on the flow pattern.

11. COMPLETE STAGE CALCULATIONS

The flow through two blade rows in relative rotation is inherently unsteady unless they are widely spaced. However, except when specifically studying unsteady effects it is usual to completely neglect the circumferential variations in flow at entry to a blade row and to assume that the interaction between two rows is confined to a spanwise variation in flow properties. This spanwise variation would usually be obtained from a throughflow type of calculation in the meridional plane and the output from such a calculation would be used as a circumferentially uniform

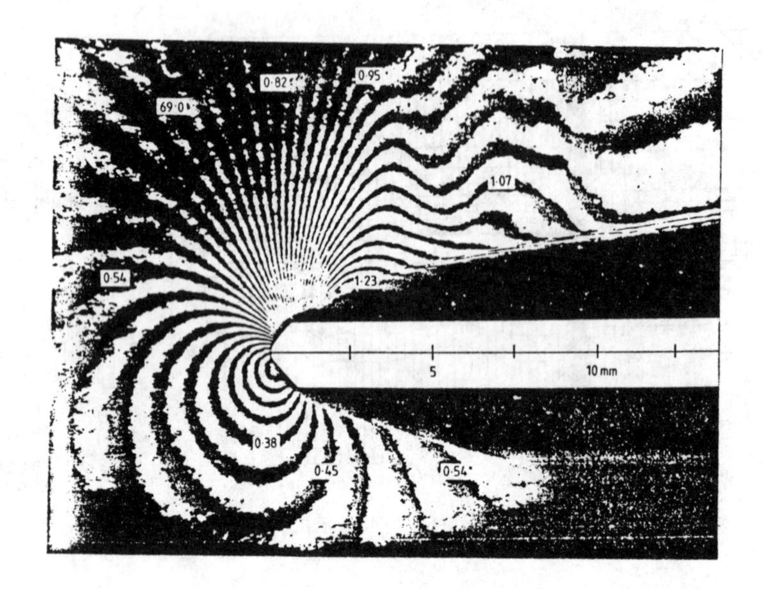

Interferogram of Blade Leading Edge.

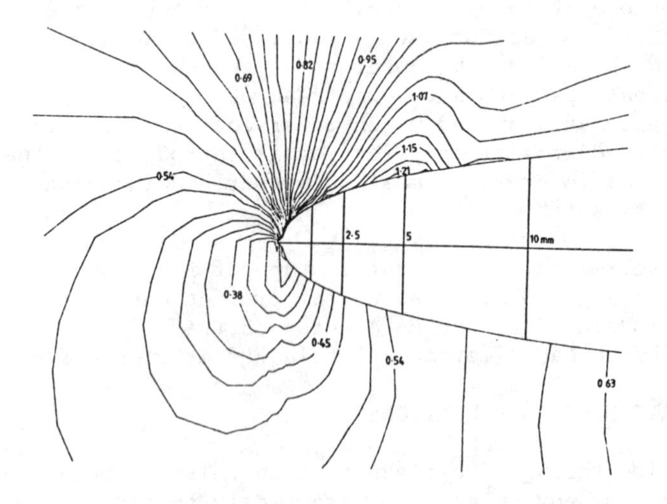

Figure 13 Time Marching Prediction of Blade Leading Edge.

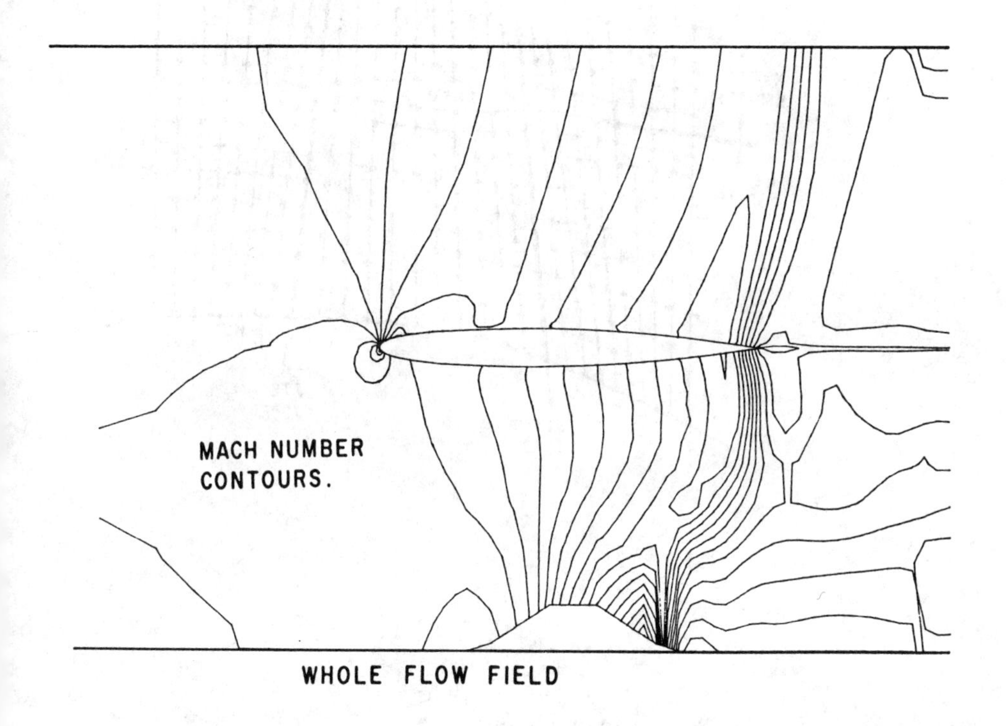

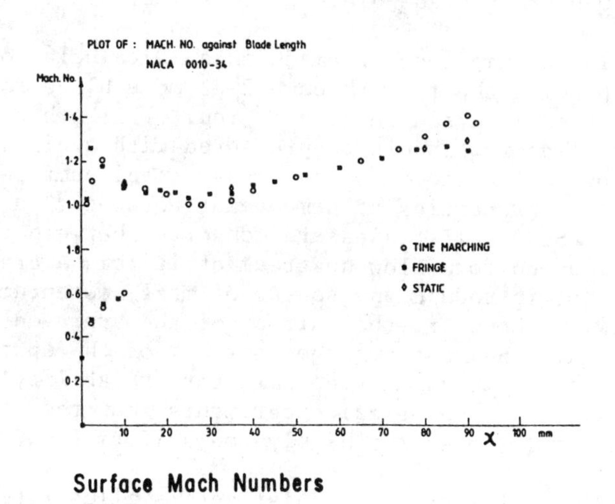

Figure 14. Aerofoil in Wind Tunnel.

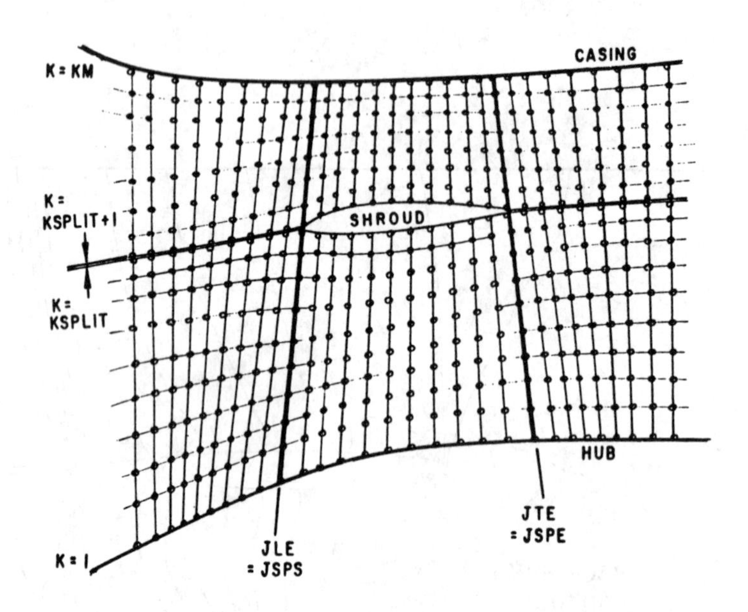

Figure 15 Grid for Part Span Shroud.

inflow to a Q3D or 3D blade-blade calculation.

If similar assumptions are made in a 3D calculation it is possible to predict the flow through 2 or more blade rows in relative rotation in a single calculation. All that is required is that each blade row should be presented with a circumferentially uniform inflow. Hence it is necessary to circumferentially average all flow properties at some point between each pair of blade rows. Since all fluxes are conserved between the averaged flow and the non-uniform flow upstream of it this averaging process does not introduce any source of mass, momentum or energy into the flow. However, the entropy of the averaged flow will not exactly equal the mass averaged entropy of the upstream flow and this is a price which must be paid for the ability to perform such calculations. Numerical experiments show that the change in entropy is small, even for highly non-uniform flow.

The grid used for stage calculations is illustrated in Fig. 17. Since, in general, there will be different numbers of blades in each row the grids will be discontinuous at the mixing

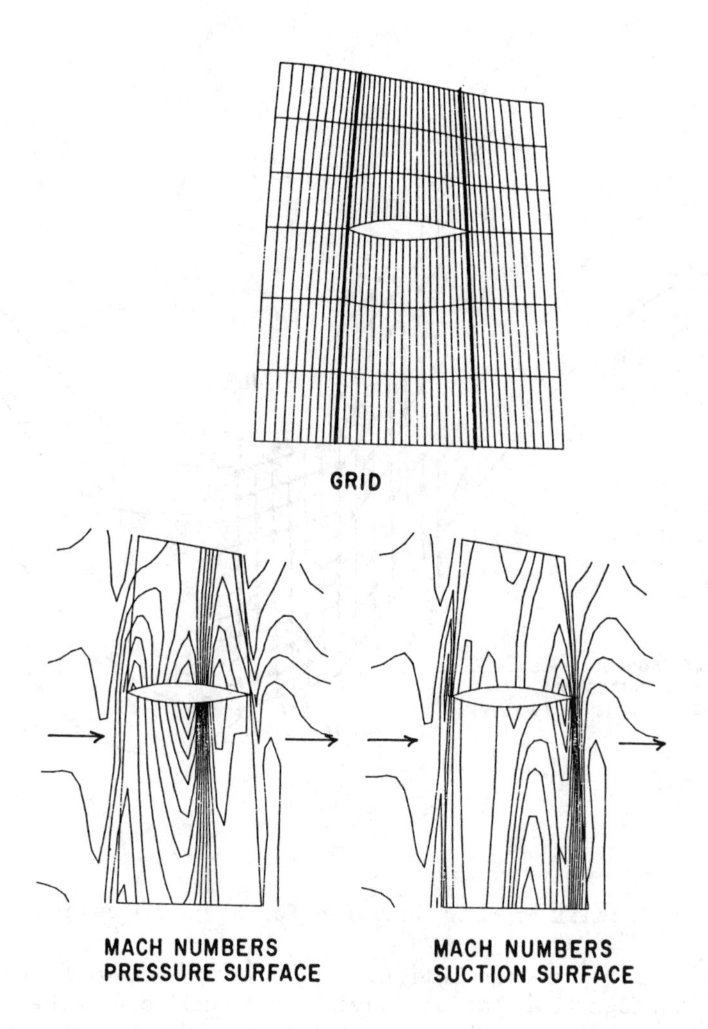

Figure 16 Fan With Shroud.

plane and the discontinuity in area of the quasi-orthogonal faces of the elements must be allowed for when satisfying the conservation equations. The only limit to the number of blade rows which can be calculated in this way is provided by computer storage limitations. The spanwise variation of flow properties at the mixing plane is not specified but is a product of the calculation exactly as it is in throughflow calculations. This constitutes one of the main attractions of the procedure since it enables a stage to be designed using a single calculation method rather than by iterating between throughflow and Q3D or 3D blade-blade methods.

An example of the predictions for the last stage of a steam turbine is shown in Fig. 18. This complex highly three dimensional flow, with Mach numbers up to 1.9, presented no special

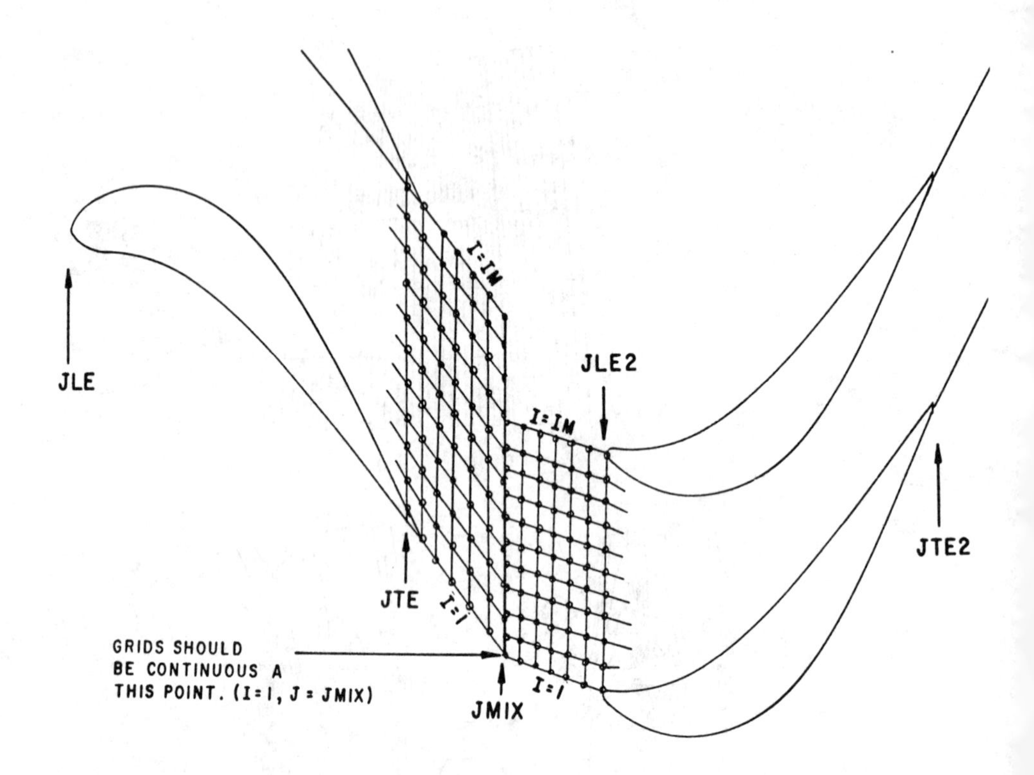

Figure 17 Detail of Grid Junction for Stage Calculation.

difficulty for the calculation. In fact, even with very coarse grids, 3D stage calculations have been found to give better predictions than throughflow calculations for stages of this type. Further examples of 3D calculations on last stage steam turbines are given in Ref. 28. The discontinuity in the contours at the mixing plane is due to the change from absolute to relative Mach numbers for the rotating blade row.

12. INCLUSION OF BOUNDARY LAYER BLOCKAGE

The major influence of viscous effects on the flow pattern in most turbomachine blade rows is known to be the effective reduction in passage area caused by boundary layer growth. This is particularly important for compressors where relative boundary layer thicknesses are greater, and in transonic flow where small area changes can produce large velocity changes.

The blockage effect of boundary layers may be readily incorporated into inviscid calculations either by displacing

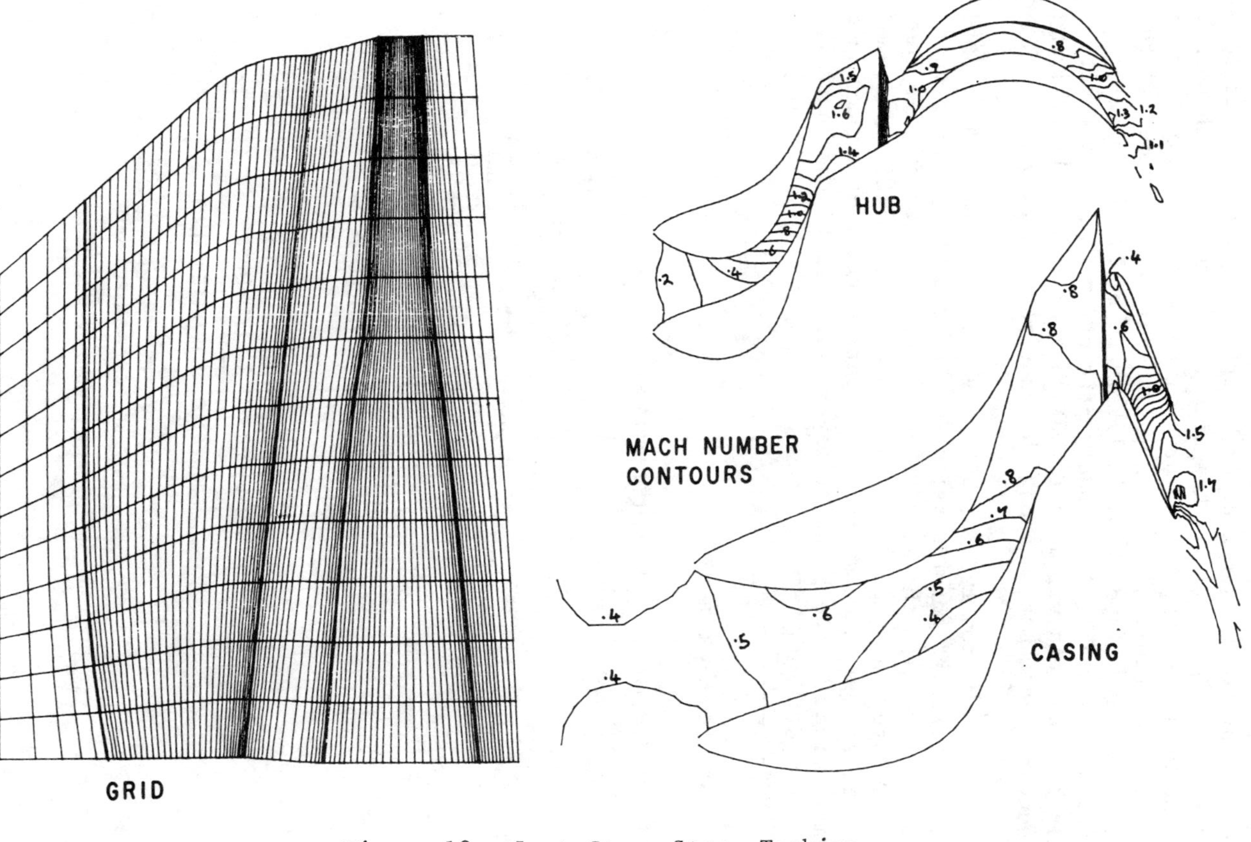

Figure 18 Last Stage Steam Turbine.

the blade surfaces by one boundary layer displacement thickness perpendicular to the surface, or by transpiration of fluid through the surface. Singh (1981) has used the former approach together with the author's original 3D program to obtain greatly improved predictions of the flow through a transonic compressor. Although the real boundary layer flow is likely to be highly three dimensional, suitable 3D boundary prediction methods are not available and Singh was able to obtain good results by applying a 2D boundary layer method along the quasi-streamlines.

A transpiration type of boundary layer displacement model has been preferred for inclusion in the present program because it does not require the grid to be re-generated every time the boundary layer is updated. In this approach fluid is assumed to flow through the blade surface at a rate sufficient to displace the mainstream by one boundary layer displacement thickness i.e.

$$\rho V_n = \frac{d(\rho V_s \delta^*)}{ds} \tag{10}$$

as illustrated in Fig. 19.

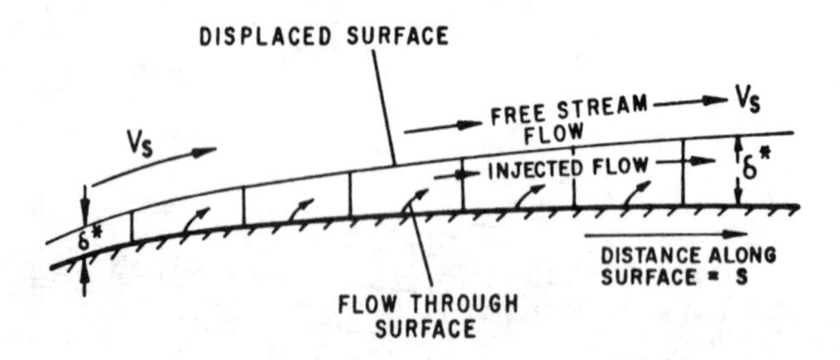

FREE STREAM IS DISPLACED BY δ^* FROM SURFACE.

VELOCITY OF FREE STREAM AND INJECTED FLUID ARE SAME - I.E BOUNDARY LAYER ASSUMED THIN.

∴ FLOW RATE OF INJECTED FLUID $= \rho V_s \delta^*$.

∴ RATE OF INJECTION PER UNIT SURFACE AREA $= \dfrac{d(\rho V_s \delta^*)}{ds}$

Figure 19 Surface Transpiration Model.

The only change needed to the inviscid 3D code is to replace the condition of zero flow through the solid boundaries with a flow as predicted by Eqn. 10. This condition may be used on both blade and endwall surfaces although the use of a 2D boundary layer prediction on the end walls must be regarded as very dubious. All that is required of the boundary layer program used is that, given values of surface velocity, surface distance and fluid properties it should return a value of displacement thickness to the main 3D program. Hence almost any 2D boundary layer program may be used. However, in view of the limitations of using a 2D boundary layer prediction in a 3D flow it is felt that there is little justification in using any but the simpler and faster methods.

It is not necessary to update the boundary layer prediction after every time step and in practice updating after every 50 time steps has been found suitable. No problems of unstable interaction between the inviscid calculation and the boundary layer calculation have been encountered although such problems are to be expected if boundary layer separation is predicted. In fact in most cases convergence of the main inviscid calculation is only very slightly delayed by the boundary layer interaction.

Figure 20 shows predictions for the NASA Rotor 33 with and without boundary layer addition in comparision with the measurements of Chima and Strazisar (1982). Although the boundary layer does not have a large effect on the flow near the tip it does greatly change the flow near the hub and the overall agreement with experiment is significantly improved. This figure also illustrates the good shock capturing properties of the method using a comparatively coarse 10 x 41 x 10 point grid with only 20 points on the blade surface. The total blockage due to blade and endwall boundary layers in this case was predicted to be about 10%.

A similar example of the use of the same program to predict the flow through a transonic fan rotor is given by Pierzga and Wood (29). They used a very fine mesh (22000 points) and were able to predict the fan pressure ratio:mass flow characteristic reasonably well despite the very crude boundary layer model.

13. CONCLUSIONS

Many published 3D flow calculation methods are developed specifically for one application and cannot be used for more general problems. The present method has evolved to the stage where the same method (in many cases the same program) can be used for all conceivable types of turbomachine and for several other applications such as flow through complex ducts, over bodies in wind tunnels and through propellers. A version, not

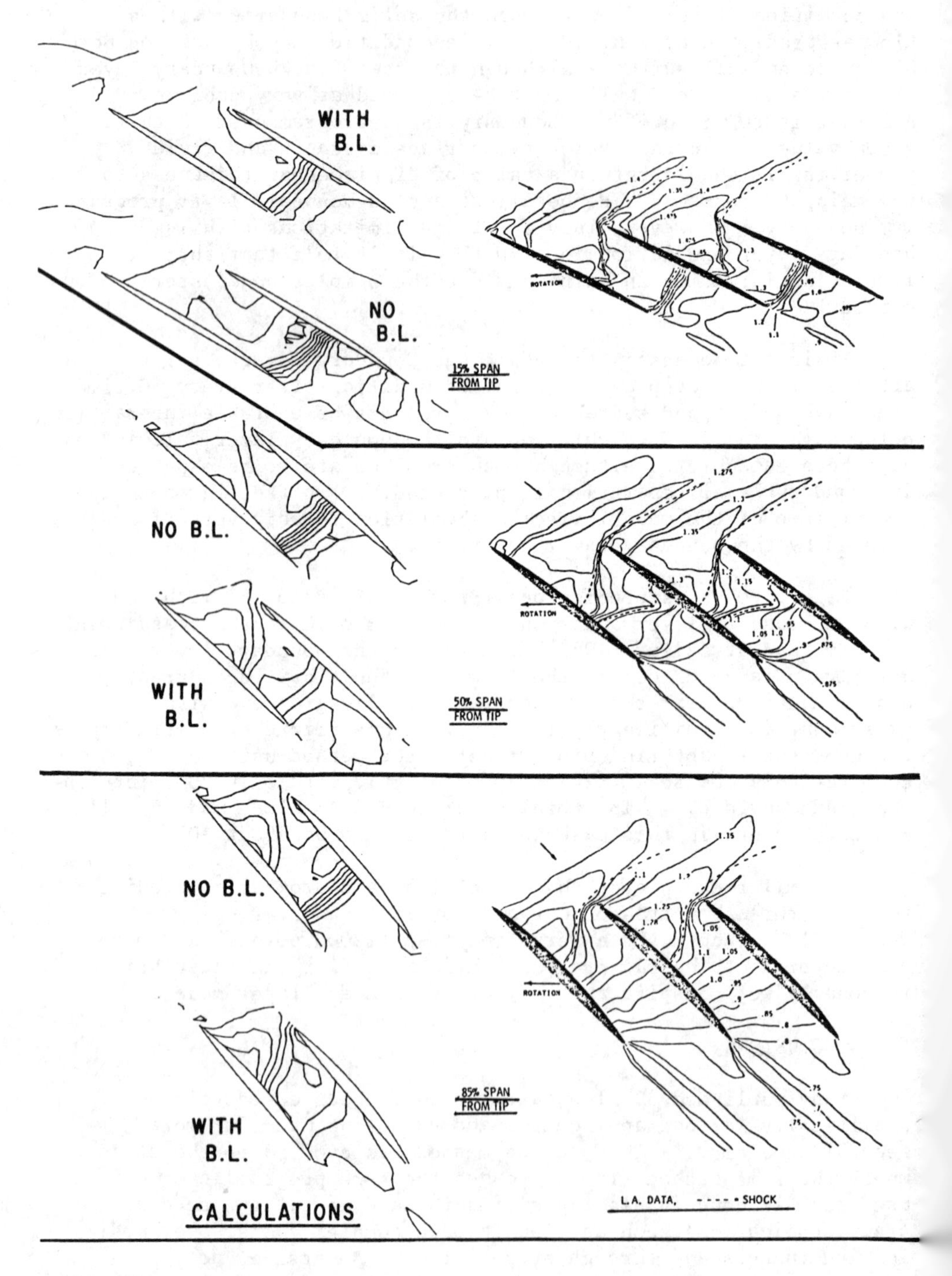

Figure 20 Comparison of Calculation and Measurements on N.A.S.A. Rotor 33.

described in this paper, has also been developed (Atkins (30)) for blade rows or ducts where the end walls cannot be regarded as surfaces of revolution.

The key feature which enables this flexibility to be achieved is felt to be the simplicity of the grid and of the numerical scheme. Although the grid is certainly not the most efficient in terms of accuracy from a limited number of points it would be difficult to cover such a wide range of geometries with a more complex grid. The simplicity of the numerical scheme makes it very fast in terms of CPU time per grid point per timestep and, perhaps more important, enables the code to be readily understood and developed by other users.

The weakest point of this, and of most other methods of solving the Euler equations in turbomachinery, is the failure to accurately capture the trailing edge shock system from transonic turbine blades. Work on this and on other developments is continuing and it is anticipated that such improvements will be progressively incorporated into the existing infrastructure of the code.

ACKNOWLEDGMENT

The development of these codes has benefitted enormously from the feedback obtained from the many users of this and of earlier 3D codes. The author would like to express his gratitude to all such users and to encourage them to keep him informed of both the successes and failures of the method.

TABLE 1

BLADE ROW	GRID USED NO. OF POINTS	ITERATIONS TO CONVERGE	TOTAL CPU TIME* MINUTES
3D Transonic Turbine Cascade Fig. 6.	10 x 49 x 10 4900	300	32
Swept Low Speed Turbine Cascade Fig.8	10 x 60 x 10 6000	370	48
Radial Inflow Turbine Fig.9	10 x 39 x 10 3900	765	65
Transonic Centrifugal Compressor Fig. 11	14 x 37 x 10 5180	735	86
2D Wing in Wind Tunnel Fig. 13	30 x 88 x 3 7920	1030	185
3D Turbine Stage Fig. 18	13 x 89 x 10 11570	825	208
Transonic Axial Fan Fig. 20	10 x 41 x 10 4100	345	31
Transonic Axial Fan with Boundary Layer. Fig. 20	10 x 41 x 10 4100	440	40

*Times on Perkin-Elmer 3230 mini-computer

REFERENCES

1. Novak, R.A., and Hearsay, R.M. A nearly three dimensional intrablade computing system for turbomachinery. Pts. I & II ASME papers 76-FE-19 and 20, 1976.
2. Katsanis, T. Use of arbitrary quasi-orthogonals for calculating flow distribution in a turbomachine. ASME J. Eng. Power, April 1966.
3. Denton, J.D. A time marching method for two and three dimensional blade to blade flow. Aer. Res. Co. R.& M 3775, 1975.
4. Bosman, C. and Highton, J. A calculation procedure for 3D time-dependent, inviscid, compressible flow through turbomachinery blades of any geometry. J.Mech Eng. Sci., Vol. 21, No. 1, 1979.
5. Thompkins, W.T. An experimental and computational study of flow in a transonic compressor rotor. Ph.D.Thesis, M.I.T. June 1976.
6. Laskaris, T.E. Finite element analysis of 3D potential flow in turbomachines. AIAA J. Vol. 16, No. 7, 1978.
7. Hirsch, C., and Lacor, C. Rotational flow calculations in 3D blade passages. ASME paper 82-GT-316.
8. Soulis, J.V. Calculation of transonic potential flow through turbomachinery blade rows. Ph.D. thesis Cambridge University September 1981.
9. Enselme, M., Brochet, J., and Boisseau, J.P. Low cost 3D flow computations using a minisystem. AIAA J. Vol. 20, No. 11, Nov. 1982.
10. Thompkins, W.T. A Fortran program for calculating 3D, inviscid, rotational flows with shock waves in axial compressor blade rows. MIT GT AND PDL report No. 162, September 1981.
11. Denton, J.D. Extension of the finite volume time marching method to 3 dimensions. VKI Lecture series. Transonic flows in turbomachinery, 1976.
12. Denton, J.D. and Singh, U.K. Time marching methods for turbomachinery flow calculations. VKI Lecture series. Transonic flows in turbomachinery, 1979.
13. Denton, J.D. An improved time marching method for turbomachinery flow calculation. ASME paper 82-GT-239, 1982.
14. Camus, J-J, Denton, J.D., Soulis, J.V., and Scrivener, C.J. An experimental and computational study of transonic 3D flow in a turbine cascade. ASME paper 83-GT-12, 1983.
15. Barber, T.J. Analysis of shearing internal flows. AIAA paper 81-0005, 1981.
16. Gotthardt, H., and Stark, U. VKI workshop on 2D and 3D flow calculations in turbine blading, 1982.
17. Dawes, W.N., and Squire, L.C. A study of shock waves in 3D transonic flow. ASME J. Fluid Eng., Vol. 104, Sept. 1982.

18. Bryanston-Cross P., and Denton, J.D. Comparison of a computed transonic flow and interferometric measurements around the leading edge of a 2D aerofoil . To be published by AIAA 1984.
19. Chima, R.V., and Strazisar, A.J. Comparison of 2D and 3D flow computations with laser anemometer measurements in a transonic compressor rotor. ASME paper 82-GT-302, 1982.
20. Singh , U.K. A computation and comparison with measurements of transonic flow in an axial compressor stage with shock and boundary layer interaction. ASME paper 81-Gr/GT-5, 1981
21. Wu, C.H. , Wang, O, and Zhu, G. Quasi 3D and full 3D flow calculations in turbomachines. ASME paper 84-GT-185, 1984.
22. Arts, T. Calculation of the three dimensional, steady, inviscid flow in a transonic axial turbine stage. ASME paper 84-GT-76, 1984.
23. Holmes, D.G., and Tong, S.S. A three dimensional Euler solver for turbomachinery blade rows. ASME paper 84-GT-79, 1984.
24. Koya, M., and Kotake, S. Numerical flow analysis of fully three-dimensional turbine cascades. ASME paper 84-GT-19, 1984.
25. Schwab,J.R., and Povinelli, L.A. Comparison of secondary flows predicted by a viscous code and an inviscid code with experimental data for a turning duct. ASME Fluids Eng. Conf. Feb. 1984.
26. Schwab, J.R., Stabe, R.G., and Whitney, W.J. Analytical and experimental study of flow through an axial turbine stage with a non-uniform inlet radial temperature profile. AIAA paper 83-1175, June 1983.
27. Povinelli, L.A. Assessment of three dimensional inviscid codes and loss calculations for turbine aerodynamic computations. ASME paper 84-GT-187, 1984.
28. Denton, J.D. 3D flow calculations on a hypothetical steam turbine last stage. VKI Lecture Series 6, 1983.
29. Pierzga, M.J., and Wood, J.R. Investigation of the three dimensional flow field within a transonic fan rotor: Experiment and analysis, ASME paper 84-GT-200, 1984.
30. Atkins, M.J. Endwall profiling in axial flow turbines. Ph.D. Thesis Cambridge University, 1984.

NOTATION

dA	Surface area of element
CF	Correction factor
c	Speed of Sound
E	Internal energy
FAC	Multiplying factor on pressure correction
h_o	Stagnation enthalpy
J	Index in streamwise direction
P	Static pressure
P_{eff}	Effective static pressure as used in the solution
RF	Relaxation factor
SF	Smoothing factor
S	Surface distance
V	Flow velocity
x) r) θ)	Cylindrical coordinate system
Δt	Time increment
Δx	Space increment
C*	Displacement thickness
ρ	Fluid density
Ω	Rotational speed

Subscripts

x) r) θ)	In coordinate directions
J	At grid point number

FINITE-DIFFERENCE SOLUTION OF BOUNDARY-LAYER FLOWS WITH SEPARATION

Tuncer Cebeci

MDC Senior Fellow - Fluid Mechanics and Heat Transfer
Douglas Aircraft Company, Long Beach, California 90846
and
Professor, Mechanical Engineering Department,
California State University, Long Beach, California 90840

1.0 INTRODUCTION

There are several finite-difference methods for solving boundary-layer flows with separation. The most widely used and popular ones are due to Crank-Nicolson [1] and Keller [2]. The latter, which has been applied to a wide range of two-dimensional and three-dimensional flows, has a number of very desirable features, as outlined in Cebeci and Bradshaw [3], and will be described here to solve the two-dimensional boundary-layer equations, namely,

$$\frac{\partial u}{\partial x} + \frac{\partial v}{\partial y} = 0 \tag{1}$$

$$u\frac{\partial u}{\partial x} + v\frac{\partial u}{\partial y} = -\frac{1}{\rho}\frac{dp}{dx} + \frac{\partial}{\partial y}\left(\nu\frac{\partial u}{\partial y} - \overline{u'v'}\right) \tag{2}$$

for both external and internal flows. For simplicity, we shall use the eddy viscosity concept ε_m to model the Reynolds shear stress term $-\rho\overline{u'v'}$,

$$-\rho\overline{u'v'} = \rho\varepsilon_m\frac{\partial u}{\partial y} \tag{3}$$

and first consider the solution of Eqs. (1) and (2) with a prescribed algebraic eddy-viscosity formulation due to Cebeci and Smith [4]. Later, in Section 5.0, we shall discuss their solution when higher turbulence models are used for $-\rho\overline{u'v'}$.

For external flows, Eqs. (1) and (2) are often solved for a prescribed external velocity distribution subject to the following boundary conditions, which for the case of no mass transfer are:

$$y = 0 \qquad u = 0 \qquad v = 0 \tag{4a}$$

$$y = \delta \qquad u = u_e \tag{4b}$$

This procedure, sometimes referred to as the standard problem, and as discussed in Section 2.0, has the disadvantage that the boundary-layer equations are singular at separation; they are not singular at separation, however, when the external velocity is computed as part of the solution by, for example, prescribing a displacement thickness or wall shear distribution. This procedure, known as the inverse problem, can be used to solve boundary-layer equations with separation, as we shall discuss in Section 3.0. Solution of internal flow problems in which the pressure drop is computed as part of the requirement to satisfy the mass flow rate can also be achieved by using the inverse boundary-layer procedure.

In many real problems, the freestream velocity distribution is unknown and must be determined from solutions of the potential-flow equations, first for the body shape and subsequently for this shape, modified by displacement thickness distributions or by a blowing velocity, $d/ds\,(u_e\delta^*)$, both obtained from solutions of the boundary-layer equations. In this case, both potential-flow and boundary-layer equations are solved interactively; we shall refer to this procedure as the interactive problem and discuss it in Section 4.0.

Equations (1) and (2) may be solved in the forms presented above or may be expressed in other forms which may be more convenient and accurate for solution. Here we shall use the definition of stream function that satisfies Eq. (1),

$$u = \frac{\partial\psi}{\partial y}, \qquad v = -\frac{\partial\psi}{\partial x} \tag{5}$$

and discuss the solution of Eqs. (1) and (2) in the following form:

$$(bf'')' - \frac{d\bar{p}}{d\xi} = f'\frac{\partial f'}{\partial\xi} - f''\frac{\partial f}{\partial\xi} \tag{6}$$

Here, with primes denoting differentiation with respect to a dimensionless y-distance Y, f and p represent a dimensionless stream function and pressure. These parameters, together with ξ and b are defined by

$$Y = \left(\frac{u_o}{\nu L}\right)^{1/2} y, \qquad \xi = \frac{x}{L}, \qquad \psi(x,y) = (u_o \nu L)^{1/2} f(\xi,Y) \tag{7}$$

$$b = 1 + \varepsilon_m^+, \qquad \varepsilon_m^+ = \varepsilon_m/\nu$$

where u_0 and L denote a reference velocity and length, respectively.

The initial conditions required for Eq. (6) can be obtained either from the use of similarity variables or from experiment. In our discussion we shall assume that they are given and will concentrate on the solutions of Eqs. (1) and (2) for standard, inverse and interactive problems.

2.0 STANDARD PROBLEM

One of the basic ideas of Keller's method is to write the governing equations in the form of a first-order system. First, derivatives of f with respect to Y are introduced as new unknown functions. With the resulting first-order equations and on an arbitrary rectangular net, Fig. 1, use is made of "centered-difference" derivatives and averages at the midpoints of the net rectangle and net segments to get finite-difference equations with a truncation error of order $(\Delta Y)^2$. The resulting difference equations, which are implicit and nonlinear, are linearized and solved by the block-elimination method described in, for example, Bradshaw et al. [5]. The description of Keller's method is presented in detail in several references (see for example, [3,5]) and for this reason, only a brief description is given here.

In terms of new variables, $u(\xi,Y)$ and $v(\xi,Y)$, Eq. (6) with the use of the Bernoulli equation, $d\bar{p}/d\xi = w(dw/d\xi)$, can be written as

$$f' = u \tag{8a}$$

$$u' = v \tag{8b}$$

$$(bv)' + w\frac{dw}{d\xi} = u\frac{\partial u}{\partial \xi} - v\frac{\partial f}{\partial \xi} \tag{8c}$$

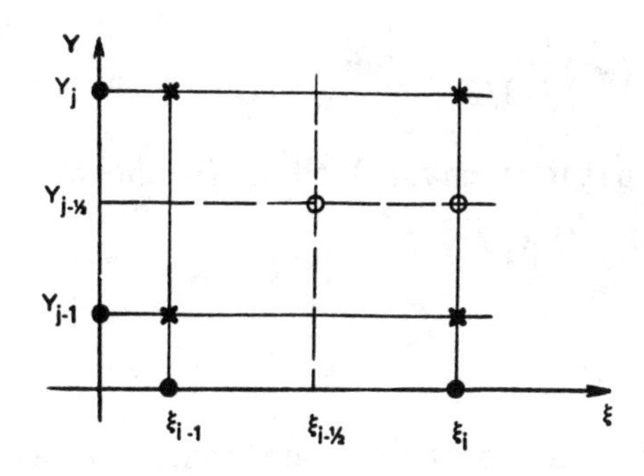

Figure 1. Net rectangle for difference approximations.

where w denotes the dimensionless velocity u_e/u_o. In terms of new variables, the boundary conditions in Eq. (4) become

$$Y = 0, \quad f = 0, \quad u = 0; \qquad Y = Y_e, \quad u = w(x) \tag{9}$$

We next consider the net rectangle, shown in Fig. 1, and denote the net points by

$$\xi_o = 0, \qquad \xi_i = \xi_{i-1} + k_i \qquad (i = 1, 2, \ldots I)$$

$$Y_o = 0, \qquad Y_j = Y_{j-1} + h_j \qquad (j = 1, 2, \ldots J) \tag{10}$$

where $k_i = \Delta\xi_i$, $h_j = \Delta Y_j$.

The difference equations that are to approximate Eqs. (8a,b) are obtained by averaging about the midpoint $(\xi_i, Y_{j-1/2})$ and Eq. (8c) about $(\xi_{i-1/2}, Y_{j-1/2})$; this gives

$$f_j^i - f_{j-1}^i - \frac{h_j}{2}(u_j^i + u_{j-1}^i) = 0 \tag{11a}$$

$$u_j^i - u_{j-1}^i - \frac{h_j}{2}(v_j^i + v_{j-1}^i) = 0 \tag{11b}$$

$$h_j^{-1}[(bv)_j^i - (bv)_{j-1}^i] - k_i[(u^2)_{j-1/2}^i - (fv)_{j-1/2}^i - f_{j-1/2}^{i-1} v_{j-1/2}^i + v_{j-1/2}^{i-1} f_{j-1/2}^i = R_{j-1/2}^{i-1} \tag{11c}$$

where

$$R_{j-1/2}^{i-1} = -h_j^{-1}[(bv)_j^{i-1} - (bv)_{j-1}^{i-1}] + k_i[(fv)_{j-1/2}^{i-1} - (u^2)_{j-1/2}^{i-1} - (w^2)_{j-1/2}^i + (w^2)_{j-1/2}^{i-1}] \tag{12}$$

Boundary conditions given by Eq. (9) become

$$f_o = 0, \qquad u_o = 0; \tag{13a}$$

$$u_J = w(x) \tag{13b}$$

Next we linearize the above nonlinear system by Newton's method as described, for example, in Cebeci and Bradshaw [3] and, dropping the superscript i for convenience, write it in the following form:

$$\delta f_j - \delta f_{j-1} - \frac{h_j}{2}(\delta u_j + \delta u_{j-1}) = (r_1)_j \equiv f_{j-1} - f_j + h_j u_{j-1/2} \tag{14a}$$

$$wu_j - wu_{j-1} - \frac{h_j}{2}(wv_j + wv_{j-1}) = (r_3)_{j-1}$$

$$\equiv u_{j-1} - u_j + h_j v_{j-1/2} \tag{14b}$$

$$(s_1)_j \delta v_j + (s_2)_j \delta v_{j-1} + (s_3)_j \delta f_j + (s_4)_j \delta f_{j-1}$$

$$(s_5)_j \delta u_j + (s_6)_j \delta u_{j-1} = (r_2)_j \tag{14c}$$

$$\delta f_o = (r_1)_o = 0, \quad \delta u_o = (r_2)_o = 0, \quad \delta u_J = (r_2)_J = 0 \tag{14d}$$

where

$$(s_1)_j = h_j^{-1} b_j + \frac{k_i^{-1}}{2}(f_j - f_{j-1/2}^{i-1}),$$

$$(s_2)_j = h_j^{-1} b_{j-1} + \frac{k_i^{-1}}{2}(f_{j-1} - f_{j-1/2}^{i-1}), \tag{15}$$

$$(s_3)_j = \frac{k_i^{-1}}{2}(v_j + v_{j-1/2}^{i-1}), \quad (s_4)_j = \frac{k_i^{-1}}{2}(v_j + v_{j-1/2}^{i-1})$$

$$(s_5)_j = - k_i^{-1} u_j, \quad (s_6)_j = - k_i^{-1} u_{j-1}$$

The linear systems, Eqs. (15), can be written as

$$A \, \delta_j = r_j \tag{16}$$

and can be solved very efficiently by the block-elimination method.

3.0 INVERSE PROBLEM

Inverse problems for boundary-layer equations have been used by a number of investigators, such as Klineberg and Steger [6], Horton [7], Carter [8,9], Williams [10], Cebeci [11], Pletcher [12], and Cebeci, Keller and Williams [13]. These studies were made by prescribing displacement thickness or wall shear distributions and calculating past the separation point using the approximation first suggested by Reyhner and Flügge-Lotz [14]. This approximation, which was necessary to overcome the stability problem associated with negative u-velocity and is referred to as FLARE (Flügge-Lotz and Reyhner) by Williams [10] neglects the $u(\partial u/\partial x)$ term in the region of negative u-velocity. The inverse method of Cebeci [11] makes use of the fact that, in thin-shear layers, the normal pressure gradient, $\partial p/\partial y$ is negligible and, as a result, treats $w(x)$ in Eq. (8c) as an unknown. This procedure, referred to as the Mechul-function approach, introduces an additional equation,

$$w' = 0 \tag{17}$$

into the equations given by Eq. (8) and an additional boundary condition, which for a specified two-dimensional displacement thickness distribution is,

$$Y = Y_e, \qquad \frac{\delta^*}{L}\sqrt{R_L} = Y_e - \frac{f_e}{w} \tag{18}$$

into the boundary conditions given by Eq. (9). This system, consisting of four equations and four unknowns, is then written in the form given by Eq. (16) and solved by the block-elimination method. Note that in the inverse problem, the terms $(u^2)^i_{j-1/2}$, $(u^2)^{i-1}_{j-1/2}$ in Eqs. (11c) and (12) now contain the FLARE approximation and are set equal to zero when $u_j \leq 0$. As long as the recirculation region is small, there is no need to improve the solutions obtained with the FLARE approximation. However, as the size of the recirculation region increases, the application of this approximation becomes increasingly inaccurate, indicating the need for a procedure in which the longitudinal convection term is represented. There are several procedures proposed for this purpose, for example those of Carter [9], Williams [10] and Cebeci [15]. One successful scheme, called the DUIT (Downstream, Upstream Iteration) procedure, is due to Williams [10] and requires several numerical sweeps through the recirculation region to reintroduce the longitudinal convective term. Thus, FLARE is used to compute an approximate solution in the recirculating region and, in successive sweeps of this region, the $u(\partial u/\partial x)$ term is progressively introduced until it is fully represented. Of course, for large regions of separation, the boundary-layer equations may be inappropriate and the Navier-Stokes equations should be solved; the size of the recirculation at which it is desirable to switch from boundary-layer to Navier-Stokes equations remains to be determined. A useful review of this subject has been provided by McDonald and Briley [16].

An alternative solution procedure, which allows the inclusion of the longitudinal convection term in the boundary-layer equations, is to use the solutions of the unsteady boundary-layer equations which are often obtained by solving the governing equations in the ξ-direction at a given time. Such spatial sweeps may be thought of as successive iterations in a procedure for solving the steady state problem corresponding, for example, to the successive sweeps of Williams' DUIT procedure. For this reason, unsteady boundary-layer calculations provide an alternative, and possibly simpler, means of computing steady separating and reattaching flows. The feasibility of this approach was explored recently by solving the two-dimensional time-dependent boundary-layer equations for a specified displacement thickness distribution [15] with the Mechul-function formulation described above.

4.0 INTERACTIVE PROBLEM

The interactive problem allows a special coupling procedure between the viscous and inviscid flow equations and in one way may be regarded as a solution to the Navier-Stokes equations with some approximations imposed by thin shear-layer assumptions. So far, two separate procedures have been developed for two-dimensional flows; the first is due to LeBalleur [17] and Carter [18] and the second to Veldman [19]. In the first, the solution of the boundary-layer equations for incompressible flows is obtained by the standard method to compute a displacement-thickness distribution $\delta_0^*(x)$. If this initial calculation encounters separation, $\delta_0^*(x)$ is obtained by extrapolation and one complete cycle of the viscous and inviscid calculation is performed. In general, this leads to two external velocity distributions, $u_{e_v}(x)$ derived from the inverse boundary-layer solution and $u_{e_i}(x)$ derived from the updated approximationto the inviscid velocity past the airfoil with the added displacement thickness. A relaxation formula is introduced to define an updated displacement-thickness distribution,

$$\delta^*(x) = \delta_0^*(x) \{1 + \omega [\frac{u_{e_v}(x)}{u_{e_i}(x)} - 1]\} \tag{20}$$

where ω is a relaxation parameter and the procedure is repeated with Eq. (20).

In the second approach, the external velocity $u_e(x)$ and the displacement thickness $\delta^*(x)$ are treated as unknown quantities and theequations are solved simultaneously by the inverse method and with successive sweeps over the airfoil surface. For each sweep, the external boundary condition for the boundary-layer equation is written as

$$u_e(x) = u_e^o(x) + \delta u_e(x) \tag{21}$$

where $u_e^o(x)$ is the inviscid velocity over the airfoil and δu_e is the perturbation due to the displacement thickness and is given by

$$\delta u_e = \frac{1}{\pi} \int_{\xi_c}^{\xi_b} \frac{d}{ds} (u_e \delta^*) \frac{ds}{\xi - s} \tag{22}$$

where the interaction is confined between ξ_a and ξ_b. A discrete approximation for the Hilbert integral involved in this equation gives

$$\delta u_e = \sum_{j=1}^{J} c_{ij} (u_e \delta^*)_j \tag{23}$$

where $[c_{ij}]$ is a matrix of interaction coefficients defining the relationship between the displacement thickness and the external velocity. In the approach used by Cebeci and Schimke [20] and Cebeci and Clark [21], Eqs. (21) and (23) are expressed in the form

$$u_{e_i} = u_{e_{o_i}}(x) + \sum_{j=1}^{J} c_{ij}[(u_e \delta^*)_j - (u_{e_o} \delta_o^*)_j] \tag{24}$$

where $u_{e_{o_i}}$ is the external velocity corresponding to a given displacement thickness distribution $\delta_{o_j}^*$.

5.0 NUMERICAL ASPECTS OF TURBULENCE MODELS

In the flow arrangements considered in the examples of Sections 7 and 8, the algebraic eddy-viscosity hypothesis described briefly in Section 1 has been used to represent the turbulent momentum flux. Alternative, and more complicated assumptions have been proposed and some of these are used to obtain the results of Section 6. These models involve the solution of additional differential equations and are intended to allow the correct representation of a wider range of flows than algebraic models. The solution of these turbulence equations introduces new numerical requirements which are considered in this section.

The boundary conditions for Eqs. (1) and (2) are either specified at the wall as given by Eq. (4a) or at some distance y_o away from the wall,

$$y = y_o, \qquad u = u_\tau[\frac{1}{\kappa} \ln \frac{y_o u_\tau}{\nu} + c], \qquad v = -\frac{u}{u_\tau}\frac{du_\tau}{dx} y_o \tag{25}$$

Here u_τ denotes the friction velocity defined by $u_\tau = (\tau_w/\rho)^{1/2}$, and κ and c are universal constants given by 0.40 and 5.2, respectively. The distance y_o is determined from a dimensionless distance $y_o^+ \equiv y_o u_\tau/\nu$ specified to be around 60. In this case the laminar shear stress term in Eq. (2) is negligible and as a result the diffusion term in Eq. (2) becomes $\partial/\partial y\ (-\overline{u'v'})$.

In general the higher-order turbulence model equations use boundary conditions away from the wall and avoid the use of wall functions that are necessary to satisfy the boundary conditions at $y = 0$. In this discussion we shall consider only those models that use this approach. Our discussion, however, is general, and the solution procedures can easily be generalized to satisfy the boundary conditions at $y = 0$.

In turbulence models which use the eddy-viscosity concept, ε_m is either given by an algebraic equation or by a differential equation or equations. In the latter case, a popular set of equations is due to Jones and Launder [22] and Hanjalic and Launder [23], who write ε_m as

$$\varepsilon_m = c_\mu k^2/\phi \tag{26}$$

and obtain the turbulence kinetic energy and dissipation rate from two transport equations for k and ϕ, respectively, written as

$$u\frac{\partial k}{\partial x} + v\frac{\partial k}{\partial y} = c_\mu \frac{k^2}{\phi}\left(\frac{\partial u}{\partial y}\right)^2 - c_k k\frac{\partial u}{\partial x} - \phi + \frac{\partial}{\partial y}\left(c_\mu \frac{k^2}{\phi}\frac{\partial k}{\partial y}\right) \tag{27}$$

$$u\frac{\partial \phi}{\partial x} + v\frac{\partial \phi}{\partial y} = 0.13k\left(\frac{\partial u}{\partial y}\right)^2 - 4.44c_k\phi\frac{\partial u}{\partial x}$$

$$- 1.90\frac{\phi^2}{k} + \frac{\partial}{\partial y}\left(c_\mu \frac{k^2}{\phi}\frac{\partial \phi}{\partial y}\right) \tag{28}$$

where $c_\mu = 0.09$, $c_k = 0.33$. These equations, in addition to those given in Eqs. (4b) and (25) are subject to the following boundary conditions:

$$y = y_0, \qquad \phi = \tau\frac{\partial u}{\partial y} - c_k k\frac{\partial u}{\partial x}, \qquad \phi = \frac{\tau^{3/2}}{\kappa y_0} \tag{29a}$$

$$y = \delta, \quad u\frac{\partial k}{\partial x} + \phi + c_k k\frac{du_e}{dx} = 0,$$

$$u\frac{\partial \phi}{\partial x} + 1.90\frac{\phi^2}{k} + 4.44c_k\phi\frac{du_e}{dx} = 0 \tag{29b}$$

In Eq. (29a), τ denotes the shear stress at $y = y_0$; it is given by

$$\tau = c_\mu \frac{k^2}{\phi}\frac{\partial u}{\partial y} \tag{30}$$

The solutions at $y = y_o$ are related to those at the wall, $y = 0$, through

$$y = y_o, \qquad \tau = u_\tau^2 - y_o\, u_e \frac{du_e}{dx} + \alpha^* \frac{du_\tau^2}{dx} y_o \tag{31}$$

where

$$\alpha^* = 0.5\{c_1(\ln y_o^+)^2 + c_2 \ln y_o^+ + c_3 + c_4/y_o^+\} \tag{32}$$

$$c_1 = 5.9488, \quad c_2 = 13.468, \quad c_3 = 13.572, \quad c_4 = -785.20$$

With $-\overline{u'v'}$ given by Eqs.(3) and (26), and with the use of the Bernoulli equation, Eq. (2) becomes

$$u\frac{\partial u}{\partial x} + v\frac{\partial u}{\partial y} = u_e\frac{du_e}{dx} + \frac{\partial}{\partial y}\left(c_\mu \frac{k^2}{\phi}\frac{\partial u}{\partial y}\right) \tag{33}$$

The Reynolds stress model equations with τ now denoting $-\rho\overline{u'v'}$ consist of transport equations for τ, $\overline{u'^2}$, $\overline{v'^2}$, $\overline{w'^2}$ and ϕ and can be written as

$$u\frac{\partial\overline{u'^2}}{\partial x} + v\frac{\partial\overline{u'^2}}{\partial y} = 2\tau\frac{\partial u}{\partial y} - \frac{2}{3}\phi - c_1\frac{\phi}{k}\left(\overline{u'^2} - \frac{2}{3}k\right) - \frac{4}{3}c_2\tau\frac{\partial u}{\partial y}$$

$$+ c_s\frac{\partial}{\partial y}\left(\frac{k\overline{v'^2}}{\phi}\frac{\partial\overline{u'^2}}{\partial y}\right) + (\Phi_{11})_w \tag{34}$$

$$u\frac{\partial\overline{v'^2}}{\partial x} + v\frac{\partial\overline{v'^2}}{\partial y} = -\frac{2}{3}\phi - c_1\frac{\phi}{k}\left(\overline{v'^2} - \frac{2}{3}k\right) + \frac{2}{3}c_2\tau\frac{\partial u}{\partial y}$$

$$+ c_s\frac{\partial}{\partial y}\left(\frac{k\overline{v'^2}}{\phi}\frac{\partial\overline{v'^2}}{\partial y}\right) + (\Phi_{22})_w \tag{35}$$

$$u\frac{\partial\overline{w'^2}}{\partial x} + v\frac{\partial\overline{w'^2}}{\partial y} = -\frac{2}{3}\phi - c_1\frac{\phi}{k}\left(\overline{w'^2} - \frac{2}{3}k\right) + \frac{2}{3}c_2\tau\frac{\partial u}{\partial y}$$

$$+ c_s\frac{\partial}{\partial y}\left(\frac{k\overline{v'^2}}{\phi}\frac{\partial\overline{w'^2}}{\partial y}\right) + (\Phi_{33})_w \tag{36}$$

$$u \frac{\partial \tau}{\partial x} + v \frac{\partial \tau}{\partial y} = \overline{v'^2} \frac{\partial u}{\partial y} - c_1 \frac{\phi}{k} \tau - c_2 \overline{v'^2} \frac{\partial u}{\partial y}$$

$$+ c_s \frac{\partial}{\partial y} \left(\frac{k\overline{v'^2}}{\phi} \frac{\partial \tau}{\partial y}\right) + (\Phi_{12})_w \tag{37}$$

$$u \frac{\partial \phi}{\partial x} + v \frac{\partial \phi}{\partial y} = c_{\phi 1} \frac{\phi}{k} \tau \frac{\partial u}{\partial y} - c_{\phi 2} \frac{\phi^2}{k} + c_\phi \frac{\partial}{\partial y} \left(\frac{k\overline{v'^2}}{\phi} \frac{\partial \phi}{\partial y}\right) \tag{38}$$

The terms $(\Phi_{ij})_w$ represent the wall proximity effect and according to Gibson and Launder [24], they, together with the constants c_1, c_2, c_s, etc., are given by

$$(\Phi_{11})_w = \frac{k^{3/2}}{\phi y} \left(c_1' \frac{\phi}{k} \overline{v'^2} + \frac{2}{3} c_2' c_2 \tau \frac{\partial u}{\partial y}\right) \tag{39a}$$

$$(\Phi_{22})_w = -2(\Phi_{11})_w , \qquad (\Phi_{33})_w = (\Phi_{11})_w \tag{39b}$$

$$(\Phi_{12})_w = \frac{k^{3/2}}{\phi y} \left(- \frac{3}{2} c_1' \frac{\phi}{k} \tau + \frac{3}{2} c_2' c_2 \overline{v'^2} \frac{\partial u}{\partial y}\right) \tag{39c}$$

$$c_1 = 1.5, \quad c_2 = 0.4, \quad c_s = 0.22, \quad c_\phi = 0.18,$$
$$c_{\phi 1} = 1.45, \quad c_{\phi 2} = 1.90, \quad c_1' = 0.5, \quad c_2' = 0.1 \tag{40}$$

Equations (14) to (16) are subject to the following "wall" boundary conditions in addition to those given by Eqs. (4b) and (11)

$$y = y_0, \quad \overline{u'^2} = 4.90\tau, \quad \overline{v'^2} = 1.03\tau, \quad \overline{w'^2} = 2.40\tau, \quad \phi = \frac{\tau^{3/2}}{\kappa y} \tag{41}$$

and, in addition to that given by Eq. (4c), to the following boundary conditions at $y = \delta$

$$u_e \frac{d\overline{u'^2}}{dx} = - \frac{2}{3} \phi - c_1 \frac{\phi}{k} \left(\overline{u'^2} - \frac{2}{3} k\right) + \frac{k^{3/2}}{\phi y} \left(c_1' \frac{\phi}{k} \overline{v'^2}\right) \tag{42a}$$

$$u_e \frac{d\overline{v'^2}}{dx} = - \frac{2}{3} \phi - c_1 \frac{\phi}{k} \left(\overline{v'^2} - \frac{2}{3} k\right) + \frac{k^{3/2}}{\phi y} \left(-2c_1' \frac{\phi}{k} \overline{v'^2}\right) \tag{42b}$$

$$u_e \frac{d\overline{w'^2}}{dx} = -\frac{2}{3}\phi - c_1 \frac{\phi}{k} (\overline{w'^2} - \frac{2}{3}k) + \frac{k^{3/2}}{\phi y} (c_1' \frac{\phi}{k} \overline{v'^2}) \tag{42c}$$

$$u_e \frac{d\tau}{dx} = - c_1 \frac{\phi}{k} \tau + \frac{k^{3/2}}{\phi y} (-\frac{3}{2} c_1' \frac{\phi}{k} \tau) \tag{42d}$$

$$u_e \frac{d\phi}{dx} = - c_{\phi 2} \frac{\phi^2}{k} \tag{42e}$$

The solution of the turbulent boundary-layer equations using the k-ϕ model equations are obtained by solving Eqs. (1), (27), (28), (33) subject to the boundary conditions given by (4b), (25) and (29). Similarly, those based on the stress model equations require the solution of Eqs. (1), (2), (34) through (38) subject to the boundary conditions given by Eqs. (4b), (25), (31), (41), (42). The boundary-layer equations that use the algebraic eddy-viscosity concept, on the other hand, are much simpler and the solution procedure consists of solving Eqs. (1) and (2) when the latter is written as

$$u \frac{\partial u}{\partial x} + v \frac{\partial u}{\partial y} = u_e \frac{du_e}{dx} + \frac{\partial}{\partial y} (\varepsilon_m \frac{\partial u}{\partial y}) \tag{43}$$

with ε_m given by empirical functions. In the eddy-viscosity formulation of Cebeci and Smith, this is achieved by defining ε_m with

$$\varepsilon_m = (\kappa y)^2 \frac{\partial u}{\partial y} \lambda_1 + 0.0168 [\int_0^\delta (u_e - u)dy]\lambda_2 \tag{44}$$

Here the parameters λ_1 and λ_2 are determined from the continuity of eddy-viscosity formulas with $\lambda_1 = 1$ and $\lambda_2 = 0$ in the inner region where $\varepsilon_m = (\kappa y)^2 \partial u/\partial y$ and $\lambda_1 = 0$ and $\lambda_2 = 1$ in the outer region where $\varepsilon_m = 0.0168 \int_0^\delta (u_e - u)dy$.

The solution of the boundary-layer equations using these turbulence models can be easily achieved by using the standard method described in Section 2.0. Before we do this, however, let us discuss some of the features of either the k-ϕ model equations or those based on the algebraic eddy-viscosity formulas. Both are nonlinear and strongly coupled to their boundary conditions which are also nonlinear. In addition, the presence of the friction velocity u_τ in the "wall" boundary conditions further complicates the solution procedure since it is unknown; it can be assumed to be known during the solution procedure, as is most commonly done, but this is not efficient and even can lead to the breakdown of the solutions in flows with strong adverse pressure gradient since friction velocity changes rapidly near separation. An accurate and efficient numerical method then, should consider carefully the linearization scheme and the manner in which u_τ is handled. The numerical method described in Sections 2.0, 3.0 and 4.0 have all these desirable features and are described for model equations based on algebraic eddy viscosity and turbulence kinetic energy and dissipation-rate (k-ϕ). The numerical features of the stress model equations are similar to those of the k-ϕ model equations and are not referred to separately.

In terms of the stream function defined by Eq. (5), and with

$$\psi' = u \tag{45a}$$

and Eq. (8b), the b-term in Eq. (2) with ε_m given by Eq. (44) can be written as

$$b = (\kappa y)^2 v\lambda_1 + 0.0168(u_e\delta - \psi_e) \tag{45b}$$

The "wall" boundary conditions given by Eqs. (25) and (31) become

$$y = y_0 \; , \quad \psi' = u = u_\tau[\frac{1}{\kappa} \ln \frac{y_0 u_\tau}{n} + c], \quad \frac{\partial\psi}{\partial x} = \frac{u}{u_\tau} \frac{du_\tau}{dx} y_0$$

$$(\kappa y_0)^2 v^2 - u_\tau^2 + y_0 u_e \frac{du_e}{dx} - \alpha^* \frac{du_\tau^2}{dx} y_0 \tag{46}$$

Since u_τ in the above equations is unknown, we use the Mechulfunction formulation discussed in Section 3.0 and set

$$u_\tau' = 0 \tag{47}$$

A common procedure would then be to solve the system given by Eqs. (8b,c), (45a) and (47) with x replacing ξ subject to the four boundary conditions given by Eqs. (4b) and (46). Since the b-term introduces additional nonlinearities into the resulting system, the usual procedure used by almost everybody is to assume this term to be known and apply Newton's method to all terms in the boundary-layer equations except for the b-term. This procedure, though convenient, has the disadvantage that this "nonproper" handling of the b-term causes the solutions to oscillate. A recent work of Cebeci, Chang and Mack [25], which makes use of the Mechulfunction approach by properly handling the linearization of the b-term, avoids these oscillations and causes the solutions to converge quadratically. In this work, the stream function at the boundary-layer edge, ψ_e, which is not known, is written as

$$s' = 0 \tag{48}$$

where s is defined by

$$s = \psi_e \tag{49}$$

and replaces ψ_e in Eq. (45b). We note that with this procedure and with Eq. (47) we have increased the system of first-order equations from three, namely, from Eq. (8b,c), (45a) to five, Eqs. (8b,c), (45a), (47) and (48).

A similar procedure can be used to express the k-ϕ model equations in the form of a first-order system; in addition to those defined by Eqs. (8b), (45a) and (47) we write

$$(c_\mu \frac{k^2}{\phi} v)' + u_e \frac{du_e}{dx} = u \frac{\partial u}{\partial x} - v \frac{\partial \psi}{\partial x} \tag{50}$$

$$(c_\mu \frac{k^2}{\phi} g)' + c_\mu \frac{k^2}{\phi} v^2 - c_k k \frac{\partial u}{\partial x} - \phi = u \frac{\partial k}{\partial x} - g \frac{\partial \psi}{\partial x} \tag{51}$$

$$(c_\mu \frac{k^2}{\phi} q)' + 0.13kv^2 - 4.44c_k \phi \frac{\partial u}{\partial x} - 1.90 \frac{\phi^2}{k} = u \frac{\partial \phi}{\partial x} - q \frac{\partial \psi}{\partial x} \tag{52}$$

where

$$k' = g \tag{53}$$

$$\phi' = q \tag{54}$$

Thus for this system of model equations we have eight first-order equations. Their boundary conditions, in addition to those defined by Eq. (4b) and (46) follow from Eqs. (29a,b) and can be written as

$$y = y_o, \quad \phi = c_\mu \frac{k^2}{\phi} v^2 - c_k k \frac{\partial u}{\partial x}, \quad \phi = (c_\mu \frac{k^2}{\phi} v)^{3/2} / \kappa y_o \tag{55a}$$

$$y = \delta, \quad u \frac{\partial k}{\partial x} + \phi + c_k k \frac{du_e}{dx} = 0,$$

$$u \frac{\partial \phi}{\partial x} + 1.90 \frac{\phi^2}{k} + 4.44 c_k \phi \frac{du_e}{dx} = 0 \tag{55b}$$

It is useful to examine the computer resources required with different turbulence models and corresponding results are provided in Table 1 and correspond to the three flows of Section 6. It is evident that the required computer storage and CPU time increase substantially with the complexity of the turbulence model and, in addition, that the increase in CPU time depends on the nature of the flow. Thus, for example, we see that the calculation of the airfoil flow with a Reynolds-stress equation model is 70 times more expensive than that of the flat plate with an algebraic stress model. The use of an efficient computer program, such as that used here, ensures that the most expensive of the calculations of Table 1 correspond to around 45s of computer time and, compared with many engineering calculations, this may seem small. The application of boundary-layer calculation methods to aid design is, however, complex in that it is likely to form part of an overall calculation which involves the solution of potential-flow equations. Experience with interactive methods for two-dimensional flows shows that some five iterations may be required to achieve convergence. In addition, it is usually necessary to perform calculations for a range of angles of attack and configurations so that the added cost associated with the higher-order turbulence models becomes increasingly important. The added costs associated with the calculation of three-dimensional flows are even greater.

Table 1
Computer Resources Required for the Three Turbulence Models

	Algebraic Eddy-Viscosity Model	Two-Equation Model	Stress-Equation Model
Computer Storage in Bytes Flow of Fig. 2 Flow of Fig. 3 Flow of Fig. 4	96K	124K	332K
Relative CPU Time			
Flow of Fig. 2	1	3	8
Flow of Fig. 3	1.7	6	30
Flow of Fig. 4	2.5	10	70

6.0 APPLICATIONS OF STANDARD METHOD

Many examples of the standard method are described in the literature, for example Cebeci and Bradshaw [3], Bradshaw et al. [5]. As a consequence, only three examples are considered here. They correspond to wall boundary-layer flows and are presented with increasing magnitude of adverse pressure gradient so as to allow assessment of the turbulence models of the previous section. The results are presented in Figures 2 to 4 in terms of variations of local skin-friction coefficient, displacement and momentum thicknesses obtained with each of the three turbulence models of the previous section. As discussed by Cebeci and Meier [26], boundary and initial conditions can influence the calculated results and it

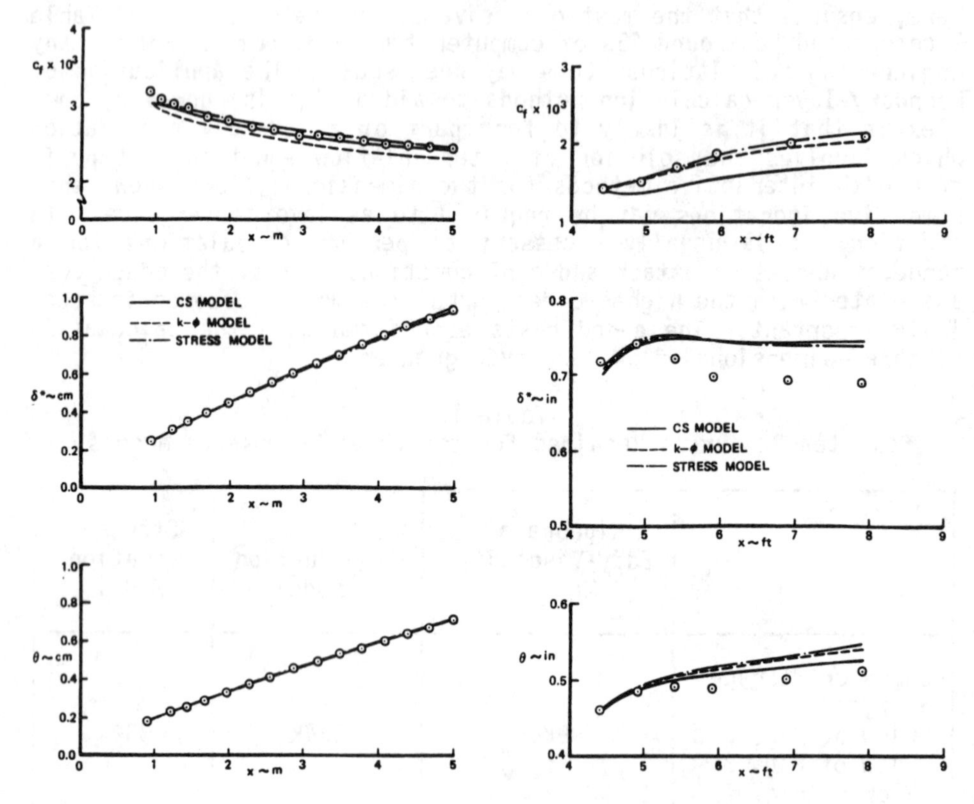

Figure 2. Results for the flat-plate flow of Wieghardt and Tillmann [27].

Figure 3. Results for the relaxing adverse pressure-gradient flow of Bradshaw and Ferriss [29].

is useful to address this question before discussing the results. In each of the three flows, the mean flow properties, that is, the initial velocity profile and wall shear-stress together with the freestream velocity distribution, were identical for the calculation with each model. Similarly, the initial distributions of normal stresses in the Reynolds stress transport model were compatible with the turbulent kinetic energy of the two-equation model and the initial shear-stress distributions were compatible with the initial assumptions for length scale and dissipation rate. It follows, of course, that each distribution of Figures 2 to 4 begins with the same origin.

The results obtained with the flat-plate flow, Wieghardt and Tillmann [27], Figure 2, can be regarded as the best achievable with each model. The variations of the integral properties are essentially identical and in very close agreement with the measurements. The local skin-friction coefficient is best predicted with the algebraic eddy-viscosity equation as might be expected since it

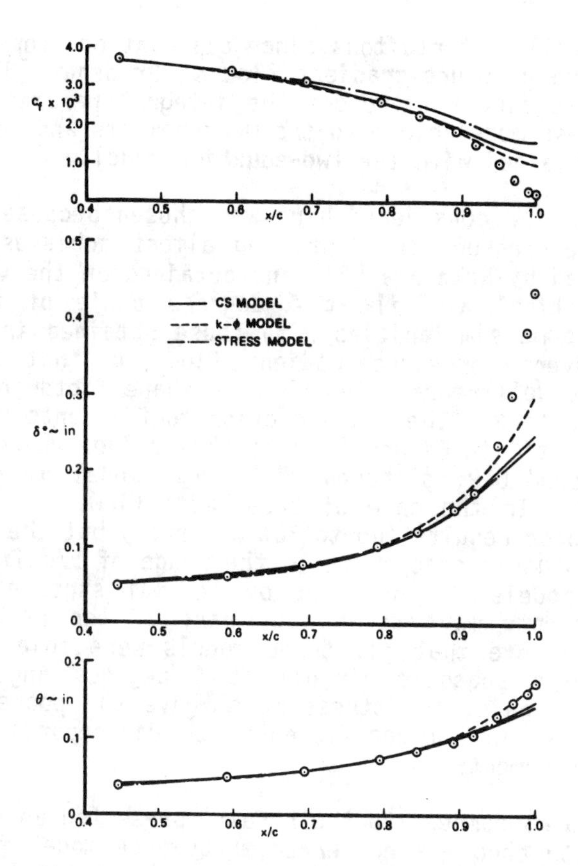

Figure 4. Results for the supercritical airfoil of Nakayama [31].

is essentially a correlation formula based on data such as this. The stress model provides results which are higher by around 3% and the two-equation model results which are lower by around 5%. It should be noted that the initial shear-stress coefficient is lower than the measurement by around 3%; it was chosen to agree with the integral properties and the Ludwig-Tillman [28] drag law rather than the measurements. It is important to remember that the difference between a skin-friction coefficient of, say, 0.0025 and 0.0024 is of the order of measurement accuracy and corresponds to 4%. Thus, the discrepancies of Figure 2 are almost within possible experimental error with all three models.

Figure 3 tells a different story in that none of the three models appears able to represent the relaxing adverse pressure-gradient flow of Bradshaw and Ferriss [29]. The algebraic eddy-viscosity equation allows the best representation of the growth of momentum thickness, but none of the models is able to represent the displacement thickness and the two transport models provide the best values of skin-friction coefficient. This result for the skin friction may well be fortuitous since calculations for the suddenly imposed adverse pressure-gradient flow of Bradshaw [30] allow all three models to almost reproduce the integral parameters with the algebraic stress model close to the measurements and some 12% below the results obtained with the two-equation model.

The third flow considered here was chosen because it involves strong adverse-pressure gradients and almost achieves separation. It was reported by Nakayama [31] and obtained on the upper surface of a supercritical airfoil at 4 degrees angle of attack. The results have some similarities with those obtained in relation to Bradshaw's adverse pressure-gradient flow, in that the integral parameters are well represented up to a shape factor of around one and perhaps up to 2. The skin-friction coefficients are, however, better represented in Figure 4 up to this value, which corresponds to x/c of around 0.9, although the stress model gives noticeably worse results. In the case of Bradshaw's flow, the stress model also yielded poor results (up to 30% in error) but the two-equation model was also unsatisfactory. In the range of x/c from 0.9 to 1, none of the models is adequate, but normal stresses and normal pressure gradients may account for part of the problem. It is interesting to note that all three models were able to represent the flow over a subsonic airfoil at 4 degrees angle of attack although, here again, the stress model gave the poorest representation of skin-friction coefficient with discrepancies up to 25% above the measurements.

There is a randomness to the results presented above which made it difficult to choose a preferred turbulence model for the range of flows considered. Perhaps the only definite conclusion is that the stress-equation model results in high values of skin-friction coefficient in the presence of adverse pressure gradients and after their relaxation.

7.0 APPLICATIONS OF THE INVERSE METHOD

7.1 An External Flow With Specified Displacement Thickness

As a first example of an inverse problem, we consider an external laminar flow in which the specified displacement thickness produces a separation bubble and provides a test to examine the ability and the accuracy of a numerical method to compute a flow with separation. In this case the solution procedure consists of solving Eqs. (8) and (17) subject to the boundary conditions given by Eqs. (9) and (18) with initial conditions, say at $\xi = \xi_0$.

Figures 5 and 6 show two flows computed employing the procedure reported by Cebeci, Keller and Williams [13], together with results obtained by Carter [9] whose numerical procedure and implementation

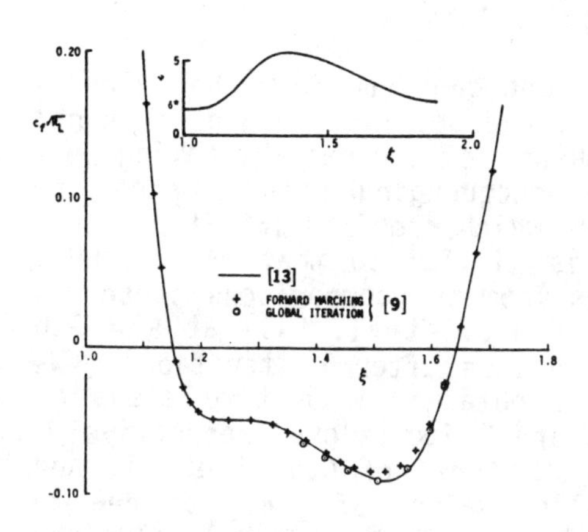

Figure 5. Local skin-friction distribution for flow A.

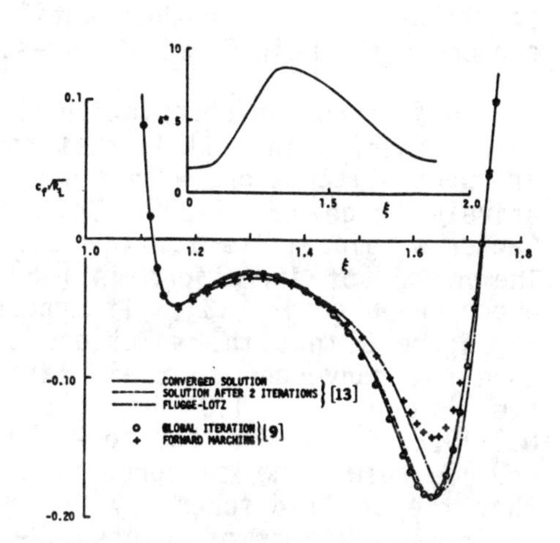

Figure 6. Local skin-friction distribution for flow B.

of the FLARE approximation is slightly different than that of Cebeci et al. Both calculations of Cebeci et al. were started at $\xi = 0$ by solving the governing equations in transformed variables for the standard problem. Then at $\xi = 1$ the method of Section 3.0 was used to solve the inverse problem with the equations expressed in physical variables defined by Eq. (7). As can be seen, both numerical methods are able to compute a separating and reattaching flow well. In Figure 5, the results of Cebeci et al., which were obtained by using only the FLARE approximation, agree well with those of Carter, indicated by forward marching. The results in Figure 5 also indicate that when the recirculation region is small, the FLARE approximation is adequate and a refined procedure indicated by global iteration which improves upon the FLARE approximation has very little effect on the FLARE-computed results.

The results in Figure 6 correspond to a more severe case of a separating and reattaching flow. The recirculation region is bigger and the computed results obtained with different implementation of the FLARE approximation differ from each other. Furthermore, in this case the FLARE approximation is not as accurate and the results obtained by the DUIT procedure and by the global iteration procedure improve the accuracy of the solutions.

The results in Figure 7 correspond to the same flow considered in Figure 6 except that now the DUIT procedure and the global iteration procedures are replaced by a new procedure based on the solution of time-dependent boundary-layer equations discussed in Section 3.0. As can be seen, the first solution generated with the FLARE approximation, is not smooth; one possible reason is that the chosen value of $\Delta\xi$ is too large. At subsequent times the solutions become smoother until by the time τ = 2, they have converged to their final steady-state values.

An interesting observation that one can draw from the calculations reported in [15] is that converged solutions can be obtained in about 5 iterations with a high degree of accuracy by taking relatively large $\Delta\tau$ steps. This is encouraging when compared to Carter's global iteration procedure which required 166 iterations. The number of iterations in [15] is similar to that of the DUIT procedure used in [13]. It appears from the comparisons presented in Figure 8 that the solutions of Cebeci et al. [13] at ξ = 1.6 have not converged to their final values after 7 iterations. We see that at τ = 1.0 the solutions obtained with time steps of 0.1, 0.2, 0.4 which lead to 20, 10 and 5 iterations, respectively, produce results which agree well with those of Cebeci et al. and that the small discrepancy at smaller values of ξ may be due to their large number of points. We also see that the final solutions of the present method (say τ = 6) do not differ significantly from those at τ = 1.0 except for the values in the vicinity of the peak c_f values, ξ = 1.6.

7.2 Calculation of Confined Jets

As a second example, we consider the solution of boundary-layer equations with boundary conditions corresponding to a ducted-jet flow, including those with jet velocities large enough to cause the wall boundary layer to separate and reattach, Figure 9. The

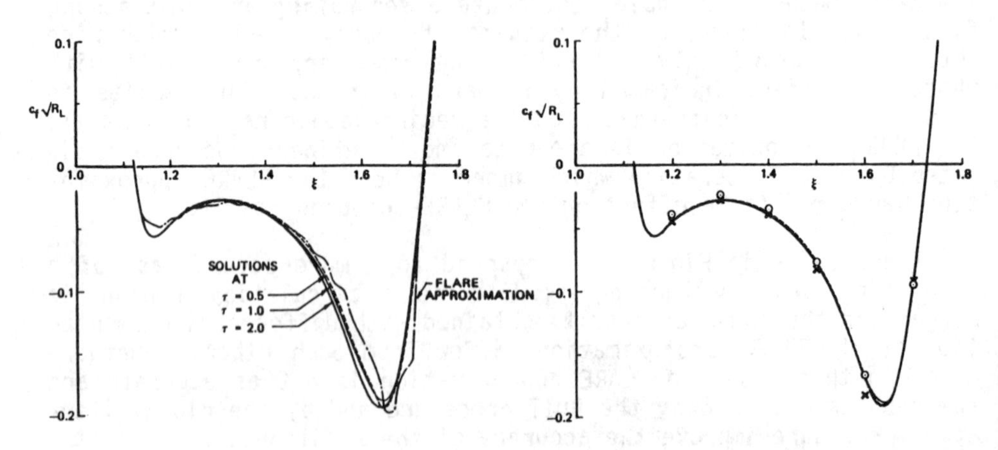

Figure 7. Variation of skin-friction values with time ($\Delta\xi$ = 0.02, $\Delta\eta$ = 0.20, $\Delta\tau$ = 0.10), after [15].

Figure 8. Comparison of computed values with those of Cebeci et al. [13]. ($\Delta\xi$ = 0.02, $\Delta\eta$ = 0.20): o, [13]; x, [15]; solutions at τ = 1.0 with $\Delta\tau$ ___, = 0.1; ---, = 0.2; —.—, = 0.4.

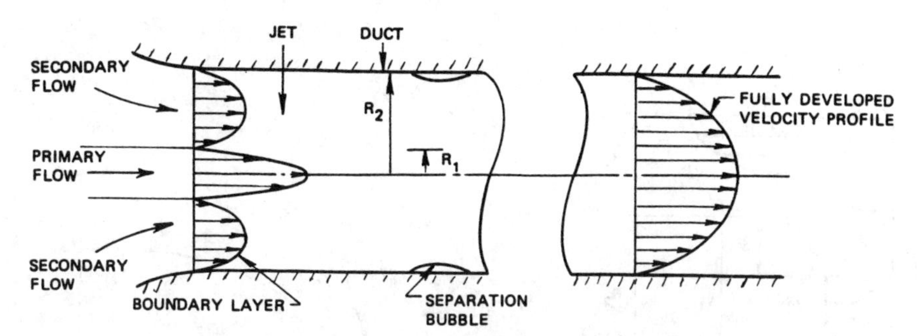

Figure 9. Diagram of ducted jet and notation.

results for laminar flows using the procedure of Section 3.0 are reported by Mean and Cebeci [32] and those for turbulent flow by Fulton, Khattab and Cebeci [33].

In internal flows, the boundary conditions corresponding to Eq. (18) can be obtained from the integral form of mass continuity, which for a two-dimensional flow can be expressed as

$$\phi_o(y)R_2 = \int_0^{R_2} u\,dy \tag{56}$$

Here R_2 denotes the half-width of the duct and $\phi_o(y)$ represents the velocity profile at $\xi = \xi_o$. Using the transformation given by Eq. (7), taking $L = R_2$, ϕ_o/u_o, Eq. (56) can be written as

$$y_c = \sqrt{R_L} \qquad f = \tilde{\phi}_o\sqrt{R_L} \tag{57}$$

and can be used as the fourth boundary condition to solve the inverse problem given by Eqs. (8), (9) and (17) for a given initial condition. In the study reported by Mena and Cebeci, the initial velocity profiles were obtained by using simple sine functions for the secondary flow and a parabolic function for the primary, as shown in Fig. 10 and as given by Eq. (58)

$$u = \begin{cases} u_{s1} \sin \frac{\pi}{2} \left(\frac{y}{\delta}\right) & \text{Region } 1 \\ u_{s1} & 2 \\ u_{s1} \sin \frac{\pi}{2} \left(\frac{R_2 - R_1 - y}{\delta}\right) & 3 \\ u_{p1} \left[1 - \left(\frac{R_2 - y}{R_1}\right)^2\right] & 4 \end{cases} \tag{58}$$

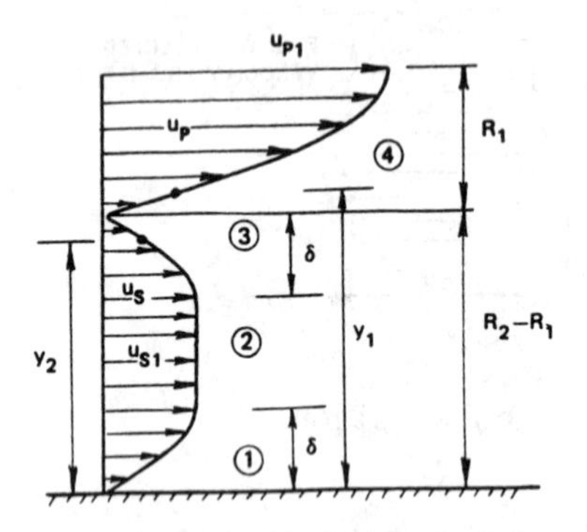

Figure 10. Sketch of the initial velocity profile.

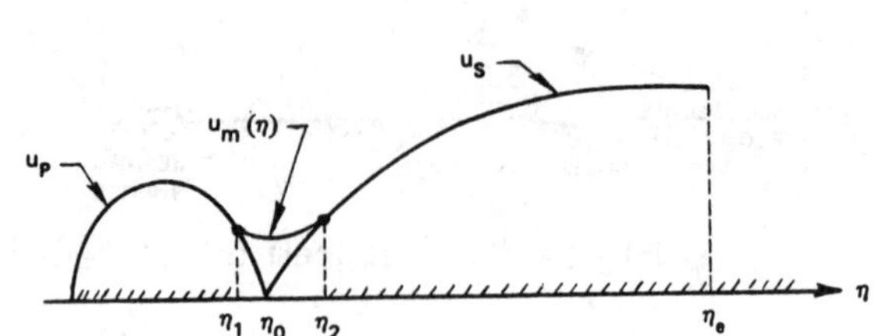

Figure 11. Notation for the blending velocity profile $u_m(\eta)$.

At the transition between the primary and secondary velocity profiles (i.e., between regions 3 and 4) a discontinuity exists. To avoid this undesirable situation, a smoothing function, known as Hermite, or osculatory interpolation (see [34, p. 192]) procedure was used and a blending velocity profile $u_m(\eta)$ defined by (see Fig. 11)

$$\psi_1(\eta) = \left(1 - 2\,\frac{\eta - \eta_1}{\eta_1 - \eta_2}\right)\left(\frac{\eta - \eta_2}{\eta_1 - \eta_2}\right)^2 ,$$

$$\psi_2(\eta) = \left(1 + 2\,\frac{\eta - \eta_1}{\eta_1 - \eta_2}\right)\left(\frac{\eta_1 - \eta}{\eta_1 - \eta_2}\right)^2 ,$$

$$\Psi_1(\eta) = (\eta - \eta_1)\left(\frac{\eta - \eta_2}{\eta_1 - \eta_2}\right)^2 \tag{59}$$

$$\Psi_2(\eta) = (\eta - \eta_2)\left(\frac{\eta_1 - \eta}{\eta_1 - \eta_2}\right)^2$$

The relation between η and y is

$$\eta = \frac{1}{2}\;R_1 + R_2 - y$$

Three series of confined jets were computed as summarized in Table 2 and described in Mena's thesis [35]. Examination of these results reveals that in laminar, confined-jet flows of the type found in injectors, with ratios of duct to jet width of 2, 4 and

8, regions of separated flow are formed on the dut walls provided the ratio of jet to secondary flow is greater than 8. The length of the region of recirculation increases and the location of separation moves upstream as the velocity ratio is increased from 8 to 10.

Table 2. Summary of test cases reported in [35]

Case	u_{p1}/u_{s1}	$R_1/(R_2 + R_1)$
1	1/1	1/2
2	3/1	1/2
3	5/1	1/2
4	8/1	1/2
5	10/1	1/2
6	1/1	1/4
7	5/1	1/4
8	8/1	1/4
9	10/1	1/4
10	1/1	1/8
11	5/1	1/8
12	8/1	1/8
13	10/1	1/8

A representative sample of results is presented in Figures 12 to 14. Figures 12 and 13 show the velocity profiles for Cases 7 and 9 and Fig. 14 the streamlines for case 9.

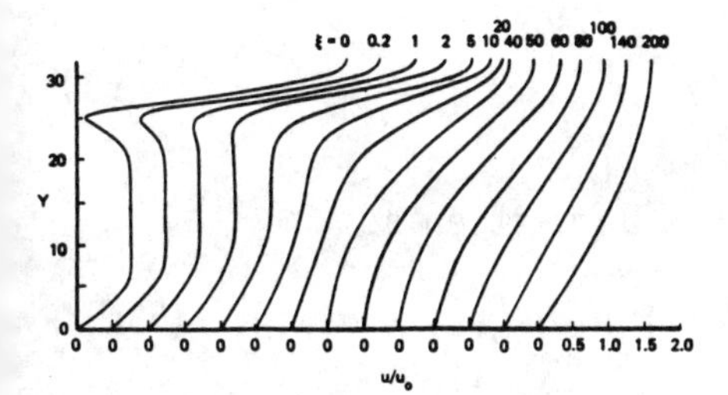

Figure 12. Velocity profile at selected duct locations for case 7.

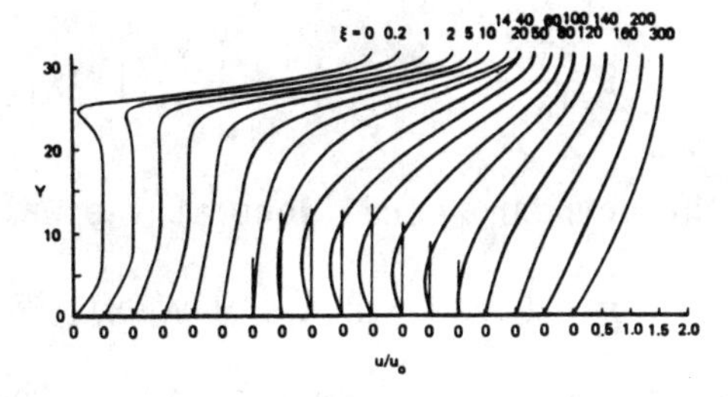

Figure 13. Velocity profiles at selected duct locations for case 9.

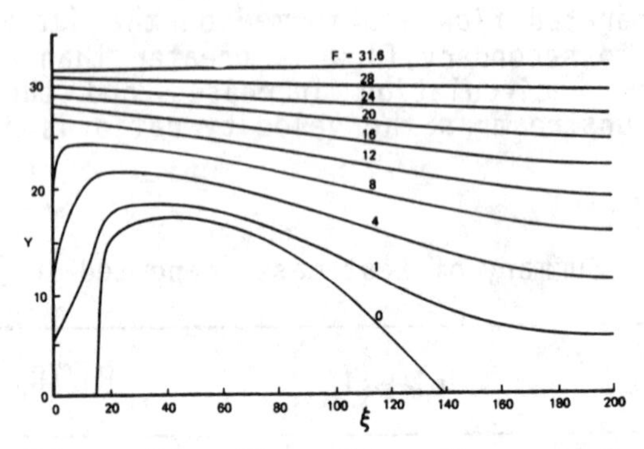

Figure 14. Streamlines for case 9.

7.3 Vertical Duct

As a third example of an inverse problem, we consider flow in a vertical axisymmetric duct where the driving force is due to forced convection, or to a combination of both forced and natural convection. For some combinations of wall temperature corresponding to wall cooling and high values of the ratio of Grashof to Reynolds number the boundary layers separate from the wall and later, at some downstream location, reattaches. The inverse method of Section 3.0 provides a good scheme to compute such a flow as was recently demonstrated by Cebeci et al. for both turbulent and laminar flows. Here, for simplicity, we shall consider only the case for laminar flow and discuss the solution of the axisymmetric boundary-layer equations including the buoyancy effects for a vertical duct. The governing equations are well known and can be written as

$$\frac{\partial}{\partial x}(ru) + \frac{\partial}{\partial y}(rv) = 0 \tag{60}$$

$$u\frac{\partial u}{\partial x} + v\frac{\partial u}{\partial y} = -\frac{1}{\rho}\frac{dp}{dx} + \frac{1}{\rho r}\frac{\partial}{\partial y}\left[r\left(\mu\frac{\partial u}{\partial y}\right)\right] + g_c\beta(T - T_o) \tag{61}$$

$$u\frac{\partial T}{\partial x} + v\frac{\partial T}{\partial y} = \frac{1}{\rho c_p}\frac{1}{r}\frac{\partial}{\partial y}\left[r\left(k\frac{\partial T}{\partial y}\right)\right] \tag{62}$$

The boundary conditions at the wall can be written as

$$y = 0; \quad u = 0, \quad v = v_w(x), \quad \alpha_0 T_w(x) + \alpha_1 T_w'(x) = \text{given} \tag{63a}$$

where $\alpha_0 = 0$ corresponds to the specified wall heat flux and $\alpha_1 = 0$ to specified wall temperature.

Depending on the shear-layer thickness, the "outer" boundary conditions may be specified either at the edge of the shear layer,

$$y = \delta, \quad u = u_e(x), \quad T = T_o \tag{63b}$$

or at the centerline of the duct,

$$y = y_c, \quad \frac{\partial u}{\partial y} = 0, \quad \frac{\partial T}{\partial y} = 0 \tag{63c}$$

In the early stages of the flow when the shear layers are developing, the core velocity increases continuously due to the growth in displacement thickness. The flow in this region, which may be called the displacement-interaction region, see Cebeci and Bradshaw [3], is physically no more than the boundary-layer flow in a mild favorable pressure gradient and it is more appropriate to use the outer boundary conditions given by Eq. (63b) than those given by Eq. (63c). On the other hand, when the thickness of the shear layer becomes comparable to the half-width or radius of the duct, it is more appropriate to use the outer boundary conditions given by Eq. (63c). Here, for simplicity, we shall only consider the case before the shear layers merge.

The presence of the dp/dx-term in Eq. (61) introduces an additional unknown to the system given by Eqs. (60) to (63). Thus, another equation is needed and is provided by the conservation of mass in integral form. For a circular pipe, mass balance gives,

$$u_o \pi r_o^2 = \int_0^{r_o} 2\pi r u dy \tag{64}$$

Equations (60) through (64) can be placed into a more convenient form by using a combination of Mangler and Falkner-Skan transformations in which we define

$$d\eta = \left(\frac{u_o}{\nu x}\right)^{1/2} \left(\frac{r}{r_o}\right) dy, \quad u = \frac{x}{r_o} \frac{1}{R}, \quad \psi = (u_o \nu x)^{1/2} f(\xi,\eta) \tag{65}$$

Then, as is discussed in [36], the governing equations with primes denoting differentiation with respect to η and θ, and w denoting dimensionless temperature and pressure, respectively,

$$\theta = \frac{T - T_o}{T_w - T_o}, \qquad w = \frac{p}{\rho u_o^2} \tag{66}$$

Eqs. (60) to (62) can be written as

$$(bf'')' + \frac{1}{2} ff'' = \xi\left(\pm\lambda\theta + f' \frac{\partial f'}{\partial \xi} - f'' \frac{\partial f}{\partial \xi} + \frac{dw}{d\xi}\right) \tag{67}$$

$$(e\theta')' + \frac{1}{2} f\theta' = \xi\left(f' \frac{\partial \theta}{\partial \xi} - \theta' \frac{\partial f}{\partial \xi}\right) \tag{68}$$

where ± denotes flow conditions in which the walls of the duct are heated (aided) and cooled (opposed). In the above equations,

$$\lambda = \frac{Gr}{R}, \qquad Gr = \frac{g_c \beta r_o^3 (T_w - T_o)}{\nu^2}, \qquad R = \frac{u_o r_o}{\nu} \tag{69a}$$

$$b = (1 - t)^2, \qquad e = (1 - t)^2 \frac{1}{Pr} \tag{69b}$$

$$t = 1 - [1 - 2(\xi^{1/2})\eta]^{1/2} \tag{69c}$$

In terms of transformed variables, the boundary conditions given by Eqs. (63a,b) can be written as

$$\eta = 0, \qquad f = f' = 0, \qquad \theta = 1 \tag{70a}$$

$$\eta = \eta_e, \qquad f' = u_e/u_o = \bar{u}_e \qquad \theta = 0 \tag{70b}$$

Similarly it can be shown that Eq. (64) with $\eta_{sp} = 1/2 \; 1/\sqrt{\xi}$ becomes

$$f(\eta_{sp}) = \eta_{sp} \tag{71}$$

Noting that $f' = \bar{u}_e$ at $\eta = \eta_e$ and that the dimensionless stream function f varies linearly with η, Eq. (71) can also be written as

$$f_e + f_e'(\eta_{sp} - \eta_e) = \eta_{sp} \tag{72}$$

The system given by Eqs. (66), (68), (70) and (72) can now be solved by using the procedure described in Section 3.0. In this case we have six first-order equations rather than four.

The method described in the previous sections is applied to laminar and turbulent flows in horizontal and vertical round ducts. When possible, computed results are compared to experimental data as well as to solutions obtained by other numerical procedures. For flows in vertical ducts, calculations are reported for various values of λ with wall conditions corresponding to heat transfer from the gas to the wall. This condition gives rise to flow separation which is a major interest to the present investigation. Calculations encompass a range of values of λ, Prandtl number, Reynolds number and include some with transition specified at different locations [36].

With cooling, depending on the value of λ, the boundary-layer flow will separate under the buoyancy effect; with increasing ξ, however, the buoyancy effect will decrease and, in the limit, the flow will become fully developed and will exhibit the characteristics of the flow corresponding to a forced convection, $\lambda = 0$. A series of runs with different values of λ indicate that the separation location ξ_s in a vertical circular pipe with wall conditions corresponding to cooling, varies with λ as shown in Fig. 15. In addition, it can be seen that the flow reattaches. As λ increases, the separation location moves forward with the size of the recirculating region increasing with λ and with reattachment location moving downstream.

Figure 16 shows the variation of $c_f R$, where the local skin friction coefficient is defined by $c_f = \tau_w/(1/2)\rho u_c^2$, along the duct for various values of λ. It is clear that at higher values of λ, in accord with results of Fig. 15, the separation bubble is large and that the numerical method is able to cope with recirculating flow without any signs of numerical difficulties. Also, distributions converge to the fully developed laminar pipe flow, a value of 8, and it should be noted that the corresponding values in the recirculation region can greatly exceed this. Figure 17 shows the corresponding distributions of Nusselt number and it can be seen that the high values of result λ in significant reductions in Nusselt number, particularly within the recirculation region. Figures 16 and 17 show that, as expected, the distance to achieve fully developed flow increases with λ. Figure 18 shows the variation of the centerline velocity with λ and confirms that the entrance velocity length increases considerably with λ. For example, when $\lambda = 0$, $\xi \sim 0.3$, while for $\lambda = 72$, $\xi \sim 1.1$, etc. The large increase in centerline velocity, and its subsequent fall, are associated with the region of recirculating flow. In the near wall region, the buoyant forces cause the local velocity to reverse and

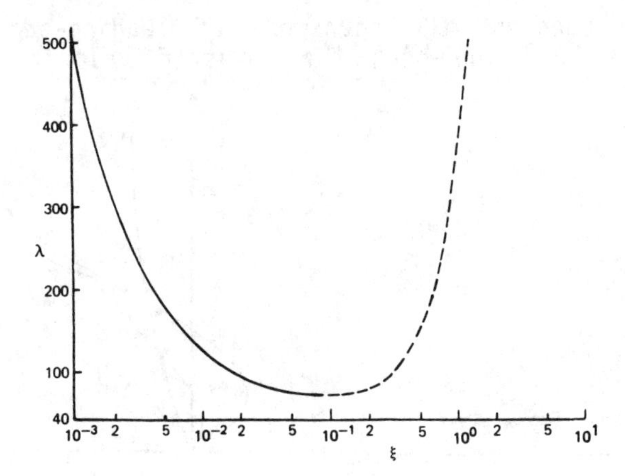

Figure 15. Separation and reattachment locations for various values of λ. Solid line denotes separation locations and dashed line reattachment location.

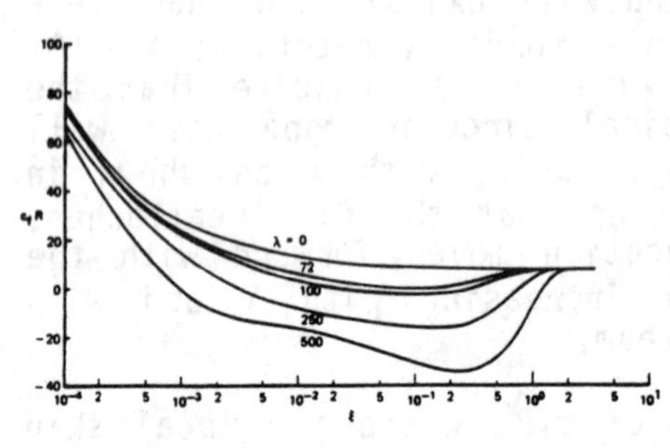

Figure 16. Effect of λ on the variation of $c_f R$ for laminar flow with separation, Pr = 0.72.

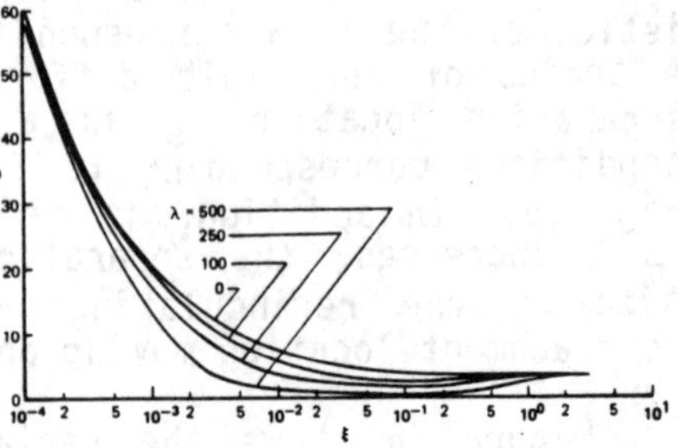

Figure 17. Effect of λ on Nusselt number for laminar flow in a circular duct, Pr = 0.72.

conservation of mass causes the forward-velocity region to accelerate. This is made clear by the profiles of Figure 19, which clearly show that the velocities at ξ of 0.0008 and 3.0 are all in the positive ξ-direction, whereas, in the region between, negative velocities occur in the near-wall region. Figure 19 corresponds to a value of λ of 600 and this has been chosen to display more clearly the effects of buoyancy on the velocity profiles.

8.0 APPLICATIONS OF THE INTERACTIVE METHOD

8.1 Separation and Reattachment Near the Leading Edge of a Thin Airfoil at Incidence

This example concerns the phenomena of leading-edge bubbles on thin airfoils [37] in which the inviscid velocity distribution

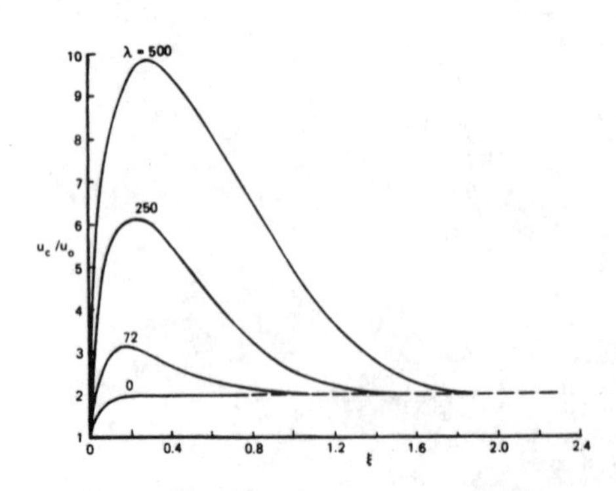

Figure 18. Effect of λ on the variation of centerline velocity of a laminar flow in a circular duct, Pr = 0.72.

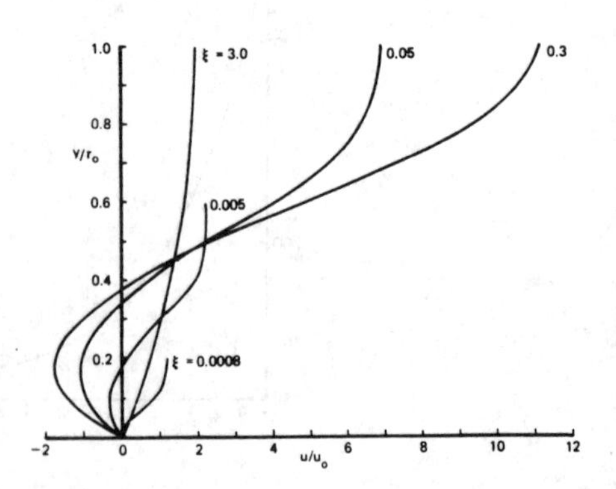

Figure 19. Velocity profiles in a circular pipe for λ = 600 and Pr = 0.72.

with ξ_o corresponding to an angle of incidence is given by

$$u_e = \frac{\xi + \xi_o}{\sqrt{1 + \xi^2}} \tag{73}$$

In this case the solution procedure is identical to that used for the inverse method except that the boundary condition given by Eq. (18) is replaced by Eqs. (21) and (23). The interactive method permits a limited amount of separated flow, as shown in Figs. 20 and 21, before it fails. A rational theory of high-Reynolds number laminar flow supports this conclusion and a likely interpretation of the theoretical results is that dramatic changes take place in the global flowfield at high angles of attack, in the region in which the interactive theory failed.

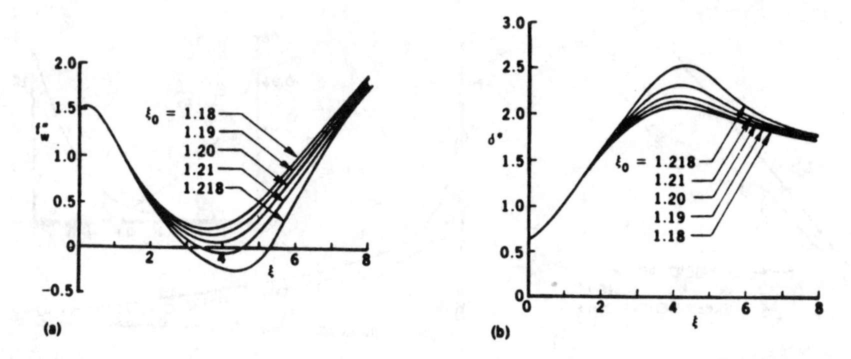

Figure 20. Results for a thin airfoil at various reduced angles of attack, ξ_o. (a) Wall shear parameter, f''_w. (b) Displacement thickness, δ^*.

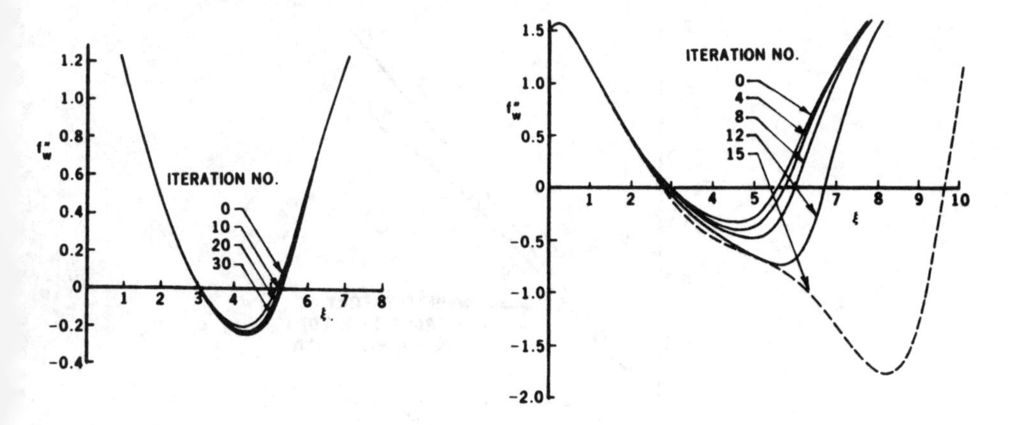

Figure 21. Variation of wall shear parameter f''_w with number of iterations. (a) ξ_o = 1.218. (b) ξ_o = 1.220.

8.2 Airfoils with Trailing Edge Separation

The interactive method of Section 4 has also been applied to the calculation of separation bubbles on single airfoils, in addition to the model problem of Section 8.1, as well as to flows with trailing-edge separation [38]. Figures 22 and 23 show the results for NACA 4412 and 0012 airfoils, respectively. As can be seen, the comparison with experiment is good up to the stall angle and a slight discrepancy with data can be attributed to the use of the FLARE approximation which becomes increasingly more important with large regions of recirculation and also due to wake effects which were neglected in the calculations.

There is no question that the interactive boundary-layer theory with attached inviscid flow can be used in all cases of separated

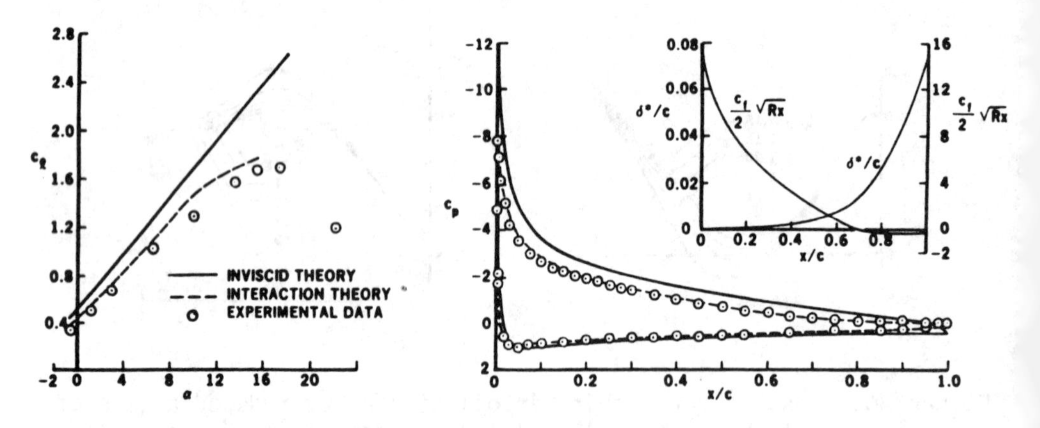

Figure 22. Comparison of results for the NACA 4412 airfoil.

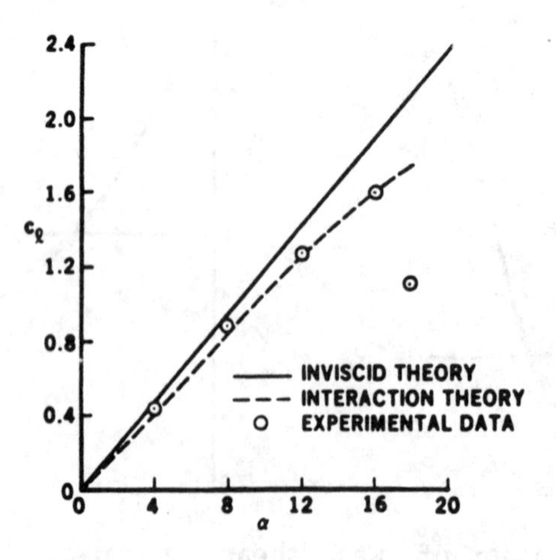

Figure 23. Comparison of results for the NACA 0012 aircraft.

flow. Either it must break down by the separated region becoming larger and larger, both longitudinally and laterally, until the assumptions of boundary-layer fail or by a sudden change in the global structure of the flow from attached to detached. It is possible that the latter situation may develop independently of boundary-layer theory, in which case the interactive theory would cease to be relevant.

1. Crank, J. and P. Nicolson. A Practical Method of Numerical Evaluation of Solutions of Partial-Differential Equations of the Heat-Conduction Type. Proceedings Cambridge Phil. Soc. 43, (1947) 50.

2. Keller, H.B. A New Difference Scheme for Parabolic Problems. Numerical Solutions of Partial-Differential Equations, Vol. II,J. Bramble, ed. (Academic Press, 1970).

3. Cebeci, T. and P. Bradshaw. Physical and Computational Aspects of Convective Heat Transfer (New York, Springer-Verlag, 1984).

4. Cebeci, T. and A.M.O. Smith. Analysis of Turbulent Boundary Layers. (Academic Press, 1974).

5. Bradshaw, P., T. Cebeci and J.H. Whitelaw. Engineering Calculation Methods for Turbulent Flow. (London, Academic Press, 1981).

6. Klineberg, J.M. and J.L. Steger. On Laminar Boundary-Layer Separation. AIAA Paper 74-94 (1974).

7. Horton, H.P. Separating Laminar Boundary Layers with Prescribed Wall Shear. AIAA Journal. 12, (1974) 1772.

8. Carter, J.E. Solutions for Laminar Boundary Layers with Separation and Reattachment. AIAA Paper No. 74-583 (1974).

9. Carter, J.E. Inverse Solutions for Laminar Boundary-Layer Flows with Separation and Reattachment. NASA TR R-447 (1975).

10. Williams, P.G. A Reverse Flow Computation in the Theory of Self-Induced Separation. Proc. 4th Int. Conf. Numer. Meth. Fluid Mech., Lecture Notes in Physics 35, (1975) 445.

11. Cebeci, T. Separated Flows and Their Representation by Boundary Equations. Mech. Eng. Rept. ONR-CR215-234-2, California State University, Long Beach (1976).

12. Pletcher, R.H. Prediction of Incompressible Turbulent Separating Flow. Journal of Fluids Engrg. 100 (1978) 427.

13. Cebeci, T., H.B. Keller, and P.G. Williams. Separating Boundary-Layer Calculations. Journal of Computational Physics 31 (1979) 363.

14. Reyhner, T.A. and I. Flügge-Lotz. The Interaction of a Shock Wave with a Laminar Boundary Layer. International Journal of Nonlinear Mech. 3 (1968) 173.

15. Cebeci, T. A Time-Dependent Approach for Calculating Steady Inverse Boundary-Layer Flows with Separation. Proceedings Royal Society London A 389 (1983) 171-178.

16. McDonald, H. and W.R. Briley. A Survey of Recent Work on Interacted Boundary-Layer Theory for Flow with Separation. In Numerical and Physical Aspects of Aerodynamic Flows II (ed. T. Cebeci) (New York, Springer-Verlag, 1984) 148.

17. LeBalleur, J.C. Couplage visqueux-Non-visqueux Analyse du probléme incluant décollements et ondes de choc. LeRech. Aerosp. 1977-6, (1977) 349. English translation E5A-476.

18. Carter, J.E. A New Boundary-Layer Inviscid Interaction Technique for Separated Flow. AIAA Paper 79-1450 (1979).

19. Veldman, A.E.P. New, Quasi-Simultaneous Method to Calculate Interacting Boundary Layers. AIAA Journal 19 (1981) 79.

20. Cebeci, T. and S.M. Schimke. The Calculation of Separation Bubbles on Interactive Turbulent Boundary Layers. Journal of Fluid Mech. 131, (1983) 305.

21. Cebeci, T. and R.W. Clark. An Interactive Approach to Subsonic Flows with Separation. In Numerical and Physical Aspects of Aerodynamic Flows II, (ed. T. Cebeci) (New York, Springer-Verlag, 1984) 193.

22. Jones, W.P. and B.E. Launder. The Prediction of Laminarization With a Two-Equation Model of Turbulence. International Journal of Heat & Mass Transfer, 15 (1972) 301.

23. Hanjalic, K. and B.E. Launder. Sensitizing the Dissipation Equation to Irrotational Strains. Journal of Fluid Engineering, Trans. ASME 102 (1980) 34-40.

24. Gibson, M.M. and B.E. Launder. Ground Effects on Pressure Fluctuations in Atmosphere Boundary Layer. Journal of Fluid Mech. 86 (1978) 491.

25. Cebeci, T., K.C. Chang and D.P. Mack. On the Linearization of Turbulent Boundary-Layer Equations. To be published in the AIAA Journal (1984).

26. Cebeci, T. and H.U. Meier. Modelling Requirements for the Calculation of the Turbulent Flow Around Airfoils, Wings and Bodies of Revolution. AGARD CP 271, Paper 16 (1979).

27. Wieghardt, K. and W. Tillmann. On the Turbulent Friction Layer for Rising Pressure. Translation in ARC, R&M 1314 (1951).

28. Ludwig, H. and W. Tillmann. Untersuchungen uber die Wandschubspannung in turbulenten reibungsschichten. Ing. Arch. 17 (1949) 288,. English translation in NACA TM 1285 (1950).

29. Bradshaw, P. and D.H. Ferriss. The Response of a Retarded Equilibrium Turbulent Boundary Layer to the Sudden Removal of Pressure Gradient. NPL Aero Rept. 1145 (1965).

30. Bradshaw, P. The Response of a Constant Pressure Boundary Layer to the Sudden Application of an Adverse Pressure Gradient. NPL Aero Rept. 1219 (1967).

31. Nakayama, A. Measurements of Attached and Separated Turbulent Flows in the Trailing-Edge Regions of Airfoils. In Numerical and Physical Aspects of Aerodynamic Flows II (ed. T. Cebeci) (New York, Springer-Verlag, 1984).

32. Mena, A.L. and T. Cebeci. Calculation of Steady Confined Jets for Two-Dimensional Plane Laminar Flows. Computer Methods in Applied Mechanics and Engineering 35 (1982) 67-86.

33. Fulton, W., A.A. Khattab and T. Cebeci. Calculation of Steady Confined Jets for Two-Dimensional Turbulent Flows. Paper in preparation (1984)

34. Isaacson, E. and H.B. Keller. Analysis of Numerical Methods. (New York, Wiley, 1966).

35. Mena, A.L. The Numerical Solution of Two-Dimensional Laminar Confined Jet. M.S. Thesis, Department of Mechanical Engineering, California State University, Long Beach, California (1981).

36. Cebeci, T., K. Lee, S. Wong and K.C. Chang. Heat Transfer in Vertical Duct Flows. Paper in preparation (1984).

37. Cebeci, T., K. Stewartson and P.G. Williams. Separation and Reattachment Near the Leading Edge of a Thin Airfoil at Incidence. AGARD CP 291 (1981) 20.

38. Cebeci, T. Problems and Opportunities With Three-Dimensional Boundary Layers. Presented at AGARD Fluid Dynamics Panel, Round Table Discussion on Three-Dimensional Boundary Layers, VKI, Brussels, 24 May 1984.

NOTATION

c_f	local skin-friction coefficient
c_p	specific heat at constant pressure
f	dimensionless stream function, Eqs. (7) and (65)
Gr	Grasshoff number, Eq. (69)
h	increment in variable Y, ΔY
k	increment in variable ξ, $\Delta\xi$; or turbulent kinetic energy; or heat conductivity
L	reference length
p	static pressure
$\bar{p}$	dimensionless static pressure, $p/\rho(u_0^2/2)$
Pr	Prandtl number
R, R_L	Reynolds number based on r_o and L, $u_o r_o/\nu$ and $u_o L/\nu$, respectively
r	radial distance
r_o	radius of the duct
s	surface distance
T	static temperature
T_o	reference temperature
u,v,w	velocity components in x-,y- and z-directions

u_o	reference velocity
u_τ	friction velocity
$-\overline{u'v'}, \overline{u'^2}, \overline{v'^2}, \overline{w'^2}$,	Reynolds stresses
x,y	orthogonal boundary-layer coordinates
Y	dimensionless y-coordinate, Eq. (7)
y_o	y at a small distance from the wall
y_o^+	Reynolds number based on y_o and friction velocity, $y_o u_\tau/\nu$
β	coefficient of thermal expansion
δ	boundary-layer thickness or finite-difference operator
δ^*	displacement thickness
ε_m	eddy viscosity
ε_m^+	dimensionless eddy viscosity, ε_m/ν
η	transformed y-coordinate, Eq. (65)
θ	dimensionless temperature, Eq. (66)
μ	dynamic viscosity
ν	kinematic viscosity
ξ	dimensionless x-coordinate, x/L
ρ	density
τ	shear stress or dimensionless time
ϕ	turbulence dissipation rate
ψ	stream function
ω	relaxation factor

Subscripts

c	centerline of the duct
e	edge of the boundary layer
o	initial value
w	wall

Primes denote differentiation with respect to Y or η.

VISCOUS-INVISCID FLOW INTERACTION METHODS

Georges MEAUZE

Office National d'Etudes et de Recherches Aérospatiales,
B.P 72, 92322 Châtillon Cedex, France

1 INTRODUCTION

The aerodynamic study of a turbomachinery needs 3D unsteady viscous flow computations.

Strictly speaking that is only possible by the resolution of the complete Navier-Stokes (N.S.) equations.

First encouraging results have been obtained lately thanks to the efforts undertaken by a lot of research workers all around the world, but some difficulties have still to be overcome.

They are of two sorts :

- First about practical applications existing computation codes are often very complex and cannot be considered as operational yet. Moreover they need very powerful computers.

- Second in especially for turbulent flow certain simplifications are necessary, and the models at our disposal don't take very well into account physical reality yet.

Improvements that will necessarily occur will make the applications of these methods more and more easy and accurate, but there is another way of describing the real behaviour of a fluid, that consists in using coupling techniques.

The basic principle harks back to split, whenever it is possible, the flow into domains of so-called perfect fluid flow where viscosity is negligible, and domains of viscous flow, where viscosity effects play a leading part. Therefore the equations governing the flow in the different domains can be different and more or less simplified.

The first step of coupling has been used for a long time. It consists in making a perfect fluid computation, this step is followed by a boundary layer computation on the airfoil or duct walls that gives the displacement thickness. Then the perfect fluid computation result is corrected to take into account the displacement thickness.

Viscous-inviscid flow coupling (or interaction) methods offers much cheaper computation possibilities than a full N.S. method.

Nevertheless we must note that they can only be applied to flows where viscous or inviscid zones can be distinguished. For instance, the case of long ducts where the viscous zones meet cannot be treated.

In other respects, it should be observed that the modellings often used for the N.S. equations are necessarily used into the viscous flow computations.

Nowadays, this coupling approach knows a lot of developings and is widely used in external aerodynamic and more recently in turbomachineries ; nevertheless, it is possible that the applications would remain advantageous and interesting only for a short or mean term. Thorough examination of the coupling approach can be found in (1) to (6).

This study proposes to enumerate the different coupling methods available and their applications to the turbomachineries, i.e. concerning internal aerodynamics point of view. It means that we will consider first of all computation domains with boundaries at a finite distance such as cascades, rows, intakes or nozzles.

To make this synthesis clear, we must consider the different possible approaches of coupling, according to the distinctive cases enumerated hereafter (not classed) and detailed successively in next paragraphs.

- Degree of simplification of equations as well for perfect as for real fluid.

- Conditions of coupling tied to the nature of the frontier on which are expressed the compatibility relations for the two flows : perfect/viscous flow.

- Kind of the coupling mode, i.e. use of direct or inverse boundary conditions on the frontier for perfect and viscous flow.

The choice of the coupling method can depend on the kind of flow on the one hand and on the computation aim (design or analysis) on the other hand.

Several examples from the literature show applications as far as possible.

Before specifying the different kinds of coupling techniques, we should first recall the terminology used sometimes, that distinguishes between a *weak interaction* and a *strong interaction* coupling.

The weak interaction corresponds to flows with a very small viscous layer (compared with the considered domain) that has a negligible effect on the inviscid flow. Then the whole field can be determined just by an inviscid flow computation, on which is superposed a boundary layer calculation that will not question flow's properties and particulary pressure distributions.

Instead we call strong interaction the cases where the viscous behaviour has an important effect on perfect fluid. Real fluid cannot be described at once just by an inviscid computation. The mutual influence of the viscous and inviscid domains requires a real coupling.

2 DISTINCTIVE CHARACTERISTICS OF THE DIFFERENT KINDS OF COUPLING

2 .1 Equations' Degree of Simplification

A lot of simplifications can be considered, as well for perfect fluid as for viscous flow. Our point is not to undertake an exhaustive synthesis of the available methods for computing perfect fluid on the one hand, viscous flow on the other hand ; we shall restrict ourselves to just recalling the different steps of simplification.

2 .1.1 Inviscid Flow

- Potential equation in incompressible flow

It's application field is limited to low speed flows, but a lot of methods are operational for complex configurations.

- Potential equation in compressible flow

It allows to deal with compressible irrotational (or considered as irrotational) flows, possibly transonic and with weak shock waves.

The flow has not to be choked (without a sonic throat). This limitation is overwhelming for internal flow computations. A lot of codes have been made, principally based on finite differences, finite volumes and finite elements methods.

- EULER equations

They represent the exact fluid mechanic equations, without viscosity. They allow to treat the general case of rotational mixed flow, i.e. comprising subsonic and supersonic zones with strong shock waves.

As often as not, the solving techniques use iterative methods, and need rather long computation times. In fact, these iterative methods include a wide variety of explicit and implicit, using all the unsteady terms (unsteady or time-marching methods) or just a part (pseudo-unsteady methods) and operate with finite differences, finite volumes or finite elements formulations. A lot of programs have been developed, especially for turbomachineries, in 2D as well in 3D flows.

Let us recall the characteristics method, restricted to supersonic flows but that remains very useful for some applications.

Finally, the choice of a perfect fluid computation method for a coupling is based not only on the application field considered by also -as we shall see- by the kind of boundary condition used for the mode of coupling.

A synthesis of the different perfect fluid computations in turbomachineries has been done in (7). Improvements have been realised since then, especially for Euler equations resolution in 2D and 3D, taking into account the coming out of new computers very efficient for their number of memories and computing speed.

2 .1.2 Viscous Flow

Thorough examination of viscous flow calculations can be found in (6). Let us recall the next classification :

- the complete N.S. equations can be used in viscous regions. Because of the small turbulence scales, the equations we are able to solve :

. either correspond to a laminar flow (very low Reynolds number)

. or are averaged equations, where turbulence global effects are represented by a model

. or are filtred equations where the large scales are taken into account and the small replaced by a model.

The use of N.S. equations for the viscous layer, joined with the Euler equations for inviscid flow as in (8) uses a coupling techniques indeed, but is still related to the resolution of the complete set of equations. The word "coupling" is generally connected with a computation where the viscous terms are schematized at a higher degree.

- the viscous layer is computed by methods based on an asymptotic development or on perturbation techniques. These techniques are very interesting, at a basic point of view, but are still rarely used (60), (9), (10), (11).

. the most widely used models employ the boundary-layer concept, that is to say Prandtl equations, that do not take into account the transverse pressure gradients.

- continuity :

$$\frac{\partial(\rho u)}{\partial x} + \frac{\partial(\rho v)}{\partial y} = 0 \tag{1}$$

- streamwise momentum :

$$\rho u \frac{\partial u}{\partial x} + \rho v \frac{\partial u}{\partial y} = - \frac{dp}{dx} + \frac{\partial}{\partial y} \left((\mu + \mu_\tau) \right) \frac{\partial u}{\partial y} \tag{2}$$

In these equations all the symbols are relative to mean quantities (Reynolds' averaging) and μ_t is the turbulent eddy viscosity defined by :

$$\mu_\tau \frac{\partial u}{\partial y} = - \rho \, \overline{u'v'}$$

As already stated, the energy equation is replaced by the Crocco's integral which writes, for adiabatic flows :

$$\frac{T}{T_e} = 1 + r \frac{\gamma - 1}{2} M_e^2 \left(1 - \frac{u^2}{u_e^2}\right) \qquad (r : \text{recovery factor})$$

These equations can be solved by finite differences techniques or, as often as not, by mean of integral methods.

Let us recall their main characteristics (as it is done in (6)) :

Integral methods

All of them use two integral equations. The first one is the Von Karman relation which can be written :

$$\frac{d\Theta}{dx} + \Theta \left(\left(2 + \frac{\delta^*}{\Theta}\right) \frac{1}{u_e} \frac{du_e}{dx} + \frac{1}{\rho_e} \frac{d\rho_e}{dx} \right) = \frac{C_f}{2} \tag{3}$$

or, if the flow at the boundary-layer edge is isentropic :

$$\frac{d\Theta}{dx} + \left(2 + \frac{\delta^*}{\Theta} - M_e^2\right) \frac{\Theta}{u_e} \frac{du_e}{dx} = \frac{C_f}{2} \quad (3')$$

The second (or complementary) equation differs according to the author's preference. Three kinds of equations are commonly used.

i) *The Mean-Flow Kinetic Energy equation* also called the *Mechanical Energy equation* (sometimes called Moment of Momentum):

This equation is obtained by multiplication of Eq. 3 by u prior to integration in the y direction, which gives :

$$\frac{d\Theta^*}{dx} + \Theta^* \left(\frac{3}{u_e} \frac{du_e}{dx} + \frac{1}{\rho_e} \frac{d\rho_e}{dx}\right) + \frac{2\Theta^{**}}{u_e} \frac{du_e}{dx} = D \quad (4)$$

or, for an isentropic external flow :

$$\frac{d\Theta^*}{dx} + \left(3 + 2\frac{\Theta^{**}}{\Theta^*} - M_e^2\right) \frac{\Theta^*}{u_e} \frac{du_e}{dx} = D$$

ii) *The Integral Continuity Equation* also called the *Entrainment Equation*

This equation results from integration of the continuity equation (1) between $y = 0$ and $y = \delta$, which gives :

$$\frac{d(\delta - \delta^*)}{dx} - \frac{(\delta - \delta^*)}{\rho_e u_e} \frac{d(\rho_e u_e)}{dx} = \frac{d\delta}{dx} - \frac{v_e}{u_e} = C_E \quad (5)$$

or, for an isentropic external stream :

$$\frac{d(\delta - \delta^*)}{dx} - (\delta - \delta^*)\,(1 - M_e^2) \frac{1}{u_e} \frac{du_e}{dx} = \frac{d\delta}{dx} - \frac{v_e}{u_e} = C_E \quad (5')$$

The coefficient C_E represents the rate at which the external flow enters the boundary-layer. It is sometimes called Head's entrainment coefficient (12). It has been demonstrated by Michel et al. (13) :

$$\frac{d\delta}{dx} - \frac{v_e}{u_e} = -\frac{1}{\rho_e u_e} \left(\frac{\partial\tau}{\partial u}\right)_{y = \delta}$$

which clearly shows that the entrainment coefficient is strongly related to turbulence properties at boundary-layer edge.

iii) *The Moment of Momentum Equation*

It is obtained by multiplication of Eq. 2 by y prior integration from $y = 0$ to $y = \delta$, which leads to the following equation :

$$\int_0^\delta \left(\rho u\frac{\partial u}{\partial x} - \frac{\partial u}{\partial y} \int_0^y \frac{\partial(\rho u)}{dx} d\eta - \rho_e u_e \frac{du_e}{dx}\right) y dy = - \int_0^y \tau dy \quad (6)$$

The moment of momentum equation may prove superior to the mean-flow kinetic energy equation due to difficulties in numerically evaluating the shear-work integral in the latter.

The set of equations constituted by Von Karman equation (3) and one of the equations (4, 5, 6) calls upon three characteristic thicknesses and two viscous terms.

So we need 3 more relations (so called closure relations) to solve the problem. Most of the time, one of the characteristic thicknesses is kept, and the two others are replaced by shape parameters. We get rid of one of the unknown by admitting that the velocity profile in the boundary layer can be described by means of a one parameter family.

Viscous terms are taken into account by turbulence models, or described by empirical formulations expressed in function of known quantities (Reynolds number, Mach number, shape factor...).

Finite difference methods

In this approach, the boundary layer equations are directly solved as often as not by an implicit finite difference technique. Basically the main difference amongst them lies in their turbulence model, i.e. the expression of shear stress $\overline{u'v'}$. They need more computing time than integral methods, but they call upon a smaller degree of empirism.

They will be more and more developed and applied while the turbulence models will be improved.

But integral method are able to give a better forecast of real phenomena nowadays, because of their high empiric degree, provided that they are applied in configurations, similar to those for which they were established of course.

A coupling is generally used for the process of strong interaction cases, that include very often separated viscous zones. However, it is well known now that the boundary layer equations have a mathematical singularity at separation when their direct formulation is used (velocity or pressure distribution assigned outside the boundary layer). This singularity doesn't exist in inverse mode. Therefore, it is necessary to abandon direct mode boundary layer calculations on behalf of inverse mode as soon as a flow separation can happen.

The flow into a turbomachine is very complex. It is necessary to consider as a preliminary flow patterns including laminar viscous layers (for instance on a leading edge or in a turbine) followed by a transition, either natural or due to a laminar separation with formation of a bubble. This laminar part is often very small, and seems to have no effect on the inviscid flow. In fact, its knowledge is of the greatest importance, for the whole turbulent viscous layer behaviour depends on it downstream. Therefore it is illusory to refine turbulent viscous effects models when similar precautions are not taken for phenomena governing the viscous layer initial development, and particularly laminar to turbulent transition with or without bubble.

Furthermore, it is necessary to foresee interactions with normal or oblique shock waves, or wakes and slip layers, or thermal effects. We must finally remind that high turbulence levels and very curved wall can be found in turbomachines. Those two phenomena, as well as Coriolis forces, have not a negligible effect on the boundary layer development, especially near the separation and in the separated zone, for the turbulent layer behaviour is affected. The computation methods taking all these effects into account are rare. Several methods are often used together to compute viscous zones. A detailed synthesis of boundary layers computation methods (mostly turbulent) being able to work in direct and inverse mode with a view to coupling, has been lately published (Agardograph 280 (6).

Some boundary layer calculation methods, integrals as well as using finite-differences, have been extended to 3 dimensions. They are enumerated in (6).

2.2 Coupling Condition

Basically, the coupling consists in computing two flows, governed by different equations, that must fulfil boundary conditions and be "compatible" on their common frontier. Pressure and flow angle evolutions must be continuous through the frontier.

To make that clear, let us place ourselves in the case of a nearly plane flow, where curvature effects are neglected and where we admit a boundary layer hypothesis. A generalization will be introduced later on. The ox axis is in the flow direction and oy is normal to it (Fig. 1).

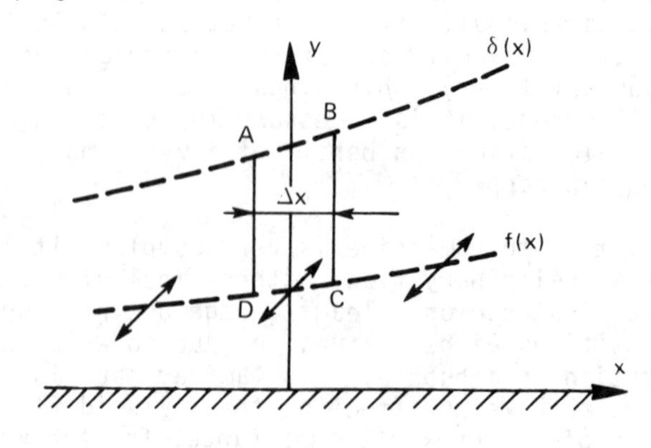

Fig. 1.

Let $\delta(x)$ be the boundary layer thickness. That is to say that viscous effects are negligible for $y > \delta(x)$. Let $f(x)$ be an arbitrary frontier line. Then the real fluid can be represented by two domains, onefor computing inviscid flow whose lower frontier is $f(x)$, and the other for computing the boundary layer stretching from the wall to $\delta(x)$. Thus the two domains may overlap.

It is obvious that the inviscid fluid, over $\delta(x)$, must be identical to the real fluid for the coupling to be representative. Particularly, when we consider the part AB of $\delta(x)$ corresponding to a given flow slice Δx, the mass flow through $\delta(x)$ must be the same for the inviscid flow as for the boundary layer.

Let us consider the control surface ABCD now. If we imagine that CD is a small wing flap imposing the inviscid flow direction locally, on the $f(x)$ frontier, it harks back to write in the overlapping zone that the mass flow variation inside the boundary layer must be equal to the one in the inviscid flow. Taking into account the displacement thickness δ^* at the one hand, admitting that the inviscid flow remains uniform in the overlapping zone (which is consistent with the boundary layer hypothesis) on the other hand, we can write :

$$\frac{d}{dx}\left(\rho_e u_e (\delta - \delta^*)\right) = \frac{d}{dx}\left(\rho_e u_e (\delta - y)\right)$$

where y is the f frontier ordinate.

The CD wing flap slope (i.e. the condition on the f frontier) gives :

$$\frac{dy}{dx} = \frac{d\delta^*}{dx} + (\delta^* - y)\frac{1}{\rho_e u_e}\frac{d(\rho_e u_e)}{dx} \tag{7}$$

This relation, called coupling relation, was first proposed by Lighthill (14). It gives the angle $\theta_f = dy/dx$ that we must impose to the inviscid flow on the f frontier, which is generally permeable. Then the coupling conditions are :

$$\left(\bar{p}(x)\right)_f = \left(p(x)\right)_{B.L.}$$

and

$$\left(\bar{\Theta}_f\right) = \left(\Theta_f\right)_{B.L.}$$

At first sight the choice of $f(x)$ is arbitrary, but usually it is either the boundary layer frontier ($f = \delta$), or the displacement frontier ($f = \delta^*$), or the wall itself ($y = 0$).

We will go again over the analysis of the Agardograph (6) about the troubles occurring according to the choice of the frontier.

Since it is assumed $\rho_e = \bar{\rho}_e$ and $u_e = \bar{u}_e$:

$$\Theta_y = \left(\frac{\bar{v}}{\bar{u}_e}\right)_y = \frac{d\delta^*}{dx} + (\delta^* - y)\frac{1}{\bar{\rho}_e \bar{u}_e}\frac{d(\bar{\rho}_e \bar{u}_e)}{dx} \tag{8}$$

From Eq. 8, we can immediately draw two alternative conclusions :

- when $y = \delta^*$, we obtain :

$$\left(\frac{\overline{v}}{\overline{u}_e}\right)_{\delta^*} = \frac{d\delta^*}{dx} \tag{9}$$

- which is a no slip condition on the displacement body

- when y = 0 one has :

$$\left(\frac{\overline{v}}{\overline{u}_e}\right)_0 = \frac{1}{\overline{\rho}_e \overline{u}_e} \frac{d}{dx}\left(\overline{\rho}_e \overline{u}_e \delta^*\right) \tag{10}$$

which corresponds to a condition of *fluid injection at the wall.*

When relations 9 and 10 are used, conditions at boundary-layer edge $\overline{\rho}_e$, $\overline{u}_e$ are generally identified with inviscid flow quantities $\overline{\rho}$, $\overline{u}$ computed either on the displacement body or on the wall. There is some inconsistency in making this identification which is justified only if the continuated inviscid flow remains practically constant in the y-direction. This is not true in strong interaction processes ; nevertheless the above identification provides a simple way to take into account static pressure variation across the boundary-layer (15). This fact is only an empirical constatation and a formulation of the "overlapping" problem free of any approximation can be made by redefining boundary-layer integral quantities from the "defect formulation" concept in (4).

Coupling equations 8, 9 and 10 are in principle equivalent within the boundary-layer approximations. The reality is more subtle and in fact they correspond to different interpretations of the viscous-inviscid interaction problem.

Use of Eq. 8 makes the problem similar to the classical *multi-domain approach* - or *"patching approach"* - in the sense that the two flows are distinct and can be considered as really "existing" on each side of the free boundary. In fact, these streams satisfy different equations (the Euler and Prandtl equations respectively) so that only continuity of quantities can be satisfied at δ ; the derivatives are discontinuous.

Coupling equations 9 and 10 imply an *overlapping* of the two streams. The boundary-layer flow is no more contiguous to the inviscid stream and the existence of the boundary-layer is essentially "felt" by the outer flow as an alteration of the inviscid no-slip condition on the body surface. This is particularly true when the coupling condition is written on the wall ; then there is no more "geometrical" constraint via δ or δ^* between the two flows. The formulation involving an overlapping between the two streams is often termed "matching" method and acquires its full significance when the defect formulation is introduced.

In fact, the determination of the similar flow (identical to the real flow) corresponds always to the association of the inviscid flow and the viscous flow on the δ frontier : this representation is obtained from inviscid flow by substracting the overlapping zone, and after by adding the viscous layer. This is particularly important when we are interested by the mass flow and momentum balance sheet in internal aerodynamic. In all respects, only the coupling on the δ frontier gives no ambiguity ; the other kinds of coupling can lead to discontinuities of some physical values on the δ frontier when the similar flow is reconstituted.

Differences between the patching and the matching approaches are immaterial in low subsonic flows and the various coupling relation give nearly identical results. However, in *supersonic* and *transonic* interacting flows, consideration of different coupling equations leads to dramatic change in the behavior of the solution. This problem, which is of outmost importance when applying the interactive concept to compute shock wave/boundary-layer interactions, is now discussed.

2 .2.1 Subcritical and supercritical boundary-layers

The coupling problem will be discussed by considering first the local boundary-layer equations in order to make it clear that this problem is inherent to boundary-layer approach and not a consequence of the use of an integral method of solution. In a second part, the problem will be formulated with integral concepts and it will be seen that similar conclusions are then reached.

2 .2.1.1 Local Analysis

Weinbaum and Garvine have for the first time established the following equation giving flow deflection $[\theta]_\delta$ at the boundary-layer edge (16) (see also (17)) :

$$\Theta_\delta = B \frac{dp}{dx} + C \tag{11}$$

where :

$$B = - \frac{1}{\gamma p} \int_\varepsilon^\delta \frac{M^2 - 1}{M^2} dy$$

$$C = \frac{1}{\gamma p} \int_\varepsilon^\delta \frac{1 + (\gamma - 1) M^2}{M^2} \frac{\partial \tau}{\partial y} dy$$

In the above expression, M is the local Mach number in the boundary-layer, p the pressure and γ the ratio of specific heats. It is assumed that Θ is zero at the wall. Equation 11 is obtained by combination of the boundary-layer equations along with the classical hypothesis : $\partial p / \partial y = 0$

For a wake flow, the inner limit ε can be set equal to zero since velocity is non zero on the axis $y = 0$ (except at a "separation" or "reattachment" station). A difficulty arises in boundary-layer flows since then B and C are singular in the limit $\varepsilon \to 0$ ($M = 0$ at $y = 0$). However, in turbulent flows, the problem can be circumvented by taking for ε the thickness $\hat{\delta}$ of the viscous sublayer, the normal velocity at $\hat{\delta}$ being asymptotically equal to zero to all orders δ^n as shown by Mellor (18). An other way to avoid the singularity is to consider a fictitious slip-velocity at the wall equal to the "wake velocity" of the Coles' composite law. In fact, behaviors to be discussed depend mainly or the more or less important "filling" of the velocity profile. For a turbulent boundary-layer, this filling is essentially represented by the wake component.

For a supersonic boundary-layer (or wake) integral B may be either positive or negative, as it is intuitively obvious if one considers the change of the sign of the integrand at the sonic point of the velocity profile. To this change in the sign of B corresponds drastically different responses of the boundary-layer in a free interaction process. To see this, let us consider a perturbation of Eq. 11 near flat-plate conditions ($dp/dx = 0$) ; then

$$\Theta_\delta - (\Theta_\delta)_{P.F.} = B \frac{dp}{dx}$$

Consequences :

- if B is *positive*, to an increase in Θ_δ, i.e. a thickening of the boundary-layer ($\Theta \simeq d\delta^*/dx$), corresponds a pressure rise. By analogy with one-dimensional perfect fluid theory, the boundary-layer is then said to be *subcritical* , in the sense that it behaves (in a global manner) like a subsonic flow.

-if B is *negative*, an increase in Θ_δ is associated with a negative pressure gradient. In this case, the boundary-layer is said to be *supercritical*, since it behaves like a supersonic flow.

The above terminology was introduced in 1952 by Crocco and Lees in their pionnering paper on viscous-inviscid interaction (19).

A laminar boundary-layer is most often subcritical, a supercritical state being encountered only in hypersonic flows or for highly cooled surface (in the second situation, very low temperature levels close to the wall entail low local speed of sound and accordingly high Mach number ; thus B is more likely to be negative). For a turbulent boundary-layer, the Mach number profile of which is much more filled, transition from subcritical state to supercritical state occurs approximately for $M_e = 1.3$ for a conventional flat-plate profile ($H_i = 1.3$). One sees that supercritical behavior is met as soon as the transonic flow regime is reached.

The essential feature of a supercritical state is that the boundary-layer cannot undergo an interaction process with smooth tendency towards separation , i.e. a process in which both the pressure p , the thickness δ^* and the shape parameter H_i increase. Such a behavior is only possible for a subcritical flow. For a supercritical boundary-layer, the onset of an interaction process leading to separation requires a preliminary transition - or jump - to subcritical conditions. Various analyses have been proposed to connect a given upstream supercritical state to the associated downstream subcritical state. These jump models have been formulated in the context of integral methods. In a manner similar to normal shock theory, they use a set of equations expressing the conservation of appropriate global quantities across the jump (20, 21, 22).

We will not comment any more about the "jump" theory, such a discontinuity in the boundary-layer evolution being artificial and physically meaningless. It must be clear that the sub and supercritical states are not "real" properties of the dissipative layer but are a consequence of the (inadequate since too simple) model adopted to depict the viscous-inviscid interaction. The subcritical-supercritical behaviors which are met when the coupling conditions are written at δ are probably a consequence of the neglect of pressure variation across the boundary-layer. As a matter of fact, in a formulation using coupling at δ, Holden (23) was able to compute smooth interactions in turbulent supersonic flows (no jump needed), provided normal pressure gradients were introduced and computed with the help of the integral y-momentum equation (these calculations were made in the context of an integral method).

Le Balleur (24) has shown that it is possible to write the coupling on the displacement body or at the wall in a manner similar to Eq. 11. By considering a continuation of the inviscid flow below the boundary δ one can write (if the pressure p is assumed constant across the boundary-layer) :

$$\overline{\Theta}_\delta - \overline{\Theta}_y = - \frac{1}{\gamma \overline{p}} \frac{d\overline{p}}{dx} \int_y^\delta \frac{\overline{M}^2 - 1}{\overline{M}^2} dy \qquad (12)$$

Within the classical boundary-layer approximations, M can be considered as constant and equal to M_e. Combination of Eqs. 11 and 12 gives :

- for the coupling on the displacement body ($y = \delta^*$) :

$$\Theta_{\delta^*} = B^* \frac{dp}{dx} + C$$

where :

$$B^* = - \frac{1}{\gamma p} \left(\int_\varepsilon^{\delta^*} \frac{M^2 - 1}{M^2} dy - \int_{\delta^*}^{\delta} \left(\frac{1}{\overline{M}^2} - \frac{1}{M^2} \right) dy \right)$$

The second integral figuring in B^* is always negative since $\bar{M}$ is always greater than M. The first integral may be positive if the boundary-layer flow is still supersonic below the displacement surface. As a consequence, B^* is less likely to be negative than B (see Eq. 11). In fact, for a flat-plate turbulent boundary-layer ($H_i = 1.3$), B^* changes sign near $M_e \simeq 2$. One sees that coupling on the displacement surface does not suppress supercritical behaviors (in turbulent flows) ; it only postpones the critical limit to higher Mach numbers.

- for the coupling at the wall (y = 0) :

$$\Theta_o = B_o \frac{dp}{dx} + C$$

with :

$$B_o = - \frac{1}{\gamma p} \int_\varepsilon^\delta \left(\frac{1}{\bar{M}^2} - \frac{1}{M^2}\right) dy$$

As it is easily seen B_o is always positive (since $\bar{M}$ is always greater than M). Thus coupling at the wall leads to a formulation of the viscous-inviscid problem in which the boundary-layer behaves always as a subcritical flow, whatever the external Mach number may be. There is no more need for an artificial jump to initiate an interaction process.

2 .2.1.2 Criticity in the Context of Integral Methods

The same above conclusions can also be drawn by consideration of an integral method of solution for the interaction problem. In a general way, the coupling relation (see Eqs. 8, 9 and 10) is expressed by mean of an ordinary differential equation involving boundary-layer golbal characteristics (δ^*, H_i) and the edge velocity (or the edge Mach number). On the other hand, x-wise variations in boundary-layer integral properties are related to change in edge conditions by ordinary differential integral equations the number of which is generally equal to two (see above). Thus, the interaction process is formulated via a system of three equations for three unknown quantities namely : a thickness, δ^* for instance, a shape parameter H_i and the velocity (or Mach number) at boundary-layer edge. This system can be written in the condensed form :

$$|A| \cdot \begin{vmatrix} \dfrac{d\delta^*}{dx} \\ \delta^* \dfrac{dHi}{dx} \\ \dfrac{\delta^*}{\bar{u}_e} \dfrac{d\bar{u}_e}{dx} \end{vmatrix} = \begin{vmatrix} \dfrac{C_f}{2} \\ f \\ \Theta_y \end{vmatrix} \qquad (13)$$

with $y = \delta$, δ^* or 0 according to the coupling relation envisaged. Function f is the Entrainement coefficient, or the Shear-work integral or the Shear stress integral depending on the "second" equation employed (see above). Coefficients of matrix A depend only of the Mach number M_e and of the velocity distribution shape parameter. Denoting by D the determinant of matrix A and applying Cramer's rule, system 13 leads to a relation of the form :

$$D \frac{1}{\bar{u}_e} \frac{d\bar{u}_e}{dx} = d_1 \Theta_y + d_2$$

One sees that a relation similar to Eq. 11 is obtained so that the same conclusions can be drawn by discussing the sign of D. When Eq. 8 is used, vanishing of D corresponds exactly to the Crocco-Lees critical point (19).

Integration of system 13 is not possible at the point where D vanishes, except if regularity conditions are locally satisfied. These conditions are obviously that $d_1 \Theta_f + d_2 = 0$ when $D = 0$. (It can be shown that if $\bar{u}_e$ is regular, δ^* and H_i are also regular(17). The critical point corresponds to a saddle-point singularity perfect fluid flow.

The existence of a critical point (with associated subcritical and supercritical states) has also important repercussions in boundary-layer calculations using the inverse mode.

One of the ways to perform an inverse calculation, consists in solving system 13 for Θ_y prescribed (as we know, see above, other inverse procedures are possible). As quoted above, integration of 13 is not possible if $D = 0$. The diagram shown in Fig. 2 gives, in the

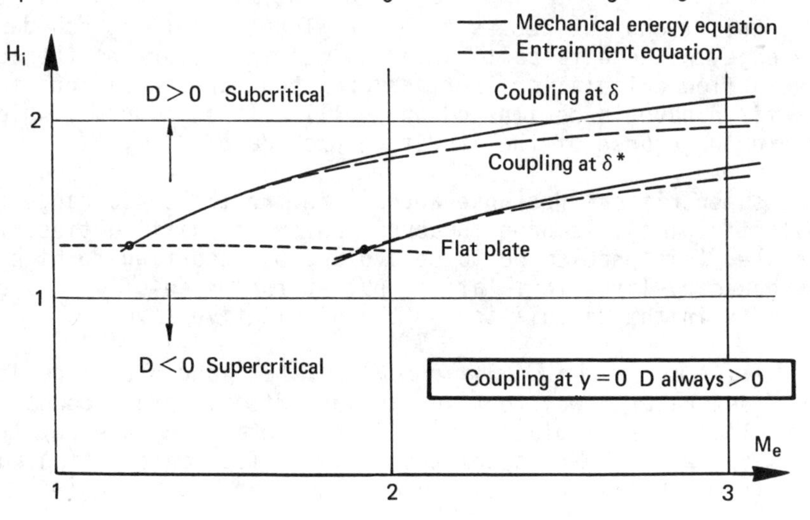

Fig. 2 – Integral coupling method. Critical curves.

plane (H_i, M_e) the locus of the points where D vanishes for the three types of inverse input Θ_y. It can be seen that the critical boundaries are not very sensitive to the "second" equation employed. Concerning an inverse integral method, the following conclusions can be drawn for a turbulent boundary-layer starting from an initial flat-plate situation :

- if Θ_δ is prescribed, supercritical behavior is met as soon as $M_e \simeq 1.3$,

- if $\Theta_{\delta^*} = \frac{d_{\delta^*}}{dx}$ (i.e. the displacement thickness) is prescribed supercritical behavior appears for $M_e \geqslant 2$,

- if $\Theta_0 = \frac{d}{dx} (\bar{\rho}_e \, \bar{u}_e \, \delta^*) \, / \, \rho_e \, u_e$ (i.e) the "perturbation mass flow" is prescribed) there is no risk of "criticity" ; the boundary-layer responds always as a subcritical flow.

The above conclusions remain valid when the boundary-layer equations are solved with a finite difference method ; as is shown by numerical experiments.

2..2.1.3 Conclusions Concerning the Coupling Formulation

To summarize the coupling problem in transonic and/or supersonic flows :

- *coupling at boundary-layer edge is not recommended* (and is now rarely used) for the following reasons :

i) this technique necessitates the location of the boundary-layer edge, hence difficulties can be met in performing the outer inviscid flow calculation. Furthermore, boundary conditions for the outer field have to be imposed on an ill defined boundary which moves in the course of the iterative procedure.

ii) supercritical response appears in the transonic range (for a turbulent regime), leading to severe numerical difficulties. In principle, this problem could be avoided by resorting to higher order boundary-layer formulations but at the price of a greater complexity in the calculation of the dissipative layers.

- *coupling on the displacement body* has long been considered as the most natural way to take viscous effects into account. Nevertheless, the displacement surface is changing at every iteration step, which complicates the inviscid flow calculation and supercritical behavior is to be expected for $M_e \geqslant 2$.

- *coupling at the wall* does not suffer of these disadvantages since :

i) the effective body geometry "seen" by the inviscid stream remains unchanged during the iterative procedure,

ii) the response of the boundary-layer is always subcritical.

However, the use of a coupling equation written as the wall may be questionable to represent viscous effects due to large separated zones. In this case, the displacement body concept seems more appropriate to correctly picture reality. Secondly, when the inviscid flow is computed with the general Euler equations (which is a necessity to compute internal flows), a problem arises to determine the entropy of the fluid entering into the computation domain when the coupling relation gives a positive mass injection.

2.2.2 Higher order methods

In the above formulation of the viscous-inviscid interaction problem, it is always tacitely assumed that the classical boundary-layer concept remains valid. According to this concept, the fluid properties in the dissipative layer (velocity, density, ...) tend towards constant values when $y \to \infty$, and accordingly the transverse pressure gradient is assumed to be zero. The limit values are identified with the inviscid flow properties along the coupling surface, which may be the outer edge δ, the displacement body or the solid body itself. This approach leads to some inconsistencies which may be the source of inaccuracies when the external inviscid flow is far from being transversally constant, as it is the case in a shock wave/boundary-layer interaction. Consequently, the pressure can no longer be considered as independent of y inside the boundary-layer. In these circumstances, coupling on the displacement body or on the surface appears as a more or less empirical way to take into account the transverse variation of the pressure. On the other hand, it seems now clear that the assumption $\partial p/\partial y = 0$ is at the origin of the artificial supercritical behavior (see above). So that the "classical" formulation leads to a situation unsatisfactory even if the prediction it gives is frequently correct.

In fact, a more rigorous formulation of the viscous-inviscid interaction concept can be made by introducing a "defect formulation (see Le Balleur (4)). Basically, this approach consists in considering the difference between the real fluid, with viscous effects near the wall, and the external inviscid fluid continued to the wall. If f designates a real flow property and $\bar{f}$ the corresponding inviscid property, one has : $\lim_{y \to \infty} (f - \bar{f}) = 0$ with

$$f = (u, v, p, \rho)$$

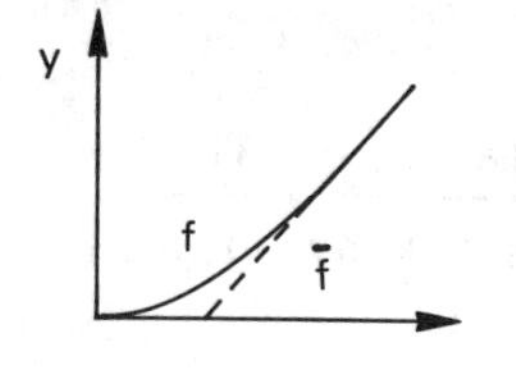

Fig. 3.

It is possible to write the full Navier-Stokes equations in terms of the difference between viscous and inviscid properties. However, for practical purposes a more simple form can be obtained by making the same order of magnitude analysis as in Prandtl equations. Then, the following set of equations is obtained :

$$\frac{\partial\,(\rho u - \overline{\rho}\overline{u})}{\partial x} + \frac{\partial\,(\rho v - \overline{\rho}\overline{v})}{\partial y} = 0$$

$$\frac{\partial\,(\rho u^2 - \overline{\rho}\overline{u}^2)}{\partial x} + \frac{\partial\,(\rho uv - \overline{\rho}\overline{u}\overline{v})}{\partial y} = \frac{\partial\,(p - \overline{p})}{\partial x} + \frac{\partial \tau}{\partial y}$$

$$0 = -\frac{\partial(p - \overline{p})}{\partial y}$$

In the classical (first order) boundary-layer theory the over-barred quantities are considered as independent of y ; now they are (fictitious) inviscid values which may vary with y. It is possible to derive from the above equations integral equations very similar to those of the first order theory provided new definitions of the integral thicknesses are introduced. These definitions take into account the variation with y of the local inviscid values.

$$\delta^* \rho_e V_e = \int_0^\infty \left(\rho\, u(x,y) - \overline{\rho}\, \overline{u}(x,y)\right) dy \quad \left(V_e^2 = u_e^2 + v_e^2\right)$$

$$(\delta^* + \Theta)\, \rho_e V_e^2 = \int_0^\infty \left(\rho\, u^2(x,y) - \overline{\rho}\, \overline{u}^2(x,y)\right) dy \text{ etc...}$$

We will not comment any more about the "Defect Formulation" which is still in a development stage.

It is also possible to improve the representation of viscous effects in high Reynolds number flows by introducing a "splitting" between a viscous component and an inviscid part at the level of the local flow variables themselves (in the viscous-inviscid interactive concept envisaged up to now, the splitting is made between regions)). There results what is called a composite representation of the pressure or of the velocity field. Such techniques have been proposed by Dodge and Lieber (25) and by Khosla and Rubin (26). Further developments about these relatively new methods would be out of the scope of the present paper.

In the following sections, we will assume that the classical boundary-layer concept remains valid, and we will mainly consider viscous-inviscid calculations in which the dissipative layers remain subcritical. This implies coupling at the wall or on the displacement body for flows the Mach number of which is not too high ($M_e \leq 2$) This is sufficient to turbomachinery applications.

2 .3 Different Coupling Modes and Corresponding Algorithms

In this paragraph, we will study the different kinds of coupling that can usefully be considered.

They depend on the one hand on the aim of the computation, and on the other hand on the flow pattern, according as it comprises a separation or not, and as the flow is supersonic or not.

The perfect flow and the viscous flow computations can both be applied either in direct or in inverse mode.

Four kinds of coupling can be distinguished a priori. But we must recall that whenever a separation occurs, boundary-layer calculations has to be done in inverse mode.

2.3.1 Coupling used for "design"

This problem corresponds to the desing of the geometry of a duct (nozzle, air intake or inter row channel) or a cascade, meeting imposed aerodynamic requirements. An inverse mode perfect flow computation is applied. Two cases have to be distinguished, according to the kind of requirements.

1) We look for a given perfect flow behaviour. For instance, we assigned a pressure evolution (or velocity or Mach number distribution) on the walls.

The choice of δ^* as coupling frontier seems to impose itself, even though the other choices can be used a priori.

Then an inverse mode perfect flow calculation and a direct mode boundary-layer calculation are simultaneously accomplished, following the algorithm hereafter :

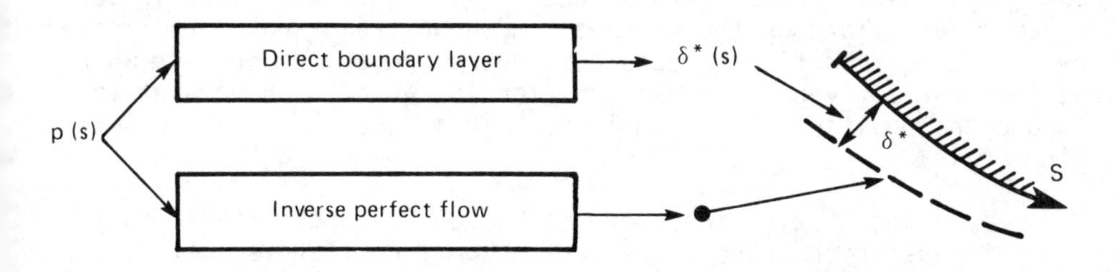

The perfect fluid yields the δ^* limit streamline. The boundary-layer calculation gives the δ^* evolution, but it can succeed only if there is no separation, else it is necessary to modify the assigned velocity distribution and to resume the process. The real geometry is easy to obtain by substracting the δ^* thickness to the computed limit streamline. This very classical kind of coupling has been used for a long time, particularly for all the cases where there is no risk for a separation to occur, such as wind tunnel nozzles.

2) We look for a given boundary-layer behavior. For instance, it is the case of flows where we want to avoid the extended separation that may occur. Then we start with an inverse boundary-layer calculation, that can even comprise a restricted separation.

From a given friction coefficient or shape factor evolution for instance, the computation yields both the velocity (or pressure) distribution, used then as a data by the inverse perfect flow calculation, and the evolution of δ^*. Therefore the computation algorithm is as follows :

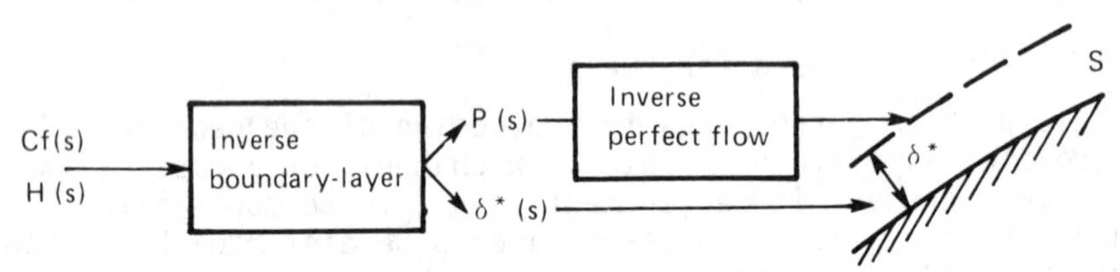

Then the wall geometry can easily be obtained. The optimization of a diffuser is one of the best illustrations of such a coupling technique. It has to be noticed that in the two cases, the problem is well set and that no iteration is needed between the perfect flow and viscous flow computations. In pratice, when taking into account the technolgical constrains often imposed for ducts or blades design, these modes are applied just on parts of the studied flow (semi-inverse design).

2 .3.2 Coupling in view of an analysis

Now, we aim at the determination of the flow established in a duct or a cascade of well known geometry, either when we give aerodynamic conditions on the upstream and downstream frontiers crossed by the flow (mass flow, pressure, temperature and so on), or when we look for a given shock pattern (for instance, when we want to impose the position of a normal shock in a supersonic cascade or air intake).

Important distinctions have to be done : whether there is a risk of separation or not, whether the perfect flow is fully supersonic or not.

2 .3.2.1 Configuration without risk of separation

We use the coupling that associates the viscous an inviscid flow computations in direct mode.

The algorithm is as follows :

The iteration loop starts with given surface boundary conditions (i.e. conditions at or near the body according to Eqs. 8, 9 or 10). These conditions are known from the previous iteration step.

An inviscid flow calculation provides a pressure distribution $\bar{p}(x)$ which is then fed into the boundary-layer calculation. This last calculation furnishes new boundary conditions Θ (Eqs. 8, 9 or 10) and the process is repeated. Such an iterative procedure is called a *fixed point iteration*, since at every station there is the following relationship between values of Θ at iterations n and n + 1 :

$$\Theta^{n+1}(x) = F(\Theta^{n}(x)) \tag{14}$$

In the above equation, **F** is an operator involving perfect fluid and boundary-layer calculations.

It is clear that this method breaks down as soon as separation occurs since the boundary-layer is always computed in the direct mode.

Let us make some remarks on the coupling in the case of unseparated flows :

- A one-dimensional analysis in a duct shows that the iterative direct-direct coupling is automatically converging. It means that there should be no need of a relaxation process but in the case of a slow down. But in fact, stability problems can occur as we shall see later on.

- We can think of other kinds of coupling. They are analysed in next paragraph. Even without any problem of stability, a relaxation can be required.

- We do not need to use an iterative perfect flow, boundary-layer computation process if the perfect fluid flow remains fully supersonic.

As a matter of fact, the inviscid fluid has an hyperbolic behavior and the boundary-layer a parabolic one. The computation can be done step by step, by directly solving the coupling relation. This is easy to apply if the perfect fluid is treated by simple waves or by the characteristics method (see some details in (6)).

2 .3.2.2 Configuration with a risk of separation

The boundary-layer has to be computed in inverse mode. Two modes of coupling can still exist, according as the perfect flow calculation is in direct or inverse mode.

2 .3.2.2.1 Inverse-inverse coupling

Now, both flows are computed by inverse methods. The iteration loop is as follows :

For a given pressure distribution $\bar{p}$ (x) (on or near the body surface, according to the retained coupling relation), an inverse inviscid flow calculation gives $\Theta(x)$ which is, in turn, prescribed to an inverse boundary-layer calculation ; hence a new pressure distribution $p(x)$..... The inverse fixed point operator is :

$$\bar{p}^{n+1}(x) = G\left[\bar{p}^{n}(x)\right]$$

In fact, the inverse boundary-layer mode is not well suited to weak interaction regions or to accelerating flows (27). So that, when computing a complete and complex flow field, (the flow past a cascade, for example) it is necessary to use alternatively direct and inverse modes. The, difficulties can be encountered to obtain a smooth transition between regions where the inviscid flow has been calculated either by direct or by inverse methods.

2 .3.2.2.2 Direct-inverse or semi-inverse coupling

For this kind of coupling, perfect flow is computed in direct mode and boundary layer in inverse mode.

According to the present procedure, the same $\Theta(x)$ distribution is fed both in the boundary-layer and in the inviscid flow calculations. Two pressure distributions are thus obtained which coincide when convergence is reached. Here a new iteration cycle is started by "guessing" a new $\Theta(x)$ distribution from the "error" $p(x) - p(x)$.

As far as applications are concerned, the advantage of such a method is that it keeps the direct mode for the perfect flow. Direct mode is rather used if the viscous layer is not separated and thin. Inverse mode is used as soon as a separation occurs (or as the dissipative flow is strongly unbalanced).

These kinds of coupling methods at the walls or on displacement frontiers δ^* have given very interesting applications, especially on cascades or wings contours. But, as we will show later on, an important problem rises when a shock boundary-layer interaction produces a separation. Basically, two cases can be distinguished according as the perfect flow near the reattachment (if there is one) is supersonic or not. As a matter of fact, if reattachment is subsonic (or if there is none, as for overexpanded nozzles), the flow pattern depends on the downstream boundary conditions.

However, in the separated zone, where a part of the flow has to be subsonic, the start of the separation, and therefore of the interaction, depends on downstream conditions, by the reattachment if there is one, or directly otherwise. The perfect flow, that includes subsonic zones separated from supersonic ones by strong shock waves, has an elliptic pattern and the physical phenomenon, which is elliptic itself, can be correctly represented by associating an elliptic perfect flow and a parabolic boundary-layer.

But the case of supersonic striking, as for a corner or for an oblique shock wave reflection, is completely different. In this case, the external flow is hyperbolic, and the physical phenomenon itself is elliptic, and therefore beyond the reach of the former coupling method, whatever mode we use. The elliptic nature of the phenomenon can only be rendered by one more condition imposed for instance at the reattachment point or dowstream to it.

In fact, the unknown is the beginning of the interaction. For example, we can proceed to series of coupling calculations and modify the point where the interaction starts, until the required condition (at the reattachment point or downstream) is satisfied.

Such a problem has been solved in different ways by several authors (28, 29, 30) for simple flow patterns (for instance by getting a flat-plate flow downstream of the interaction).

For further details, one can refer to the synthesis done in (6).

The condition downstream of the reattachment point is not easy to obtain for a complex flow pattern.

It has to be noticed that in the case of a subsonic reattachment, depending on downstream conditions, in which the coupling process doesn't lead to problems a priori, can be considered in two different ways :

- Either a downstream condition (most frequently the static pressure) is given : then the separated flow location will be a computation result.

- Or we impose the beginning of the interaction, and the inverse-inverse mode calculation gives the downstream conditions.

This last possibility leads to interesting applications, mostly in internal aerodynamics, when strong shocks occur, because we often try to compute flow patterns corresponding to well determined locations of shock waves (air intakes, or supersonic blade cascades).

Anyway, those two methods areequivalent, as far as the description of different working points is concerned.

The three above coupling techniques are said to be *explicit* in the sense that the boundary-layer and the inviscid stream are computed in turn, the one after the other. Some supersonic methods were in fact *implicit* coupling procedures, since in these methods the two streams were determined simultaneously (see (6)). Extension of the implicit procedure to elliptic external flows, with a view to obtain higher convergence rates, has been recently proposed by Veldam (31, 32). In essence his method is as follows in the case of a strictly incompressible flow. The outer velocity distribution $\bar{u}_e(x)$ is computed by using Cauchy's integral which involves the displacement thickness distribution. This integral constitutes the interaction law. Discretization of the interaction law results in an algebraic relation involving δ^* and $\bar{u}_e$ at every grid point i (i = 1, I) along the body surface. This relation is added to the discretized boundary-layer equations to obtain a system which is solved at each streamwise station x_i (i varying from from 1 to I). Due to the fact that the interaction law contains values of δ^* downstream of the computation station x_i (the problem is here elliptic), it is necessary to perform several upstream-downstream sweeps in order to properly account of the ellipticity of the problem. The essential feature of the present method is to use an interaction law (or coupling equation) at iteration number n, and for station x_i, which involves both δ^*_i and $\bar{u}_{ei}$ at the same iteration number.

This is in contrast to :

- direct methods (explicit as well as implicit) where $\bar{u}_e^{(n)}$ is computed from $\delta^{*(n-1)}$,

- inverse methods (explicit as well as implicit) where $\delta^{*(n)}$ is computed from $\bar{u}_e^{(n-1)}$.

Such a quasi-simultaneous procedure avoids difficulties incurrent when either fully direct (as in Werle and Vatsa's method (33, 34) or fully inverse modes are used.

2 .3.2.3 Convergence properties of direct and inverse mode

The next analysis is extracted from (6).

2 .3.2.3.1 Fixed point methods

The relationship implicit in Eqs. 14, 15 and can be viewed conceptually, in simplified form, as representing curves or traces in the p, Θ space, such as depicted in Fig. 4, 5, 6. This graph may also be interpreted as the situation at one particular point of the computation grid along the coupling surface. The two curves represent respectively :

- relation between $\bar{p}$ and $\bar{\Theta}$ satisfying the inviscid flow equations

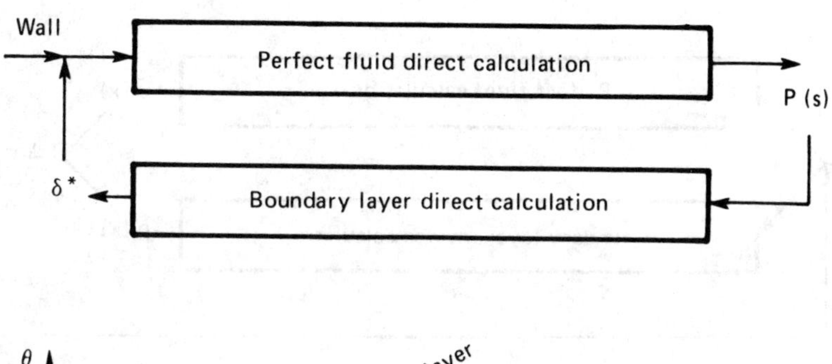

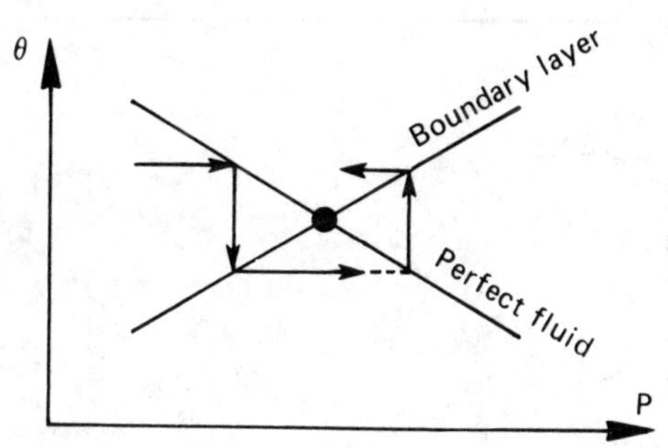

Fig. 4 — Direct-Direct mode iteration path.

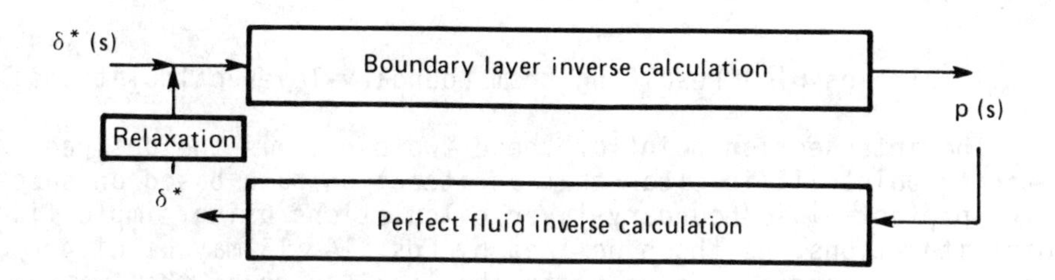

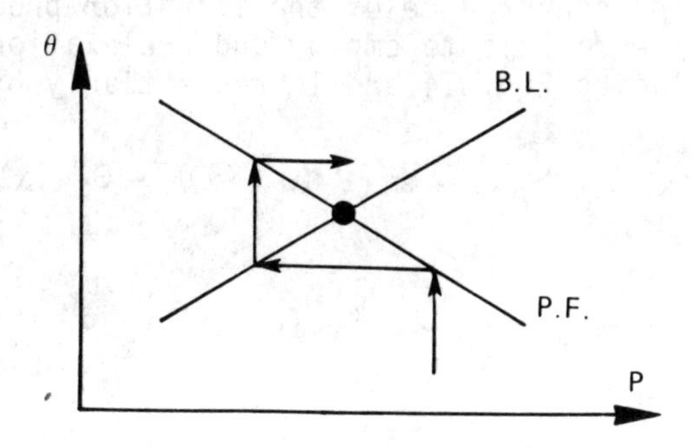

Fig. 5 — Inverse-Inverse mode iteration path.

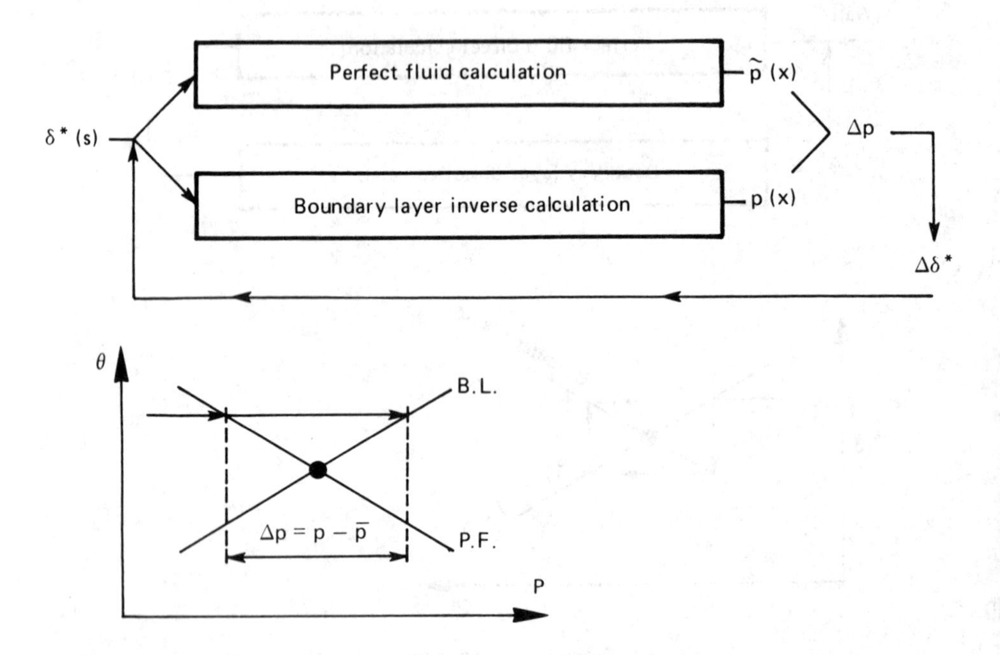

Fig. 6 – Semi inverse mode iteration path.

- relation $p(\Theta)$ resulting from boundary-layer calculations.

The intersection point of these two curves is the desired matching point. It is clear that an iteration path based on successive perfect-fluid/boundary-layer calculations using simple fixed point iterations, as those defined by Eqs. 14, 15 may be either converging or diverging according to the local shape of the two response" curves (see diagrams in Fig. 7). The classical and well known method to insure the convergence of the iteration process or to enhance its convergence rate is to employ underrelaxation. The process consists in replacing Eqs. 14 and 15 respectively by :

$$\Theta^{n+1}(x) = \Theta^{n}(x) + \omega\left(F(\Theta^{n}(x)) - \Theta^{n}(x)\right)$$

$$\bar{p}^{n+1}(x) = \bar{p}^{n}(x) + \omega\left(G(p^{n}(x)) - p^{n}(x)\right)$$

where the underrelaxation coefficient ω is most often determined empirically from trial and error. Effect of underrelaxation in the plane p, Θ is shown in Fig. 8.

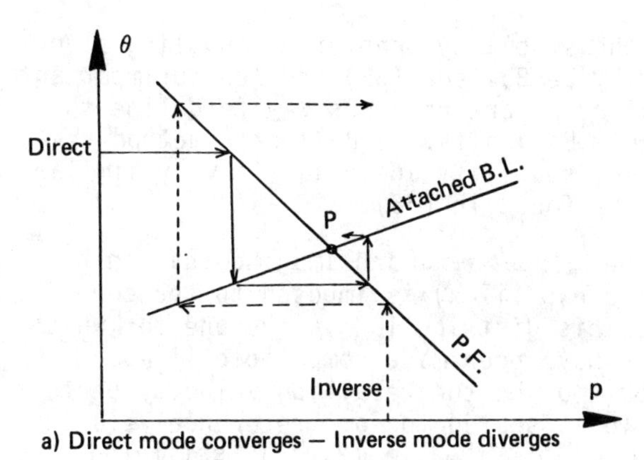

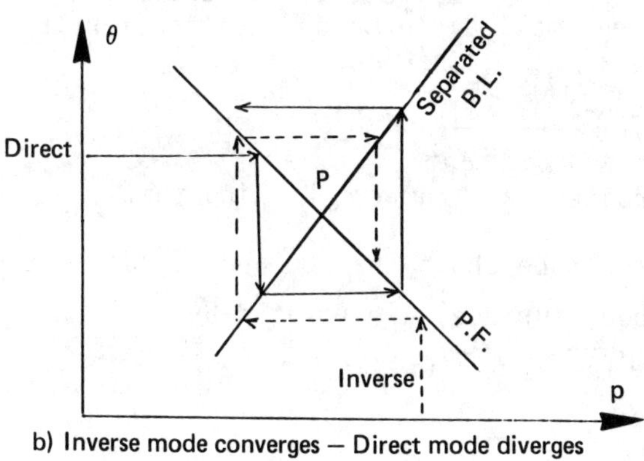

Fig. 7 – Convergence properties of Direct and Inverse modes.

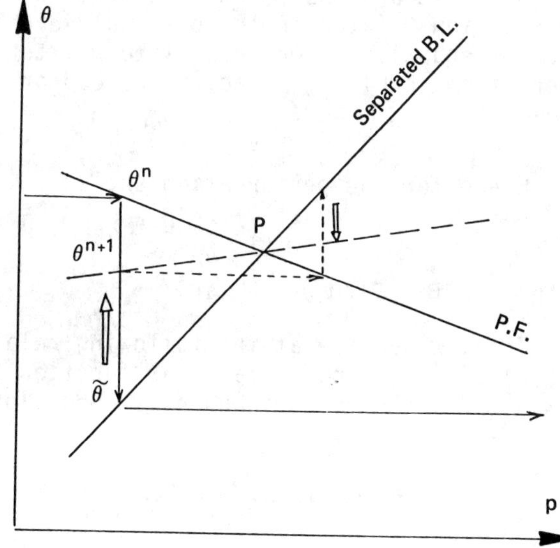

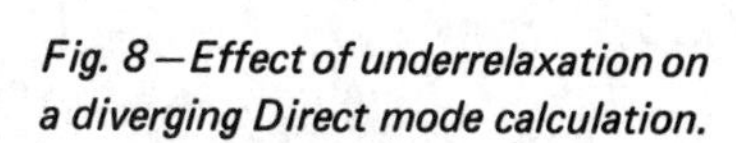
Fig. 8 – Effect of underrelaxation on a diverging Direct mode calculation.

A rational approach to the stability problem of coupling algorithms was published in 1978 by Le Balleur (35) and led to important results allowing, in particular, a more rigorous way to define the optimum relaxation coefficient. Basically, Le Balleur's method relies on an approximate, linear, stability analysis. (a very similar analysis of this problem can be found in (36).

Let us first consider the *direct mode* and imagine that an harmonic perturbation $\delta\Theta(x,o) = \varepsilon \exp(i\lambda x)$ is imposed to the converged $\Theta^*(x,o)$ distribution. This distribution is the one for which the outer and the inner flows have been made compatible (i.e. satisfy Eqs. 7).Let $\delta\bar{u}(x,o)$ be the resulting velocity perturbation of the inviscid flow ($\bar{u}$ is scaled to some reference velocity, say the velocity at upstream infinity U_∞). $\delta\bar{u}(x,o)$ is estimated with the help of the linearized small disturbance equation for the perturbation potential $\phi(x,y)$:

$$(1 - \bar{M}^2)\frac{\partial^2\phi}{\partial x^2} + \frac{\partial^2\phi}{\partial y^2} = 0$$

where $\bar{M}$ is the local unperturbated Mach number. Knowing that :

$\delta\bar{\Theta} = \frac{\partial\phi}{\partial y}$ and $\delta\bar{u} = \frac{\partial\phi}{\partial x}$, it can be shown that $\delta\bar{\Theta}$ and $\delta\bar{u}$, along the boundary $y = 0$ (i.e. on the body surface) are related by :

$$\delta\bar{\Theta}\ (x,0) = i\sqrt{1 - \bar{M}^2}\ \delta\,\bar{u}\ (x,0)$$

if the flow is locally *subsonic*,

$$\delta\bar{\Theta}\ (x,0) = -\sqrt{\bar{M}^2 - 1}\ \delta\bar{u}\ (x,0)$$

if the flow is locally *supersonic*.

Now, let us consider the response of the boundary-layer to the perturbation $\delta u(x,o)$ in the external velocity. This response is given by equation 11 which will be written here in a slightly different form involving velocity instead or pressure :

$$\Theta = B'\frac{d\bar{u}}{dx} + C$$

Hence for the perturbation angle :

$$\delta\Theta\ (x,0) = i\,\lambda\,\tilde{B}\,\delta\,\bar{u}\ (x,0) \tag{17}$$

where $\tilde{B} = B'/\bar{u}$

Thus, we arrive at the following relations giving the response $\delta\Theta(x,o)$ of the boundary-layer as function of the perturbation $\delta\bar{\Theta}(x,o)$ of the converged boundary condition :

$$\delta\Theta\ (x,0) = \frac{\lambda\tilde{B}}{\beta}\delta\Theta\ (x,0) \qquad \text{if } M < 1$$

$$\delta\Theta\ (x,0) = -\frac{i\lambda\ \tilde{B}}{\beta}\ \delta\Theta\ (x,0) \qquad \text{if } M > 1$$

where : $\beta = \sqrt{|\bar{M}^2 - 1|}$

Letting $\mu_D = \frac{\lambda\tilde{B}}{\beta}$ if $M < 1$ or $\mu_D = -\frac{i\lambda\tilde{B}}{\beta}$ if $M > 1$ (μ_D is the amplification coefficient of the coupling mechanism), one can see that the classical chain iteration will converge only if $|\mu_D| < 1$ condition which corresponds to a damping of the oscillations. The restriction $|\mu_D| < 1$ must be satisfied for every wave number λ. If the perturbation is not made of a simple sinusoid, it can be decomposed into simple harmonics by Fourier analysis. The condition $|\mu_D(\lambda)| < 1$ must be satisfied for all the wave numbers λ contained in the spectrum of the perturbation. In the reality, due to the x-discritization of the computation methods, the numbers λ constitute a finite set of discrete values lying in the range.

$$\frac{\pi}{L} < \lambda < \frac{\pi}{\Delta x}$$

where Δx is the mesh size and L the length of the computation domain.

A very similar analysis can be made for the inverse problem by considering a perturbation $\delta\bar{u}(x,o)$ of the external velocity. Now, the boundary-layer response is given by :

$$\delta u\ (x,0) = \mu_I\ \delta\ \overline{u}\ (x,0)$$

It is readily verified that μ_D and μ_I are such that :

$$\mu_D\ \mu_I = 1$$

The above analysis leads to very useful conclusions regarding the convergence properties of the direct and inverse modes :

- in direct iteration, a small mesh size (λ_{max} is large) tends to degrade the convergence properties by increasing the amplification factor $|\mu_D(\lambda_{max})|$. The converse is true for an inverse iteration : convergence is deteriorated by an enlargement of the computation domain. This latter finding is confirmed by calculations made by Melnik and Chow (37).

- in "well" attached flows, $\tilde{B}$ is small (it is recalled that $\tilde{B}$ depends only of the shape of the velocity distribution in the boundary-layer and of the edge Mach number). Consequently, the direct mode is more appropriate to compute these flows. On the other hand, when the boundary-layer is separated or strongly destabilized B is large and the inverse mode becomes preferable. The above situations are illustrated in Fig. 7. When the slope at the matching point P of the boundary-layer response curve is small (Fig. 7a) i.e. when the flow is attached, the direct mode is more likely to give convergence. On the other hand, when this slope is increased, which corresponds to a destabilized boundary-layer, the inverse mode has a natural tendency to converge (see Fig. 7b).

Meauzé and Délery (38) have developed a coupling method to compute the flow in a transonic channel in which shock/boundary-layer interactions with large separation occur. The aim of this research is the extension of coupling techniques to the prediction of the flow in supersonic axial compressors. Due to the necessity to determine a flow with choking conditions, the outer inviscid stream is computed by solving the full Euler equations with a time marching method (32). This method can be worked out either in the direct mode (i.e. with slip condition on the body surface) or in the inverse mode (i.e. with prescribed pressure along free boundary (40). The turbulent boundary-layer is computed by a direct-inverse finite difference method (41). The coupling conditions are written on the displacement body in order to satisfy the mass flow conservation in the channel. The iteration procedure can be performed according to the fully inverse mode or to the semi-inverse mode.

In situations where the Inverse mode tends to diverge, an underrelaxation coefficient is computed at every grid point situated on the coupling surface. Reasoning is made in the (δ^*, M_e) plane. The boundary-layer edge Mach number M_e is in fact equivalent to the pressure p of Fig. 5. Let m_{PF} and m_{BL} be respectively the slopes of the Perfect Fluid and Boundary-Layer response curves at the (desired) matching point P. If in the vicinity of P these response curves are assumed to be rectilinear, it is clear that convergence of the fixed point iteration is insured provided that $m_{PF}/m_{BL} > -1$. In these circumstances, it can be readily demonstrated that the iteration is made converging by choosing the relaxation coefficient ω in such a way that :

$$\omega < \frac{2\, m_{PF}}{m_{PF} - m_{BL}}$$

At every iteration cycle, the slopes m_{PF} and m_{BL} are determined by assuming rectilinear response curves. The perfect-fluid slope is evaluated by considering that locally the inviscid stream is one-dimensional with a Mach number $\overline{M}_{PF}$ equal to the average Mach number of the two-dimensional inviscid flow. Then, by applying the equation for mass flow conservation, one obtains :

$$\frac{d\,\overline{M}_{PF}}{d\delta^*} = \frac{\left(1 + \frac{\gamma - 1}{2}\overline{M}^2_{PF}\right)}{\left(1 - \overline{M}^2_{PF}\right)} \; \frac{\overline{M}_{PF}}{(A - \delta^*)}$$

where A is the cross section of the channel.

The slope m_{BL} relative to the boundary-layer is computed by considering a simplified Von karman equation where the skin friction is neglected (in a manner similar to the method used by Carter (42).

This equation is written :

$$\frac{d\Theta}{dx} + \Theta\,(H + 2 - M_e^2)\,\frac{1}{1 + m_e}\,\frac{1}{M_e}\,\frac{dM_e}{dx} = 0$$

where : $m_e = \frac{\gamma - 1}{2} M_e^2$

By considering the approximate relation $H = H_i + \alpha M_e^2$ where $\alpha = 0.4$, one has :

$$\frac{d\delta^*}{dM_e} = \frac{\delta^*}{M_e}\left(\frac{2\,\alpha\,M_e^2}{H} - \frac{1}{1 + m_e}\,(H + 2 - M_e^2)\right) \qquad (18)$$

In the above equation, M_e is the local Mach number at the boundary-layer edge and H is provided by the boundary-layer calculation.

Equation 18 is also employed in the Semi-Inverse mode to guess the new δ^* distribution from the mismatch ΔM_e between the perfect-fluid and the boundary-layer calculations.

2 .3.2.3.2 Newton method

As it was suggested by Brune, Rubbert and Nark (43) convergence of the coupling iteration can also be achieved by using Newton method. Let us consider a computation grid on the coupling surface the points of which are characterised by index i. Any perturbation $\Delta\bar{\Theta}$ of the boundary condition for the inviscid flow will produce changes in pressure $\Delta\bar{p}$ at every point i. These changes can be expressed in a linearized form :

$$\{\Delta\bar{p}_i\} = (P_{ij})\{\Delta\bar{\Theta}_j\}$$

where the P_{ij} are the inviscid flow influence coefficients.

Similarly, one can write a linearized expression for the changes in boundary-layer deviation Θ due to changes in surface pressure p :

$$\{\Delta\Theta_i\} = (B_{ij})\,\{\Delta P_j\}$$

The situation at iteration number n being defined by :

$\{\bar{P}_i\}^n, \{\bar{\Theta}_i\}^n, \{P_i\}^n, \{\Theta_i\}^n$, the problem is to estimate values of pressure and deviation at iteration (n + 1) such that :

$$\{\bar{P}_i\}^{n+1} = \{P_i\}^{n+1}$$

$$\{\bar{\Theta}_i\}^{n+1} = \{\Theta_i\}^{n+1}$$

Thus, we arrive at the following system :

$$\{\Delta\bar{p}_i\} = (P_{ij})^n \{\Delta\bar{\Theta}_j\}$$

$$\{\Delta\Theta_i\} = (B_{ij})^n \{\Delta p_j\}$$

$$\{\bar{p}_i\} + \{\Delta\bar{p}_i\} = \{p_i\}^n + \{\Delta p_i\}$$

$$\{\bar{\Theta}_i\} + \{\Delta\bar{\Theta}_i\} = \{\Theta_i\}^n + \{\Delta\Theta_i\}$$

which allows, in principle, the calculation of the perturbation values $\{\Delta\bar{p}_i\}$, $\{\Delta\bar{\Theta}_i\}$, $\{\Delta p_i\}$ and $\{\Delta\Theta_i\}$.

The above process is depicted schematically in Fig. 9. Conceptually, the perturbation equations (18,19) define tangent lines to the inviscid flow and boundary-layer solution curves. In essence, the method is seen to be equivalent to approximating thesesolution curves by straight lines locally tangent to the starting points and solving for the point at which these tangent lines intersect.

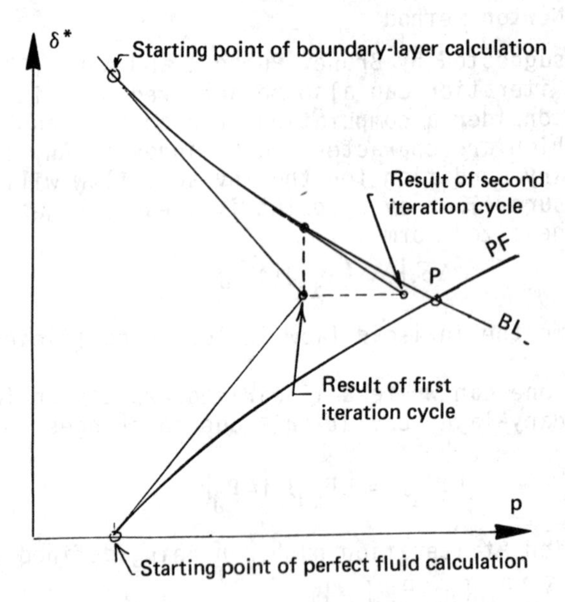

Fig. 9 – Newton method for coupling problem (after Brune et al. [43]).

If non linear effects in P_{ij} and B_{ij} are moderate (i.e. if the solution curves are nearly straight) convergence of the method may be rapid. However, each cycle is very complex since it requires inversion of full matrices. It is why rather few examples of this approach have been reported for transonic flow calculations (44).

2.3.2.4 Semi-Inverse techniques

2.3.2.4.1 Fixed point methods

Now the problem is to "guess" a new $\Theta(x)$ distribution from the difference (or "error") in pressure $\Delta p = p(x) - \bar{p}(x)$ in such a way that Δp cancels at convergence. This kind of iteration procedure was first proposed by Kuhn and Nielsen (45) in the computation of transonic shock wave/turbulent boundary-layer interactions. At that time, they use a rather empirical way to iterate on the $\Theta(x)$ distribution. A more rational examination of the convergence properties of the Semi-Inverse algorithm can be made by resorting to Le Balleur's stability analysis.

Let Θ^* and $\bar{u}^*(x,o) = u(x,o)$ be the converged values on the coupling boundary and $\Theta^n(x,o)$, $\bar{u}^n(x,o)$, $u(x,o)$ the corresponding values at iteration number n (it is recalled that in the Semi-Inverse procedure $\bar{\Theta}^n(x,o) = \Theta^n(x,o)$, barred quantities being relative to the inviscid flow). Considering the subsonic case, the local linear analysis leads to the following relations :

- for the *inviscid flow* (see Eq. 16)

$$\Theta^n(x,0) - \Theta^*(x,0) = \frac{B}{\lambda}\left(\frac{1}{\bar{u}^n}\frac{d\bar{u}^n}{dx} - \frac{1}{u^*}\frac{du^*}{dx}\right)$$

- for the *boundary layer flow* (see Eq. 17)

$$\Theta^n(x,0) - \Theta^*(x,0) = B^*\left(\frac{1}{u^n}\frac{du^n}{dx} - \frac{1}{u^*}\frac{du^*}{dx}\right)$$

Combination of the two above equations gives :

$$\frac{1}{\bar{u}^n}\frac{d\bar{u}^n}{dx} - \frac{1}{u^n}\frac{du^n}{dx} = \left(\frac{\lambda}{\beta} - \frac{1}{B^*}\right)\left[\Theta^n(x,0) - \Theta^*(x,0)\right]$$

If convergence is to be achieved at the $(n+1)^{th}$ iteration, we must have : $\tilde{\Theta}^{n+1}(x,o) = \Theta^*(x,o)$, hence :

$$\tilde{\Theta}^{n+1}(x,0) - \Theta^n(x,0) = \frac{B^*\beta}{\lambda B^* - \beta}\left(\frac{1}{u^n}\frac{du^n}{dx} - \frac{1}{\bar{u}^n}\frac{d\bar{u}^n}{dx}\right) \quad (20)$$

A similar calculation gives for the supersonic case :

$$\tilde{\Theta}^{n+1}(x,0) - \Theta^n(x,0) = \frac{\beta B^{*2}}{\lambda^2 B^{*2} + \beta^2}\left(\frac{1}{u^n}\frac{d^2u^n}{dx^2} - \frac{1}{\bar{u}^n}\frac{d^2\bar{u}^n}{dx^2}\right)$$

A relaxation coefficient must generally be introduced and the $(n+1)^{th}$ values are taken to be :

$$\Theta^{n+1}(x,0) = \Theta^n(x,0) + \omega\left[\tilde{\Theta}^{n+1}(x,0) - \Theta^n(x,0)\right]$$

It is to be noticed that this Semi-Inverse iteration method requires only the knowledge of the error on the pressure gradient, so that integration of u(x) (or p (x) is not necessary. This property may facilitate a zonal switch between Semi-Inverse iteration which is more appropriate for attached flows.

More recently, Carter (42, 46) (see also application of the method in (47)) has proposed a rather simple updating procedure which takes the following form when coupling conditions are expressed on the displacement body : the updated displacement $\delta^{*(n+1)}$ is deduced from the mismatch of the viscous and inviscid velocities by :

$$\delta^{*\,n+1} = \delta^{*\,n} \left(\frac{u^n}{\bar{u}^n}\right)$$

The updating procedure including a relaxation coefficient can be written :

$$\delta^{*\,n+1} = \delta^{*\,n} + \omega\, \delta^{*\,n} \left(\frac{u^n}{\bar{u}^n} - 1\right) \tag{21}$$

A similar expression can be obtained if one considers the simplified Von Karman equation where the skin friction has been omitted. In that case, one can write :

$$\Delta\delta^* = -\left(H_i + 2\right) \frac{\delta^*}{u} \Delta u$$

thus, if :

$$\Delta u = \bar{u}^n - u^n$$

$$\Delta\delta^* = \delta^{*\,(n+1)} - \delta^{*\,(n)}$$

one gets :

$$\delta^{*\,(n+1)} = \delta^{*\,(n)} + \left(H_i + 2\right) \delta^{*\,(n)} \left(\frac{u^n}{\bar{u}^n} - 1\right) \tag{22}$$

Comparison of Eqs. 21 and 22 shows a close resemblance in relating an increment in u to that in δ^*. Estimation of $\delta^{*(n+1)}$ by Eq. 22 is not in principle entirely rigorous since the momentum integral equation expresses change in u and δ^* in the streamwise direction, whereas those in Eq. 22 refer to change between successive iteration cycles. Carter found that overrelaxation could be used in Eq. 21 ($\omega > 1$) to accelerate the convergence of the iteration process. The similarity between Eqs. 21 and 22 offers an explanation since, even with overrelaxation, it was observed in these calculations that :

$$\omega < H_i + 2$$

H_i being always greater than 1.

It can be easily demonstrated that, for subsonic flows, Carter 's and Le Balleur's approaches are essentially equivalent. Differentiating Eq. 21 with respect to x gives (after some approximations) :

$$\frac{d\delta^{*(n+1)}}{dx} - \frac{d\delta^{*(n)}}{dx} = \omega\, \delta^{*(n)} \left(\frac{1}{u^n}\frac{du^n}{dx} - \frac{1}{\bar{u}^n}\frac{d\bar{u}^n}{dx}\right)$$

The above equation is identical to Eq. 20 applied to coupling on the displacement body (in this case $\Theta(x,o) = \frac{d\delta^*}{dx}$) provided that :

$$\omega\delta^{*(n)} = \frac{B^*\beta}{\lambda B^* - \beta}$$

In supersonic or transonic flows, Carter uses coupling on the body surface. As seen before, then the effect of the boundary-layer is felt by the outer inviscid flow as mass bleed (positive or negative) along the body surface. In this case, the boundary-layer calculation of the iteration cycle, is made by specifying the perturbation mass flow : $m = \rho u \delta^*$, and the same updating procedure as for coupling on the displacement body is employed ; it gives :

$$m^{(n+1)} = m^{(n)} \frac{u^n}{\bar{u}^n}$$

2.3.2.4.2 Application of Newton method

Newton iteration method can also be used to enhance convergence of the Semi-Inverse algorithm. Most often, the method is employed in a much more simple form that the initial version proposed by Brune et al (43). For example, Gordon and Rom (48) have devised a matching procedure based on the assumption that the relation between $\{\Delta\bar{p}_i\}$ and $\{\Delta\bar{\Theta}_i\}$ and the one between $\{\Delta\bar{p}_i\}$ and $\{\Delta\Theta_i\}$ are two dimensional. This means that for each station $\{\Delta\bar{p}_i\}$ depends only on $\{\Delta\Theta_i\}$ (and not on $\{\Delta\Theta_j\}$, $j \neq i$) and $\{\Delta p_i\}$ depends only on $\{\Delta\Theta_i\}$. The procedure consists essentially in keeping only the diagonal of matrice $[P_{ij}]$ and $[B_{ij}]$. The guessed displacement thickness (coupling is expressed on the displacement body) has to be underrelaxed rather strongly to prevent oscillations.

In 1976, Alziary de Roquefort (49) has proposed an updating procedure in which only two diagonals in the boundary-layer influence matrix B_{ij} (i and i-1) were retained. The method worked satisfactorily for laminar shock-wave/boundary-layer interactions.

More recently, Ardonceau (50) (see also Ardonceau, Alziary and Aymer (51)) has published a method which worked efficiently both for (supersonic) laminar and turbulent shock/boundary-layer interaction. Its principle consists in representing the perfect fluid and the boundary-layer sensitivity functions by :

- for the *perfect fluid* :

$$\overline{P}^{\,n+1}_{i-\frac{1}{2}} = \overline{P}^{\,n}_{i-\frac{1}{2}} + A\overline{p}_{\infty} \left(\overline{\Theta}^{\,n+1} - \Theta^{\,n}\right)_{i-\frac{1}{2}} \tag{23}$$

which is the linearized Prandtl-Meyer law

- for the *boundary-layer* :

$$\left(\frac{dp}{dx}\right)^{n+1}_{i-\frac{1}{2}} = \left(\frac{dp}{dx}\right)^{n}_{i-\frac{1}{2}} + S\left(\Theta^{\,n+1} - \Theta^{\,n}\right)_{i-\frac{1}{2}} \tag{24}$$

where the influence coefficient $S = \left(\frac{\partial \text{ grad } p}{\partial \Theta}\right)_{i-\frac{1}{2}}$ is approximately by $\left(\frac{\partial p}{\partial \delta^*}\right)_i$ and estimated numerically.

Discretization of Eqs. 23 and 24, where $\overline{p}^{\,n+1}$ is set equal to p^{n+1}, results in a recursive relation allowing the calculation of p^{n+1}. This calculation is started from the most downstream station where the pressure is prescribed as the downstream boundary condition of the viscous-inviscid interaction problem. Thereafter, the updated displacement thickness distribution is computed from 23 or 24 (in the case where $\Theta = \frac{d\delta^*}{dx}$). An underrelaxation of p^{n+1} is re-required when separation occurs.

3. SUBSONIC COUPLING APPLICATIONS

Viscous-inviscid interactions in transonic and supersonic flow always leads to shock boundary-layer interaction ; they are illustrated in part II.

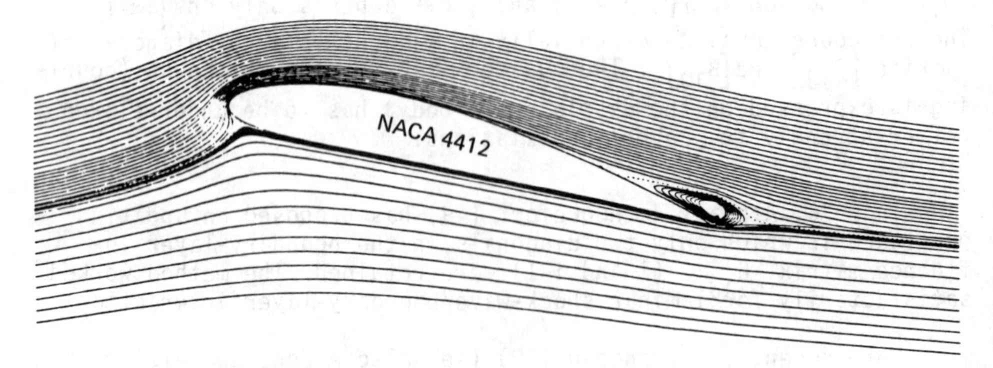

Fig. 10 – Streamlines on subsonic airfoil (NACA 4412, V = 20 m/s, α = 13.6°, R = 1.5 x 10⁶). [38].

Nowadays, many complex configurations have been treated by coupling methods, in subsonic external flows. We reproduce here only some recent applications.

Results on 2D airfoils showed on Fig. 10 to 12 are obtained by Le Balleur (52) and reproduced in (53). The semi-inverse mode is used for these calculations.

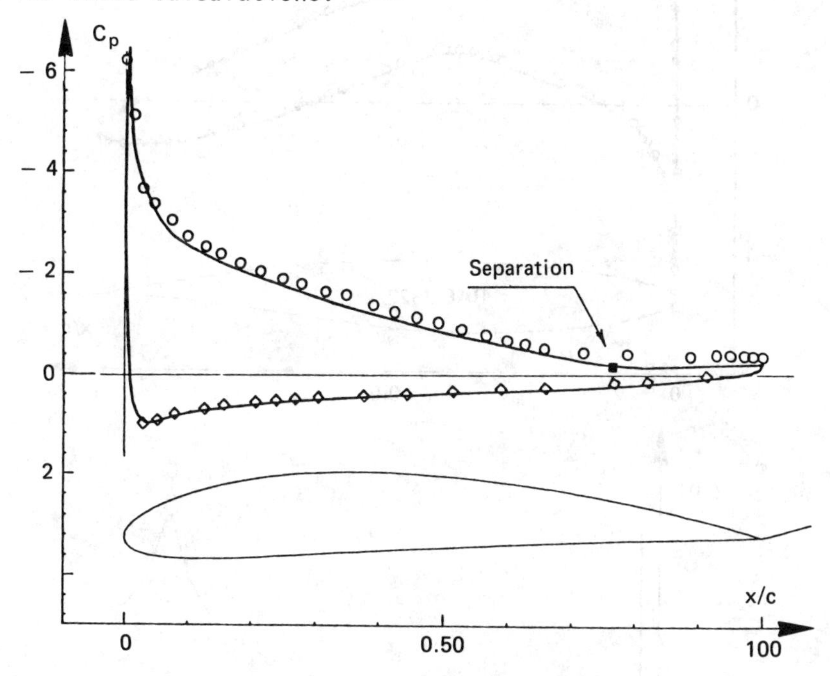

Fig. 11 – Pressure distribution (NACA 4412, V = 20 m/s, α = 13.6°, R = 1.5 x 10⁶). [38].

The inviscid flow is computed in direct mode by a finite difference technique solving the full potential equation, the viscous layer is computed in direct or inverse mode by an integral method which can be applied to the laminar or turbulent boundary-layer and to the wake.

A more complex configuration concerning a three elements airfoil was calculated by Neron and Le Balleur (54). The main results are reported on Fig. 13 and 14 (see (55) for more details).

Applications on cascades are more rare. Calvert (56) has used the inverse mode to compute separated zones. The inviscid flow is calculated by a time marching method. Viscous layer is solved by an integral method. Nowadays, some recent applications have been calculated. They are reported in (53). Some of them are reproduced on Fig. 15 and 16.

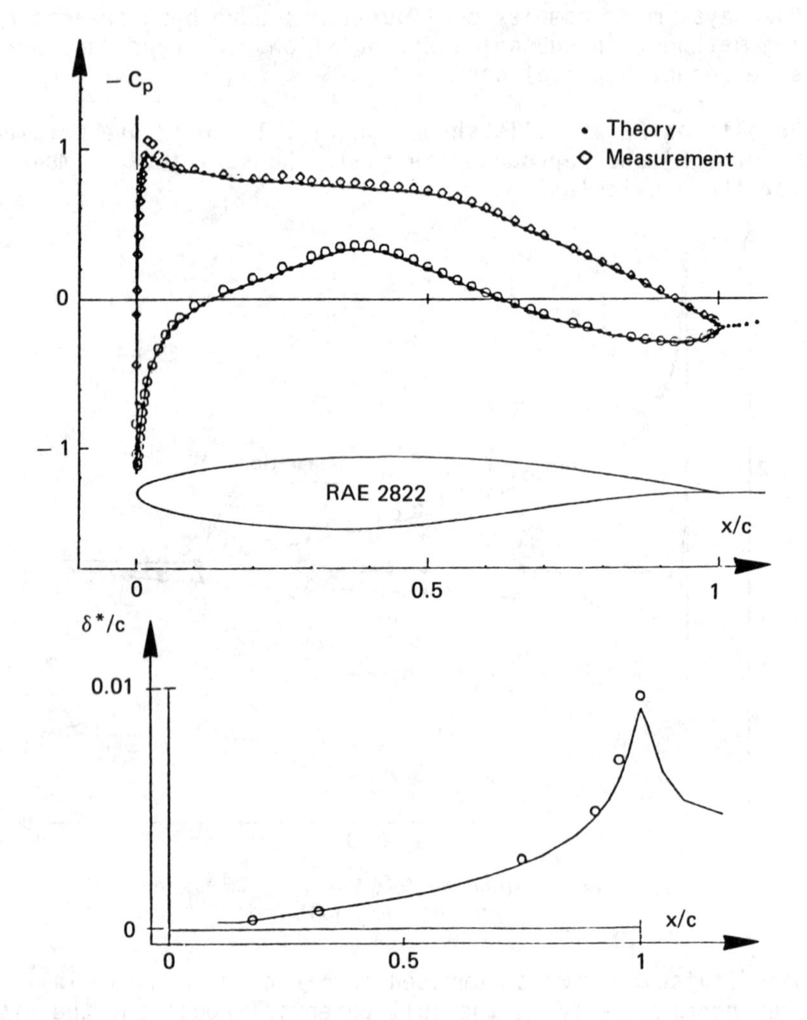

Fig. 12 – Pressure and displacement thickness distribution (RAE 2822, $M = 0.676$, $\alpha = 2.40°$, $R = 5.7 \times 10^6$).

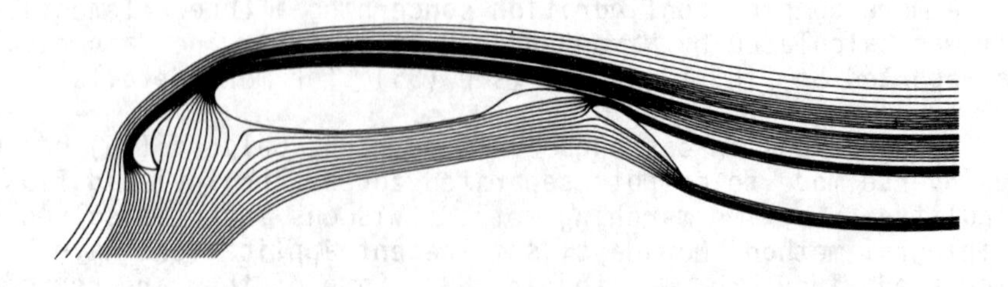

Fig. 13 – Calculated streamlines on three elements airfoil with large separated zones.

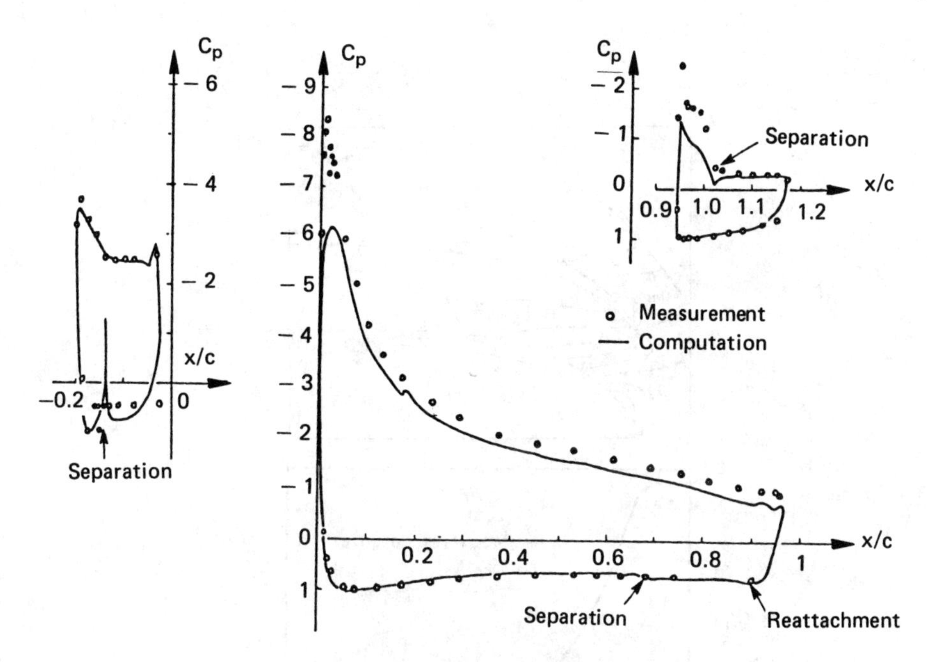

Fig. 14 – Pressure distribution on the three elements ($\alpha = 12°$, $\delta_{volet} = 40°$, $R \cong 2 \times 10^6$). [38].

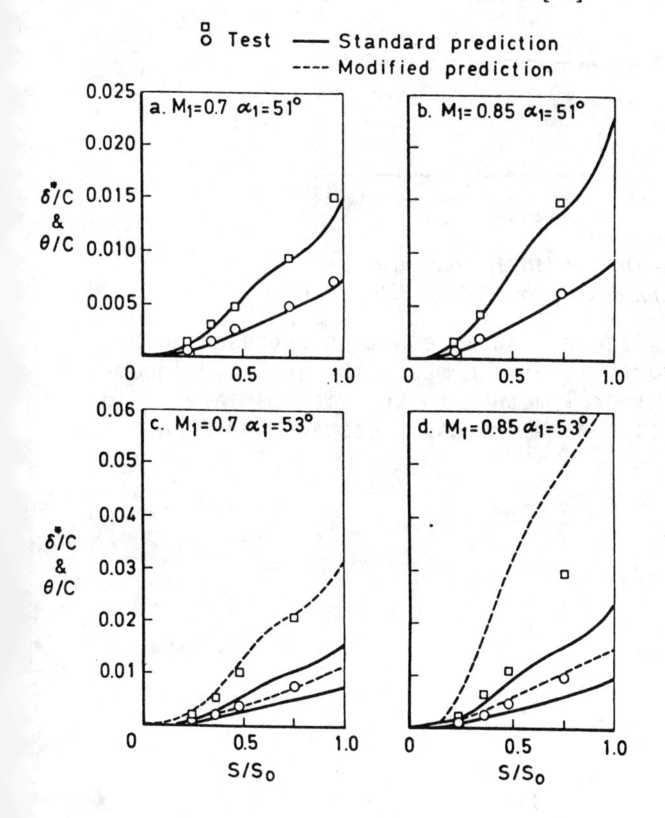

Fig. 15 – ONERA 115 suction surface boundary layer growth (Calvert [38]).

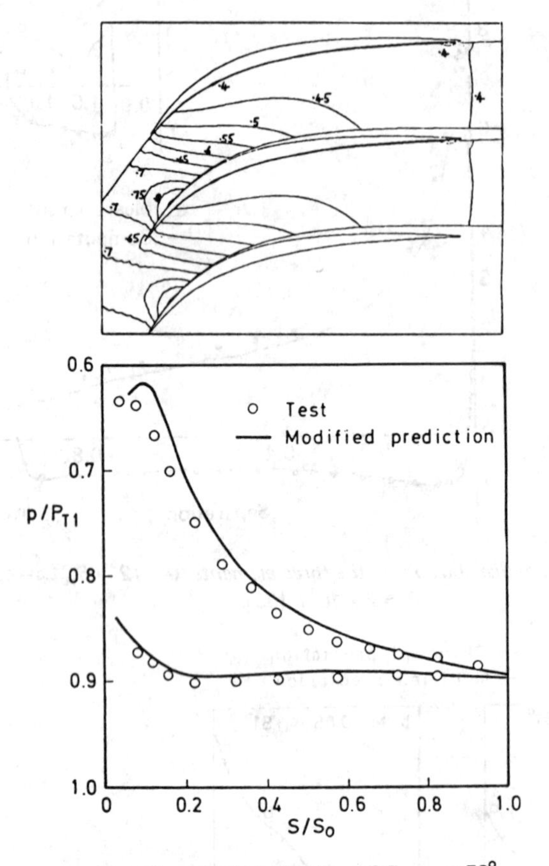

Fig. 16 – ONERA 115 $M_1 = 0.7$, $\alpha_1 = 53°$ modified prediction (Calvert [38]).

Psarudakis (53) uses the potential equation represented by continuous distributed vorticity to compute the inviscid incompressible flow and an integral method to the viscous layer. Fig. 17 and 18 show two results for a compressor cascade and a turbine cascade.

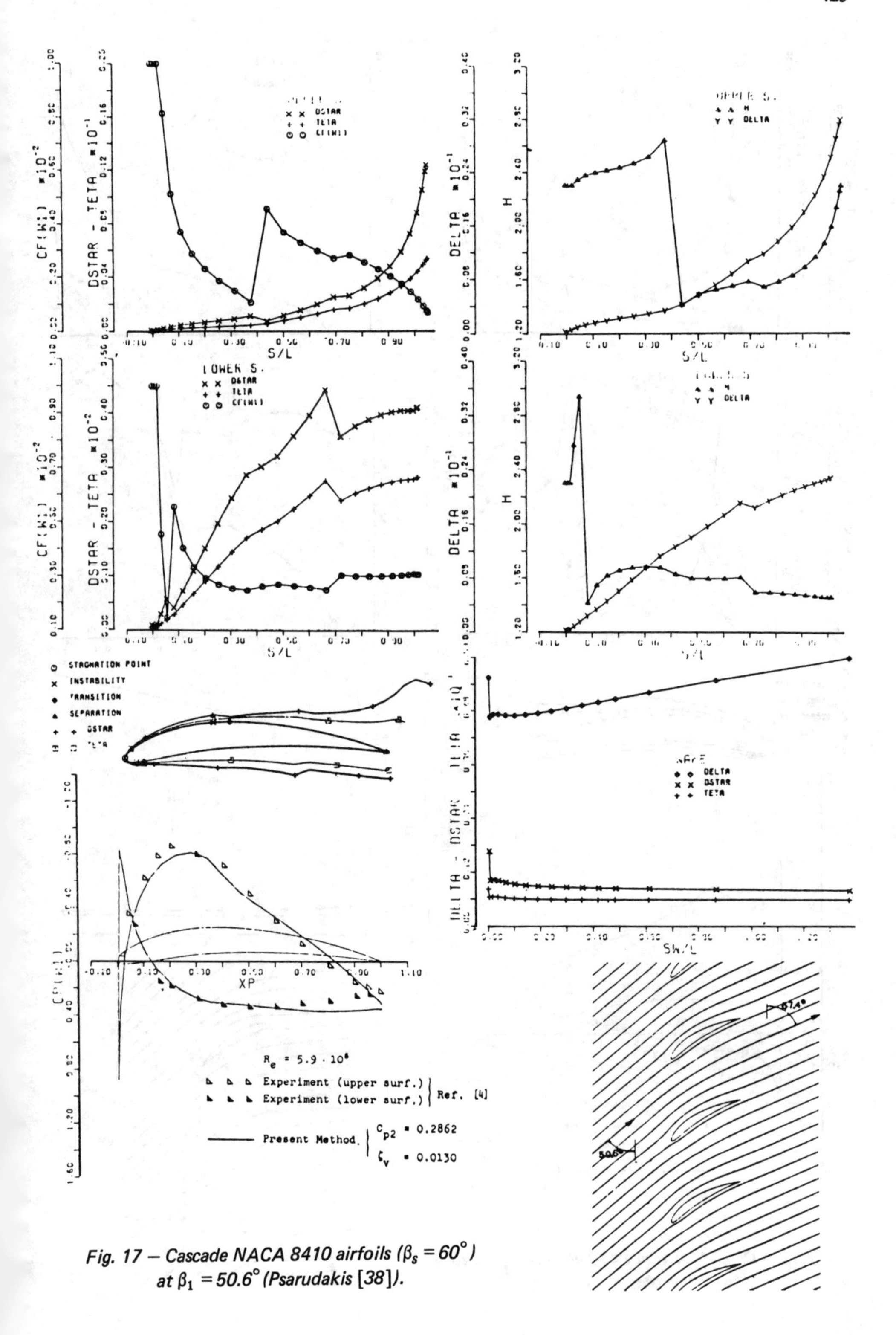

Fig. 17 – Cascade NACA 8410 airfoils ($\beta_s = 60°$) at $\beta_1 = 50.6°$ (Psarudakis [38]).

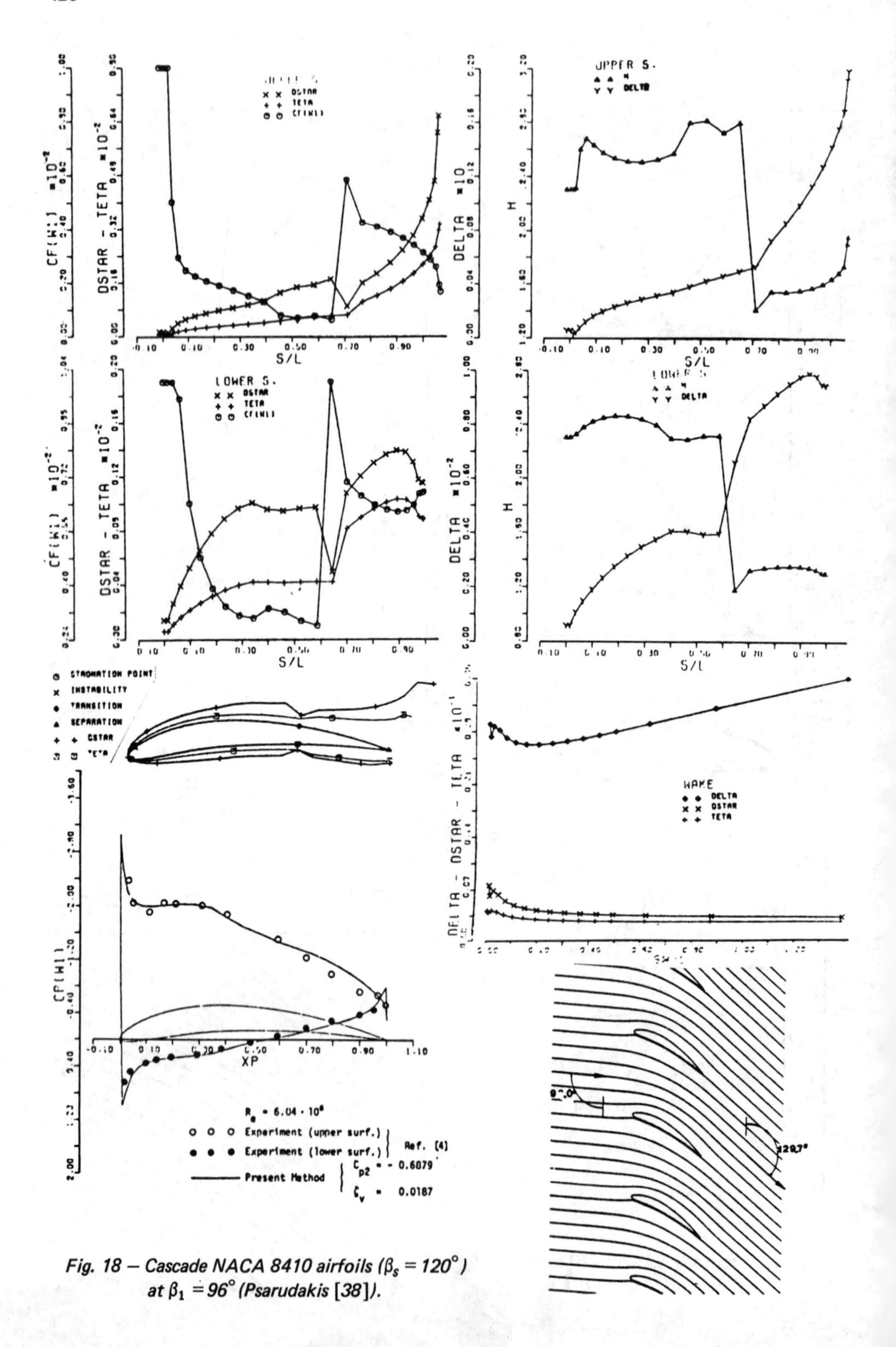

Fig. 18 – Cascade NACA 8410 airfoils ($\beta_s = 120°$) at $\beta_1 = 96°$ (Psarudakis [38]).

ACKNOWLEDGEMENTS

The author wishes to sincerely thank J. Délery who helped him to write this paper.

NOMENCLATURE

Cf	Skin friction coefficient
H	Boundary layer shape factor δ^*/Θ
M	Mach number
P	Static pressure
Pi	Total pressure
r	Recovery factor
T	Static temperature
Ti	Total temperature
u,v	Velocity components
x,y	Coordinate directions
γ	Ratio of specific heats
δ	Boundary layer thickness
δ^*	Displacement thickness
Δ	Difference
λ	Wave number
ρ	Density
Θ	Momentum thickness
Θi	Angle
τ	Friction
μ	Viscosity coefficient
ν	Kinematic viscosity
ω	Relaxation coefficient

Subscript

e	external inviscid flow
i	incompressible or isentropic

REFERENCES

1. Le Balleur, J.C. Calculs couplés visqueux-non visqueux incluant décollements et ondes de choc en écoulement bidimensionnel. VKI/AGARD LS-94 on "Three dimensional and unsteady separation at high Reynolds number" (1978).

2. Le Balleur, J.C. Calcul des écoulements à forte interaction visqueuse au moyen de méthodes de couplage. AGARD-CP-291 (1980).

3. Lock, R.C. and Firmin, M.C.P. Survey of techniques for estimating viscous effects in external aerodynamics. RAE. Tech. Memo. Aero 1900 (1981).

4. Le Balleur, J.C. "Viscid-inviscid coupling calculations for two and three dimensional flows. VKI LS-1982-6 on "Computational Fluid Dynamics" (1982).

5. AGARD Conferences proceedings no 351 "Viscous effects in Turbomachines" (1983).

6. Agardograph no 280 (to be published).

7. Through flow Calculations in axial Turbomachines. AGARD A.R. 175 (octobre 1981).

8. Cambier, L., Ghazzi, W., Veuillot, J.P., and Viviand, H. A multi-domain approach for the computation of viscous transonic flows by unsteady type method.
Recent Advances in Numerical methods in fluids - vol. 3 : Viscous flow computational methods. Editor Habashi. Pineridge Press.

9. Stewartson, K. and Williams, P.G. Self-induced separation. Mathematika 20, pp. 98-108, (1973).

10. Inger, G.R. and Mason, W.H. Analytical theory of transonic normal shock/turbulent boundary-layer interaction. AIAA Journal, vol. 14, no 9, pp. 1266-1272 (1976).

11. Melnik, R.E. Turbulent interaction on airfoils at transonic speeds. Recent developments. AGARD - CP - 291 (1980).

12. Head, M.R. Entrainment in the turbulent boundary-layer. A.R.C. - R. & M. 3152 (1958).

13. Michel, R., Quémard, C. and Durant, R. Application d'un schéma de longueur de mélange à l'étude des couches limites d'équilibre. ONERA N.T. n° 154 (1969).

14. Lighthill, M.J. On displacement thickness. J. Fluid Mech., vol. 4, Part 2, pp. 383-392 (1958).

15. Le Balleur, J.C. and Mirande, J. Etude expérimentale et théorique du recollement bidimensionnel turbulent incompressible. AGARD - CP - 168 (1975).

16. Weinbaum, S. and Garvine, R.W. An exact treatment of the boundary-layer equations describing the two-dimensional viscous analogy of the one dimensional inviscid throat. AIAA Paper no 68-102 (1968).

17. Carrière, P., Sirieix, M. and Délery, J. Méthodes de calcul des écoulements turbulents décollés en supersonique. Progress in Aerospace Sciences, vol. 16, n° 4, pp. 385-429 (1975).

18. Mellor, G.L. The large Reynolds number asymptotic theory of turbulent boundary-layers". Int. J. Engineering Sci., vol. 10, pp. 851-873 (1972).

19. Crocco, L. and Lees, L. A mixing theory for the interaction between dissipative flows and nearly isentropic streams. JAS, vol. 19, no 10, pp. 649-676 (1952).

20. Crocco, L. Consideration on the shock/boundary-layer interaction. Proc. Centennial of Brooklyn Poly. Inst. (1974).

21. Klineberg, J.M. Theory of laminar viscous-inviscid interactions in supersonic flow. Ph. D. Thesis, California Institute of Technology (1968).

22. Hunter, L.G. and Reeves, B.L. Results of a strong interaction wake-like model of supersonic separated and reattaching turbulent flows. AIAA Journal, vol. 9, no 4, pp. 703-712 (1971).

23. Holden, M.S. Theoretical and experimental studies of the shock-wave/boundary-layer interaction on curved compression surfaces. Proc. of the Symposium on Viscous Interaction Phenomena in Supersonic and Hypersonic Flow, Hypersonic Research Laboratory, Aerospace Research Laboratories, 7-8 May 1969.

24. Le Balleur, J.C. Couplage visqueux-non visqueux : analyse du problème incluant décollements et ondes de choc. La Recherche Aérospatiale, n° 1977-6, pp. 349-358 (1977)

25. Dodge, P.R. and lieber, L.S. A numerical method for the solution of Navier-Stokes equations for a separated flows. AIAA Paper no 77-170 (1977)

26. Hkosla, P.K. and Rubin, S.G. A composite velocity for the compressible Navier-Stokes equations. AIAA Paper no 82-0099 (1982).

27. Ardonceau, P.L. and Alziary de Roquefort, T. Direct and inverse calculation of laminar boundary-layer solution. AIAA Journal, vol. 18, no 11, pp. 1392-1394 (1980).

28. Garvine, R.W. Upstream influence in viscous interaction problems. The Physics of Fluids, vol. 11, pp. 1413-1423 (1968).

29. Neyland, V. Ya. Upstream propagation of disturbances in hypersonic boundary-layer interactions". (in Russian) Akad. Nauk. SSSR, Izv. Mekh. Gaza., no 4, pp. 44-49 (1970).

30. Werle, M.J., Hankey, W.L. and Dwoyer, D.L. Initial conditions for the hypersonic shock/boundary-layer interaction problem. AIAA Journal, vol. 11, no 4, pp. 525-530 (1973).

31. Veldman, A.E.P. A numerical method for the calculation of laminar incompressible, boundary-layer with strong viscous-inviscid interaction. Report NLR TR 79023U (1979).

32. Veldman, A.E.P. New, quasi-simultaneous method to calculate interacting boundary-layers. AIAA Journal, vol. 19, no 1, pp. 79-85 (1981).

33. Bertke, S.D., Werle, M.J. and Polak, A. Finite difference solutions for the interacting supersonic turbulent boundary-layer equations, including separation effects. Department of Aerospace Engineering, University of Cincinnati, Cincinnati, Ohio. Report AFL - 74-4-9 (1974)

34. Werle, M.J., Polak, A., Vatsa, V.N. and Bertke, S.D. Finite difference solution for supersonic separated flows. AGARD - CP - 168 (1975).

35. Le Balleur, J.C. Couplage visqueux-non visqueux : méthode numérique et application aux écoulements bidimensionnels transsoniques et supersoniques. La Recherche Aérospatiale, n° 1978-2, pp. 67-76 (1978).

36. Wigton, L.B. and Holt, M. Viscous-inviscid interaction in transonic flow. AIAA Paper no 81-1003 (1981).

37. Melnik, R.E. and Chow, R. "Asymptotic theory of two dimensional trailing edge flows. NASA SP-347, p. 197 (1976).

38. Meauzé, G. and Délery, J. Calcul de l'interaction onde de choc-couche limite par emploi de méthodes inverses. AGARD/PDP 61-A Specialists Meeting on Viscous Effects in Turbomachines. Copenhag, Danmar, 1-3 June 1983 - AGARD - CP - 351.

39. Viviand, H. and Veuillot, J.P. Méthodes pseudo-instationnaires pour le calcul d'écoulements transsoniques. Publication ONERA n° 1978-4 (1978).

40. Meauzé, G. Méthode de calcul inverse pseudo-instationnaire. La Recherche Aérospatiale, n° 1980-1, pp. 23-30 (1980).

41. Délery, J. and Le Balleur, J.C. Interaction et couplage entre écoulement de fluide parfait et écoulement visqueux. ONERA RSF n° 4/7073 (1980).

42. Carter, J.E. A new boundary-layer iteration technique for separate flow. AIAA Paper no 79-1450 (1979).

43. Brune, G.W., Rubbert, P.E. and Nark, T.C. A new approach to inviscid flow boundary-layer matching. AIAA Paper no 74-601 (1974) see also AIAA Journal, vol. 13, no 7, pp. 936-938 (1975).

44. Thiede, P.G. Prediction method for steady aerodynamics loading on airfoils with separated transonic flow. AGARD-CP-204 (1976).

45. Kuhn, G.D. and Nielsen, J.N. Prediction of turbulent separated boundary-layers. AIAA Paper no 73-663 (1973).

46. Carter, J.E. Viscous-inviscid interaction analysis of transonic turbulent separated flow. AIAA Paper no 81-1241 (1981)

47. Whitfield, D.D., Swafford, T.W. and Jacocks, J.L. Calculation of turbulent boundary-layer with separation and viscous-inviscid interaction. AIAA Journal, vol. 19, no 10, pp. 1315-1322 (1981).

48. Gordon, R. and Rom, J. Transonic viscous-inviscid interaction over airfoils for separated laminar and turbulent flows. AIAA Journal, vol. 19, no 5, pp. 545-552 (1981).

49. Alziary de Roquefort, T. Couplage fort et couplage faible entre couche limite et écoulement extérieur. 13ème Colloque de l'Association Aéronautique et Astronautique de France, Ecully, nov. 1976.

50. Ardonceau, P. Etude de l'interaction onde de choc-couche limite supersonique. Thèse de Docteur ès Sciences Physiques, Université de Poitiers (1981).

51. Ardonceau, P., Alziary, T. and Aymer, D. Calcul de l'interaction onde de choc/couche limite avec décollement. AGARD - CP - 191 (1980).

52. Le Balleur, J.C. Viscid-Inviscid Coupling Calculations for Two and Three-dimensional flows. Lecture series 1982-04, Von Karman Institute, Computational fluids dynamics, Belgium, (March 1982).

53. AGARD - CP - 351.

54. Le Balleur, J.C., Néron, M. Calcul d'écoulements visqueux décollés sur profils d'ailes par une approche de couplage - AGARD - CP - 291, Paper 11 (1981).

55. Porcheron, B. and Thibert, J.J. Etude détaillée de l'écoulement autour d'un profil hypersustenté. Comparaison avec les calculs. AGARD - CP - 365.

56. Calvert, W.J. and Herbert, M.V. An inviscid-viscous interaction method to predict the blade-to-blade performance of axial compressors. Aeronautical Quarterly, vol. XXXI, Part 3 (August 1980).

SHOCK BOUNDARY-LAYER INTERACTION

Georges MEAUZÉ

Office National d'Etudes et de Recherches Aérospatiales,
B.P 72, 92322 Châtillon Cedex, France

1 INTRODUCTION

This second part deals with the particular problem of shock-boundary-layer interaction.

The consequences of this phenomenom are very important especially when separation occurs. Viscous zones and, consequently, losses increase with a strong intensification of turbulent fluctuations. Thus the phenomenon of shock/boundary-layer interactions are harmful to the correct working of parts, especially in internal aerodynamics, i.e. air intakes, compressor or turbine blades and nozzles.

There are a lot of shock-wave configurations (as far as a boundary-layer interaction is concerned) with a boundary-layer on a wall. All of these can be separated into two types which have already been distinguished for the application of coupling techniques.

i) The transonic interaction corresponds to the case where the shock-wave would be normal to the wall without a boundary-layer. The downstream flow is subsonic.

ii) In the supersonic interaction, the downstream flow remains supersonic. This occurs for an oblique shock-wave reflection or in the beginning of the oblique shock-wave in the presence of a compression corner.

The external configuration of the flow (inviscid flow) is very different, but we can observe quite the same phenomena in the boundary-layer interaction, especially the pressure distribution.

Figures 1 to 4 show the schemes of configurations usually encountered in air intakes, compressors and turbines, cascades and nozzles, in transonic and supersonic regimes.

Before the presentation of results concerning the application of coupling computations to shock/boundary-layer configurations, let us recall some generalities concerning phenomena appearing in these interactions. To do so we have used Délery's synthesis (1) limiting ourselves to cases where the Mach number is less than 2. This covers to most of the cases in turbomachinery applications.

2 GENERALITIES CONCERNING THE MECHANISMS OF SHOCK/BOUNDARY-LAYER INTERACTION

An attached boundary-layer is basically a thin region of thickness δ, where the cross velocity distribution varies from the external velocity u_e to zero on the wall. The Mach number varies consequently from M_e to zero. Without strong interaction, the static pressure remains constant along the normal direction to the wall. So, the boundary-layer can be assumed to be a quasi-parallel flow where the entropy changes at each streamline (rotational flow).

When an incoming shock-wave crosses a boundary-layer, the stress friction resulting from the laminar viscosity and from the turbulent shear stress, has in fact no effect on the phenomenon for the major portion of the thickness δ because the deceleration is very high. This behaviour has been completely demonstrated by both experience (2), (3) and theory (4) . Thus an incoming shock-wave in a boundary-layer can be assumed to be an inviscid fluid phenomenon.

This scheme permits a forecasting of the flow configuration but it should be observed that using a strictly inviscid model leads to unreaslistic behavior close to the wall.

Lighthill (4) has shown that viscosity terms must be taken into account to satisfy the wall condition.

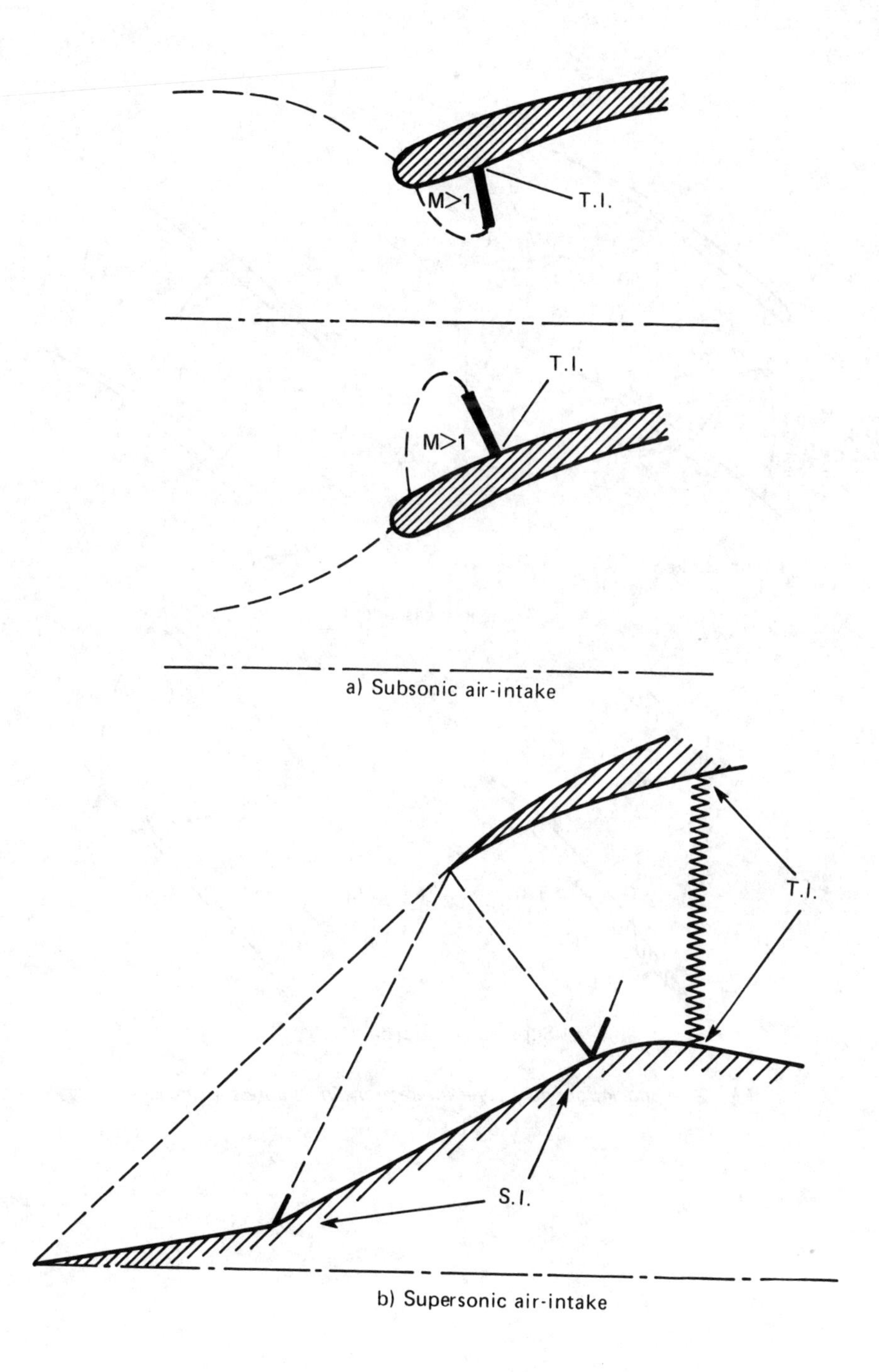

Fig. 1 — Shock boundary layer interactions in air-intakes.

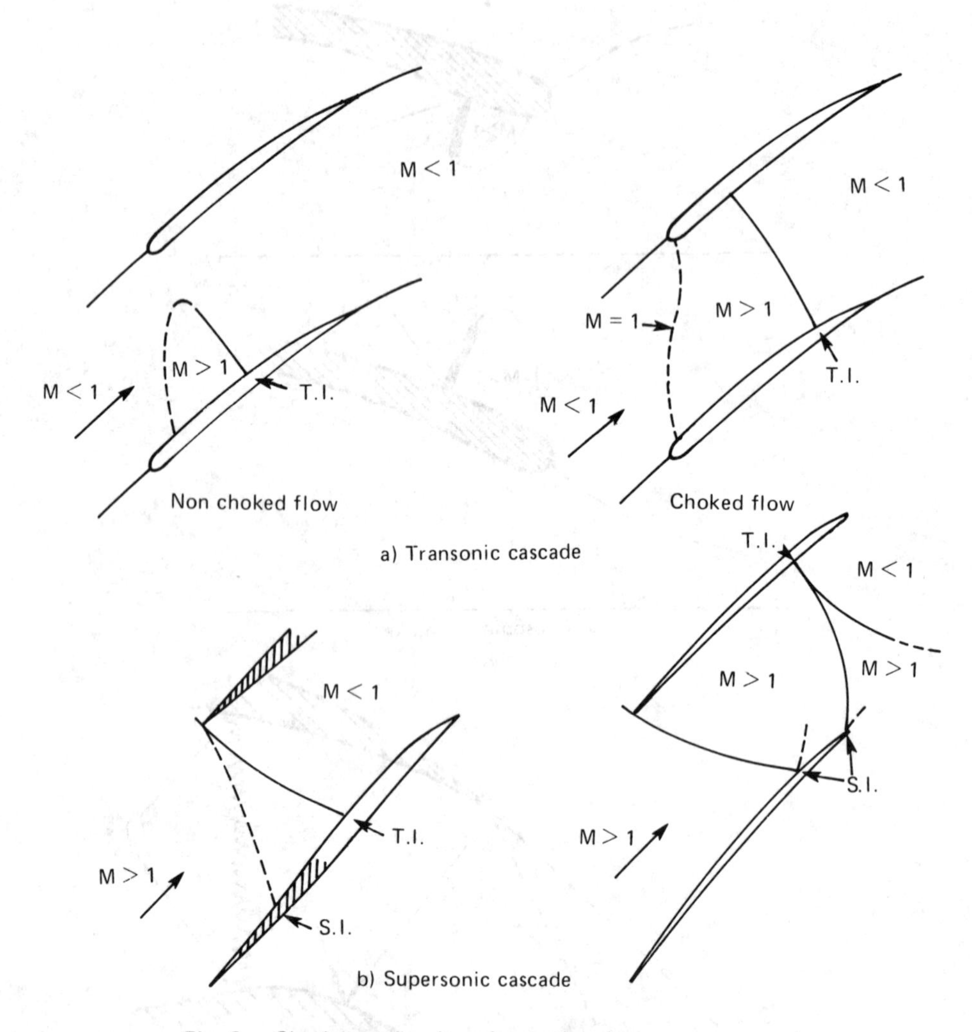

Fig. 2 – Shock boundary layer interactions in compressor cascades.

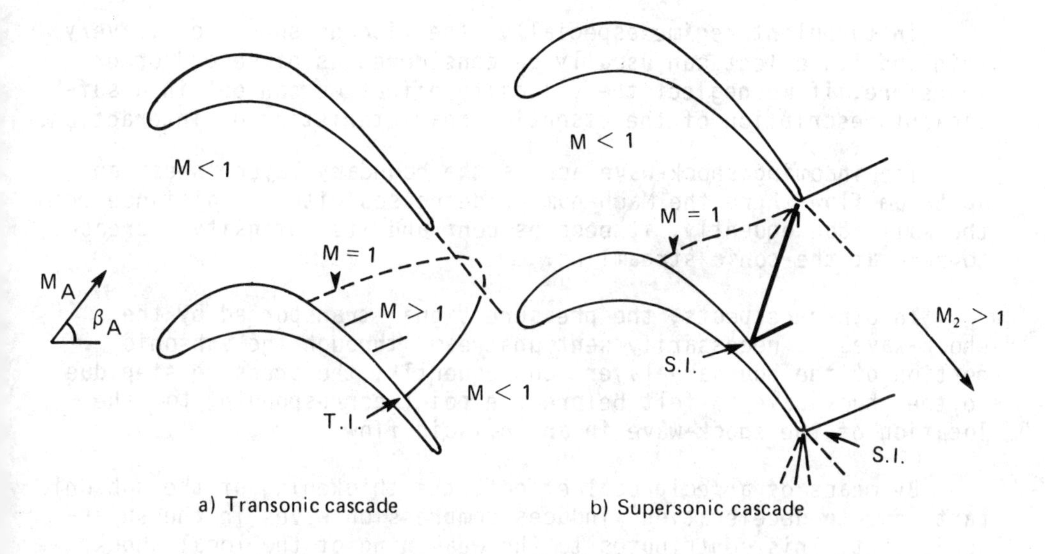

Fig. 3 – Shock boundary layer interactions in turbine cascades.

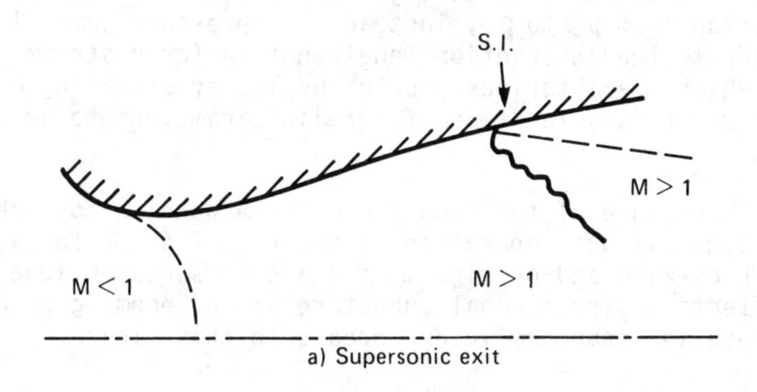

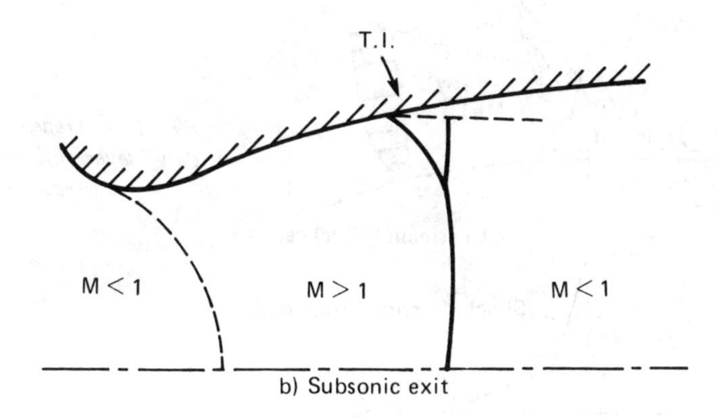

Fig. 4 – Shock boundary layer interaction in nozzles.

In turbulent regime especially, the viscous sub-layer is very thin and its effect can usually be considered as of second order. Therefore, if we neglect the viscosity effect we can obtain a sufficient description of the essential characteristics of interaction.

The incoming shock-wave across the boundary-layer "sees" an upstream flow where the Mach number decreases with the distance from the wall. Consequently, it becomes bent and its intensity decreases to zero at the sonic streamline.

In other respects, the pressure signal transported by the shock-waves is necessarily sent upstream through the subsonic portion of the boundary-layer. Consequently, the pressure step due to the shock-wave is felt before the point corresponding to the location of the shock-wave in an inviscid flow.

By means of a reciprocal effect, the thickening of the subsonic part (due to deceleration) induces compression waves in the supersonic part. This contributes to the weakening of the local shock.

This mechanism leads to a gradual evolution of the pressure distribution from p_o to p_1, instead of a pressure jump. This in turn leads to the interaction length notion (or upstream influence length) which characterises greater or lesser spreading of the compression which is a function of certain parameters to be recalled later on.

The structure of the flow-field corresponding to turbulent interactions without separation is shown on fig. 5 for a transonic normal shock-wave and on fig. 6 for a supersonic oblique shock-wave reflection. The gradual curvature of an incoming shock-wave (C_o) across the boundary-layer appears in this figure.

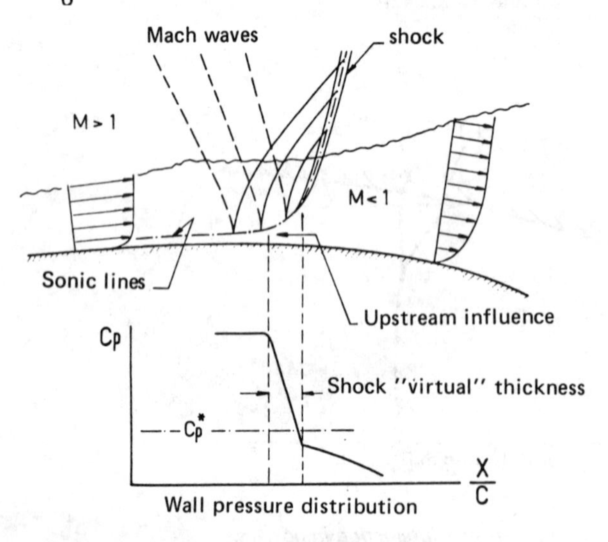

Fig. 5 — Transonic shock boundary layer interaction without separation.

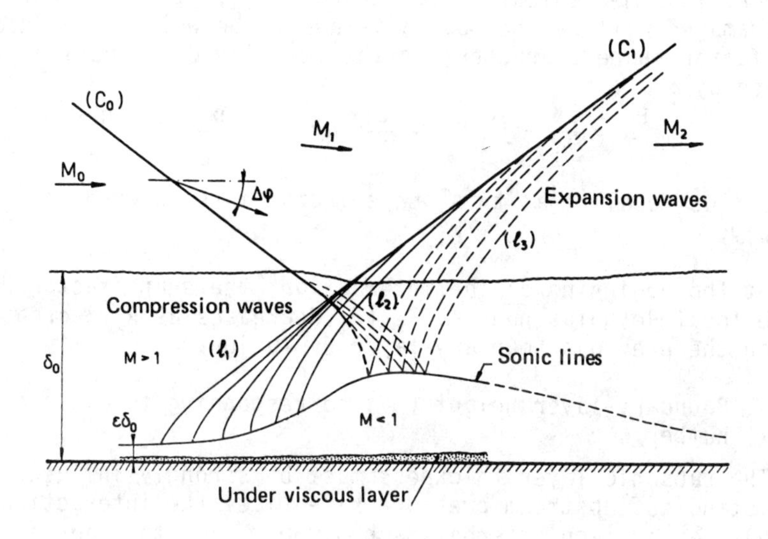

Fig. 6 – Supersonic shock reflexion without separation.

The subsonic layer growth with an initial thickness $\varepsilon\delta_0$, generates the compression waves (l_1) which reduce the intensity of (C_0). Their focalisation creates the reflected shock (C_1).

The refraction of the wave (l_1) and (C_0) during their propagation into the rotational supersonic layer $(1 - \varepsilon)\ \delta_0$ induces a secondary system of waves (l_2). Their reflection on the sonic line gives the expansion waves (l_3) just behind (C_1). The figure also shows the growth of the subsonic layer. Normally this growth can be contracted because of the acceleration of the boundary-layer lower part by the effect of viscosity forces which becomes more and more important.

In fact, the interaction between a boundary-layer and a shock-wave induces very complex mechanisms which are not very well understood.

We can well imagine that velocity and pressure distribution in the boundary-layer just before the interaction zone strongly contribute to the extension and severity of the interaction.

Briefly speaking, three factors occur :

i) Mechanical energy carried out with initial profile

The less mechanical energy there is (i.e., empty profile), the more damaged will be the boundary-layer. The well known incompressible factor shape characterises the velocity distribution. It is defined by :

$$H_i = \frac{\delta *}{\Theta_i} = \int_0^\delta \left(1 - \frac{u}{u_e}\right) dy \Big/ \int_0^\delta \frac{u}{u_e}\left(1 - \frac{u}{u_e}\right) dy$$

Let us recall that this shape factor is high when the profile is empty.

At the beginning of the interaction, the shape factor depends on the local Reynolds number, R_δ (H_1 decreases as R_δ increases) and also on the previous boundary-layer evolution.

ii) Boundary-layer height (δ^*) corresponding to a sonic Mach number

The subsonic layer thickness must be strongly influenced by the distance of upstream pressure rise (i.e. the interaction length). For a given external Mach number M_e on the boundary-layer thickness δ, the normalised height y^*/δ is a function of the Mach number distribution $M(y)$ which depends on both the velocity profile (i.e. H_i) and the thermic configuration on the wall, the latter being an important factor of the sound velocity distribution. For instance, with a cooled wall, Mach numbers are higher in the lower part.

The sonic line location in an adiabatic flow ($M_1 = 2$) is given in Fig. 7 for different turbulent velocity profiles represented by the Coles composite law (5). The same diagram for a laminar boundary-layer (Fig. 7) shows that the values of y/δ are much higher. This explains the large interaction domains in laminar flow.

iii) Viscosity effect

Chapman's free interaction theory gives an idea of the influence of viscosity.

The pressure rise in a separated boundary-layer can be written as :

$$\Delta p_D \simeq \left(M_0^2 - 1\right)^{-1/4} C_{f_0}^{\;1/2} q_0$$

Where q_0, M_0, Cf_0 are respectively the dynamic pressure, the Mach number and the skin friction at the beginning of the interaction. So if the Reynolds number increases, Δp_D decreases as does the interaction length.

The reality is a compromise between opposite influences. In fact, if the Reynolds number $R\delta_0$ is less than 10^5, or if we are in laminar regime, viscosity is the most important factor.

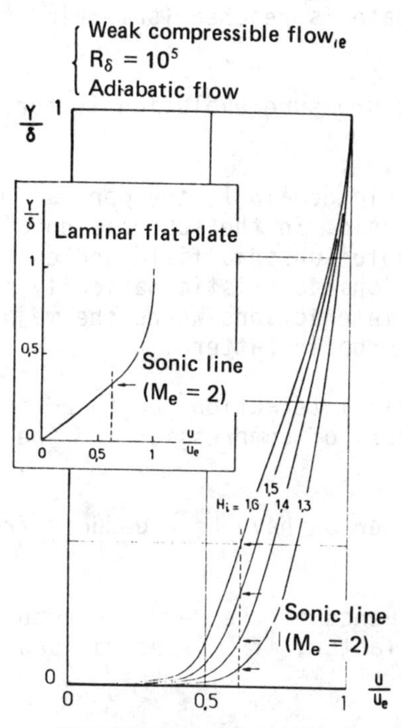

Fig. 7 — Velocity distribution in a turbulent boundary layer.

3 INTERACTION PROPERTIES - INTERACTION LENGTH IN TRANSONIC FLOW WITHOUT SEPARATION

Figure 8 shows an interferometric visualisation of an interaction. The Mach number just in front of the shock-wave is close to 1.3. The beginning of the compression waves inside the boundary-layer can be barely seen. These waves focalise into a unique quasi normal shock. Its thickening in the picture is due to the lateral wall boundary-layer interaction.

Such visualisations and static pressure distributions show that the influence domain can be divided into two parts, as indicated on Fig. 8.

Fig. 8 — Transonic interaction. Definition of the supersonic interaction length.

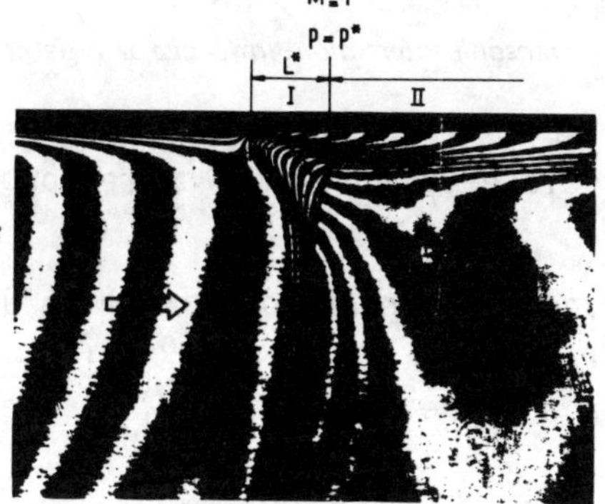

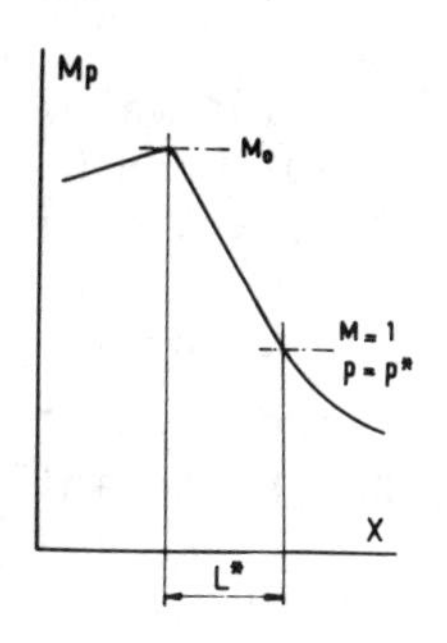

i) A first domain I corresponds to a quick but continuous supersonic compression until the sonic state is reached (same kind of compression as simple waves).

ii) In the second domain II, the pressure evolution is more moderate.

It is clear that in the supersonic domain I, the general properties of the interaction are intrinsic. In the subsonic domain II, the flow structure depends on the whole subsonic field and especially on downstream effects. This characteristic basically distinguishes subsonic from supersonic interactions where the major part of the flow-field follows a hyperbolic pattern.

For this reason, a characteristic interaction length L^* is defined as the distance between the start of compression and the sonic state.

The correlations presented further on have been deduced from a lot of experimental results (6) .

As shown on Fig. 9.a, the influence of the Reynolds number is weak for a given value of the shape factor, if L is normalised with the displacement thickness δ_0^* at X_0.

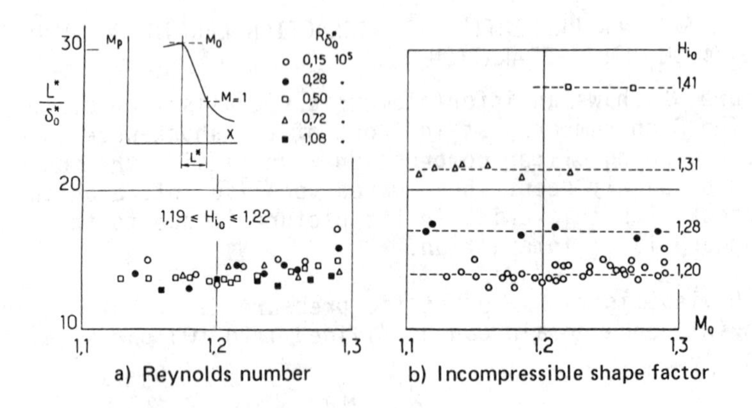

Fig. 9 – Transonic interaction. Influence of initial conditions.

The value of $L^*/_{\delta^*}$ is quasi independent of the Mach number, except when M_0 is close to 1.3 which corresponds to the emergence of separated flow.

L^*/δ_0^* is strongly dependent on H_{i_0} which doubles when H_{i_0} changes from 1.2 to 1.4. A simple empirical correlation gives well enough the influence of Reynolds number, upstream Mach number and shape factor (Fig. 10). It is given by:

$$\frac{L^*}{\delta_0^*} \frac{1}{Hi_0 - 1} = 70$$

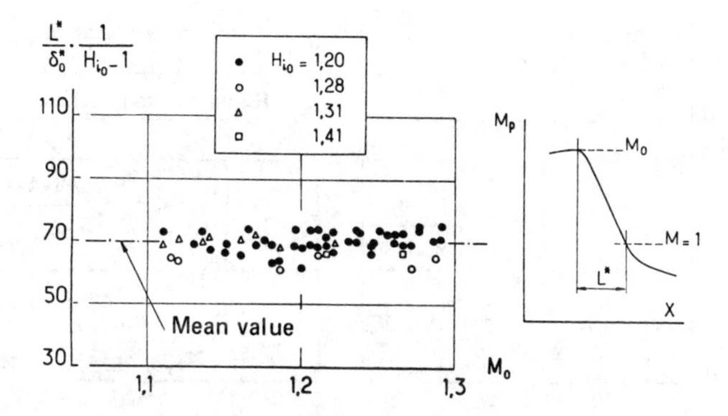

Fig. 10 – Transonic interaction. Correlation for the supersonic interaction length.

3.1 Supersonic interaction length

The influence length L_0 is usually defined as the distance between the beginning of the interaction (X_0) and the point where the shock-wave would strike the wall in inviscid flow (Fig. 11).

It is difficult to determine with good accuracy the actual beginning of the interaction from wall pressure distributions. Most authors on this subject take a conventional starting point obtained by the extrapolation to the wall of the quasi linear part of the pressure evolution as indicated on fig. 11.

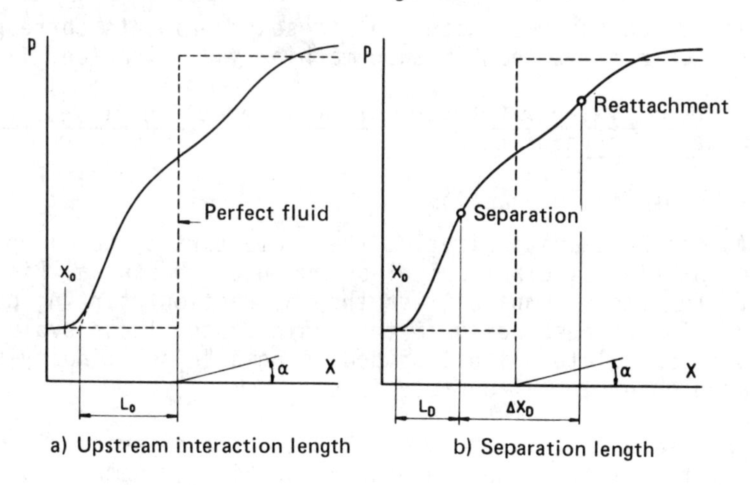

Fig. 11 – Supersonic interaction characteristics lengths.

Fig. 12 shows the interaction length variations as functions of main parameters : upstream Mach number, Reynolds number, corner angle...

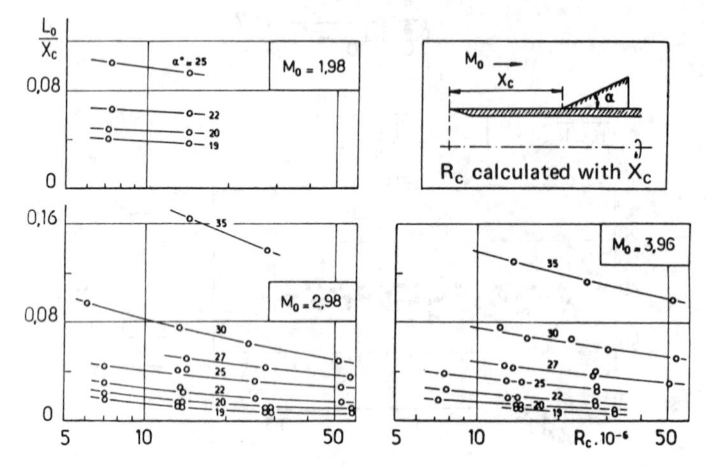

Fig. 12 — Upstream interaction length of main parameters.

The study of these results (from (7)) shows similar tendencies as found in transonic interaction.

i) For fixed values of Mach number M_0 and Reynolds number R_c, L_0 increases with α, i.e. with the perturbation intensity.

ii) If α and R_c are fixed, L_0 decreases as M_0 increases.

iii) If α and M_0 are fixed, L_0 decreases as Reynolds number.

It is necessary to note that these experiments correspond to higher Mach numbers than in turbomachinery application.

3.2 Boundary-layer behavior in an interaction without extended separation

- Transonic interaction

It can be characterised on the whole part by the jump of certain quantities as displacement or momentum thickness. Figure 13 illustrates these jumps between the interaction starting point and the location corresponding to the sonic state. These evolutions have been calculated with a method using a "discontinuity analysis" and based on empirical correlations.

The hachured zone shows the limit where separation occurs as given by this method. We can observe on Fig. 13 some major tendencies.

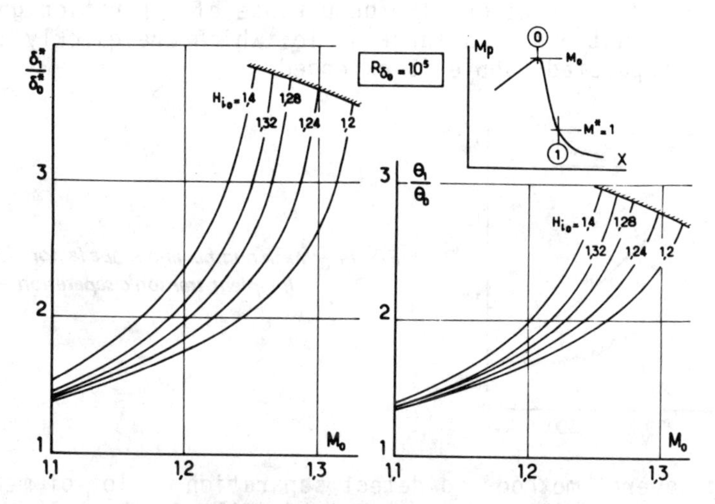

Fig. 13 – No separated transonic interaction. Characteristics thickness jumps.

i) The jump amplitude of δ^* and Θ , increases with the Mach number M_0, i.e. with shock intensity. The displacement thickness increases in a quasi exponential manner when separation occurs.

ii) The δ^* or Θ jump is more large than the initial shape factor H_{i_0} is high.

iii) Calculation have shown that the influence of the Reynolds number $R\delta_0$ is weak when H_{i_0} is fixed. The strong influence of $R\delta_0$, sometimes observed is due in fact to the H_{i_0} change resulting from $R\delta_0$ variation.

3.3 Conditions under which separation occurs - Transonic incipient separation

Usually, incipient separation is defined as the situation where the minimum of the wall skin friction τ evolution, is just equal to zero. The increase of the shock intensity leads to a configuration where the sign of τ changes and the separated zone corresponds to negative values of τ . The size of the separated bubble is very small. In fact, in a turbulent regime this bubble is unsteady and thus it is not easy to define incipient separation. This problem was discussed in detail by Simpson et al. (8).

The shape factor evolution for incipient separation is presented on fig. 14. We can see that H_i has a maximum value close to 2.6. This corresponds quite well to the usual value of 2.5 for a turbulent separation point. Downstream H_i quickly decreases and tends to the flat-plate level. We can also observe that longitudinal turbulence intensity reaches a higher level than in the case of a non-separated interaction. The occurrence of separation generates turbulent fluctuations on a large scale, which are quickly amplified if the separated bubble is extended.

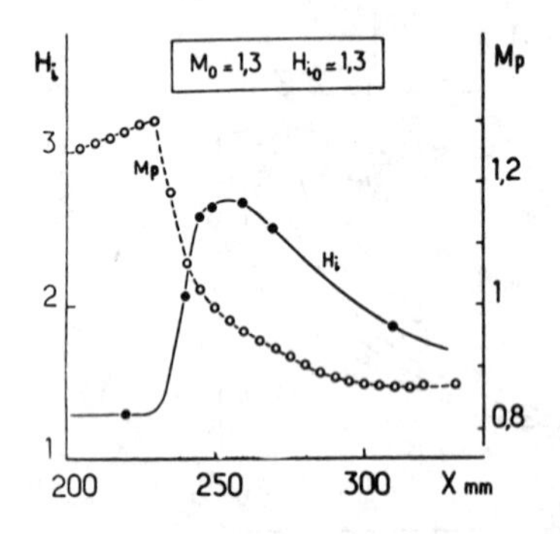

Fig. 14 – Mach number and shape factor distribution. Incipient transonic separation.

Using several methods to detect separation, a lot of measurements taken on airfoils and on contoured walls in channels have made it possible to draw the separation limit line on fig. 15 as function of the two parameters (M_0 and H_{i_0}) that seem to cause major effects. Let us recall that this limit is only available for basic interactions, i.e. non-porous and athermal wall. The Reynolds number effect is in fact taken into account by its influence on H_{i_0} but this is true only for a flat-plate boundary-layer.

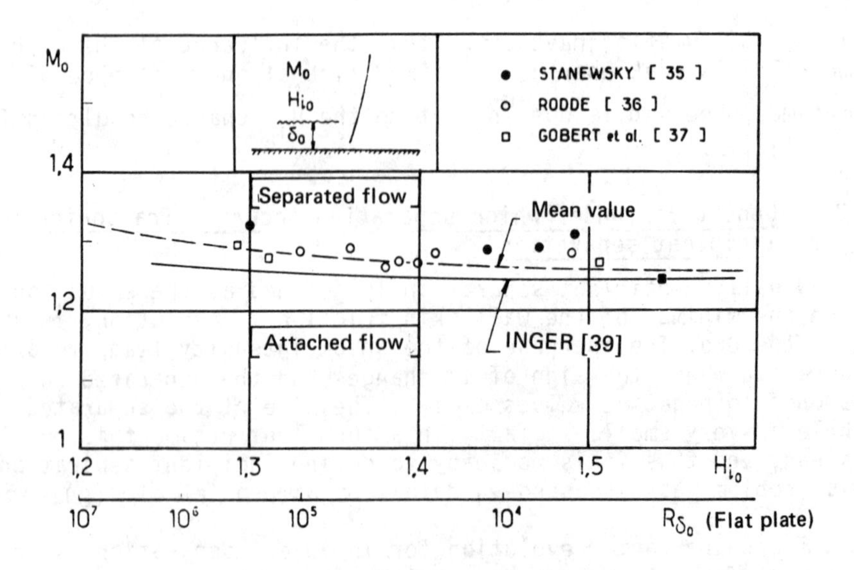

Fig. 15 – Transonic incipient separation. Adiabatic wall.

The empirical separation limit presented on fig. 15 is compared to the prediction given by Inger's model (9). It uses a small perturbations theory applied to the triple deck model of Lighthill (4). Inger's theory is in quite good agreement with experiments.

3.4 Supersonic incipient separation

Some years ago, a lot of experiments were performed to determine incipient separation in a supersonic or hypersonic flow. Usually, the results present the limit angle α_s of the corner corresponding to incipient separation as a function of the Reynolds number $R\delta_o$ and Mach number M_o. α_s can be easily replaced by the corresponding pressure ratio P_s/P_o using Rankine-Hugoniot equations.

Some tendencies can be observed on the fig. 16 where the major part of published results are plotted.

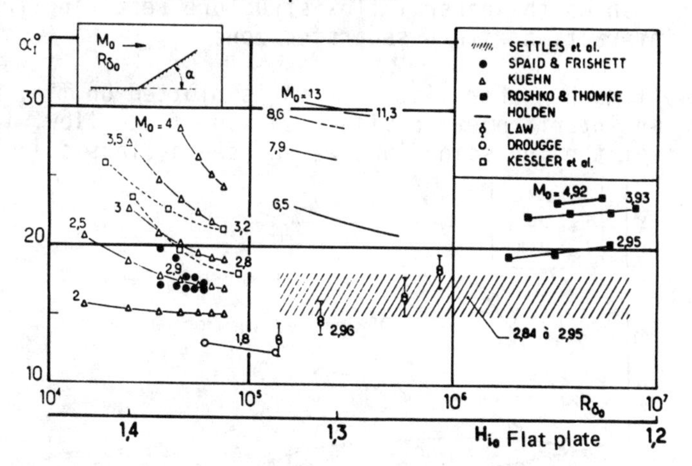

Fig. 16 – Supersonic incipient separation. Adiabatic wall.

i) The value of α_i increases as the upstream Mach number increases.

ii) For low Reynolds numbers, α_i decreases as $R\delta_o$ increases.

iii) But for high Reynolds numbers, α_i increases slightly as $R\delta_o$ increases.

Because of this slight influence and questionable experiments, some authors believe that incipient separation might be independent of the Reynolds number.

Several simplified methods have been proposed for predicting supersonic or hypersonic incipient separation. Some of them have been deduced from Chapman's free interaction theory and can be applied to low Reynolds number flows (10, 11).

Other neglect viscosity (12, 13, 14) and thus they can be applied only for high Reynolds numbers.

Several entirely empirical prediction methods exist as well (7, 15).

3.5 Interaction with large separation

The situation is more complex when the shock intensity is sufficient to induce an extended separation. Then the dissipative effects are essential.

Fig. 17 shows the general flow structure resulting from a transonic interaction with a separated zone.

The wall Mach number distribution is plotted on fig. 18, where we can see an interferometric visualisation of the flow. In the external part of the inviscid flow, we can see a classic lambda shock system.

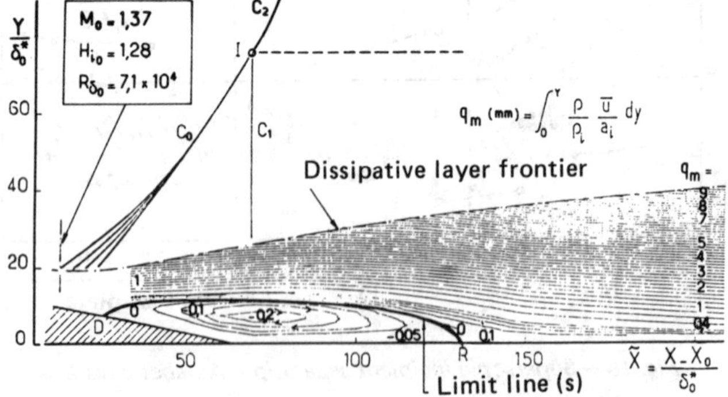

Fig. 17 – Separated transonic interaction. Shock system and dissipative layer streamlines.

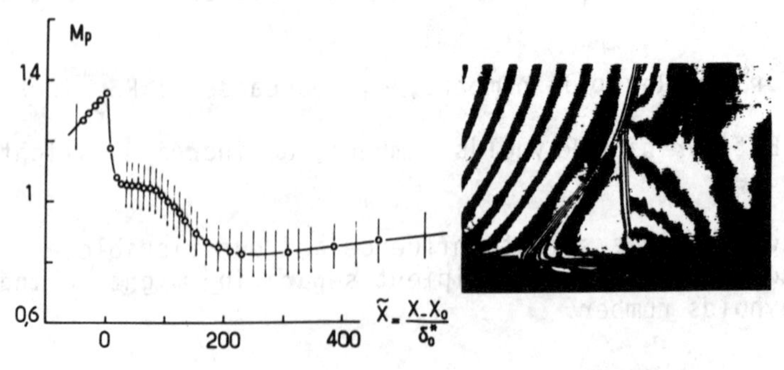

Fig. 18 – Separated transonic interaction. Mach number distribution.

A "weak" oblique shock-wave (C_0) starts at the beginning of the separation. The supersonic zone after (C_0) is terminated by a "strong" oblique shock wave (C_1) (in fact of weak intensity).

The two shocks (C_0) and (C_1) converge at a triple point (I) where the "strong" oblique shock (C_2) begins.

Streamlines of the mean flow in the dissipative layer are also shown on fig. 17. We can see a recirculation bubble restricted by the (S) limit streamline issued from the separation point D and terminated at the reattachment point R.

The reflection of a supersonic oblique shock wave is shown on fig. 19. The separation begins far from the theoretical point corresponding to the inviscid flow. The corner due to the separation creates compression waves which focalise to build the reflected shock (C_1). The oblique shock (C_0) impinges on the separated dissipative layer and is reflected into an expansion waves system. This induces a continuous static pressure evolution. Then the main flow deviates to the wall and reattachment occurs at the point R. We have the same kind of bubble restricted by the limit streamline (S) as in a transonic interaction.

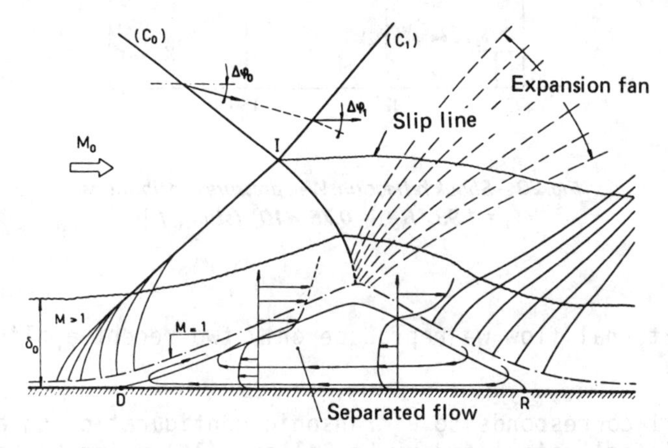

Fig. 19 — Shock reflection with boundary layer separation. Schematic representation of the flow field.

Some wall pressure distributions measured with an upstream Mach number of 1.93 are given on fig. 20.

We see that the size of the separated bubble increases as does the pressure ratio p_2/p_0.

4 SHOCK BOUNDARY-LAYER COUPLING APPLICATIONS

A lot of applications have been made more in external flow than in internal flow.

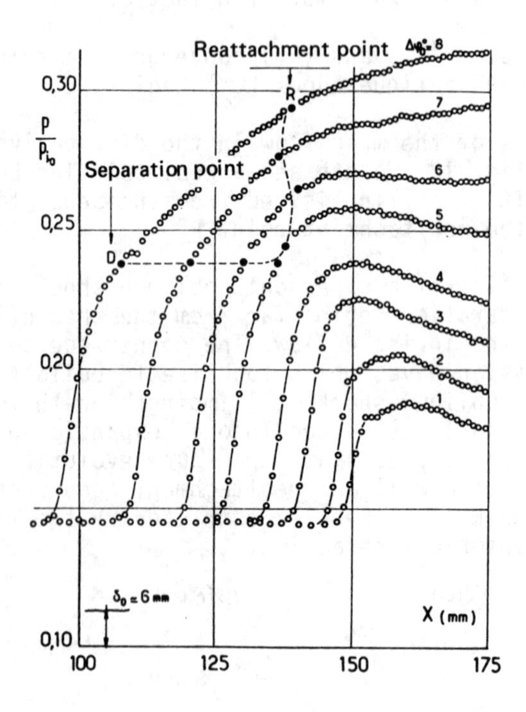

Fig. 20 – Shock reflection. Wall pressure distributions
$M_0 = 1.93$, $R_{\delta_0} = 0.75 \times 10^5$ (after [1]).

For external flow we reproduce only two recent applications on airfoils.

Fig. 21 corresponds to a transonic configuration on a supercritical airfoil calculated by Le Balleur (16) using his semi-inverse method.

The same airfoil has been calculated by Melnik et al. (17) and Whitfield et al (18).

The result is illustrated on fig. 22.

For internal flow, ducts or nozzles are often studied because on the one hand, the geometry is simple and on the other hand, measurements are easy.

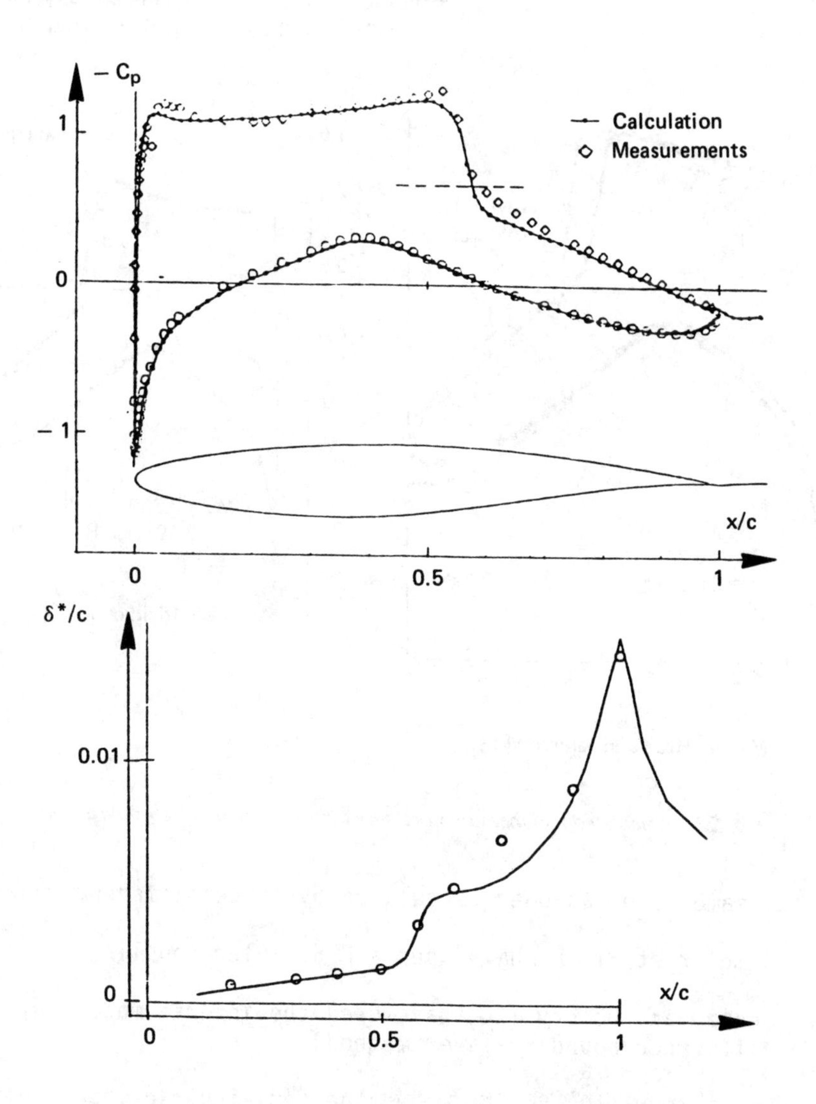

Fig. 21 – Calculated and measured pressure distribution and displacement thickness on a transonic airfoil. (RAE 2822, $M = 0.730$, $\alpha = 3.19°$, $R = 6.5 \times 10^6$).

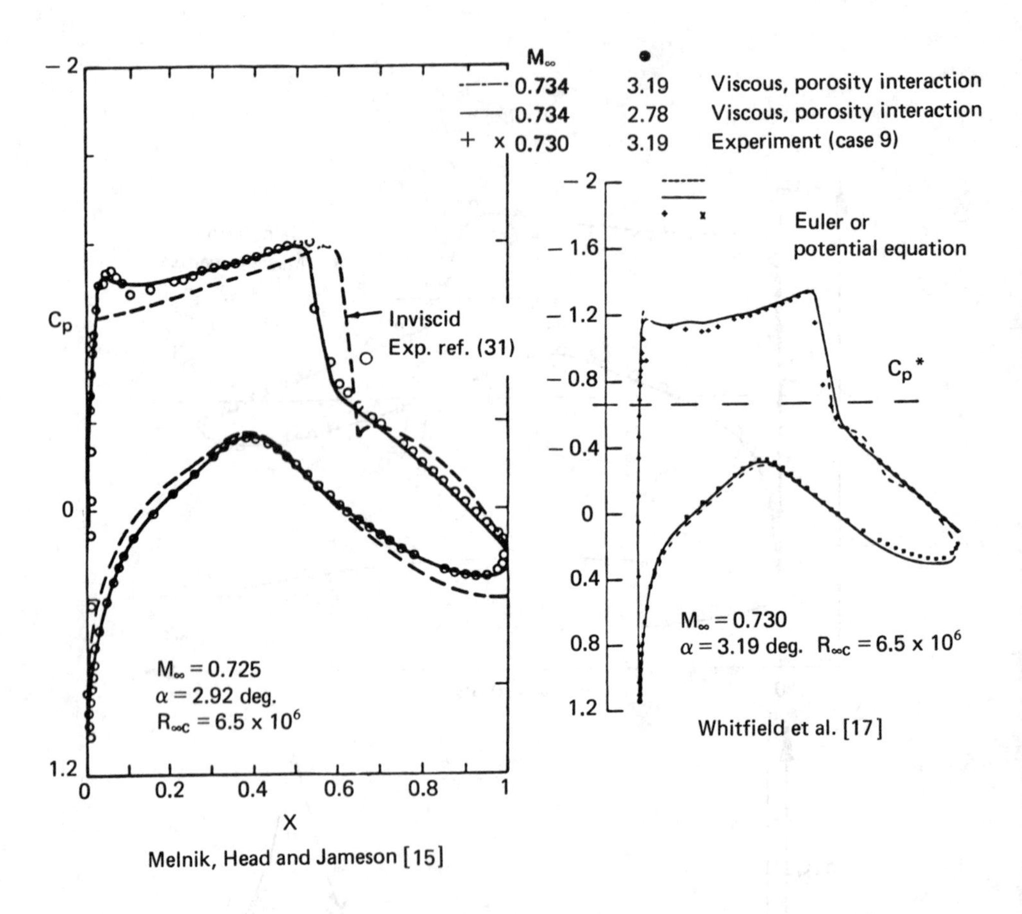

Fig. 22 — Calculated and measured pressure distribution on a transonic airfoil.

The same case has been calculated by three different ways.

. Cambier et al (19) have used a N.S. Euler coupling.

. Meauzé and Délery (20) have used the inverse mode (Euler and finite difference boundary-layer method).

. Le Balleur (16, 21) has used the semi-inverse mode (Euler and integral boundary-layer method).

Results are reported on fig. 23.

On fig. 24 and 25, we can see the result of the inverse mode applied for two different shock locations in a symmetrical transonic nozzle, obtained by the inverse mode (20).

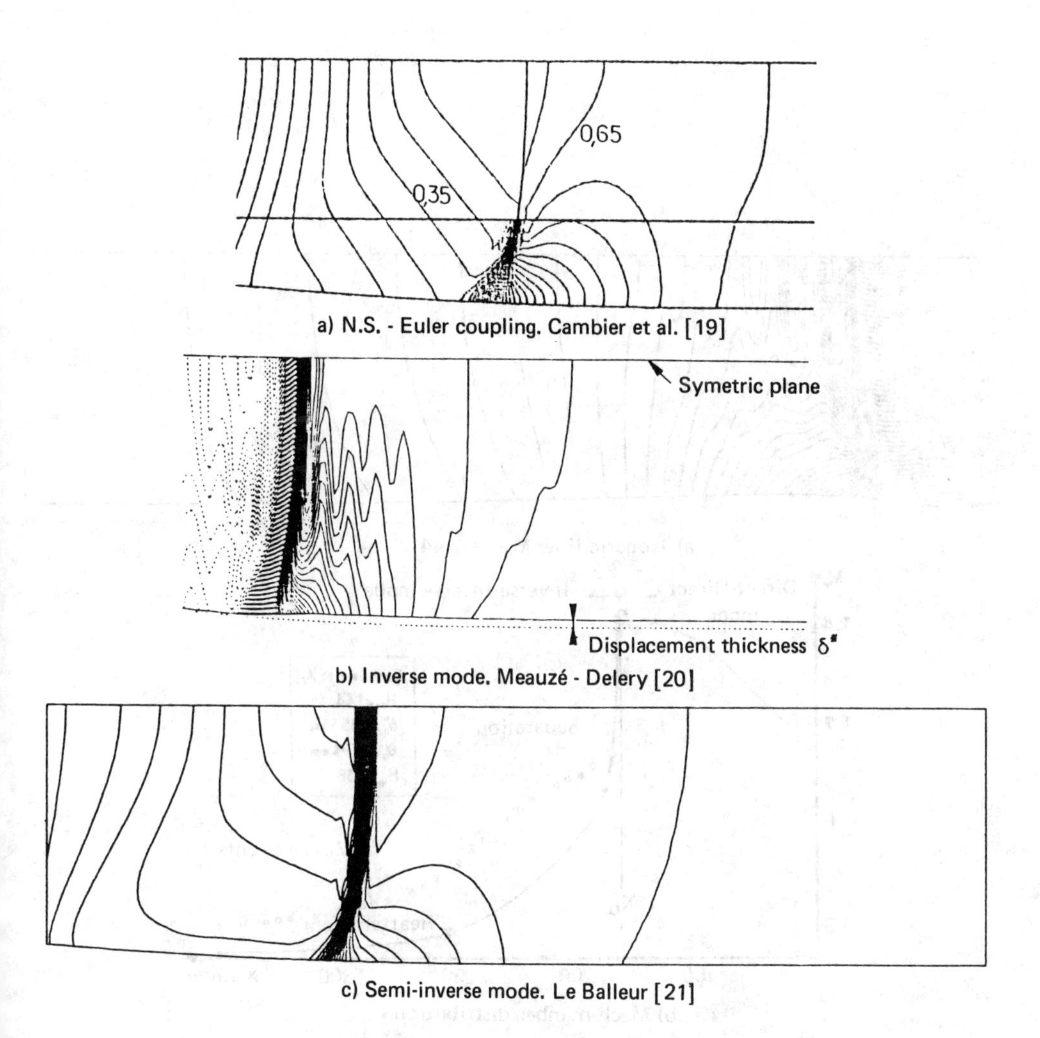

Fig. 23 – Calculation of isobaric lines. Transonic shock boundary layer interaction in a nozzle.

A configuration with a double transonic interaction (on two walls) has been also calculated, with the inverse mode on the lower wall and the semi-inverse mode on the upper wall (fig. 26). The same configuration has been computed by Le Balleur with semi-inverse mode applied on both walls.

Inverse mode is very useful to compute also the supersonic shock boundary-layer interaction in an overexpanded nozzle (without reattachment). Fig. 27 shows the result corresponding to different locations of the beginning of the interaction (two different back pressures).

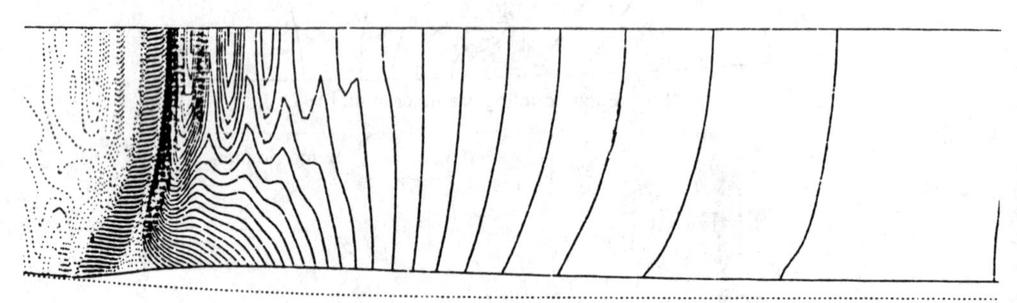

a) Isobaric lines $M_{e_0} = 1.44$

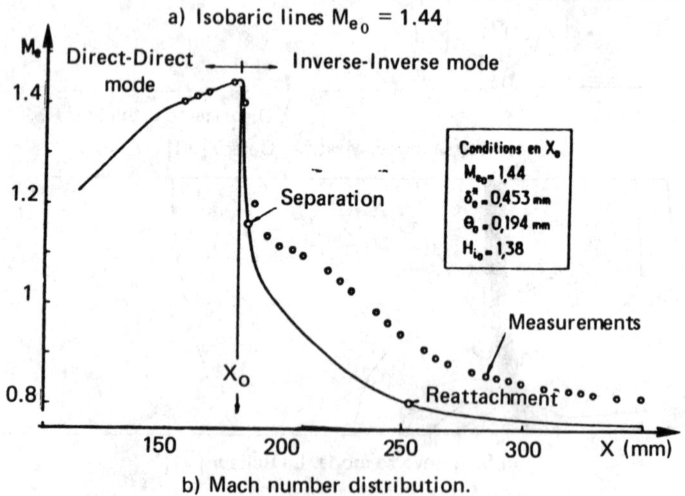

b) Mach number distribution.

Fig. 24 – Calculated and measured transonic shock boundary layer interaction in a nozzle (M = 1.44).

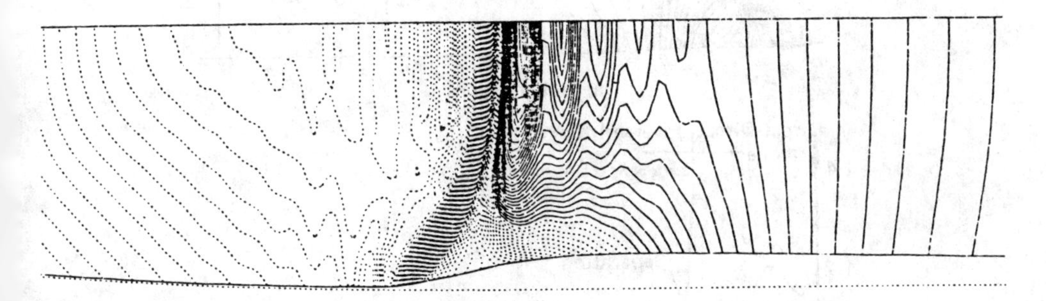

a) Isobaric lines $M_0 = 1.55$

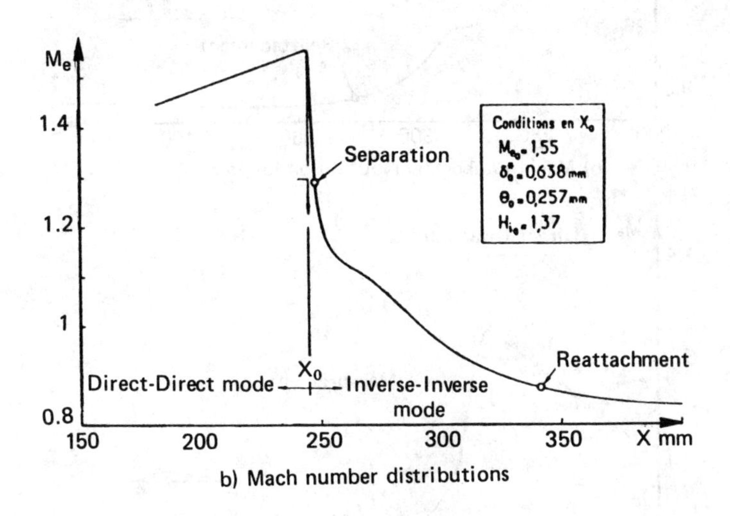

b) Mach number distributions

Fig. 25 – Transonic boundary layer interaction in a nozzle (M = 1.55).

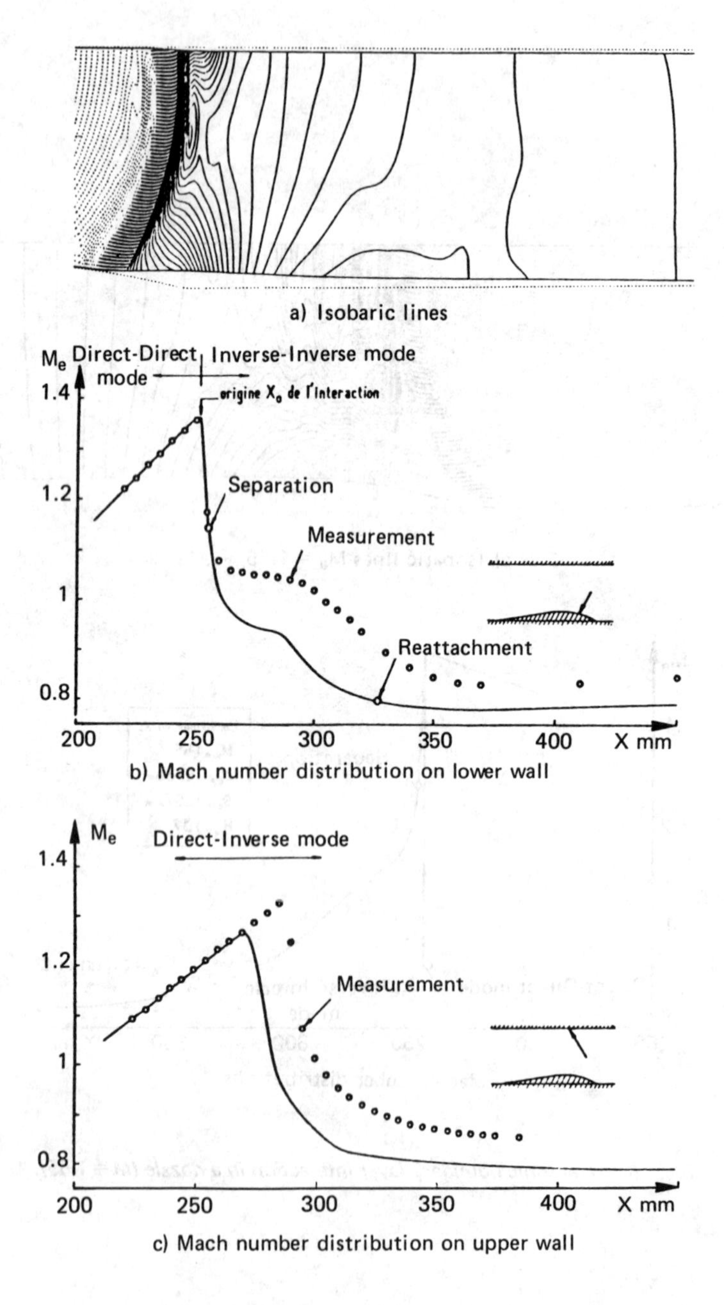

Fig. 26 – Double transonic interaction in a nozzle.

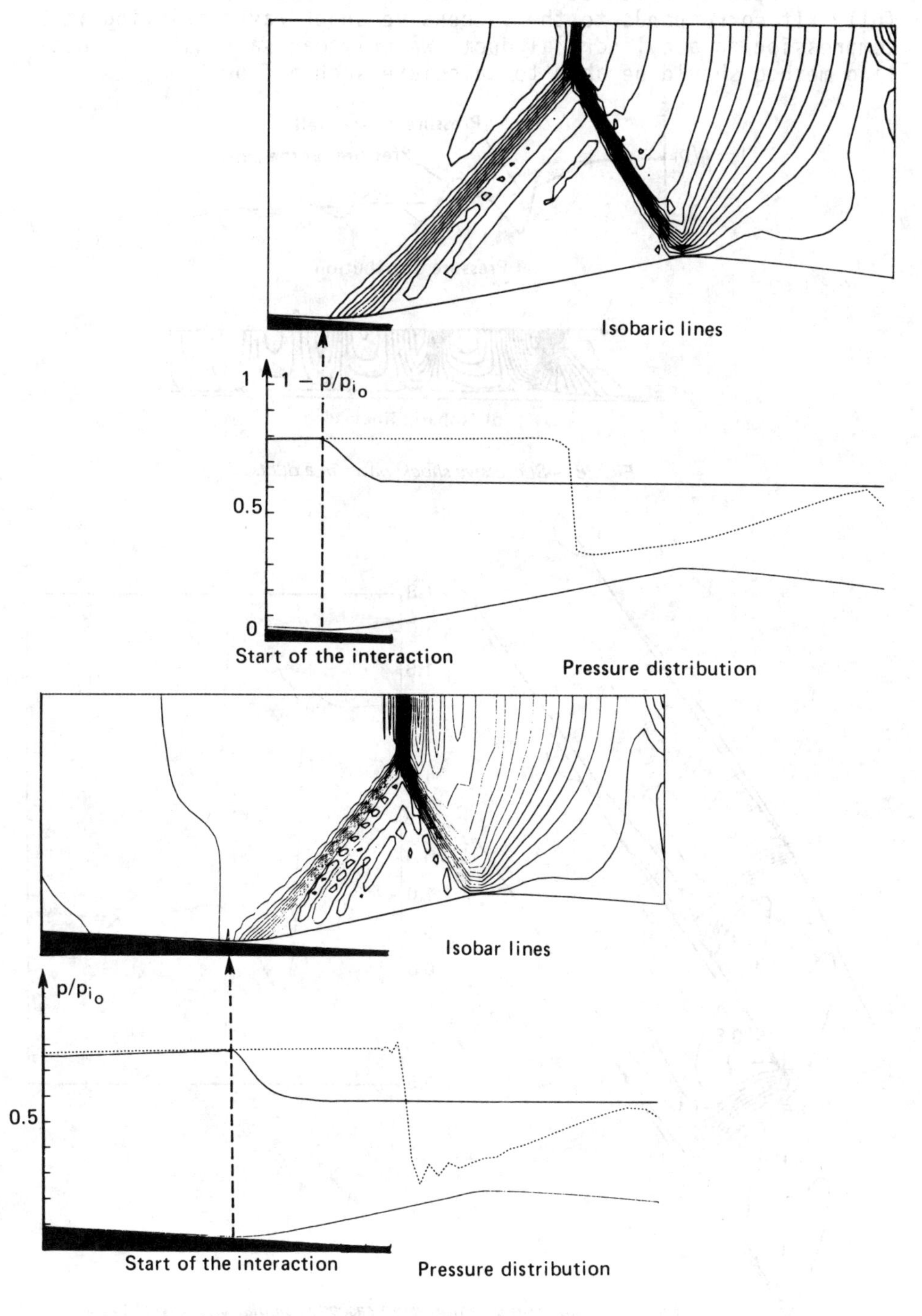

Fig. 27 – Supersonic shock boundary layer in an over expanded nozzle (without reattachment).

The result presented on fig. 28 must be considered very carefully. It corresponds to the successive shock-waves existing in a compression in a cylindrical duct. We only can say that the coupling method should be able to calculate such a flow.

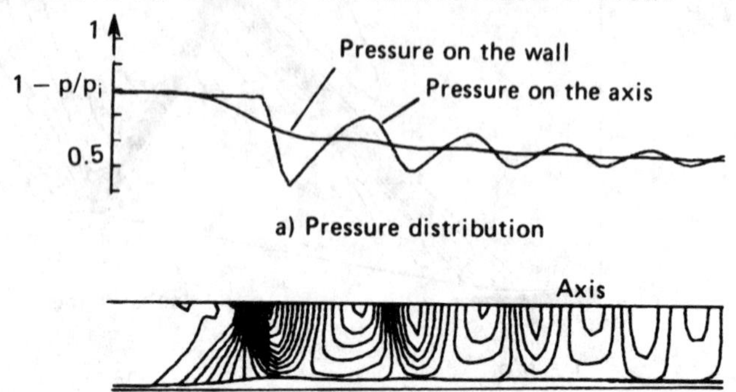

Fig. 28 – Successive shock waves in a duct.

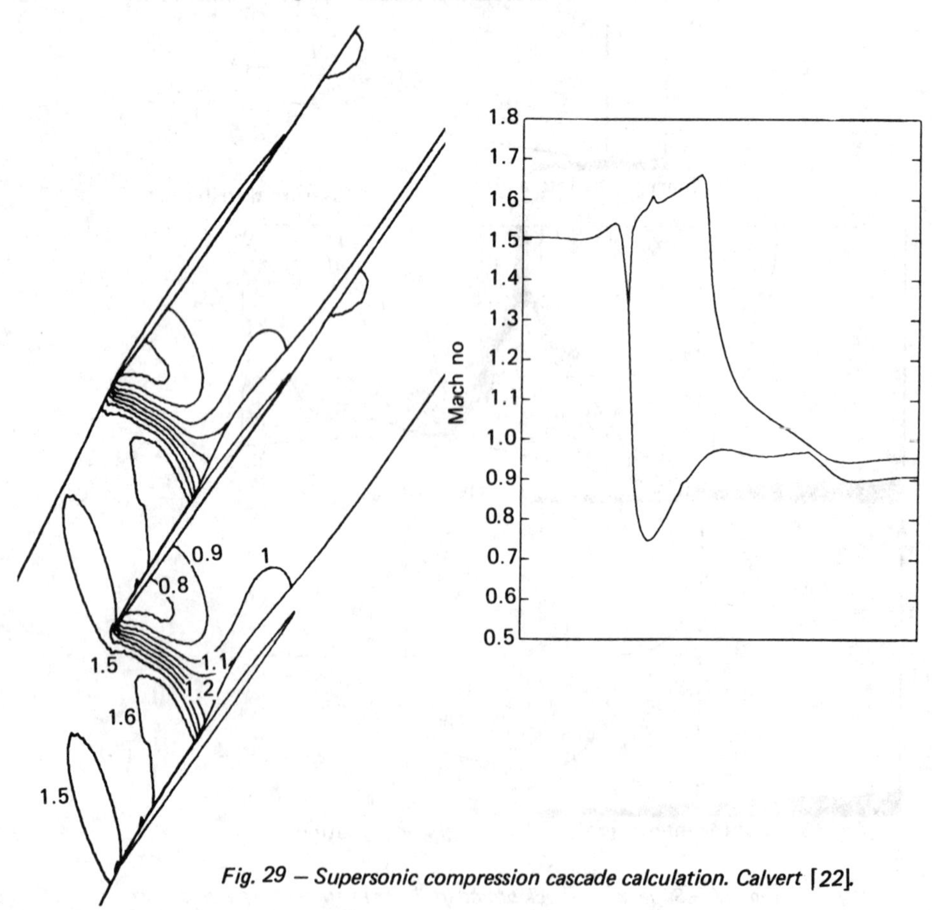

Fig. 29 – Supersonic compression cascade calculation. Calvert [22].

The application of coupling to shock boundary-layer interaction on cascades are rather rare.

We reproduce on fig. 29, 30, 31, 32 some supersonic compressor cascade configurations calculated by Calvert (22, 16).

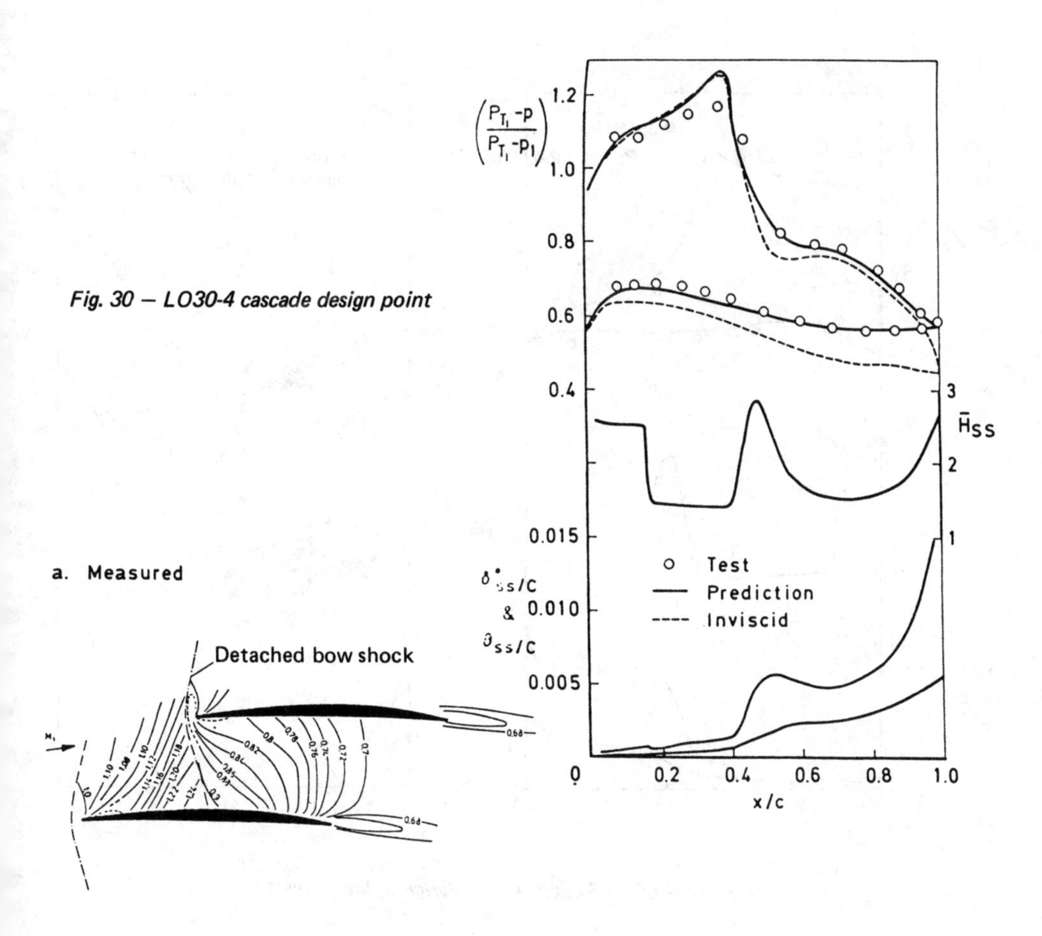

Fig. 30 – LO30-4 cascade design point

a. Measured

b. Predicted

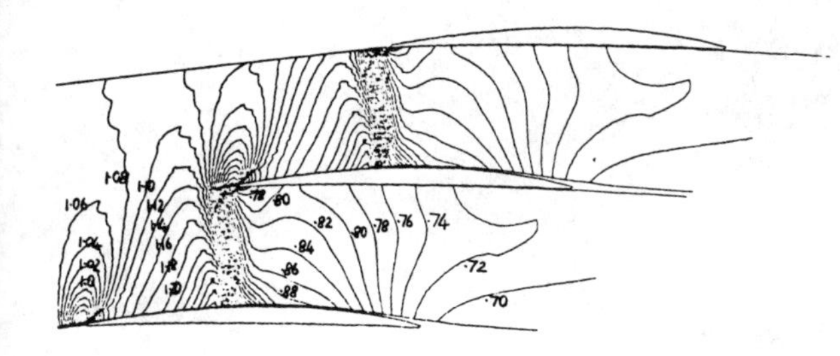

Fig. 31 – LO30-4 cascade Mach number contours at design point.

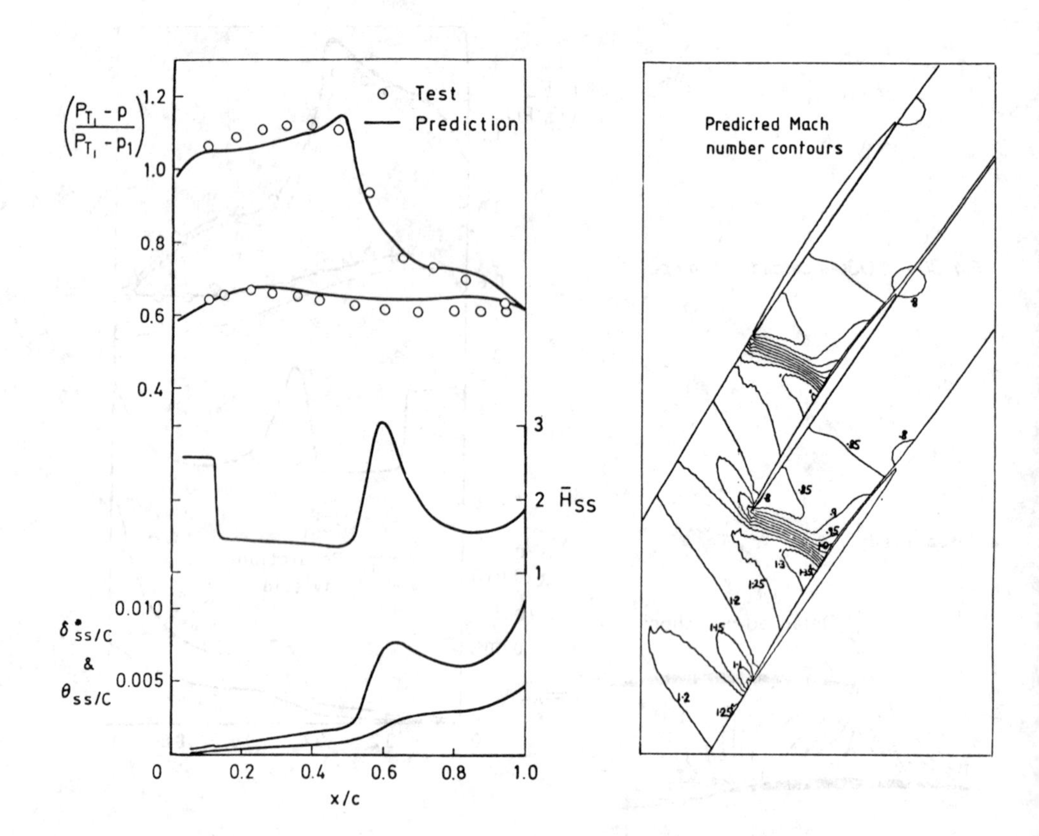

Fig. 32 — LO30-6 cascade near design point. Calvert [22].

ACKNOWLEDGEMENTS

The author wishes to sincerely thank J. Délery who helped him to write this paper.

NOMENCLATURE

a	Speed of sound
C_i	Shock waves
Cp	Pressure coefficient
Cf	Skin friction coefficient
H	Boundary-layer shape factor
l	Mach waves
L	Length
M	Mach number
P	Static pressure
Pi	Total pressure
q	Dynamic pressure
R	Reynolds number
U	Velocity
x,y	Coordinate directions
α	Angle
δ	Boundary-layer thickness
δ^*	Displacement thickness
Δ	Difference
Θ	Momentum thickness
ρ	Density

REFERENCES

1. Délery J. L'interaction onde de choc-couche limite turbulente et son contrôle. AGARD - CP - 365.

2. Délery, J. et Masure, B. Action d'une variation brusque de pression sur une couche limite turbulente et application aux prises d'air hypersoniques. La Recherche Aérospatiale, n° 129, pp. 3-12 (1969).

3. Roshko, A. and Thomke, G.J. Supersonic Turbulent Boundary-Layer Interaction with a Compression Corner at Very High Reynolds Number. Mc Donnell-Douglas. Paper 10163 (mai 1969).

4. Lighthill, M.J. On Boundary-Layers Upstream Influence. II Supersonic Flows with Separation. Proc. Roy. Soc, A 217, pp. 478-507 (1953).

5. COLES, D.E. "The Law of the Wake in Turbulent Boundary-Layer. J. Fluid Mech., vol. 1, Part 2, pp. 191-226 (1956).

6. Sirieix, M., Délery, J. and Stanewsky, E. High Reynolds Number Boundary-Layer Shock-Wave Interactions in Transonic Flow. Lecture Notes in Physics 148, Springer Verlag (1981).

7. Roshko, A. and Thomke, G.J. Flare Induced Interaction Lengths in Supersonic, Turbulent Boundary-Layers. Mc Donnell-Douglas, MDAC Paper WD 2416 (Dec. 1974). See also AIAA Journal, vol.14, no 7, pp. 873-879 (July 1976).

8. Simpson, R.L., Strickland, J.H., and Barr, P.W. Features of separating turbulent boundary-layer in the vicinity of separation. J. Fluid Mech. Vol. 79, Part 3, pp. 553-594 (1977).

9. Inger, G.R. Transonic Shock/Turbulent Boundary Layer Interaction and Incipient Separation on Curved Surfaces. AIAA Paper 81-1244 (1981).

10. Erdos, J. and Pallone, A. Shock/boundary-Layer Interaction and Flow Separation. Heat Transfer and Fluid Mechanics Institute Procs., Stanford University Press (1962).

11. Roshko, A. and Thomke, G.J. Correlations for Incipients Separation Pressure. Douglas Aircraft Co., DAC - 59800 (1966).

12. Reshotko, E. and Tucker, M. Effect of a Discontinuity on a Turbulent Boundary-Layer Thickness Parameters with Application to Shock-Induced Separation. NACA TN-3454 (1955).

13. Gadd, G.E. Interaction between Normal Shock-Waves and Turbulent Boundary-Layers. ARC R & M no 3262 (1961).

14. Elfstrom, G.M. Turbulent Separation in Hypersonic Flow. Imperial College of Sciences and Technology, i.e. Aero Report 71-16 (Sept. 1971) ; see also Turbulent Hypersonic Flow at a Wedge Compression Corner. J. Fluid Mech. Vol. 53, Part 1, pp. 113-127 (1972).

15. Korkegi, R.H. Comparison of Shock-Induced Two and Three-Dimensional Incipient Turbulent Separation. AIAA Journal, vol. 13, no 4, pp. 534-535 (April 1975).

16. AGARD - CP - 351

17. Melnik, R.E., Mead, H.R. and Jameson, A. A multi-grid method for the computation of viscid/inviscid interaction on airfoils. AIAA Paper no 83-0234 (Jan. 1983).

18. Whitfield, D.L., Thomas, J.L., Jameson, A. and Schmidt, W. Commutation of transonic viscous-inviscid interacting flow. 2nd Symposium on Numerical and Physical Aspects of Aerodynamics Flows, Long-Beach (Calif.), 17-20 Jan. 1983.

19. Cambier, L., Ghazzi, W., Veuillot, J.P., and Viviand, H. A multi-domain approach for the computation of viscous transonic flows by unsteady type method.
Recent Advances in Numerical methods in fluids - vol. 3 : Viscous flow computational methods. Editor Habashi. Pineridge Press.

20. Meauzé, G. and Délery, J. Calcul de l'interaction onde de choc-couche limite par emploi de méthodes inverses. AGARD/PDP 61-A Specialists Meeting on Viscous Effects in Turbomachines. Copenhag, Danmar, 1-3 June 1983 - AGARD - CP - 351.

21. Le Balleur, J.C. Numerical flow calculation and viscous-inviscid interaction techniques. Recend Advances in Numerical Methods in Fluids, vol. 3, Viscous Flows Computational Methods, edited by W.G. Habashi, Pineridge Press, Swansea, U.K., 1983.

22. Calvert, W.J. An inviscid-viscous interaction treatment to predict the blade-to-blade performance of axial compressors with leading edge normal shock-waves. ASME Paper 82 - GT - 135 (1982).

[illegible] Interaction between [illegible] Shock Waves and Turbulent Boundary Layers [illegible] ARC R & M no. 3[illegible] ([illegible])

[illegible] Turbulent Separation [illegible] Aero [illegible] Flow [illegible] Compression Corner [illegible] Fluid Mech. [illegible]

15. Korkegi, R.H., "[illegible] of Shock-Induced Two- and Three-Dimensional [illegible]", AIAA Journal, Vol. 13, no. 4, pp. 534-535 (1975).

17. [illegible] and [illegible] [illegible] AIAA [illegible]

[illegible] Separation [illegible] Physics [illegible] of Aerodynamic [illegible]

[illegible] A [illegible]

[illegible]

[illegible]

[illegible]

22. [illegible] Paper [illegible]

MEASUREMENT TECHNIQUES

FUNDAMENTALS OF FLOW FIELD MEASUREMENT TECHNIQUES

H.B. Weyer

German-Dutch Wind Tunnel

INTRODUCTION

The basic flow in turbomachines is highly three-dimensional and unsteady; it covers - so far in a single blade row - the whole flow regime of subsonic, mixed-flow, and supersonic velocities as it is found in modern transonic compressors and turbines. Further imponderabilities aggravating the understanding and treatment of turbomachine flows are added by the existence of wall boundary-layers, of shock-boundary-layer interaction, of secondary flow vorticity, and of clearance flows, all superimposed on the basic flow.

The still urgent need to improve the performance and reliability of the turbomachines, and to reduce the development costs requires permanent great efforts to accomplish the knowledge on the actual flow processes in turbomachines. Concerning the experimental side of the activity there are two primary aspects to be considered:

- First of all an accurate analysis of the axisymmetric steady-state flow is needed to estimate the mean flow characteristics of each blade row, and to evaluate the overall performance and efficiency.

- On the other hand a detailed investigation of the three-dimensional and unsteady flow phenomena is necessary to accomplish - at least step by step - the physical flow modelling that is absolutely required for any adequate theoretical flow treatment or any further performance improvement.

Following this scope of experimental necessities the paper deals with some important domains of turbomachine testing:

- Conventional testing of unsteady flows by probes and wall taps will be covered first placing particular emphasis upon integrating effects on probing, upon the effects of flow turbulence and blade wake cutting. Finally, unresolved problems of temperature and flow angle measurement will be briefly described.

- Measurements of unsteady flow quantities will be treated next, pointing primarily to the capability, the accuracy, and to the operating limits of high-response sensors.

- Optical methods have become very attractive for flow studies in turbomachines because they do not disturb the flow in almost narrow blade channels and allow relatively easy access to even rotating components. The laser velocimetry most promising among current optical techniques will be covered in detail by R. Schodl within the succeeding paper "Optical Techniques for Turbomachinery Flow Analysis". On the other hand the fundamentals of flow visualization, of Schlieren Technique, of Raman and fluorescent spectroscopy will be briefly described in the second part of this paper, putting main emphasis upon the prospects of these techniques for turbomachinery flow field testing.

PART I: CONVENTIONAL TESTING OF UNSTEADY FLOW IN TURBOMACHINERY

MEASUREMENT OF AVERAGE FLOW QUANTITIES

Unsteady flows in turbomachines occur predominantly within and behind the rotors; frequency and strength of those fluctuations depend primarily upon the speed and blade loading levels. In order to characterize the axisymmetric steady-state flow of a rotor blade row mean flow data of the mass-, momentum-, or energy-averaged type are required; however, there is no relevant measuring procedure available to get these data directly from the unsteady flow field, e.g. by means of probes.

Usual total pressure or temperature probes are believed to read time-averaged flow quantities that are so far equivalent to the corresponding area-weighted data in periodic rotor flow. In unsteady flows, however, probes always tend to be subject to remarkable errors the magnitude of which depends on the probe design, on the strength, and on the type of the flow fluctuations.

Figure 1.1 demonstrates - as an example - the erroneous reading of a pitot-probe that was used to measure the steady-state total pressure downstream of a state-of-the-art transonic axial flow compressor rotor. The data were sampled a quarter of chord downstream of the rotor exit plane at various relative blade heights. The probe error is plotted against rotor speed, peak efficiency points on the speed lines have been selected to illustrate the

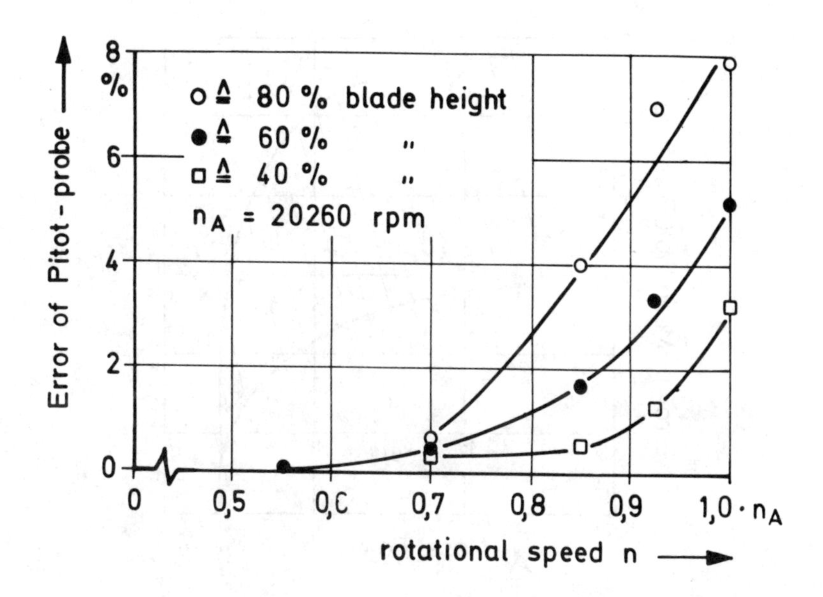

Fig. 1.1 Error of Conventional Pitot-Probe Measurement a Quarter of Chord Downstream of a Transonic Axial Compressor Rotor (Peak Efficiency)

experimental finding. The error increases with increasing rotational speed, because the downstream total pressure fluctuations intensify due to the increase of flow velocity, to the associated appearance of shock waves in the rotor blading, and to the occurrence of severe boundary-layer separation on the blades. The maximum pitot-probe error observed at full speed near blade tip amounts to about 8% of the correct time-weighted value, that was measured by an oil-filled probe described in detail in a succeeding chapter.

Figure 1.2 points to the even more serious problem of accurate mean wall pressure measurement at the tip of transonic rotors. The rotor used for these tests is characterized by very strong shock waves within the blade channel leading to considerably high wall pressure fluctuations (up to 70% of mean value) at rotor tip. The data shown were taken at 15% chord downstream of rotor inlet plane, using two different types of wall taps - sharp-edged and well rounded, respectively. The error in wall pressure reading - shown in function of rotor mass flow at maximum speed - is slightly negative near choke line and grows positive with increasing rotor throttling up to 10% near surge line. Here, the type of pressure fluctuation (wave shape in periodic flow) decides upon the error

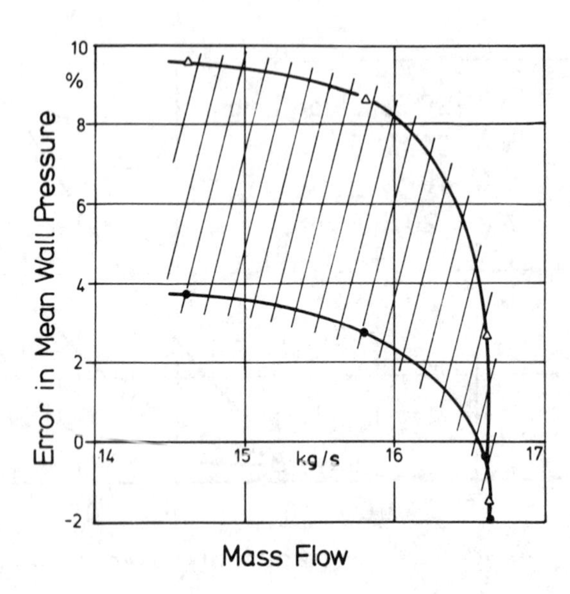

Fig. 1.2 Error of Outer Wall Static Pressure Measurement in a Transonic Rotor Using Conventional Pneumatic Techniques as a Function of Compressor Throttling (90% Design Speed)

being negative or positive; this is outlined in detail in a succeeding chapter. The hatched area localizes the probable scattering of the pressure reading due to random construction of the wall taps.

Figure 1.3 illustrates the situation of total temperature measurement in the unsteady flow behind rotors. Usual shielded stagnation probes were used to measure the mean total temperatures behind a transonic rotor; the experimental data were then reduced to relative temperature rise and compared to equivalent data evaluated from laser velocimetry results. The discrepancies found are primarily caused by an inaccurate reading of the conventional temperature probes in the unsteady rotor discharge flow. In Figure 1.3 the data are shown as a percentage of the correct time-averaged rotor temperature rise and plotted against mass flow at design speed. The various symbols represent the data at blade heights of 43%, 66%, and 88%.

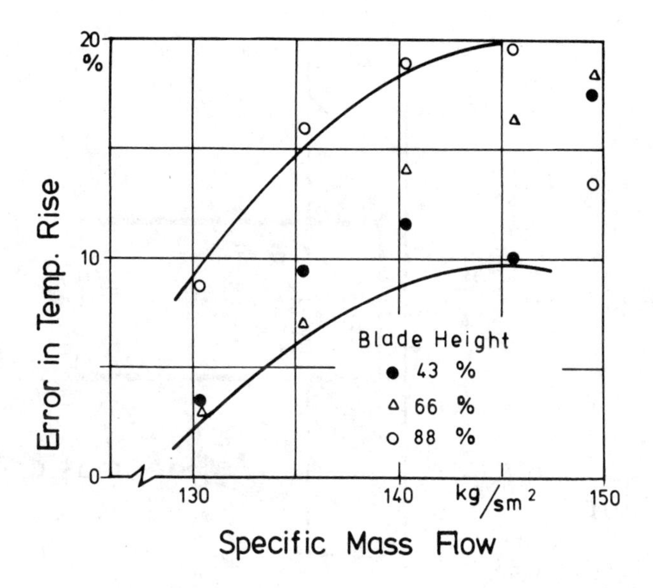

Fig. 1.3 Error of Stagnation Temperature Measurement Behind a Transonic Rotor Using Conventional Probes as Function of Compressor Throttling (100% Design Speed)

Effects of Probe Averaging in Unsteady Flow

The determination of mean quantities in unsteady flow using customary measuring techniques is based on flow integration within the probe system. For physical reasons this integration usually leads to an unbalance between the time-weighted value looked for and the actual probe reading. For a detailed explanation pressure measurement techniques as pitot-probes or wall taps will be considered that are connected to inert instrument via pneumatic lines. The head of such a probe is shown very schematically in Figure 1.4. Inside the probe the steady-state mean pressure $\overline{P}_p$ exists balancing the unsteady pressure p(t) at probe entrance. Due to compressibility effects an alternating mass flow takes place within the probe orifice, and steady-state operation now requires that this mass flow, integrated over one period in periodic flow or over a long time interval in just fluctuating flow, has to be zero.

The instantaneous mass flow within the orifice reveals to be a function of the pressure drop $p(t) - \overline{P}_p$ across probe entrance, of the density on both sides of the orifice, and of the construction of the orifice, that means of the flow coefficients for both in-

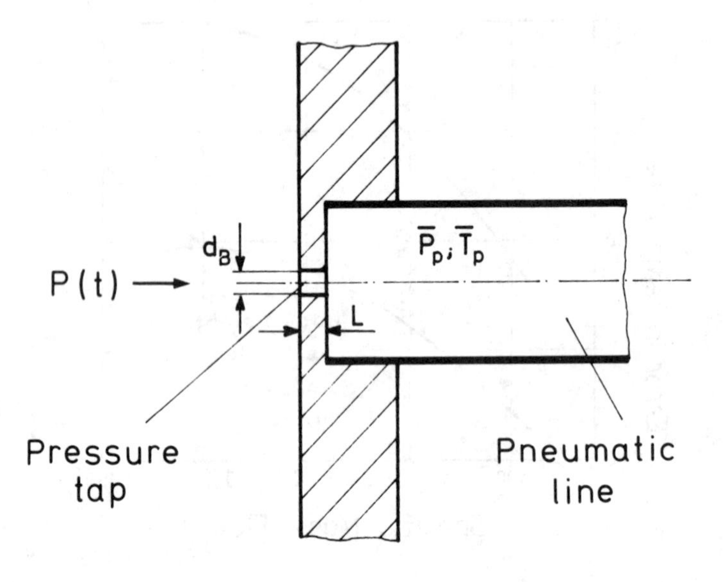

Fig. 1.4 Sketch of Conventional Pressure Probe - Wall Tap and Pneumatic Line.

and outflow. This dependence on various parameters implies a couple of serious non-linearities that directly cause systematic deviation of the probe reading $\overline{P}_p$ from the correct time-weighted value $\overline{P}$ (references 1.1, 1.2, 1.3, 1.4, 1.5).

The periodic pressure fluctuation at probe entrance (Figure 1.4) is selected to study fundamentally this systematic "integrating error" and to get a rough idea on its probable magnitude; for this it is also assumed that the alternating flow within the orifice is quasi steady-state, a hypothesis that is fulfilled as far as the axial length of the orifice is small compared to the wave length of the pressure fluctuation (reference 1.5).

Figure 1.5 demonstrates the effect of pressure amplitude on the "integrating error" with a symmetrical rectangular and a sinusoidal pressure wave being present. P_{max}/P_{min} designates the ratio of highest to lowest pressure during the oscillation thus being equivalent to the unsteady pressure strength. As expected the error increases with increasing amplitude, strongest for the rectangular type of oscillation. Figure 1.6 shows the effect of unsymmetry of the pressure wave on the magnitude of the "integrating error". For a rectangular wave of variable amplitude the error is plotted

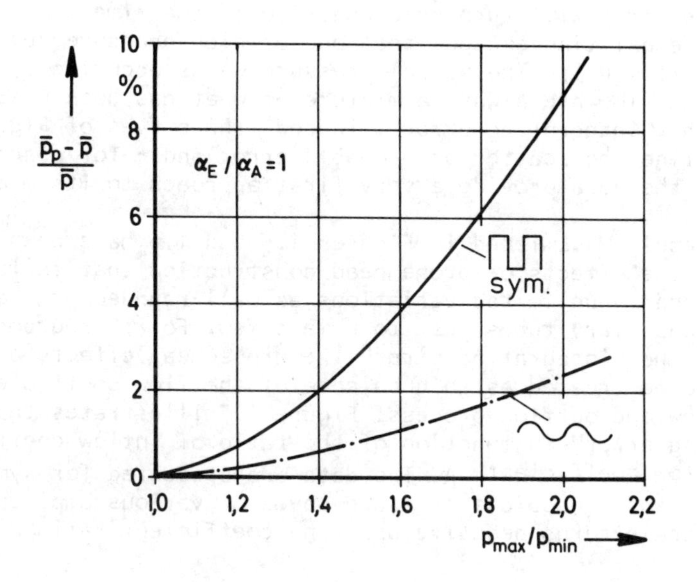

Fig. 1.5 Estimated "Integrating Error" of Conventional Pressure Probe as Function of Unsteady Pressure Strength for Rectangular and Sinusoidal Pressure Waves

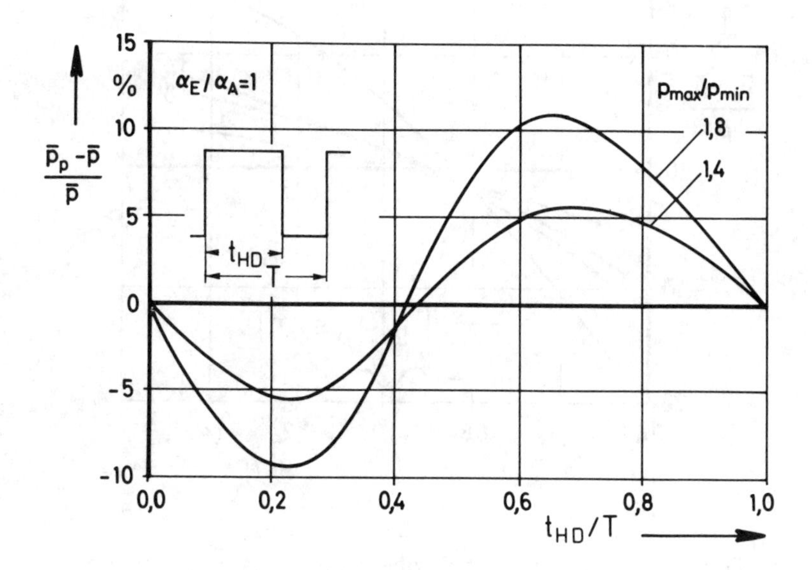

Fig. 1.6 Continued; Error as Function of Pressure Wave Shape (Unsymmetric Rectangular Waves)

against the time ratio t_{HP}/T that describes that fraction of period T where the high pressure P_{max} is present at probe entrance. Obviously the error will turn even negative if the time fraction with low pressure existing exceeds that one of high pressure during the period ($t_{HD}/T < 0.5$). The actual pressure waves occuring in unsteady turbomachine flow are always a mixture of what has been used here to estimate the "integrating error"; indeed, the curves of Figures 1.5 and 1.6 define the scatter of probable error and - for practical purposes - the data provide a very first approach to its magnitude.

The results submitted in Figures 1.5 and 1.6 have been evaluated neglecting the effects of probe head construction that in fact is a matter of nearly unlimited variations as well-rounded or sharp-edged orifices, capillary tubes just to name a few. For the purpose of estimating the "integrating error" the probe head effects are believed to be treated as an unbalance of the flow coefficients of probe inflow and outflow process. Figure 1.7 illustrates the "integrating error" in function of the ratio of inflow coefficient α_F to outflow coefficient α_A. The data are presented for symmetric rectangular and sinusoidal pressure waves of various amplitudes. The error again turns negative when the coefficient ratio drops below 0.85.

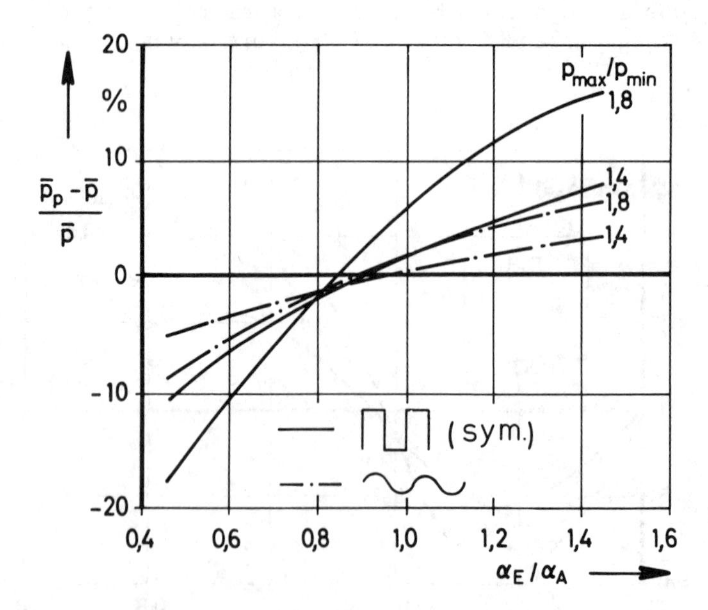

Fig. 1.7 Continued; Effect of Probe Head Construction on "Integrating Error"
($\alpha_E \hat{=}$ Inflow Coefficient; $\alpha_A \hat{=}$ Outflow Coefficient)

So far, the systematic error associated with pressure amplitude and wave form is quite easy to be calculated provided the unsteady pressure profile (dynamic component only) is available. It is hard, however, to take properly into account the probe head effects because there is almost no way to determine the flow coefficients or to guarantee that no instantaneous effects will occur at the entrance of probes with capillary tubes the length of which is not small compared to the wave length of the pressure oscillation. The broad variety of probe head designs in use today has revealed to bring about a large uncontrollable scattering of the experimental data; consequently, usual pneumatic techniques have probed to be inappropriate for the measurement of correct time-weighted pressures at least in strongly fluctuating flows (for more details see reference 1.5).

Hydraulic Probe for Correct Mean Pressure Measurement

All the previously described effects on probe reading in unsteady flow will be eliminated completely if one succeeds in suppressing the alternating mass flow at probe entrance, for example by filling up the whole probe system with an incompressible fluid. A probe based on this hydraulic concept is shown schematically in Figure 1.8. A capillary tube - 0.35 mm in diameter, 50 mm long - is connected to an adapter housing as usual pressure transducer of

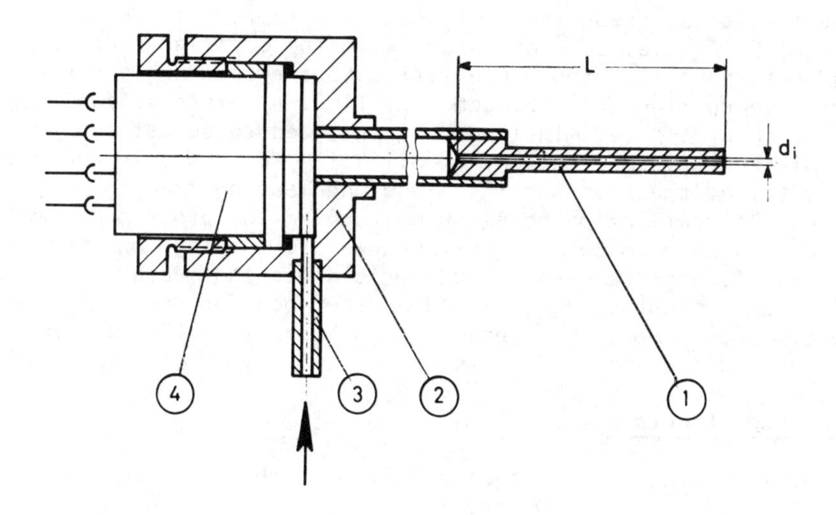

Fig. 1.8 Scheme of Hydraulic Pressure Probe; 1 ≙ Capillary Tube, 2 ≙ Adapter, 3 ≙ Oil Supply, 4 ≙ Pressure Transducer

low natural frequency. The system is completely filled with silicon oil extending to the capillary's open end that is exposed to the unsteady pressure. The concept of capillary and highly viscous oil has the twofold effect of attenuating the pressure fluctuations on their way to the transducer's place, and of stabilizing the oil column against the strong unsteady forces at probe entrance.

With the help of momentum equation it is easily proved that the nearly steady-state pressure ahead of the transducer equals the time integral of the fluctuating pressure to be measured. The slight movement of the oil column - due to the oil's weak compressibility and to the structure's elasticity - and the unsteady friction forces associated with the movement do physically not effect the integration process (reference 1.5).

In extensive experimental studies it has been found that a capillary tube with an inner diameter of 0.35 mm and a length of 50 mm combined to a silicon oil with a viscosity of 120 cSt at room temperature is the best compromise to meet the requirements of most turbomachine applications. The probe is able to meet high pressure amplitudes, any fluctuation frequency higher than 200 cps, and ambient temperatures of at least 200 centigrades in uncooled version. However, in order to guarantee absolutely safe probe operation it is recommended to add a very small amount of oil - about 10^{-7} litres per hour - to the system (at point 3 - Figure 1.8), thus making sure that the oil column will always reach up to the capillary's open end.

The hydraulic technique is appropriate to measure wall static as well as total pressures; a typical pitot-probe is shown in Figure 1.9. A shielded probe head (with the capillary tube inside) is used to have the probe insensitive to the flow angle variations normally occurring in turbomachinery flow. A reference capillary tube of several meters in length is provided to adjust and to control the very small amount of oil being continuously added to the system at the transducer's place and leaving the probe head through the ventilation holes. When running the probe two simple corrections have to be made to the pressure reading, one for any height difference between transducer's place and measuring locus (oil column weight component) and another one for the small pressure drop of about 5 mb that is needed to push the oil through the capillary tube.

Turbulence Effects

Besides pressure fluctuations flow turbulence also effects pressure measurement, however, in a physically more complex way. The unsteady pressures associated with turbulent flow initiate the probe "integrating error" as described before, although this effect is believed to become important primarily at very high turbulence

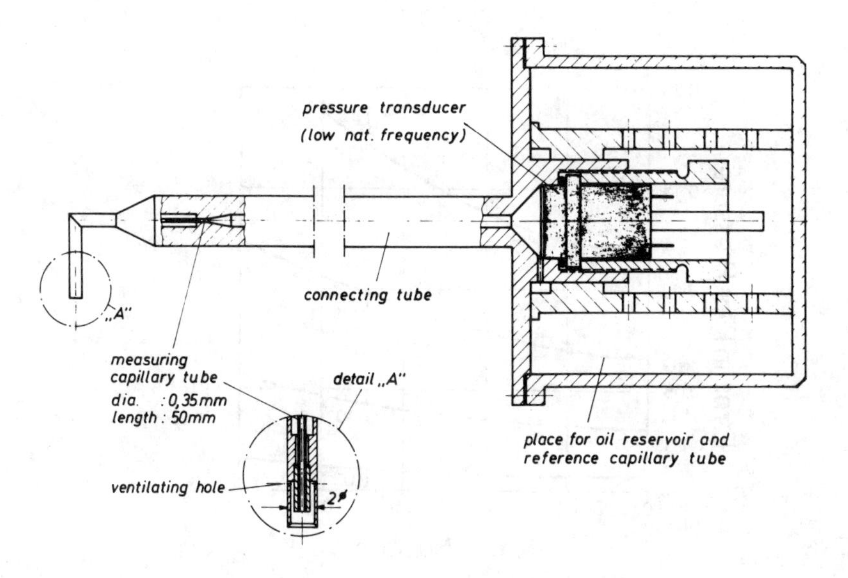

Fig. 1.9 Design Details of Hydraulic Pitot-Probe

levels (> 20%). The "integrating error" might be superimposed by an additional error due to the kinetic energy of the turbulent movement. Concerning wall static pressures the effect of turbulence is that the turbulent flow component is stagnated at the downstream edge of the wall tap thus falsifying the pressure reading. Shaw (reference 1.6), however, has proved experimentally that this error is always negligible as far as the wall tap diameter is small, about 1 mm and below.

The effect of turbulent kinetic energy on time-weighted total pressures has been found to be considerable under certain circumstances, as shown in Figure 1.10. The data have been estimated theoretically based on complete isentropic stagnation of turbulent flow at probe entrance assuming that the probe head dimensions are small compared to the turbulent eddies; the effect of turbulent flow angle fluctuations on probe reading is neglected (references 1.7, 1.8). In Figure 1.10 the "turbulence error" referred to the mean kinetic pressure is plotted in function of Mach number of compressible flow with the turbulence level as a parameter. This plot permits a very rough estimation of the "turbulence error" that is thought to be unavoidable even with the oil-filled probe.

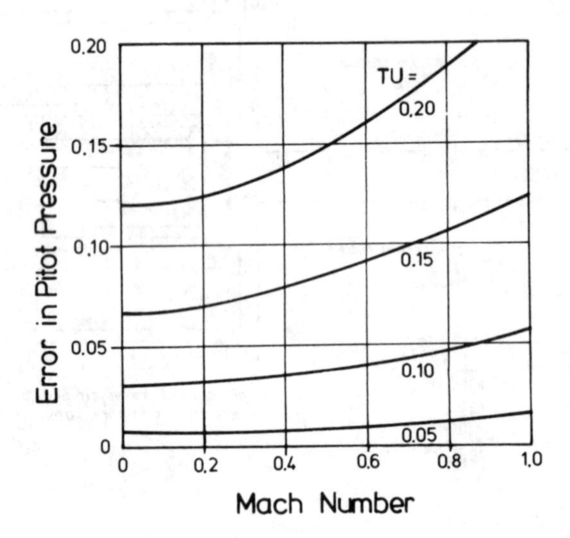

Fig. 1.10 Effect of Flow Turbulence on Stagnation Pressure Measurement Depending on Flow Mach Number and Turbulence Degree

Blade Wake Cutting

The arrangement of blade rows in turbomachines always rises the problem of "blade wake cutting" characterizing the influence of upstream wakes on the flow inside succeeding blade rows. "Wake cutting" considerably contributes to the flow unsteadiness and noise emission of turbomachines, but also requires appropriate test procedures to be applied as shown by the following simple approach: The flow model of stator-rotor combination is illustrated in Figure 1.11 (reference 1.9). The stator blade wakes are cut by the rotor, pass through the rotor blade channels, and move downstream - as separated segments - within the field marked by the dashed lines. Since this field is steady-state from the absolute frame of reference only those probes located inside the field will detect the stator blade wakes, those outside will not. Thus a careful probe positioning even downstream of rotors is required to analyse the flow properly.

Figure 1.12 (reference 1.10) illustrates the flow model of rotor-stator combinations; the rotor wake fluid entering the stator has a slip vector towards the pressure side of the stator blades as shown by the velocity triangle in the right hand graph. Thus, the

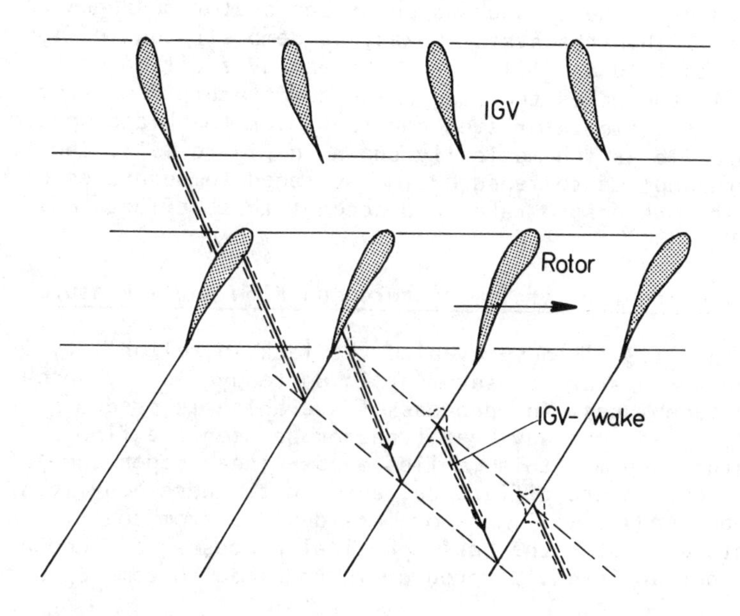

Fig. 1.11 Model of "Blade Wake Cutting" in Stator (IGV)-Rotor Combinations (Ref. 1.9)

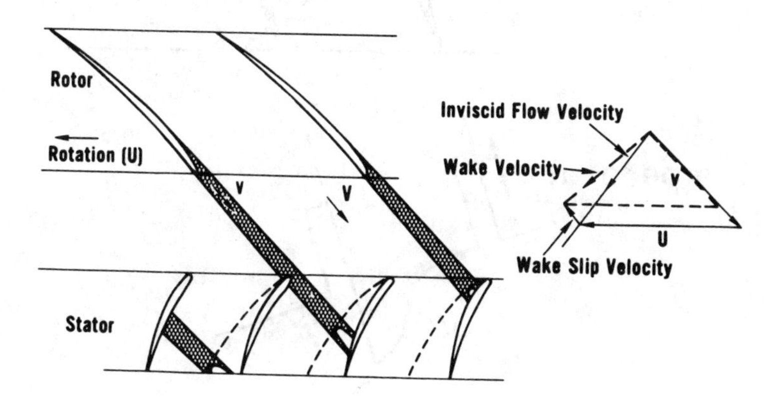

Fig. 1.12 Model of "Blade Wake Cutting" in Rotor-Stator Combinations (Ref. 1.10)

rotor wakes are shifted towards the pressure side with the stator blades interrupting this transport. The rotor wake fluid is accumulated near the pressure side of the stator passages and tends to appear within the stator wakes, as demonstrated in Figure 1.13 (references 1.10 and 1.11). For the energy addition of rotor is highest in the wakes the stagnation temperature downstream of the stator is not smooth or even constant over the blade spacing but has a profile as schematically shown in Figure 1.13. The test procedure applied to measure time-averaged temperatures of pressures behind stators has to take into account these effects of "blade wake cutting".

Unresolved Problems of Temperature and Flow Angle Measurement

Figure 1.3 already revealed the lack in reliability of time-averaged temperature measurement in unsteady flow by unconventional testing techniques. The process of averaging is predominantly determined by an unsteady heat transfer between the flow and the temperature sensor; this implies a non-linear dependence of heat flux on temperature difference, and due to these non-linearities the probe reading will more or less deviate from the time-averaged temperature. Indeed the basic physical processes are known qualitatively, but an intensive program is required to come up with

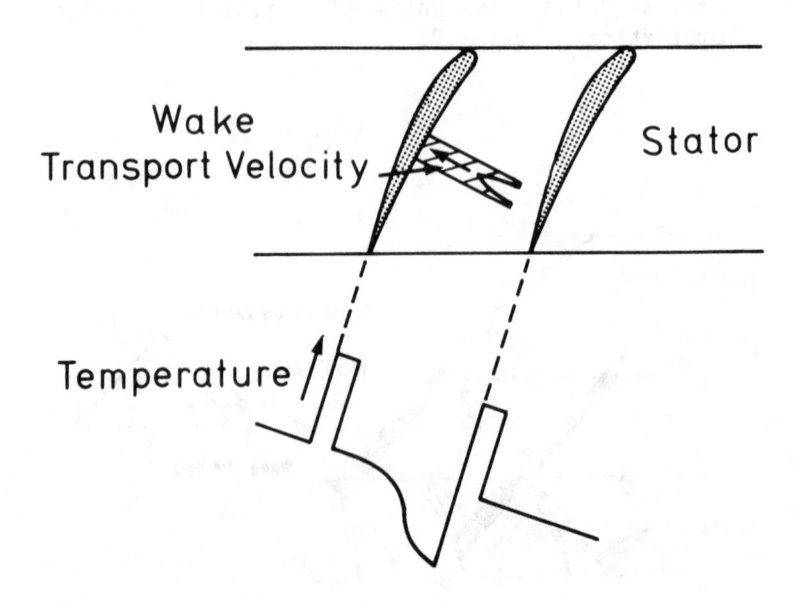

Fig. 1.13 Transport of Rotor Wake in Stator Blading and Accumulating of Wake Material at Stator Blade Pressure Side (Ref. 1.10)

reliable procedures to correct for the averaging error or to develop a new type of probe to read the temperature data needed. The situation of unsteady flow angle measurement is quite similar to that of pressure measurement, because flow angle analysis is based on pressures, i.e. on a mixture of flow static and total pressure, where two pressure readings to be balanced are subject to the "integrating error" described before. No work has been done so far to clarify the fundamentals and to develop probes correctly averaging unsteady flow angles.

MEASUREMENT OF UNSTEADY FLOW QUANTITIES

The basic frequency of flow oscillations around turbomachinery rotors corresponds to the blade passing frequency whereas the amplitude of fluctuations depends primarily on the rotor design, i.e. on the flow velocity and loading levels; particularly shock waves appearing in the blade passages at supersonic velocities give rise to heavy flow oscillations. Pressure, temperature, and flow angle are the main unsteady quantities to be investigated for proper flow analysis. The following statements will be concentrated on unsteady pressure measurement.

Unsteady Pressure Measurement

Today, for unsteady pressure measurements transducers are available based on piezo and semi-conductor components. Their natural frequency is sufficiently high compared to the blade passing frequency, at least in quite a number of turbomachine applications. A choice of relevant transducers is shown in Figure 1.14. All the transducers have proved to be almost unattenuated systems that means the output signal will tend to overshoot if the actual pressure frequency exceeds about one third of the natural frequency; thus, in order to eliminate any dynamic effects it is recommended to limit the transducer's application to the lower 30%-range of their "eigenfrequency".

Due to their size (less than 2 mm in diameter) the semiconductor transducers are most advanced to be used in turbomachinery even to be built into pitot-probes allowing for unsteady total pressure measurement behind rotors.

Before illustrating in detail the capability of the semiconductor transducers I would like to discuss briefly those properties that might have some severe impact on the transducer's operation: frequency response and sensitivity to ambient temperature variations and to mechanical acceleration. In principle, they allow to measure fluctuating absolute pressures, however, they have proved to be very sensitive to ambient temperature variations giving rise to an almost severe zero drift, as demonstrated in Figure 1.15. The

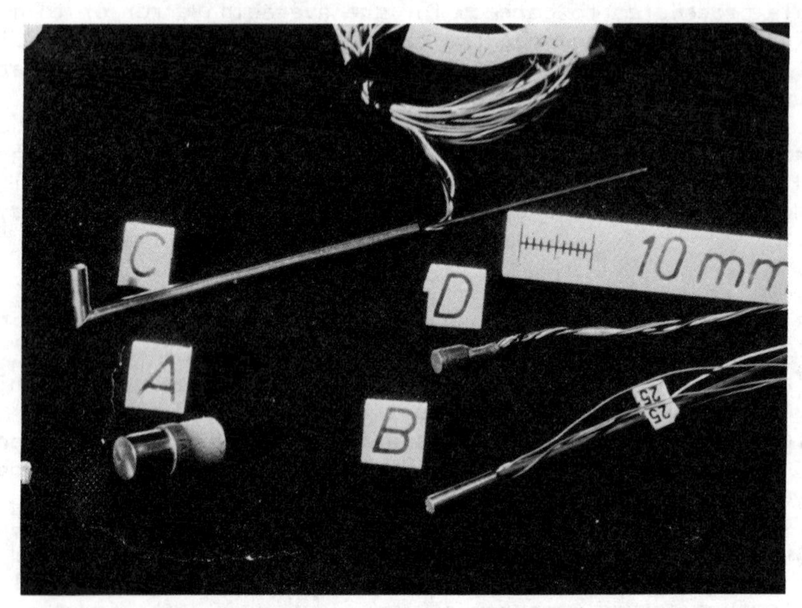

Fig. 1.14 Collection of High-Response Pressure Transducers Appropriate for Turbomachine Application

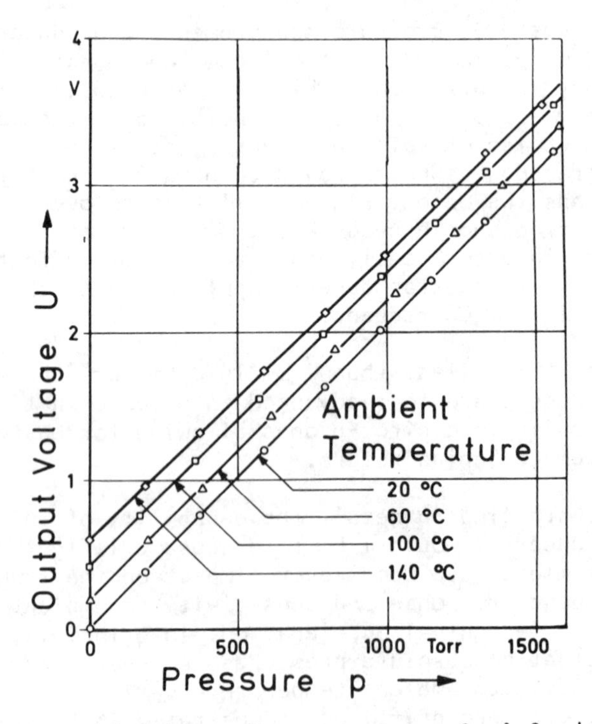

Fig. 1.15 Effect of Ambient Temperature on Semi-Conductor Pressure Transducer Output Voltage

transducer's output voltage is plotted in function of pressure at different ambient temperatures; a 5% full scale drift is observed with a 30 degree temperature rise. As evident in Figure 1.15 the zero balance is strongly effected by ambient temperature variation, whereas the pressure sensitivity keeps quite uneffected as necessary for accurate unsteady pressure analysis.

Using cooling adapters always necessary in temperature environments beyond 370 K it is possible to reduce considerably the transducer's thermal sensitivity, however, under actual turbomachinery test conditions with strong spatial and temporal temperature gradients being present one has to account for a thermal zero shift of at least 5% full scale. Thus, semi-conductor transducers are capable of measuring the dynamic pressure component with satisfying accuracy (~ 0.5 to 1% f.s.) but they are not able to identify properly the pressure level.

Several investigations on the frequency response of the transducers (references 1.12, 1.13, 1.15) have shown that the dynamic calibration factor varies less than about ± 1.% from the static value, thus proving that steady-state calibration meets the demands of unsteady pressure measurement. Furthermore, the semi-conductor transducers show very little sensitivity to mechanical acceleration resulting from using silicon for the diaphragm's material being 4 times less in density than steel normally used to build the diaphragm (reference 1.12). If adapters made of materials of high damping effectiveness (as teflon or delerin) are additionally applied the influence of mechanical acceleration will be reduced thus that machine vibrations have not to be considered as a critial factor.

Effects of Transducer Installation

Full profit of the transducer's dynamic properties will be obtained if the transducers are installed such that the unsteady pressure is directly supplied to the diaphragm (references 1.14 and 1.15). Concerning wall pressure measurements it is recommended to have the transducers flush-mounted with the casing inner wall, as shown in Figure 1.16. For stagnation pressure measurements the transducer should be directly built into the probe head with the shield extending the diaphragm by a few tenth of a millimeter (Figure 1.16) in order to have the probe insensitive - as far as necessary - to the flow angle variations within the discharge flow of turbomachine rotors. Tests on the effects of shield positioning have proved that the shield may not extend beyond 0.5 mm, otherwise strong flow vortices appear at the probe leading edge due to the high angles of attack that occur during blade wake passing. The vorticity deteriorates considerably the pressure signal and, therefore, should be carefully avoided by probe design.

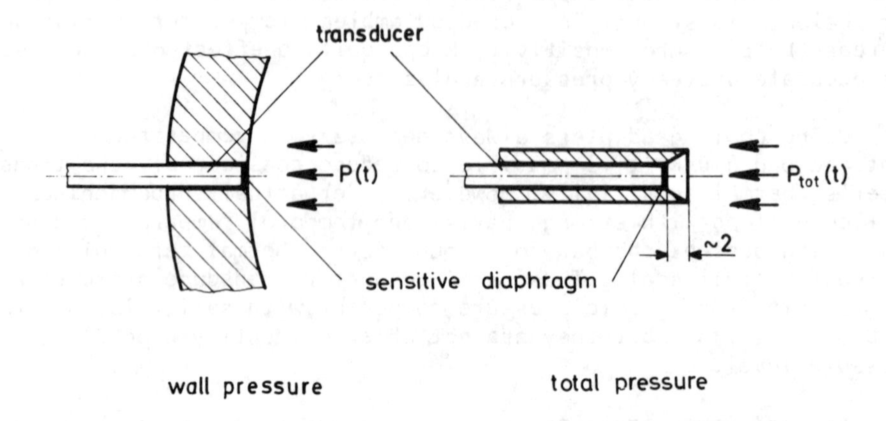

Fig. 1.16 Scheme of Transducer Installation for Wall Static and Total Pressure Measurement

The direct type of transducer installation is limited to a.o. transverse pressure fluctuations the wave length of which is large compared to the transducer's diameter. As illustrated in Figure 1.17 the transducer always supplies an integral of the momentary pressure

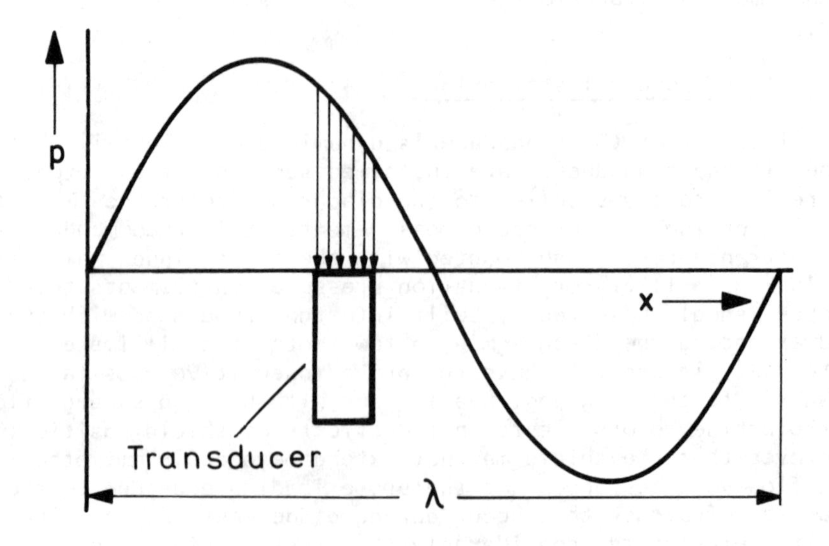

Fig. 1.17 Averaging Effect of Transducer Size on Unsteady Pressure Measurement

distribution across its face thus falsifying the actual pressure gradients as in shock waves. On the other hand, reducing this effect by use of small pressure taps ahead of the transducer will rise the problem of unsteady pressure attenuation within the pneumatic transmission system. Furthermore, such systems are used whenever the measuring locii are not accessible for a direct transducer installation. However, the frequency response of pneumatic lines is rather limited, and for practical purposes it has been found that an accurate correction of amplitude damping and phase lag within pneumatic transmission systems is very difficult. A lot of relevant work has been done on evaluating the frequency response of pneumatic systems, particularly by Taback (reference 1.16), Iberall (reference 1.17), Rohmann and Grogan (reference 1.18), Bergh and Tijdeman (reference 1.19), Healy and Carlson (reference 1.20).

Unsteady Pressure Measurements in Compressors

Figure 1.18 illustrates appropriate installation of semiconductor transducers for unsteady wall and total pressure measurements in compressors. The left hand sketch shows a wall pressure transducer built into a water cooled adapter to protect the pickup in high ambient temperature environments of centrifugal compressors (up to 550 K). The right hand sketch shows an uncooled pitot probe with the transducer built in.

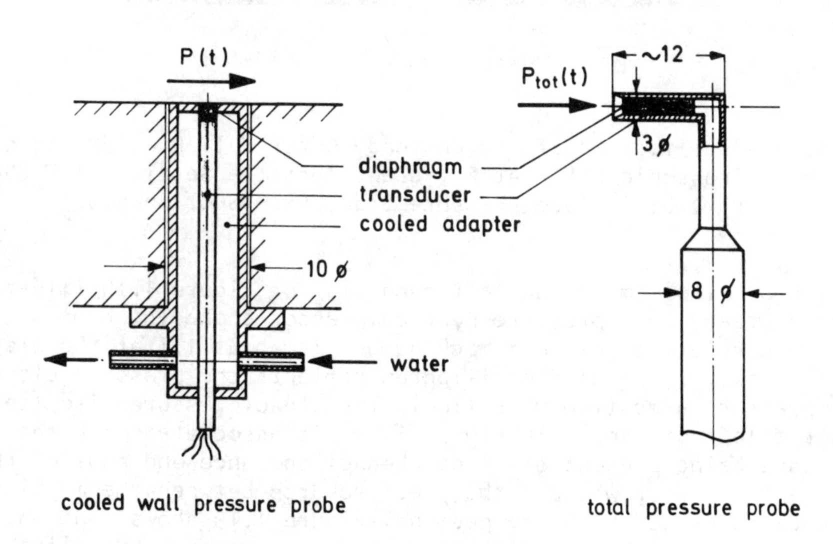

Fig. 1.18 Effective Installation of Semi-Conductor Transducers for Unsteady Wall and Total Pressure Measurements (Dimensions in mm)

The subsequent figures present some results of unsteady pressure measurement in axial and radial flow compressors. Figure 1.19 shows oscillograms of the oscillating wall pressure at the outer casing of a transonic compressor with a rotor design pressure ratio of 1.6. The pressure data measured 10% of chord downstream of rotor leading edge (Figure 1.20) represent different throttle settings of the compressor at 90% design speed; all the oscillograms cover a bit less than two pressure periods.

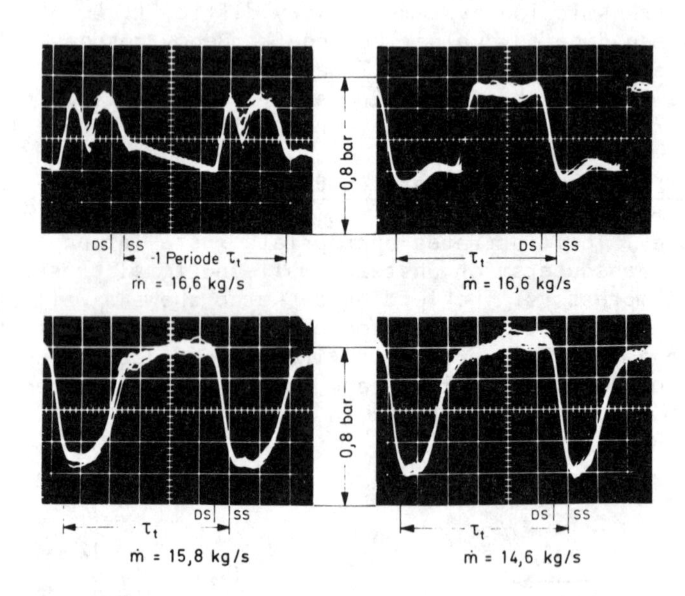

Fig. 1.19 Oscillograms of the Unsteady Outer Wall Pressure in a Transonic Rotor at Different Throttle Settings (15.000 rpm, DS ≙ Blade Pressure Side, SS ≙ Suction Side)

The oscillogram on the left hand side of Figure 1.19 illustrates the unsteady wall pressure near compressor choke (high mass flow); the relative upstream Mach number is about 1.1 at the blade tip. The lower right hand oscillogram reflects the pressure close to compressor surge (low mass flow). The steep pressure rise (towards the left of center in Figure 1.19) is associated with the shock wave being present at blade channel entrance and passing the pressure transducer whereas the pressure drop between the markings DS and SS is caused by blade passing. Figure 1.19 shows that the pressure wave form varies with increasing compressor throttling as does the pressure amplitude. In particular, the area of low pressure on the blade suction gets smaller, indicating the detached bow shock travelling upstream due to the increasing back pressure. The pressure profiles representing always the same blade channel are the result of superposing the transducer's output signal over 25 rotor revolutions. A small data scattering is observed primarily due to unsteadiness of the rotor relative flow field; the shock wave fluctuates a little around its mean position and thus causes weak pressure fluctuation of low frequency (~ rotation).

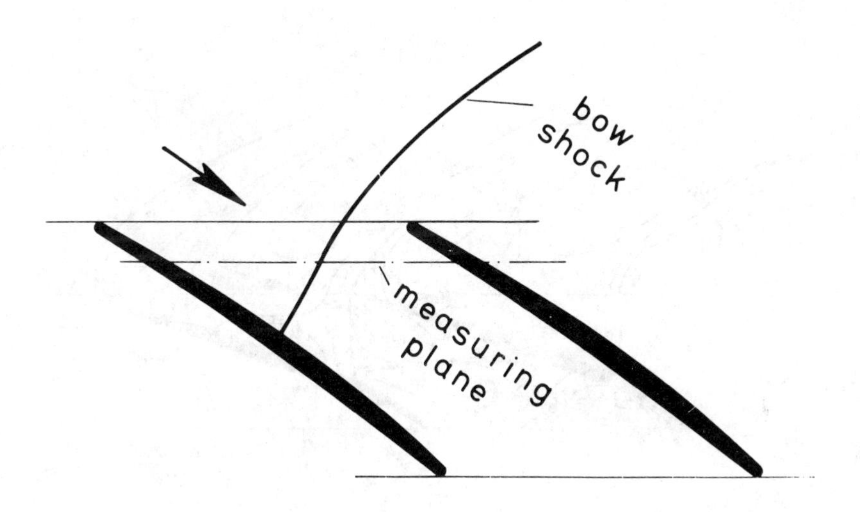

Fig. 1.20 Position of Semi-Conductor Transducer With Respect to Rotor Blade Tip Section

Using lots of transducers properly distributed over rotor tip and applying the correct time-averaging pressure probe described before allows to analyse in detail the static pressure field at rotor tip as shown in Figure 1.21 that presents data obtained in a second transonic axial flow rotor. Isobars of local static to inlet absolute total pressure p/p_{tot} are plotted versus the blade tip section. At low back pressure (figure at the top) the inlet bow shock and additionally an oblique shock system inside the blade channel resulting from bow shock reflection become evident. The flow is decelerated throughout the blade passage from Mach number 1.36 to 0.84 undergoing a turning of 8.9° (blade turning: 12.0°). Thus, in the rear part of just downstream of the blade passage a normal shock must be present decelerating the flow to subsonic velocity. However, at that station, the pressure measured does not make the normal shock to be clearly identified. With increasing back-pressure (figure at the bottom) the bow shock travels upstream and its internal branch becomes normal. No further shock system appears inside of the blade channel, and the internal pressure rise is due to subsonic diffusion. The flow Mach number now decreases from 1.30 at inlet to 0.62 at rotor outlet whereas the flow turning angle is 8.4^{o}, only half a degree lower compared to the weak-throttled case (further results see a.o. references 1.21 and 1.22).

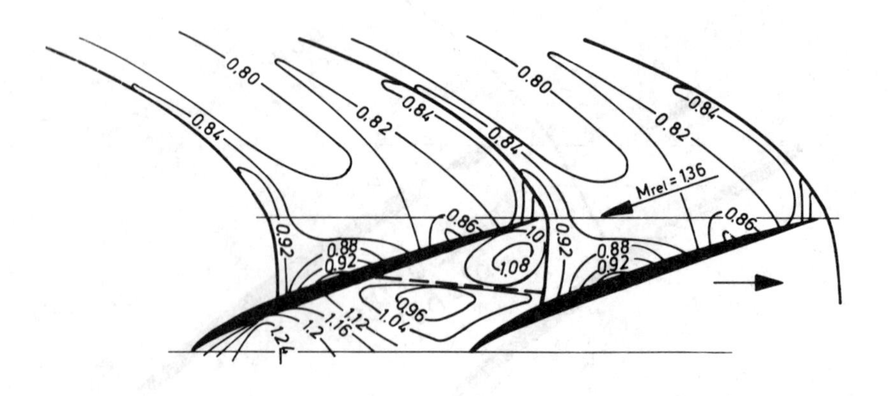

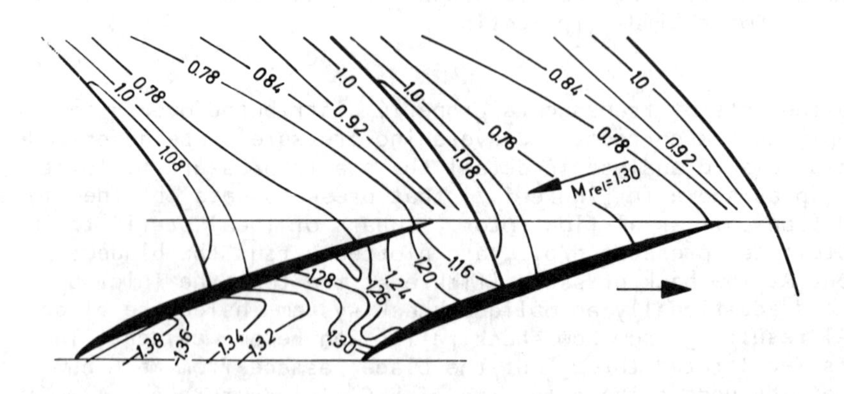

Fig. 1.21 Isobars of Outer Wall Static Pressure (referred to Absolute Inlet Total Pressure) in a Transonic Rotor at Low and High Back-Pressure (π_{stat} = 1.37 and 1.47 respectively)

Figure 1.22 presents an example of oscillating wall pressure measurement in a centrifugal compressor with an impeller having 20 radially ending blades. On the left hand side the installation and position of several semi-conductor transducers are demonstrated; on the right hand side oscillograms of the typical periodic wall pressure at stations I, II, III and IV are shown. Each oscillogram covers somewhat more than two pressure periods. A vertical scale unit corresponds to a pressure of 0.12 bar. The rotor speed is 14.000 rpm, the mass flow amounts to 7.9 kg/s.

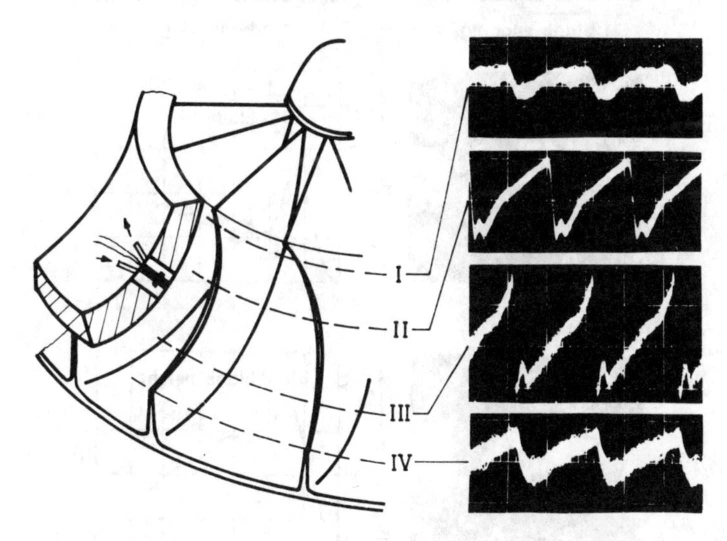

Fig. 1.22 Oscillograms of Outer Wall Static Pressure in a Radial Flow Compressor (Vertical Unit ≙ 0.12 bar; 14.000 rpm, 7.9 kg/s) and Corresponding Semi-Conductor Transducer Positioning

At rotor inlet (I) there is a relatively weak pressure fluctuation; because of the low rotor speed the flow velocity towards the rotor is far below the speed of sound. Further downstream (II and III) a saw-tooth pressure profile appears with a steep pressure drop over the rotor blades and practically linear pressure rise from blade suction to pressure side. The pressure amplitude grows within the forward section of the rotor reaching a maximum of 0.29 bar (III) and drops again near rotor outlet (VI) which is probably caused by flow separation and/or flow mixing behind the rotor.

Figure 1.23 demonstrates, by a few selected examples, the unsteady total pressure at the outlet of the second transonic axial compressor rotor running near peak efficiency at design speed (~ 20.000 rpm). The data were gathered using a pitot-probe with a semi-conductor transducer built into the probe head.

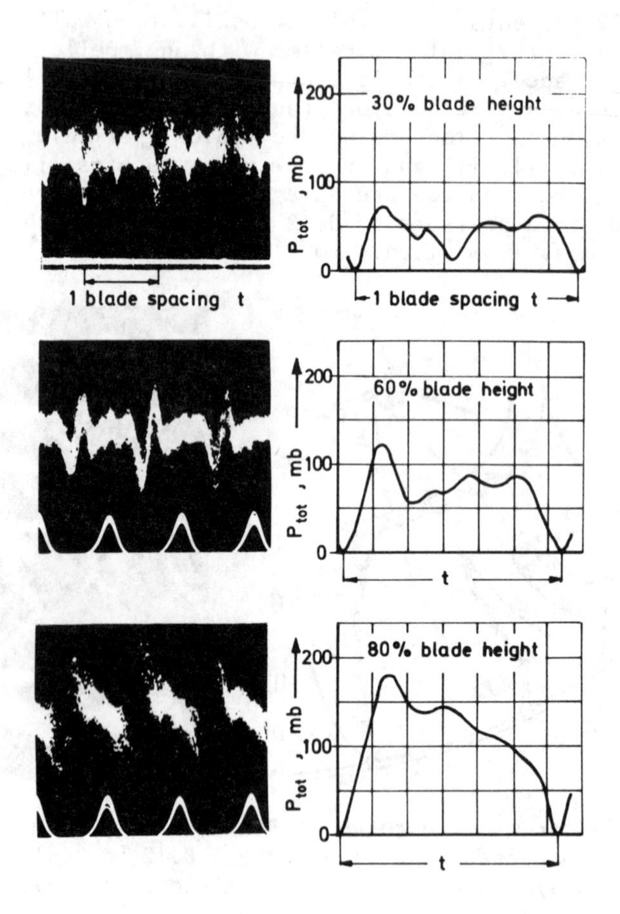

Fig. 1.23 Unsteady Total Pressure (Dynamic Component!) Behind a Transonic Axial Flow Rotor at Design Speed (~ 20.200 rpm)

The top row of graphs illustrates the fluctuating total pressure at 30%, the middle one at 50%, and the lower one at 80% of the rotor blade height. The oscillograms on the left again represent the pressure pulsation behind the same blade passage over 25 rotor revolutions; the intense signal noise is caused by the unsteady character of the relative rotor flow. The strongest irregular fluctuations are found within the blade wakes. With the help of the synchronized sampling technique, the transducer signals are freed from noise and submitted as statistically averaged periodic distributions in the diagrams on the right hand side.

The probes shown in Figures 1.9 and 1.18 have been used to investigate in detail the unsteady total pressure field at rotor outlet. The measuring plane was located a quarter of chord downstream of the rotor trailing edge. Simultaneous radial traverse of both probes was performed to study also the spanwise variation of total pressure. The probe with built-in transducer was used to read the dynamic pressure component whereas the correct mean pressure was measured by means of the oil-filled probe. Superposition of both probe readings revealed the desired absolute total pressure as a function of the circumferential and radial coordinate. The test conditions were the same as selected for the outer wall static pressure analysis (Figure 1.21).

Figure 1.24 a-c presents the total pressure related to the inlet total pressure. The data are plotted as isobars over a bit more than one blade passage. In order to facilitate the interpretation of the results the blade passage boundaries are shown in a very simplified form with straight lines and with radii at hub and tip, respectively. The blade trailing edge marked by the thick dash-pointed line is extrapolated from the rotor outlet plane to the measuring plane accounting for the variation of the circumferential velocity component along the streamlines and assuming that the relative flow angle does not vary considerably over the blade spacing.

At low back pressure (near maximum mass flow) a quite large region of relatively high pressure appears near the blade pressure side extending over about 80% of the blade height; approaching the blade tip this region bends against the direction of rotation as the extrapolated trailing edges do as a result of the spanwise distribution of the circumferential velocity component. The most striking phenomenon is observed near the hub where zones of low and high effective energy addition exist close together. The low work region where the pressure ratio does not exceed unity is primarily caused by flow separation at the hub and by secondary flow. A strong clockwise rotating secondary flow vortex is present in the corner between blade suction side and hub. Another zone of low pressure ratio occurs at blade tip and is also a result of flow separation already known from hot-wire measurements in transonic compressors (references 1.23 and 1.24). Generally, in separated flow regions low mass flow and, thus, small axial velocity components exist leading often to large swirl components and therefore to relatively high energy addition. Nevertheless, the total pressure may be low in those regions because of tremendous flow losses.

With increasing back pressure the total pressure ratio also increases until its maximum of 1.8 is achieved near blade tip at minimum mass flow (15.8 kg/s). Furthermore, with higher back pressures the region of relatively high total pressure rise extends more and more over the blade pitch and span. Regarding the flow phenomena at the hub no significant variation becomes evident from the figures, except that the region of low total pressure decreases considerably; however, the pressure ratio near its center differs hardly from unity. On the other hand, the area of low total pressure at blade tip which happens obviously at the low back pressure case (Figure 1.24 a) disappears continuously with reduced mass flow.

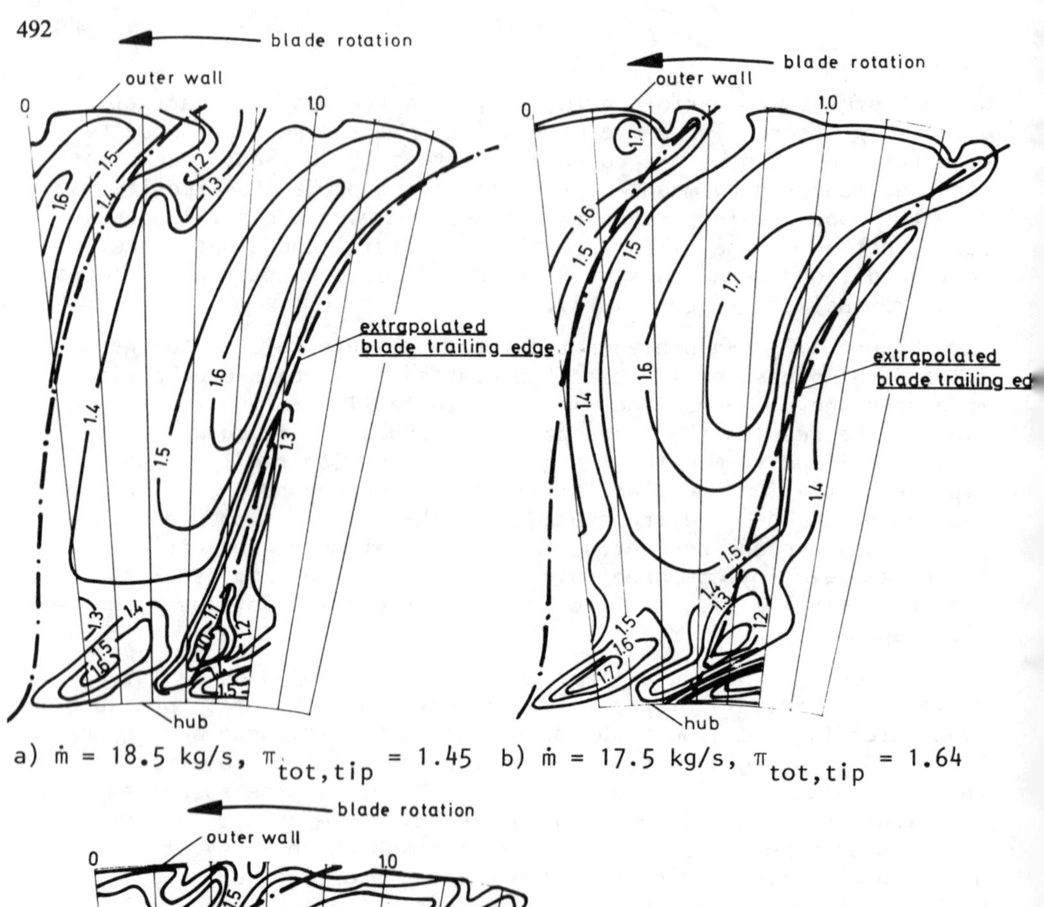

a) $\dot{m}$ = 18.5 kg/s, $\pi_{tot,tip}$ = 1.45 b) $\dot{m}$ = 17.5 kg/s, $\pi_{tot,tip}$ = 1.64

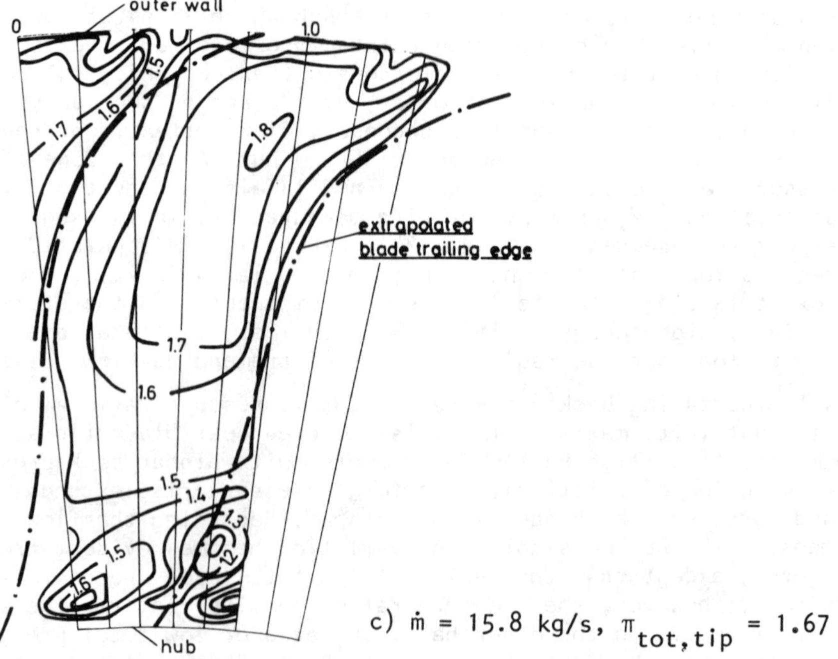

c) $\dot{m}$ = 15.8 kg/s, $\pi_{tot,tip}$ = 1.67

Fig. 1.24 Isobars of Total Pressure Ratio Behind a Transonic Rotor Over One Blade Passage at Design Speed (~ 20.200 rpm)

REFERENCES

1.1 Nesbitt, M., "The Measurement of True Mean Pressures and Mach Numbers in Oscillatory Flow", N.G.T.E. Memorandum No. M 180 (1953).
1.2 Grant, H., Kronauer, R., "Presssure Probe Response in Fluctuating Flow", Pratt & Whitney Research Report No. 139 (1954).
1.3 Johnson, R., "Averaging of Periodic Pressure Pulsations by a Total-Pressure Probe", NACA-TN 3568 (1955).
1.4 Fowle, A., "Measurement of Nonstationary Pressure with a Pitot Probe", Massachussetts Institute of Technology, Dissertation (1948).
1.5 Weyer, H., "The Determination of Time-Weighted Average Pressures in Strongly Fluctuating Flows, Especially in Turbomachines", ESRO-TT-161 (1975).
1.6 Shaw, R., "The Influence of Hole Dimensions on Static Pressure Measurements". J. Fluid Mech. 7, 550-564, Cambridge Univ. Press (1960).
1.7 Goldstein, S., "A Note on the Measurement of Total Head and Static Pressure in a Turbulent Stream", Proceedings of the Royal Soc. of London A155, 570-575 (1936).
1.8 Schodl, R, "Einfluss turbulenter Geschwindigkeitsschwankungen auf die Druckmessung in Strömungen", DFVLR-IB 352-78/1 (1978).
1.9 Smith, L.L., "Wake Dispersion in Turbomachines", Journal of Engineering for Power, September 1966, p. 688 (1966).
1.10 Mikolajczak, A.A., "The Practical Importance of Unsteady Flow", AGARD-CP-177 (1976).
1.11 Kerrebrock, J.L. and Mikolajczak, A.A. "Intra-Stator Transport of Rotor Wakes and Its Effect on Compressor Performance", Journal of Engineering for Power, October 1970, p. 359.
1.12 Brosh, A., "Miniature Pressure Transducers for Wind Tunnel Models", Proc. Instrument Soc. of America, 17th Aerospace Instrumentation Symposium, Las Vegas, Nevada, May 10-12, 1971; B. Washburn (ed.), Pittsburgh (1971), p. 209-216.
1.13 Simpson, J. and Gatley, W., "Dynamic Calibration of Pressure Measuring Systems", McDonnel Aircraft Company, MDC 70-010, 16th Aerospace Instrumentation Symposium of America, Seattle, Washington, 11-13 May 1970.
1.14 Willmarth, W.W., "Unsteady Force and Pressure Measurement", Annual Review of Fluid Mechanics, Vol. 3 (1971) ed. by M. van Dyke, W. Vincenti and J. Wehansen.
1.15 Bynum, D., Ledford, R. and Smotherman, W., "Wind Tunnel Pressure Measuring Techniques", AGARDograph No. 145 (1970).
1.16 Taback, J., "The Response of Pressure Measuring Systems to Oscillating Pressures", NACA TN 1819 (February 1949).
1.17 Iberall, A., "Attenuation of Oscillatory Pressures in Instrument Lines", NBS-Journal of Research, RP 2115, Vol. 45 (July 1950).
1.18 Rohmann, C. and Grogan, E., "On the Dynamics and Pneumatic Transmission Lines", ASME-Paper No. 56-SA-1 (1956).

1.19 Bergh, H. and Tijdeman, H., "Theoretical and Experimental Results of the Dynamic Responses of Pressure Measuring Systems", NLR-TR F.238 (1956).
1.20 Healy, A. and Carlson, R., "Frequency Responses of Rectangular Pneumatic Transmission Lines", ASME-Paper 60-WA/Fles-5 (1969).
1.21 Weyer, H.B. and Hungenberg H.-G., "Analysis of Unsteady Flow in a Transonic Compressor by Means of High-Response Pressure Measuring Techniques", AGARD-CP-177 (1976).
1.22 Miller, G.R. and Bailey, E.E., "Static-Pressure Contours in the Blade Passage at the Tip of Several High Mach Number Rotors", NASA TM X-2170 (February 1971).
1.23 Fessler, T. and Hartman M., "Preliminary Survey of Compressor Rotor-Blade Wakes and Other Flow Phenomena with a Hot-Wire Anemometer", NACA RM E56A13 (1956).
1.24 Lewis, G., Tysl, E. and Fessler, T., "Analysis of Transonic Rotor-Blade Passage Loss with Hot-Wire Anemometers", NACA RM E58C04 (1958).

PART II: OPTICAL METHODS OF FLOW MEASUREMENT IN TURBOMACHINERY

Conventional testing techniques, as pressure and temperature probes, so far widely used in turbomachinery R and D, are quite limited to flow studies in the non-rotating components; similar investigations within the rotors are extremely difficult primarily due to mechanical problems at high rotational speeds. Therefore, optical measuring methods have always attracted the experimentalists because they offer a real chance to investigate at least some important aspects of the complex flow in rotating components without giving rise to flow disturbances or mechanical difficulties. Figure 2.1 summarizes the most appropriate optical techniques being already in use or momentarily under development for turbomachine applications.

REVIEW OF EXISTING OPTICAL METHODS

Flow Visualization

When investigating the flow around models in wind tunnels the experimentalists very soon decided not only to rely upon surface and total pressure measurements but to visualize the flow by injecting traces of small particles into the upstream flow. For instance, they use smoke in gaseous, hydrogen bubbles or dye traces in liquid flows, being able to follow the streamlines or flow paths, to analyze the spatial extention of separated flow areas, and to observe distinct flow vortices.

Coloured liquid traces moving along the walls with the flow are used to identify the streamlines of flow paths just on the model surfaces, or the surfaces are painted with thermosensitive or sublimating (azobenzene) materials to get informations on the wall temperature distribution and the boundary-layer development, particularly on the location of transition and separation. The fundamentals of the various techniques are described to some extent in reference 2.1

TECHNIQUE	PRINCIPLE	ANALYSIS
VISUALISATION	SMOKE INJECTION, LIQUID TRACING, SURFACE PAINTING	• STREAMLINES • AREAS OF FLOW SEPARATION • BOUNDARY-LAYER TRANSITION
SCHLIEREN- AND SHADOWGRAPH METHODS	REFRACTION OF LIGHT RAYS	• DENSITY GRADIENTS • SHOCK WAVES • BOUNDARY-LAYER THICKNESS • FLOW SEPARATION
HOLOGRAPHIC INTERFEROMETRY	PHASE SHIFT OF COHERENT LIGHT	• DENSITY GRADIENTS • SHOCK WAVES
LASER VELOCIMETRY	LIGHT SCATTERING BY SMALL PARTICLES	• FLOW VELOCITY • FLOW DIRECTION • FLOW TURBULENCE
RAMAN SPECTROSCOPY	LIGHT SCATTERING BY MOLECULES (RAMAN EFFECT)	• GAS DENSITY • GAS TEMPERATURE • SPECIES CONCENTR.
FLUORESCENT TECHNIQUES	FLUORESCENT LIGHT EMISSION	• GAS DENSITY

Fig. 2.1 Survey of Non-Intrusive Optical Measuring Techniques Appropriate for Turbomachine Application

At least, the streamline or flow path visualization by injecting particles is restricted to flows of very low turbulence and consequently of low velocity, because increasing turbulence intensifies the diffusion of the particles into the surrounding flow, and thus leads to a very rapid outwash of the traces at higher speeds. There is one example found for the dye trace technique being applied to study the incompressible hydraulic flow in an axial turbomachine. The tests were carried out at ONERA in France (reference 2.2), and Figure 2.2 presents one of the results obtained.

The rotor shown is running at 24 rpm; the flow velocity upstream of the blade row amounts to 0.2 m/s. Due to the low turbulence level at low Reynolds number (10^4 based on blade chord) discrete vortices shedding from the blades appear within the wakes.

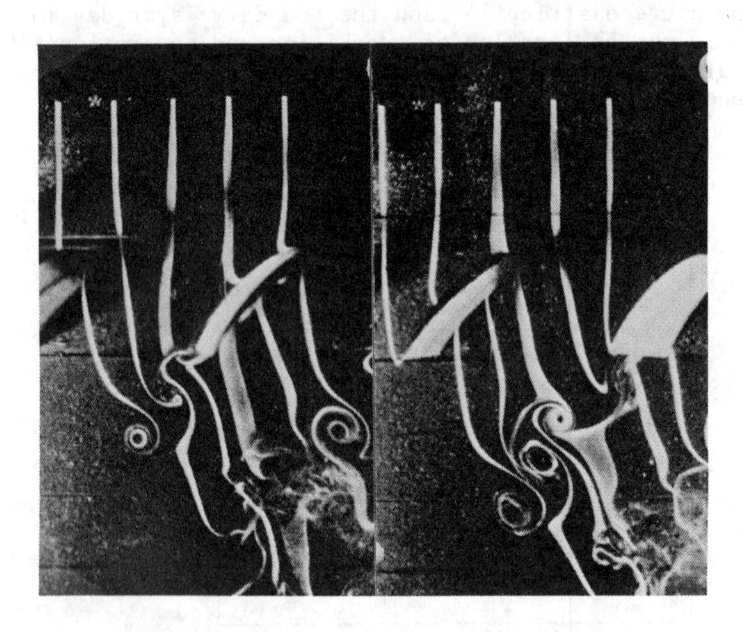

Fig. 2.2 Flow Visualization With Dye Traces in an Axial Turbomachine (Reference 2.2)

A critical examination of the techniques briefly described herein reveals that they - because of principal restrictions to low speed flows - are of no major importance for experimental flow studies in modern high speed turbomachinery, except perhaps the particle injection method (e.g. smoke) to analyse the extensions of flow separation and the surface painting technique to locate blade boundary-layer transition and separation (references 2.3, 2.4).

Schlieren- and Shadowgraph Methods

The schlieren- and shadowgraph methods are based on the wave character of light which means that a light ray travelling through a transparent but with respect to its refractive index inhomogeneous medium will be deflected from its original direction. Refractive index inhomogeneities always occur in compressible flows due to the inherent density variations. If a light ray passes through a test section with compressible flow as shown in Figure 2.3, it will be bent at an angle α which depends on the refractive index gradient, i.e. on the density gradient normal to the light path. The shadowgraph system measures the light ray shift from A to B on the observing plane and therefore refers to the second derivative of the refractive index while the schlieren system measures directly the angle α and thus responds to variations of the refractive index gradient. Further basic informations are given in reference 2.1.

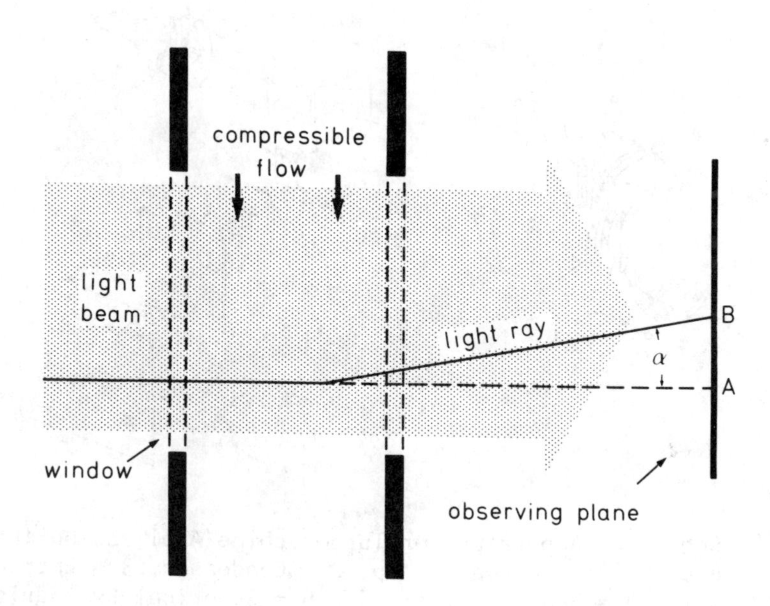

Fig. 2.3 Principle of Light Refraction of Schlieren- and Shadowgraph Methods (Reference 2.1)

In wind tunnels, the schlieren- and shadowgraph methods have proved to be very useful to study qualitatively density variations within the compressible flow field around models. Particularly shock waves in supersonic flows, shock/boundary-layer interaction, boundary-layer thickness growing and separation may be visualized by either the schlieren- or the shadowgraph technique. Their application to transonic and supersonic turbomachine testing, however, raises principal difficulties since they integrate along the line-of-sight and therefore are generally limited to two-dimensional flows. Turbomachine flow is highly three-dimensional and will approach two-dimensionality only if the hub-to-tip ratio is extremely high as in rotating cascades where the blades remain identical and untwisted over the whole blade span. Test rigs of this type are available a.o. at ONERA and TH Aachen; there they are widely used to investigate qualitatively the structure, periodicity and stability of shock waves in supersonic rotors. This excellent basic research work is reported e.g. in references 2.5 to 2.9.

Figure 2.4 (reference 2.6) illustrates the schlieren system developed at ONERA and applied to supersonic axial flow turbomachinery. A flash lamp is used for light source and operated stroboscopically to allow for rotor flow measurements.

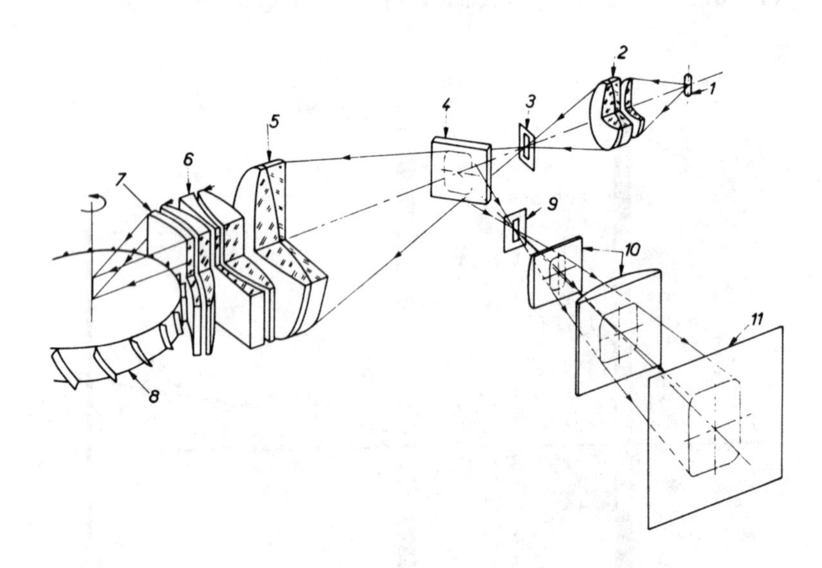

Fig. 2.4 Schlieren Apparatus for Turbomachine Application (Reference 2.6): 1 = Flash Lamp, 2 = Condenser, 3 = Entrance Slit, 4 = Beam Splitter, 5 - 6 = Spherical-Cylindrical Lens System, 7 = Window, 8 = Moving Blade Row, 9 = Schlieren Slit, 10 = Cylindrical Lenses (Crossed), 11 = Observing Screen

The main component of the optical arrangement is the spherical-cylindrical lens system 5 - 6 that transforms the spherical light wave front formed at the entrance slit 3 into a cylindrical wave front being concentric to the machine axis. Hence, the light rays enter the blade row on radial paths, are reflected at the polished metal hub, and travel back to the beam splitter 4 and finally to the observation screen 11. The shock wave visualization in the high speed rotors (up to 12000 rpm) requires a special fast shutter camera associated with the flash lamp. An ultra-fast electronic shutter was found that keeps the exposure times well below 50 μsec and enables to obtain satisfactory schlieren pictures. (For further details see reference 2.6). Figure 2.5 presents a schlieren photograph selected from the results published in reference 2.5. The picture represents the flow in a supersonic compressor running in freon at 6000 rpm. The bow shock at the blade entrance is clearly visible with its external branch travelling upstream. Its internal oblique branch meets a normal shock in the rear part of the blade that decelerates the supersonic flow to subsonic velocity and interferes with blade suction side boundary-layer as can be seen from the λ-shape of the shock. This type of measurement revealing an excellent flow periodicity allows to study in detail the shock movement and the flow stability in function of back-pressure (references 2.5, 2.6 and 2.8). Similar results were obtained at ONERA in diffusers of supersonic radial flow compressors (e.g. reference 2.6).

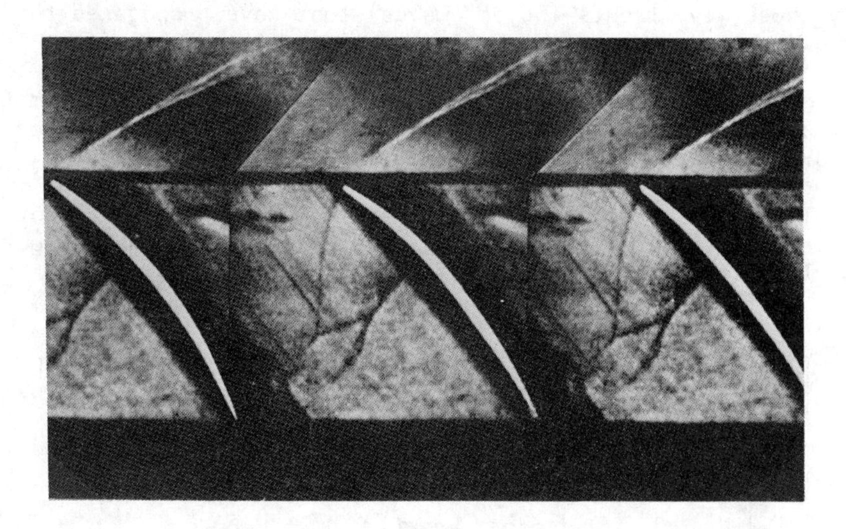

Fig. 2.5 Schlieren Picture of Shock Waves in a Supersonic Axial Flow Rotor (6000 rpm, Freon), (Reference 2.5)

Holographic Interferometry

While the schlieren- and shadowgraph methods are based on the light deflection in inhomogeneous flows interferometry uses the associated phase alteration to measure quantitatively flow density fields. In order to make visible the light phase alteration two coherent light beams are used: an undisturbed reference beam and an object beam that propagates through the flow field of interest there undergoing phase alterations due to the flow density inhomogeneities. Both the light beams are superimposed on the observation screen interfering with each other and generating the well-known fringe patterns. Conventional interferometry of the Mach-Zehnder or Michelson type is strictly limited to the application in two-dimensional flows for integrating along the line-of-sight.

Holography is an optical storage technique that allows to store and reconstruct the coherent wave fronts proceeding from a three-dimensional object, even a flow field, thus that a real three-dimensional image of the object results. Simple photographic techniques may be applied for hologram storage; its analysis must not follow on-line with the measurement, but can take place at any later time. (For more basic information see reference 2.10.)

The introduction of holography has widely expanded the possibilities of interferometry for flow field diagnostics even in turbomachinery. The holographic interferometry offers important advantages against the conventional type because the holographic system compares only those wave fronts (object waves) that have propagated through a flow test section at two different times. One prime effect is that the test cell windows and all optical components do not need to be of high optical quality. The two object waves proceeding from the test section at the two different times are stored on the same hologram and are reconstructed simultaneously.

The basic principles of holographic interferometry do not differ from those of classical interferometry as shown in Figure 2.6. The reference beam is always necessary to establish and to

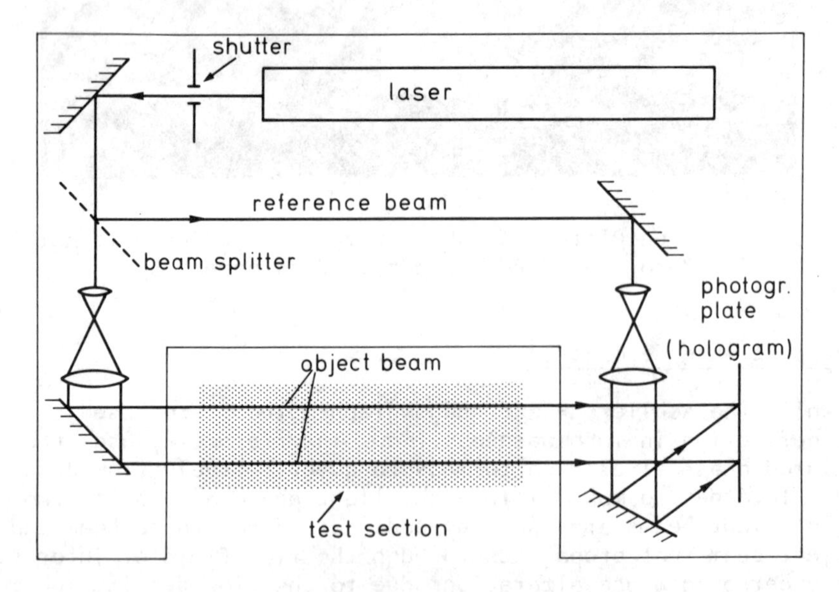

Fig. 2.6 Principle of Holographic Interferometry

reconstruct the hologram. The object wave propagates through the test section and is mixed with the reference beam on the photographic plate creating the hologram. In recent years many holographic interferometers have been developed; the method that revealed to be most appropriate for turbomachine applications is the double exposure interferometry. Two exposures are made; the first one of a comparison flow field and the second one of the same test section with the flow of interest present. Thus, on the photographic plate there are two holograms stored. The simultaneous reconstruction of the two object waves produces the interference between the comparison and the test case. The interference fringe pattern provides all the informations requested on the density of the three-dimensional flow field. A detailed description of the fundamentals of holographic interferometry is given a.o. in references 2.11 and 2.12.

A couple of years ago, NASA initiated an extensive program to apply pulsed laser holographic interferometry to the analysis of shock patterns in the outer span regions of axial flow transonic compressor rotors (reference 2.13). Two holographic approaches were made to two separate existing transonic rotors (references 2.14 and 2.15). The first holographic technique utilizes diffuse light reflected from the centerbody just ahead of the rotor inlet as is schematically illustrated in Figure 2.7 (from reference 2.14). The object beam enters the test section through a small window, is reflected from the polished hub, passes outward through the blade tip section, through a second window, and finally reaches the holographic plate. The reference beam is contained within the associated holocamera which is not shown in the figure.

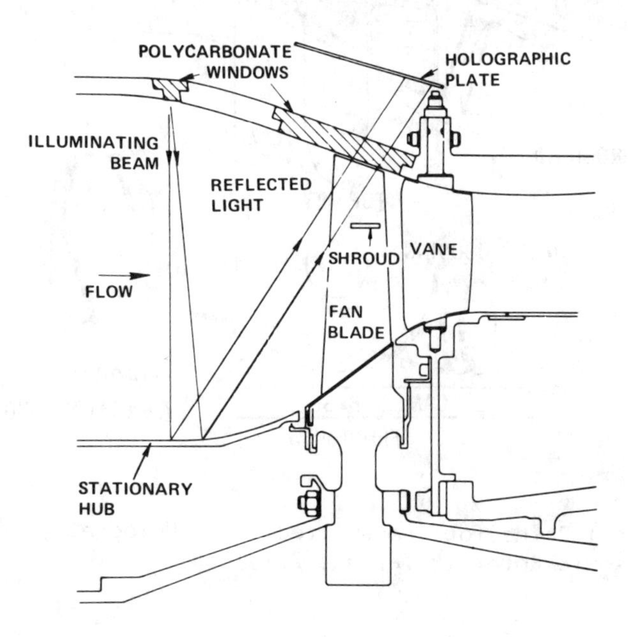

Fig. 2.7 Optical paths of Reflected Light Holography in a Transonic Fan Rig

The technical details of the camera are given in reference 2.14 and will not be discussed herein. The rotor that has been tested by means of this interferometer is a highly loaded high-tip speed fan using precompression type blades for the outer blade sections. The design tip speed is about 550 m/s, the rotor pressure ratio amounts to 2.3.

The second holographic technique transmits the object beam diagonally across the compressor inlet as demonstrated schematically in Figure 2.8 (from reference 2.13). Diffuse light enters the test section through the large window in front of the rotor, passes by the centerbody, through the blade tip region, and leaves through the window over the blade tip onto the photographic plate. Further details of the technique particularly of the holocamera used are given in references 2.15 and 2.16. With this method a low-loading high-tip speed rotor designed for a tip speed of 490 m/s and a pressure ratio of 1.5 was tested.

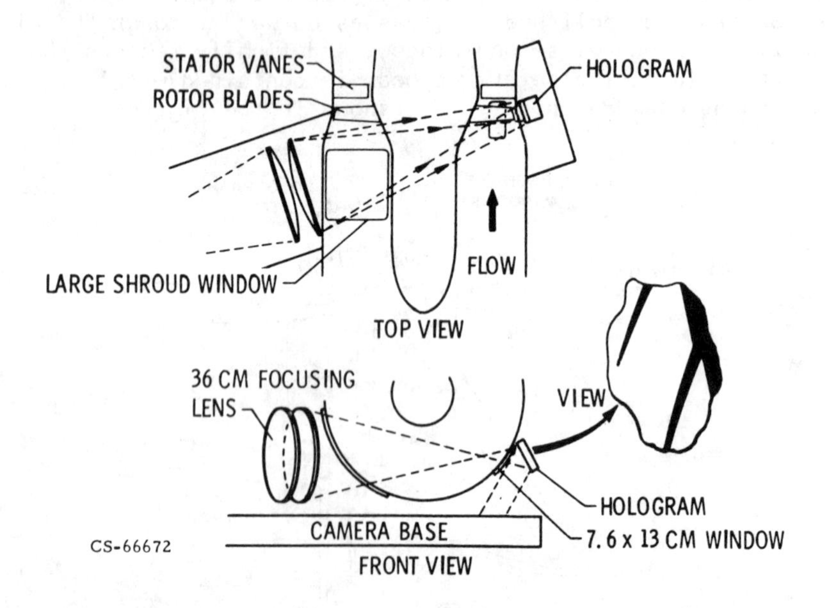

Fig. 2.8 Optical Paths for Transmitted Light Holography in a Transonic Rotor (Reference 2.13)

In order to get the holograms, in each case the rotor operating point is set and then the laser is pulsed twice at a time interval of about 10 and 5 microseconds, respectively. During the interval the density field within the blade passage moves about 10 or 5% of the blade spacing. The double exposure provides a

g. 2.9 Reconstructed Double-Exposure Reflected Light Hologram of Fan Flow (100% Design Speed, 10 μsec Pulse Separation)(Ref. 2.1)

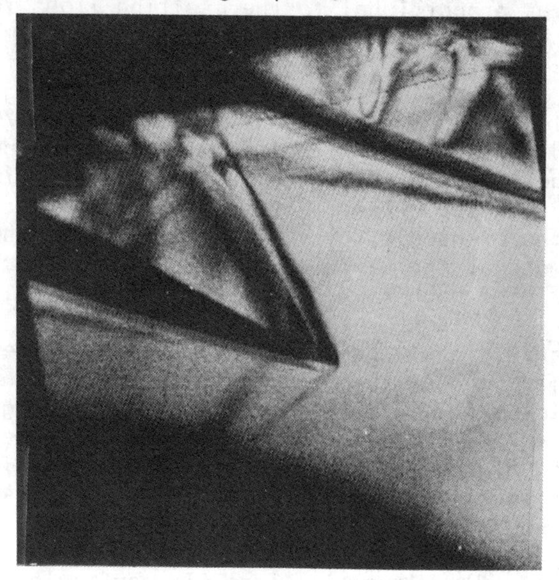

g. 2.10 Reconstructed Rapid Double-Exposure Transmitted Light Hologram of Fan Flow (100% Design Speed, 5 μsec Pulse Separation)(Ref. 2.16)

Today Raman spectroscopy has progressed so far that combustion studies in laboratory environments can be performed without major difficulties and at acceptable accuracy levels. However, the current state-of-the-art does not permit to apply the technique immediately to turbomachinery flow studies. Further extensive work has to be done first to develop adequate lasers (continuous high power or pulsed with high repetition rate) and to find out efficient methods of improving the signal-to-noise ratio. As far as the handling and the accessibility to turbomachinery flow is concerned it would be desirable to have the spectrometer operating in the back-scatter mode if that is possible at all.

Gas Fluorescence

The table in Figure 2.1 finally points to some promising methods of flow density measurement by fluorescence. The well-known electron beam fluorescence (reference 2.24) which is a versatile technique for obtaining quantitative data from two- and three-dimensional flows, however, requires basically a low density test environment, and its application is therefore usually restricted to rarefied gas flows. At MIT a new fluorescent gas techniques has been developed to study the three-dimensional flow in a transonic compressor rotor (references 2.25 and 2.26). A fluorescent gas, Biacetyl, is added to the Argon-Freon test gas in small fractions and excited by a flashlamp-pumped dye laser illuminating selected blade-to-blade planes within the rotor (for hardware details see reference 2.25). The wave length of the laser must be in the Biacetyl absorption band (370 - 450 nm), the fluorescent radiation then is of slightly longer wave length than that of the laser, and its intensity behaves proportional to the exciting light intensity and to the gas density, but independent of gas temperature. The fluorescence is very fast (10^{-8} sec) and of low quantum efficiency ($\sim 10^{-7}$), thus the emitted light cannot be directly photographed, but must be amplified by an image intensifier (reference 2.26), however, then yielding useful density informations from the whole blade-to-blade plane selected. A typical result is given in Figure 2.13 (reference 2.26) which presents the relative density pattern near blade tip of the transonic rotor. The density plot shows a well defined bow shock at the blade channel entrance, which is to be expected at the inlet Mach number of 1.17.

The accurate analysis of the "density images" requires several corrections to be made concerning the non-uniformity of the laser beam and the detector system (image intensifier and photographic film). Finally an overall precision of $\pm$ 2% in relative density is expected by applying high quality components (e.g. intensifier) and by careful calibration of the system.

Indeed, the fluorescent gas technique seems to be very promising for turbomachinery flow investigations, in particular because

differential interferogram of this density variation, which becomes visible by reconstructing the hologram. Typical results are shown in Figures 2.9 and 2.10 (references 2.13 and 2.16) representing the shock waves in the tip region of the two different rotors at 100% speed. In principle the holograms contain complete informations on the three-dimensional density field within the flow volume that is illuminated by the object wave.

A critical review of holographic interferometry shows that this technique may have great advantages for turbomachine applications: First of all, three-dimensional flow data can be obtained by a single measurement. Furthermore, imperfections of the test section windows and of the optical components do not affect the measuring accuracy. Less optical precision means a relatively simple and cheap apparatus which is easy to handle. However, there are still unsolved problems, too: The accessibility of interferometry to turbomachinery flow fields is rather limited. The quantitative analysis of the results is difficult. On the other hand, machine vibrations will disturb the measurement as far as the time interval between two exposures is not small compared to the vibrating period.

Laser Velocimetry

Laser velocimetry which in general measures fluid velocities indirectly by determining the velocity of particles moving with the flow field has become already a routine tool for flow studies in turbomachinery. Fundamentals of laser anemometry and various techniques appropriate for turbomachinery testing will be covered in great detail in the succeeding paper of R. Schodl.

Raman Spectroscopy

The full description of flow fields in turbomachinery requires - besides the local velocity data - additional informations on the local pressure, density or temperature. While the pressure cannot be determined without disturbing the flow by probes, Raman spectroscopy offers some promising prospects for non-intrusive measurements of gas density and temperature. The technique has already proved successful in flame studies (references 2.17 to 2.20).

Raman spectroscopy is based upon the phenomenon of frequency change when light is scattered by molecules. The interference between photons and molecules affects the vibrational and rotational state of the molecules as illustrated in Figure 2.11 (reference 2.21). The frequency shift depends on the type of molecule (type of gas) and is positive as well as negative, defined as Stokes (Raman) and Anti-Stokes scattering, respectively. The figure points also to the Rayleigh scattering which designates that fraction of scattered light which is modulated in its direction whereas the frequency remains equal to that of the exciting light. Now, Raman spectroscopy exploits the effect of frequency shift suppressing the incident light as well as the Rayleigh scattering by using adequate optical filters.

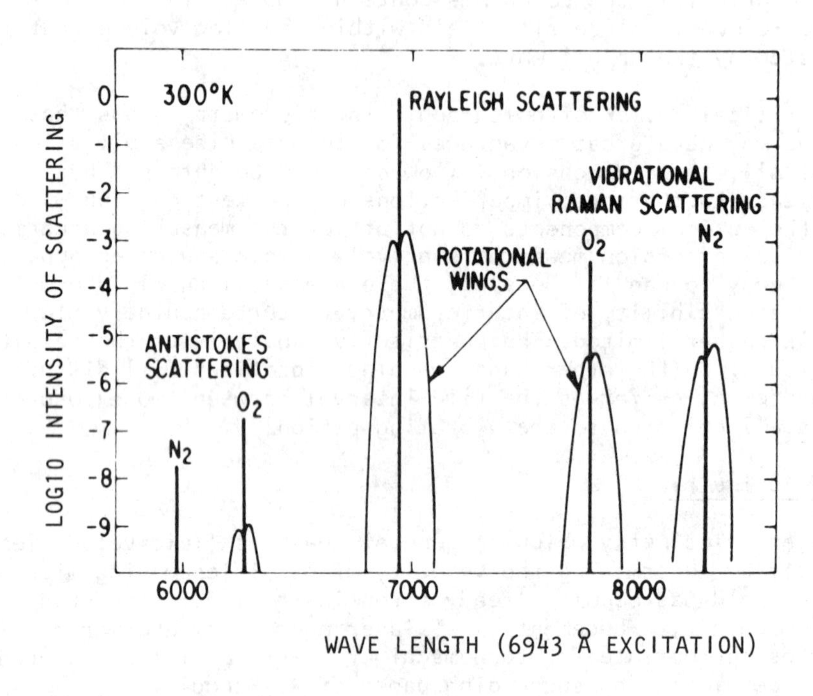

Fig. 2.11 Raman and Rayleigh Scattering of Air at Room Temperature

The temperature measurement then is based upon the fact that the intensity distribution of the Stokes and Anti-Stokes lines reflects the Boltzmann distribution of the molecular energy levels. The ratio of two arbitrary intensities is a measure for the gas temperature whereas the absolute intensity of each individual line (Stokes or Anti-Stokes) is a measure for the gas density. Complete descriptions of the basic principles of Raman spectroscopy are given in references 2.21 to 2.23.

Density measurements in turbomachinery flow ask for high accuracy; the error should not exceed 1% of the actual density because - at least in subsonic axial flow compressors - only weak density variations occur that should be able to be determined. The accuracy mentioned is attainable by using high excitation energy (around 1 W sec), which is supplied today by Q-switch ruby lasers (pulse duration ≅ 10 nsec), when rotor flow testing is considered. However, the low repetition rate of these lasers - about one pulse per second - rises the problem of extremely high measuring time; an accurate investigation of the density in turbulent flows requires around 1000 individual measurements at each probe location amounting to an overall time of 1000 sec per probing. Chopped continuous lasers having in general lower peak power cannot be employed because either a long integration time of the photon counting system (in stators) or frequently repeated measurements (in rotors) are necessary to arrive at the excitation energy level of 1 W sec needed to achieve an acceptable accuracy. Thus, the information on the turbulent character of the flow is widely lost.

The other severe problem of Raman spectroscopy is the extremely low signal-to-noise ratio that remains several orders of magnitude below that of the laser velocimetry (Mie scattering). Due to this unfavourable situation the spectrometer cannot be operated in the back-scatter mode, and alternative light path orientations have to be considered as illustrated in Figure 2.12. The laser light enters the turbomachine downstream of the rotor, is turned 90° by a mirror, propagates upstream through the rotor blading, and is finally absorbed in the inlet tubing. The spectroscopy set-up is aligned radially permitting an observing angle of 90° which is convenient with respect to the background noise. However, with this light path concept large shadowed flow areas will result from blade camber and twist, while the unavoidable mirror will disturb the flow essentially.

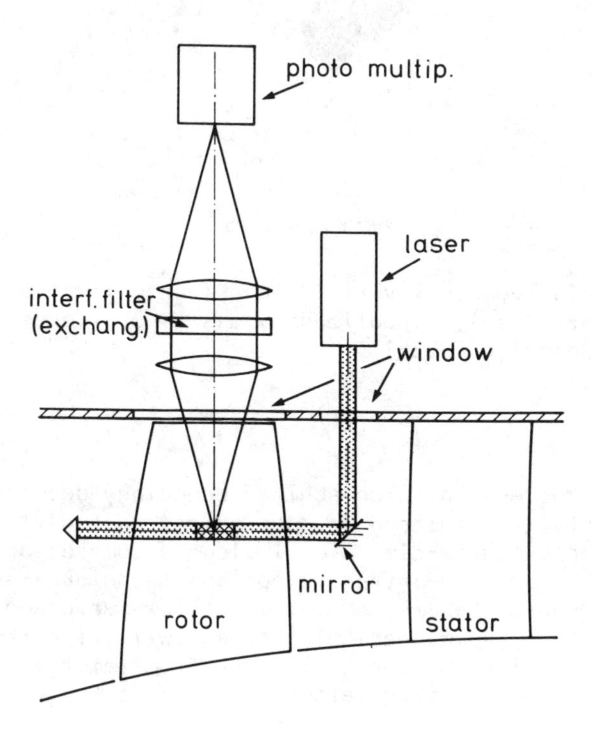

Fig. 2.12 Sketch of Raman Spectroscopy Applied to an Axial Flow Turbomachine Rotor

one individual measurement yields density data from a full blade-to-blade flow area. But further development is still needed to overcome some basic problems as the extremely low quantum efficiency, the short life times, and possible quenching effects in air/tracer mixtures. Handling problems could arise from explosiveness, corrosiveness, and toxicity of the fluorescent media. Additional efforts should be adressed to minimizing the shadowed areas in the blade channels and to refining the various system's components (optics, electronics) for the benefit of accuracy.

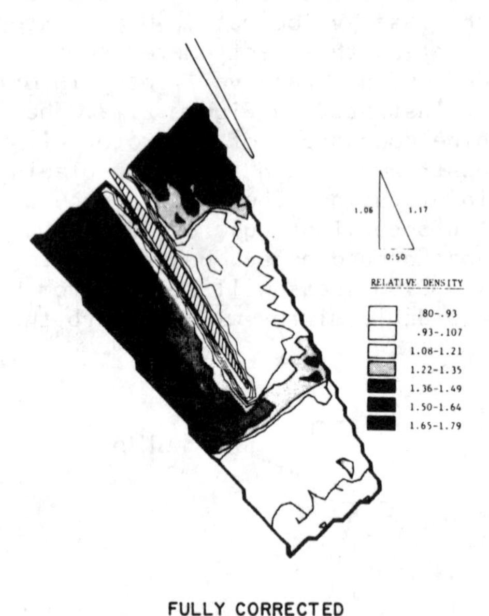

Fig. 2.13 Relative Density Field in a Transonic Compressor Rotor (88% Span), Visualized by Gas Fluorescence (M_1 = 1.17) (Reference 2.26)

CONCLUSIONS

Future progress in turbomachine technology depends upon how the experimentalists succeed in completing the today's knowledge about the complex three-dimensional flow in compressors and turbines. In doing so, non-intrusive optical techniques are of great importance because they render detailed flow measurements even in high-speed rotors. No mechanical problems will rise from rotating probes, no flow disturbances will occur from immerging probes. The critical review of existing methods has led to the following conclusions:

- Flow visualization by tracing is of minor importance for experimental flow research in modern turbomachines for being restricted to low turbulent flows. Smoke injection and surface painting, however, may be of some interest for special investigations on separated flow areas (jet/wake pattern), on surface heat exchange, and on boundary-layer development.

- Schlieren- and shadowgraph techniques have proved worthwhile methods to study shock wave structures in two-dimensional supersonic flows. As far as turbomachinery research is concerned their application is limited to high hub-to-tip ratio machines, e.g. rotating cascades.

- The holographic interferometry in principle is suitable for gathering complete density data in three-dimensional flows by a single probing. The double exposure technique thereby guarantees that imperfections of the optical components and of the test section windows do not affect the accuracy of the measurement. Main problems associated with the application to turbomachinery flow are: limited accessibility to the flow in blade rows, extensive quantitative analysis of the test data, difficult spatial orientation of the results, and the effects of machine vibrations on accuracy.

- Laser velocimetry already has become a routine tool for flow investigations in turbomachines, manifesting its great potential for turbomachinery research and development. Today practically all major gasturbine manufacturers and research laboratories are involved in adapting the laser velocimetry to flow studies in all types of turbomachines. The current activities are primarily addressed to improving the signal-to-noise ratio by refining the optical and electronic components, to clarifying still open questions of flow seeding and particles' following in strongly accelerated flows. Some further activities aim at the development of three-component velocimeters.

- Raman spectroscopy has progressed so far that combustion studies in laboratory environments can be performed without major difficulties and at acceptable accuracy levels. The current state-of-the-art, however, does not permit to apply the technique immediately to turbomachinery flow studies. Further work has to be done to develop adequate lasers (continuous high power or pulsed with high repetition rate) and to discover efficient methods of improving the signal-to-noise ratio. As far as the handling and the accessibility to turbomachinery flow is concerned it would be desirable to have the spectrometer operating in the back-scatter mode if that is possible at all.

- The fluorescent gas technique seems to be very promising for turbomachinery flow investigations; however, further activities are still needed to overcome some basic problems as the extremely low quantum efficiency, the short life times, and possible quenching effects in air/tracer mixtures. Handling problems as explosiveness, corrosiveness, and poisonousness of the fluorescent media must be resolved, too. Additional efforts should be addressed to minimizing the shadowed areas in the blade channels and to refining the various system's components (optics, electronics) for the benefit of accuracy.

REFERENCES

2.1 Merzkirch, W., "Flow Visualization", Academic Press, New York and London, 1974.

2.2 Werlé, H. and Gallon, "Adaptation de la Cave Hydraulique à la Visualisation de l'Ecoulement dans les Turbomachines", La Recherche Aérospatiale, No. 1, 1972, pp. 15-21.

2.3 Rygh, P.J. and Martin R.E., "The Use of Azobenzene to Provide Visual Indication of Supersonic Boundary-Layer Transition", Progress Report No. 20-335, Jet Propulsion Laboratory California, Institute of Technology, Pasadena, California, 1957.

2.4 Heilmann, W., "Experimentelle und grenzschichttheoretische Untersuchungen an ebenen Verzögerungsgittern bei kompressibler Strömung, insbesondere bei Änderung des axialen Stromdichteverhältnisses und der Zuströmturbulenz", DLR-FB 67-88, 1967

2.5 Fabri, J., "La Visualisation de l'Ecoulement dans un Compresseur Axial Supersonique", L'Aeronautique et l'Astronautique, No. 32, 1971, pp 3-11.

2.6 Fertin, G., "Recording of Schlieren Pictures for Flow Visualization in Rotating Machines", ONERA, T.P. No. 1396, 1974.

2.7 Le Bot, Y. and Larguier, R., "Charactérisation Experimentale des Ecoulements Internes dans les Turbomachines", Revue Française des Mécanique, No. 57, 1976, pp. 5-34.

2.8 Gallus, H.E., Bohn, D. and Broichhausen, K.-D., "Measurement of Quasi-Steady and Unsteady Flow Effects in a Supersonic Compressor Stage", ASME-Paper 77 Gt-3, October 1977.

2.9 Bohn, D., "Untersuchung zweier verschiedener axialer Überschallverdichterstufen unter besonderer Berücksichtigung der Wechselwirkung zwischen Lauf- und Leitrad", Dissertation TH Aachen, December 1977.

2.10 Collier, R.J., Burckhardt, C.B. and Lin, L.H., "Optical Holography", Academic Press, New York, 1971.

2.11 Erf, R.K., ed., "Holographic Nondestructive Testing", Academic Press, 1974.

2.12 Trolinger, J.D., "Laser Instrumentation for Flow Field Diagnostics", AGARD-AG 186, March 1974.

2.13 Benser, W.A., Bailey, E.E. and Gelder, T.F., "Holographic Studies of Shock Waves within Transonic Fan Rotors", NASA-TMX-71430, 1974.

2.14 Hartmann, R.G., Burr, F.J., Alwang, W.G. and Williams, M.C., "Application of Holography to the Determination of Flow Conditions within the Rotating Blade Row of a Compressor", NASA CR-121112, July 1973.
2.15 Wuerker, R.F., Kobayashi, R.J. and Heflinger, L.O., "Application of Holography to Flow Visualization within Rotating Compressor Blade Row", NASA CR-121264, 1973.
2.16 Dodge, P., "Laser Holography, Optical Design and Set-Up", Advanced Testing Techniques in Turbomachines, VKI-Lecture Series, April 1975, von Karman Institute for Fluid Dynamics, pp. 1-115.
2.17 Goulard, R., ed., "Combustion Measurements, Modern Techniques and Instrumentation". Academic Press, New York, San Francisco, London, 1976.
2.18 Ledermann, S., "Modern Diagnostics in Combustion", AIAA Paper No. 76-26, January 1976.
2.19 Setchell, R.E., "Time Measurements in Turbulent Flames Using Raman Spectroscopy", AIAA Paper No. 76-28, January 1976.
2.20 Ledermann, S., Celentano, A. and Glaser, J., "Flow Field Diagnostics", AIAA-Journal, Vol, 17, No. 10, October 1979, pp. 1106-1110.
2.21 Lapp, M., Penner, C.M., ed., "Laser Raman Gas Diagnostics", Plenum Press, New York, 1974.
2.22 Szymanski, H., ed., "Raman Spectroscopy, Theory and Practice", Plenum Press, New York, 1967.
2.23 Veret, C., "Review of Optical Techniques with Respect to Aero-Engine Applications", AGARD-LS-90, 1979.
2.24 Maquire, B., "Quantitative Visualization of Low Density Flow Fields Using the Electron Beam Excitation Technique", ICIASF '69 Record, IEEE No. 69 C 19 - AES, 79, 1969.
2.25 Epstein, A.H., "Quantitative Density Visualization in a Transonic Compressor Rotor", Thesis, MIT, September 1975.
2.26 Kerrebrock, J.L., "Flow in a Transonic Compressor Rotor", Proceeding of the SQUID Workshop on Transonic Flow Problems in Turbo-Machinery, Naval Postgraduate School, Monterey, California, 1976, pp. 485-513.

AERODYNAMIC DEVELOPMENT OF AXIAL TURBOMACHINERY BLADINGS
- Testing Techniques and Steady State Performance Measurements -

C.H. Sieverding

von Karman Institute for Fluid Dynamics

1. LINEAR CASCADE TESTING

1.1 Definition of cascade model

In the case of an annular blade row with cylindrical endwalls, the corresponding linear cascade is directly obtained by developing a cylindrical cut through the blading onto a plane surface. This simple cylindrical cut neglects of course that : (a) the streamline in the meridional plane is in general not a straight line and (b) the initial axisymmetric stream surface becomes twisted when passing through the blade row.

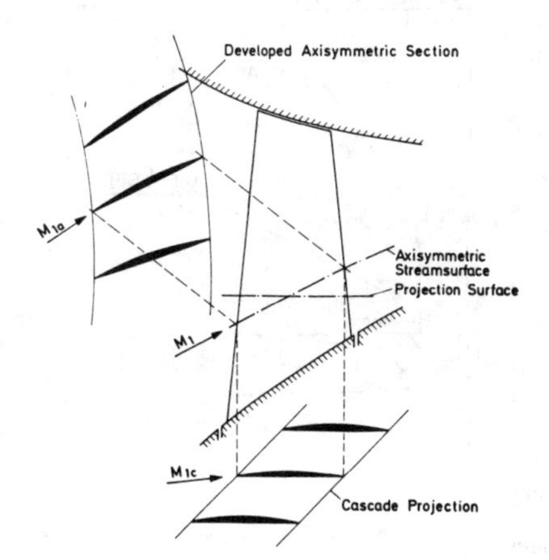

FIG. 1 - BLADE SECTIONS FOR STRAIGHT CASCADES [1]

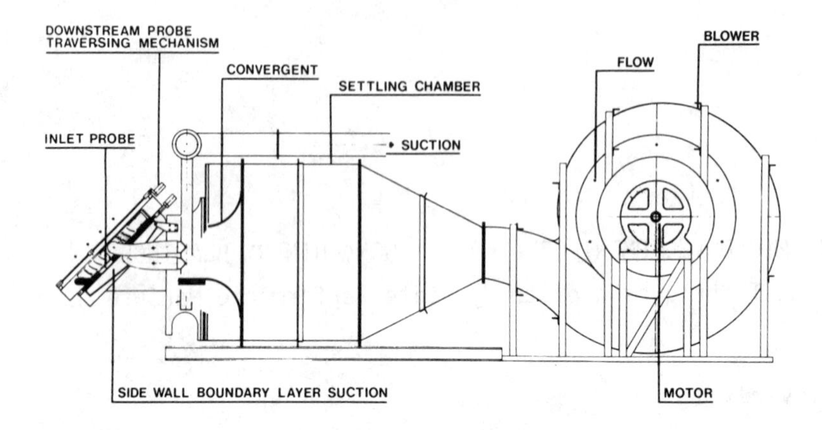

FIG. 2 - NASA LOW SPEED WIND TUNNEL

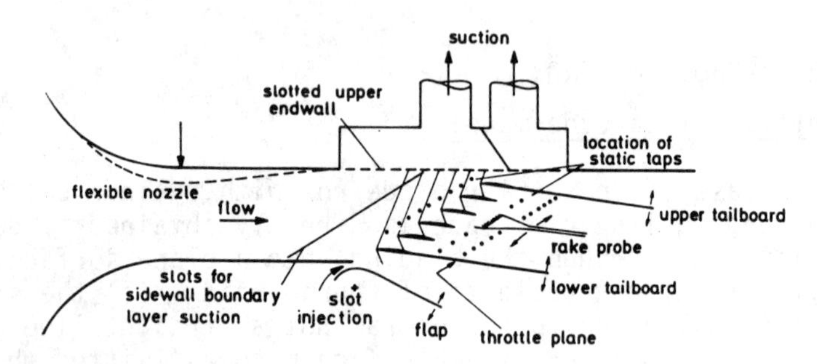

a - Test section of DFVLR continuous-closed loop transonic compressor cascade tunnel

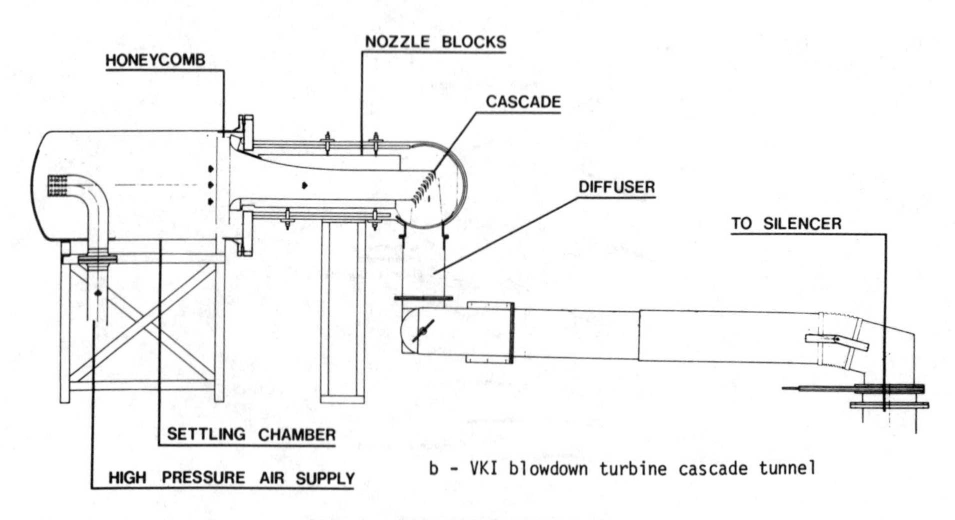

b - VKI blowdown turbine cascade tunnel

FIG. 3 - HIGH SPEED CASCADE FACILITIES

The endwalls of annular cascades are, however, in general neither cylindric nor parallel. Under these circumstances, it is much more difficult to define appropriate blade sections for straight cascades, see figure 1 [1]. There is no general agreement as to whether the blade sections should be defined on a cylindrical or a conical surface, or along a streamline. At subsonic velocities and moderate radius change, the cylindrical cut seems to be most widely applied. At supersonic flow and increased radius change, preference is given to blade sections from conical surfaces. The requirement of a constant pitch in straight cascades results of course in a significant change of passage geometry. Hence the two dimensional performance of these sections cannot be used without correction factor to evaluate the performance of the corresponding stage. In view of these limitations and due to the progress in predicting two dimensional blade performances, systematic two dimensional cascade tests are in regression. Instead, considerable effort is spent to use linear cascade tunnels for investigating three dimensional flow effects like stream tube convergence and divergence, and sweep and dihedral effects.

1.2 Simulation of principal flow parameters

1.2.1 Mach number. The Mach number range in axial turbomachines extends typically from $M = 0.2$ to 2.5. Depending on the Mach number range cascade wind tunnels are split into two groups : low speed and high speed wind tunnels. Low speed wind tunnels are usually open loop continuous running facilities operating at 40-60 m/s. These facilities can provide reliable performance data for compressor and turbine stages as long as the blade velocity does not exceed sonic velocities; i.e., for $M_{1,cr}$ ($M_{2,cr}$) ≤ 0.7 to 0.85 (M_1 - inlet Mach number for compressor blades, M_2 - outlet Mach number for turbine blades). For $M < M_{cr}$ the losses are said to be constant. A typical example for this type of wind tunnels is the NACA low speed cascade tunnel in figure 2 which served to establish the NACA loss correlation for compressors.

High speed cascade wind tunnels may be (a) of the continuous closed loop type or (b) of the intermittent blowdown (exhaust to atmosphere) or suction (from atmosphere) type. A closed loop continuous cascade facility for transonic and moderate supersonic inlet velocity is presented in figure 3a [2]. The upper wall is slotted and equipped with a suction device. This allows the operation in the transonic regime and reduces the shock reflection from the upper wall back into the cascade. The flexible front part of the upper wall allows to transform the inlet duct into a convergent-divergent nozzle providing uniform supersonic upstream flow conditions. The pressure ratio across the cascade is set by a downstream throttling device. So the downstream flow conditions are varied at constant inlet Reynolds number.

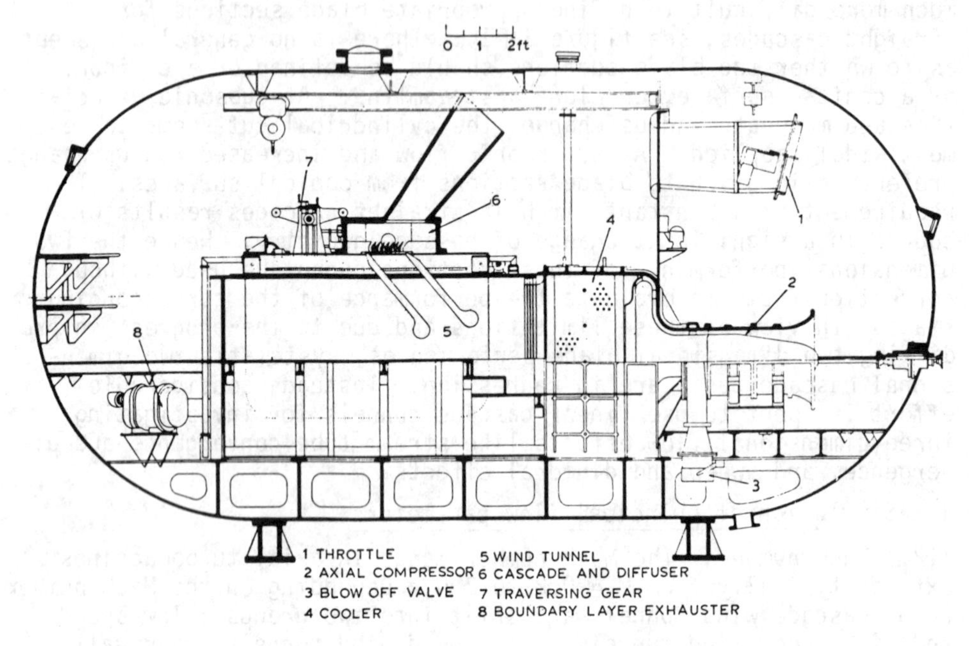

a - GEC variable density tunnel

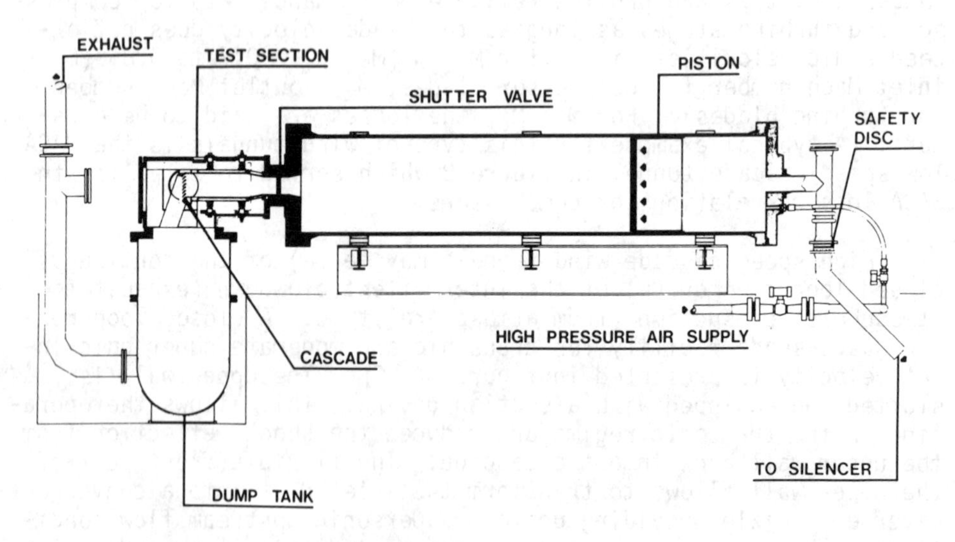

b - VKI light piston tunnel

FIG. 4 - VARIABLE DENSITY CASCADE TUNNELS

Contrary to compressor cascade facilities, turbine cascade tunnels are often intermittent blowdown or suction tunnels and this for two reasons : (a) compared to compressor cascades, the control of the flow conditions (periodicity and two dimensionality) in turbine cascades is much less elaborate and therefore less time consuming; i.e., the total mass flow per test point is much smaller; (b) testing of turbine cascades with outlet Mach numbers up to M_2 = 1.8 and higher is not exceptional (blade sections for last stage of steam turbines). Hence, high pressure air storage reservoirs and vacuum tanks are best suited to cover short duration high pressure ratio needs. A typical example of a blowdown tunnel is the small VKI high speed cascade tunnel in figure 3b. Suction tunnels are in operation at DFVLR-AVA Göttingen and at Mitsubishi. The latter one is a shock tunnel with running times of $\sim$500 milliseconds [4]. A major disadvantage of all these tunnels is the lack of an independent Mach number/Reynolds number variation.

1.2.2 Reynolds number - Turbulence. The Reynolds number range in turbomachines extends from about 2×10^4 to 5×10^7. Most loss correlations indicate that Reynolds number effects become negligible for Re $> 2\times10^5$. This number seems to be on the low side for low turbulence cascade tests of turbine blades with little rear suction side diffusion. Since most turbine cascade tests are performed at Reynolds numbers between 10^5 and 5×10^5 it is likely that a certain number of tests are conducted in a critical Reynolds number range. Variable density tunnels are best suited to overcome this problem. They are usually closed circuit tunnels with continuous operation. An early example of this type of cascade tunnel is the EE (at present GEC) variable density tunnel in figure 4a [3]. This facility was put in operation in 1953. It has been upgraded since to higher pressure ratios (originally it was laid out for pressure ratios up to 2.5) and allows testing at pressure levels between 0.1 and 4 bars with pressure ratios up to 4.

The isentropic light piston tunnel is a newcomer between the variable density tunnels, figure 4b. This short duration wind tunnel operates on the prinicple of isentropic recompression of the test gas by a light weight piston driven by compressed air until the desired total pressure is reached. The flow through the test section is initiated by a fast operating shutter valve, triggered by the output of a fast response pressure transducer measuring the total pressure ahead of the piston. The pressure level in the test section and dump tank on one side and the high pressure cylinder on the other side can be set independently from each other before starting the compression. Originally developed for heat transfer measurements this tunnel can also be used for aerodynamic performance measurements as demonstrated by Oldfield et al [5].

In absence of any artificial turbulence devices most cascade

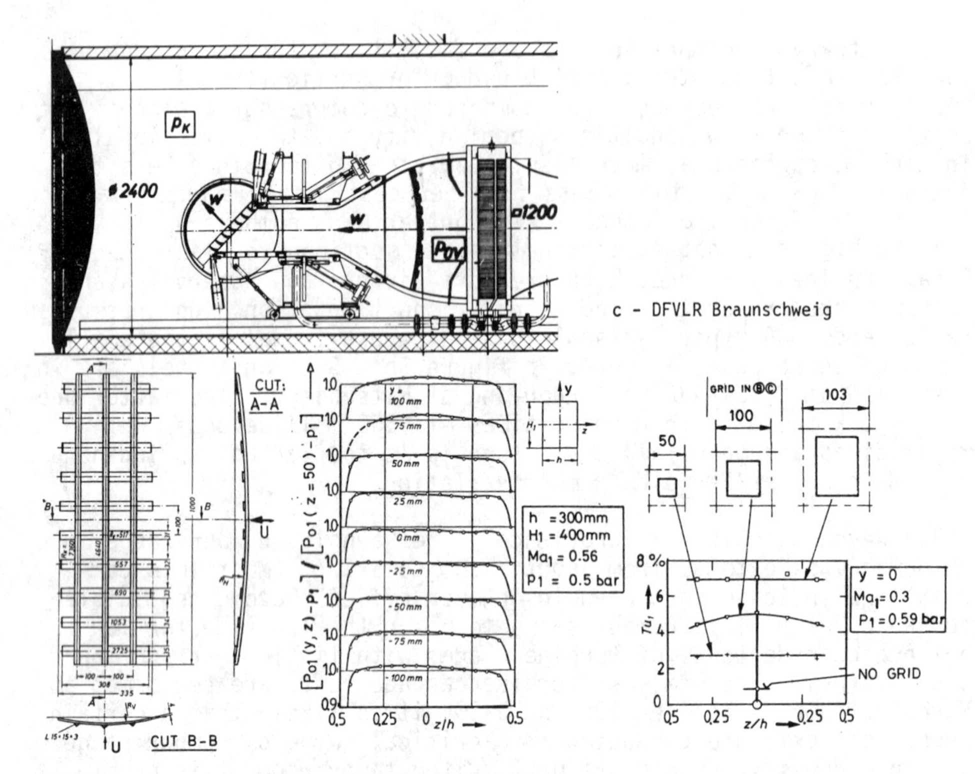

FIG. 4 - VARIABLE DENSITY CASCADE TUNNELS (cont'd)

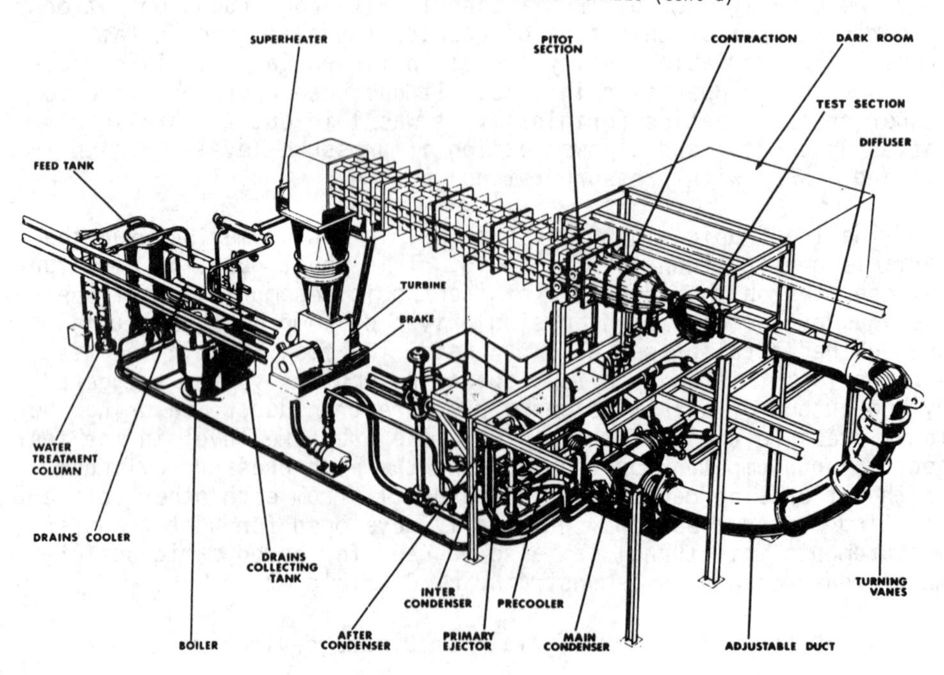

FIG. 5 - CERL WET STEAM TUNNEL

tunnels work with about 1% natural turbulence level. This is far from the turbulence levels in turbines and compressors of Tu = 5 to 15% (RMS values). Using special turbulence grids in the upstream duct of cascade tunnels, maximum turbulence levels of 5% to 6% can be obtained maintaining a homogeneous flow. Figure 4c shows the geometry and the characteristics of some turbulence grids used in the high speed variable density tunnel of the DFVLR Braunschweig [6]. For a given grid, the turbulence level varies with contraction, Mach number and Reynolds number. The turbulence generated with a fixed grid simulates only very poorly the real character of turbulent flows in turbomachines.

1.2.3 Blade cooling. Aerodynamic tests for the investigation of the effect of blade cooling on the blade performance require the correct simulation of both the density and density-velocity ratios between main flow and coolant flow. However, cold flow tunnels in general do not dispose of a sufficient variation of the temperature of the coolant gas. This problem can be overcome by using air-CO_2 mixtures for the coolant flow. This procedure was successfully used at VKI.

1.2.3 Wet steam. Steam turbine blade sections are frequently investigated in air tunnels. This might be sufficient to evaluate the benefits of design modifications on a relative basis. However, the lack of a proper simulation of the specific heat ratio, Reynolds number and wetness fraction is a serious drawback. There are several ways in obtaining the desired wetness fraction : (a) by using water spray jets in the settling chamber [7], (b) by extracting heat from the live steam line through a steam cooler [8], or (c) by using steam which has previously expanded in a steam turbine. The latter method is used in the CERL wet steam tunnel presented in figure 5 [9] for which the test section inlet stage may be varied between approximately 10°C superheat and 5% wetness. Drop sizes produced by the turbine depend on the wetness and cannot be varied independently. Nevertheless, experiments can be carried out in non equilibrium flows using droplets of mean diameter 0.2 to 2 μm by varying the turbine speed and hence wetness.

1.3 Two dimensional cascade flow

1.3.1 Periodicity. Depending on the velocity range it is more or less difficult to establish periodic inlet conditions. A trivial case is a cascade with axial sonic inlet conditions, figure 6a [10]. The leading edge shocks are parallel to the leading edge and the inlet flow is not only periodic but also uniform. The case of a high supersonic flow with a subsonic axial component is described in figure 6b [11]. Periodic conditions exist only within the triangle J.K.L. At low supersonic and transonic inlet Mach numbers, porous upper walls and suction are used to reduce the reflection of the leading edge shocks back into the cascade (Fig. 3a).

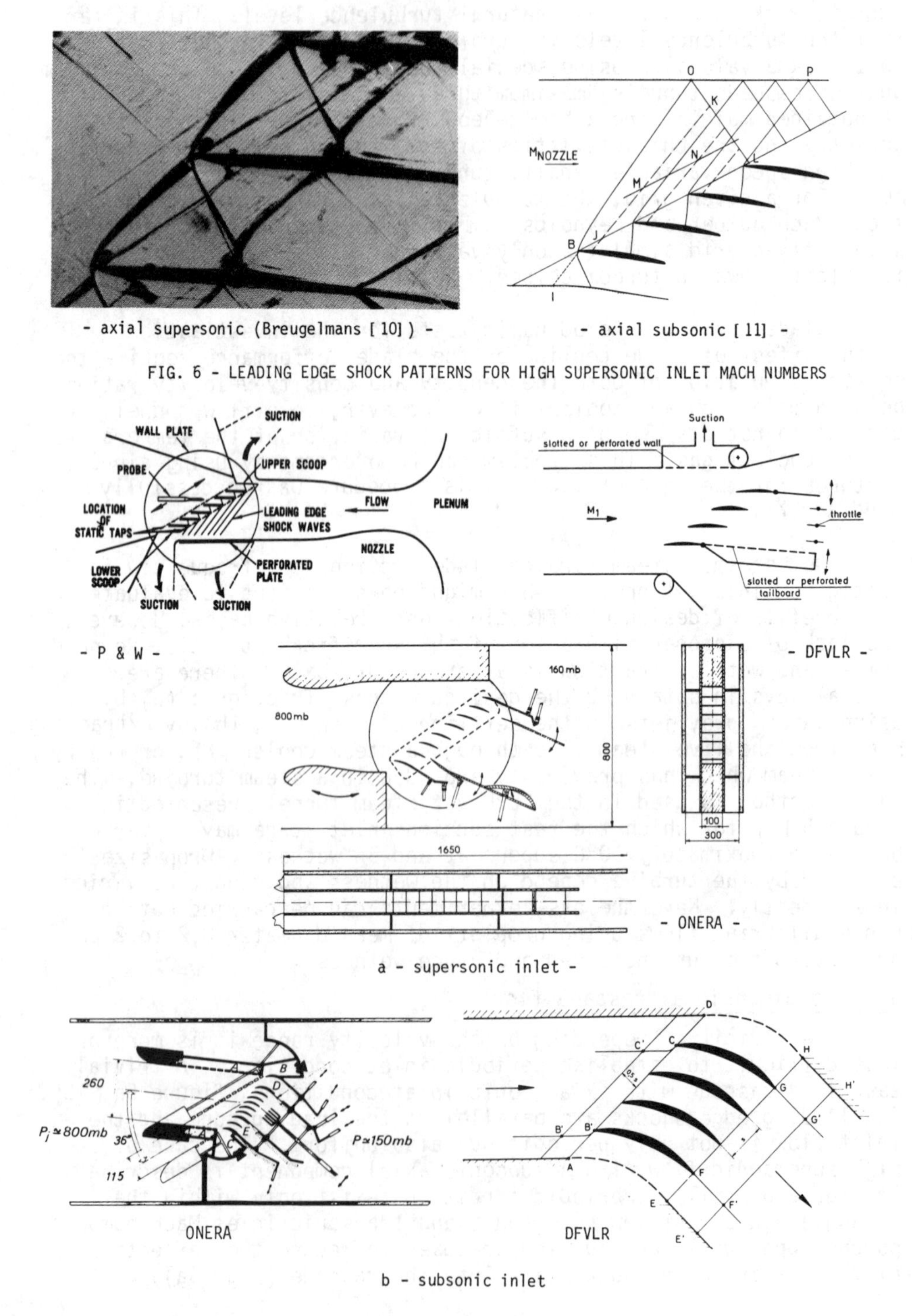

- axial supersonic (Breugelmans [10]) -

- axial subsonic [11] -

FIG. 6 - LEADING EDGE SHOCK PATTERNS FOR HIGH SUPERSONIC INLET MACH NUMBERS

- P & W -

- DFVLR -

- ONERA -

a - supersonic inlet -

ONERA

DFVLR

b - subsonic inlet

FIG. 7 - COMPRESSOR CASCADE FLOW CONTROL SYSTEMS

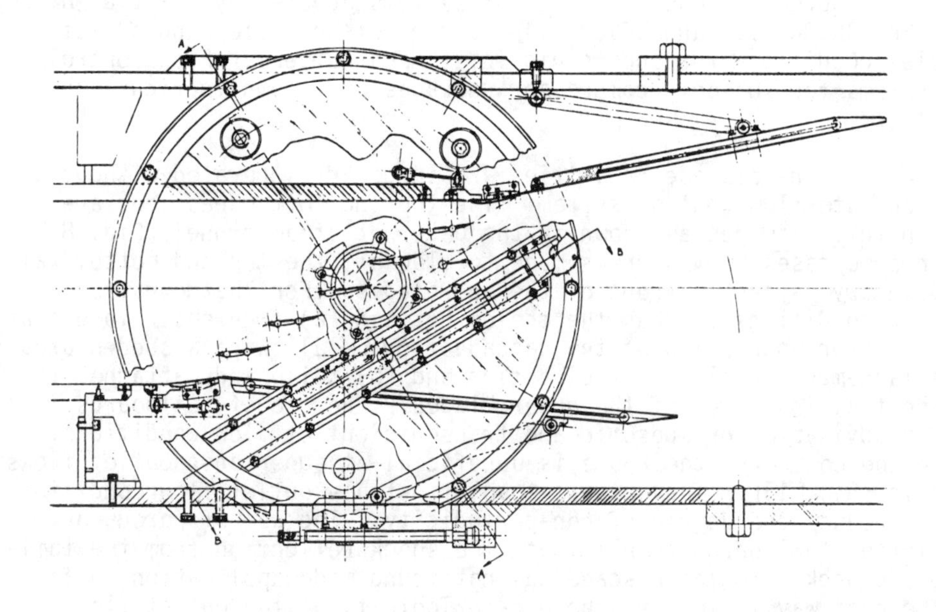

FIG. 8 - TEST SECTION OF CERL WET STEAM CASCADE TUNNEL

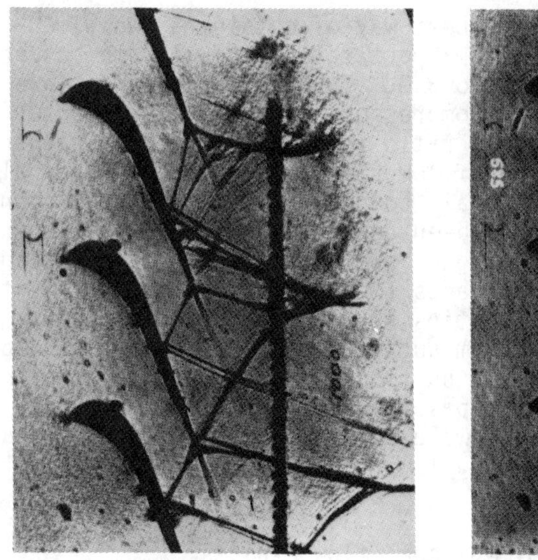

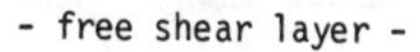

- free shear layer - - tailboard -

FIG. 9 - SUPERSONIC CASCADE OUTLET FLOW FIELD WITH FREE LAYER OR TAILBOARD AT CASCADE EXTREMITY. $M_2 \cong 1.2$ (Sieverding [15])

Figure 7a shows a variety of systems adopted by Pratt & Whitney [12], ONERA [13] and DFVLR [14] to control the inlet and outlet flow conditions for supersonic inlet Mach numbers. Flow control systems for subsonic compressor cascades are shown in figure 7b [11,13].

Turbine cascade facilities are in general operated without any elaborate flow control system. The top and bottom passages are often full passages as shown in the CERL wet steam tunnel (Fig. 8)[9]. In some cases provision is made to blow off the top and bottom wall boundary layers in front of the end blades. For the downstream flow conditions one has the choice between (a) free shear layers at the outer boundaries of the cascade flow by allowing a sudden area enlargement in pitchwise direction and (b) tailboards attached to the trailing edges of the outer blades. The use of tailboards is not advisable for subsonic and transonic outlet flow conditions. On the contrary, their use is beneficial at supersonic outlet flows (Fig. 9), [15]. They assure absolute stable outlet flow conditions and reduce the degree of non-periodicity. The use of porous or slotted tailboards to eliminate the shock reflection from the tailboard back into the cascade has not found wide application so far. The best way to improve the non-periodicity of the outlet flow is a high number of blades, such that disturbances introduced from the cascade extremities have time to attenuate before they reach the blades to be measured.

1.3.2 Two dimensionality. The best way of obtaining two dimensional flow conditions is a high aspect ratio. Usually the aspect ratio varies between $h/C = 1.0$ to 3.0. Turbine cascades get away with smaller aspect ratios than compressor cascades. For transonic turbine cascades the significant ratio is height to throat rather than height to chord. Testing of compressor cascades requires always a close control of the endwall boundary layer. Several boundary layer suction systems have been developed. The endwall boundary layer may be sucked off through slots upstream of the cascade and porous walls across the cascades (see Fig. 2). This allows to obtain axial velocity-density ratios ADVR=1 based on mid-span measurements. However, a uniform suction across the blade passage is not ideal since the amount of suction varies locally due to the pressure variation in the blade passage. At ONERA satisfactory two dimensional flow conditions were obtained by removing the endwall boundary through slots along isopressure lines within the blade passage (Fig. 10) [16].

1.4 Investigation of three dimensional flow aspects

The stream lines in the meridional plane of a compressor form in general convergent stream tubes. Attempts have been made in recent years to simulate this situation in straight cascade

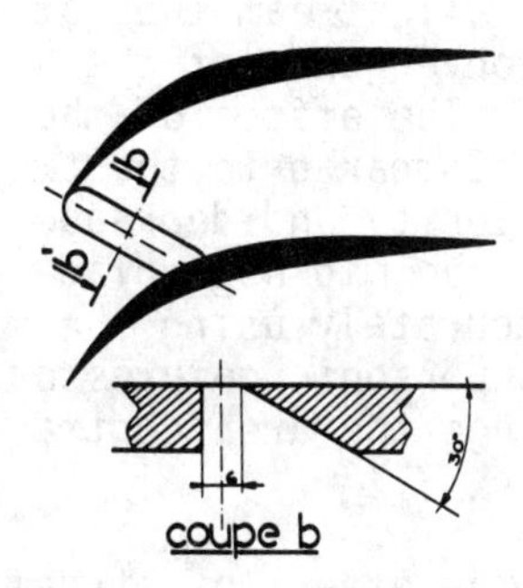

FIG. 10 - ENDWALL BOUNDARY LAYER CONTROL USED AT ONERA [17]

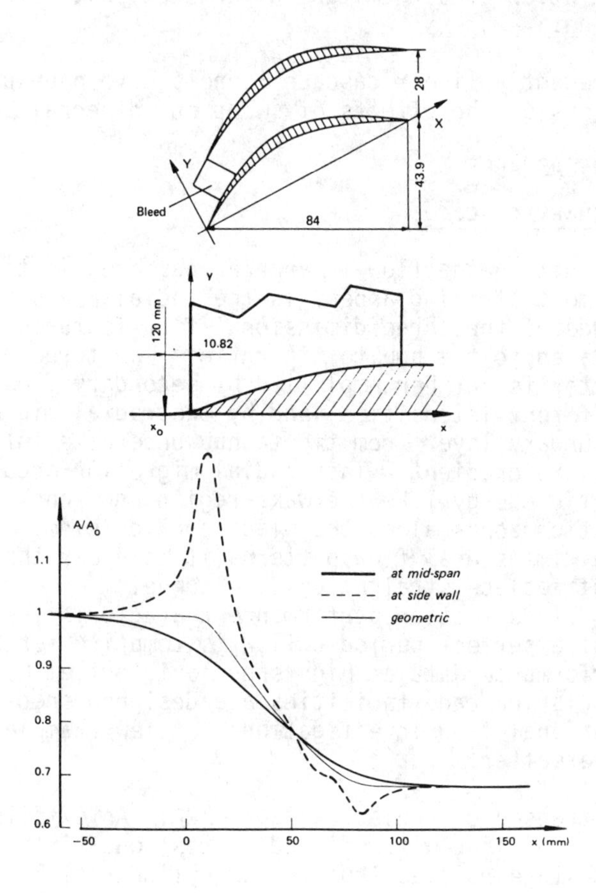

FIG. 11 - EVOLUTION OF MID SPAN STREAM SHEET THICKNESS IN A CASCADE WITH CONVERGING SIDE WALLS (Meauze [17])

facilities by appropriate shaping of the side walls. The evolution of the convergence of a small stream tube at mid span is, however, different from the area convergence of the side walls as shown by Meauze in figure 10 [17]. The effect of the convergence extends upstream of the cascade. Compared to the two dimensional case, it increases the absolute velocity and decreases the angle of attack. The author has shown that the mid span blade pressure distributions can be predicted quite accurately using the mid span convergence calculated beforehand. Supersonic compressor cascade tests with movable side walls to change the area contraction are described by Schreiber [18].

The significance of the effect of stream tube divergence on the blade performance of high turning low reaction blades was demonstrated by Forster [19], but a systematic investigation of the proper simulation of stream tube divergence in turbine cascades is still lacking.

More recently linear cascade tunnels have been used for basic investigations of the effects of sweep and dihedral [20,21].

2. ANNULAR CASCADES

2.1 Stationary cascades

Be definition the flow in annular cascades is three dimensional and its most striking aspect is the radial pressure gradient. The magnitude of the three dimensional flow features are of course closely related to the hub to tip ratio. The three dimensional flow character is further amplified by secondary flows at hub and tip (of different intensities) and by the radial migration of low momentum boundary layer from tip to hub under the influence of the radial pressure gradient. This radial migration occurs in regions of low kinetic energy like the wake region and shock induced boundary separation zones along the blade suction side. In view of all these three dimensional flow patterns it is clear that annular cascades cannot replace straight cascade tunnels for obtaining two dimensional blade section performance characteristics, although the advantage of a perfect periodic flow is tempting for obtaining blade section performance data at high supersonic outlet Mach numbers. In general annular cascade facilities are designed specifically for three dimensional flow investigations. A few examples will be described hereafter.

2.1.1 MEL transonic annular cascade [22]. An existing supersonic straight cascade loop was modified to test one-sixth scale models of the last stage nozzles from the LP cylinder of 500 MW turbines. The annular test section in figure 12 from [22] shows a blade row with a hub diameter of 0.306 m and a tip diameter varying from 0.506m at the leading edge to 0.59 at the trailing edge. The flow enters

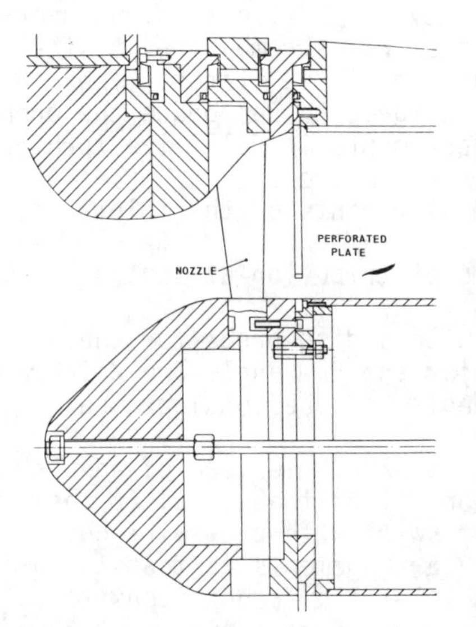

FIG. 12 - MEL TRANSONIC ANNULAR CASCADE FOR TESTING LAST STAGE STEAM TURBINE DIAPHRAGM BLADES

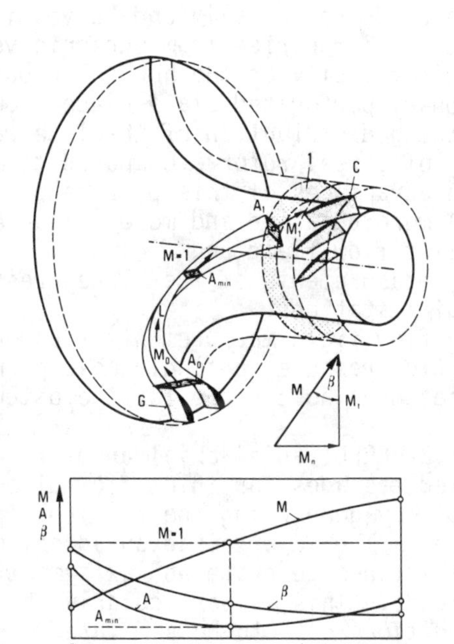

G radial preswirl vanes
1 axial test section
C test cascade

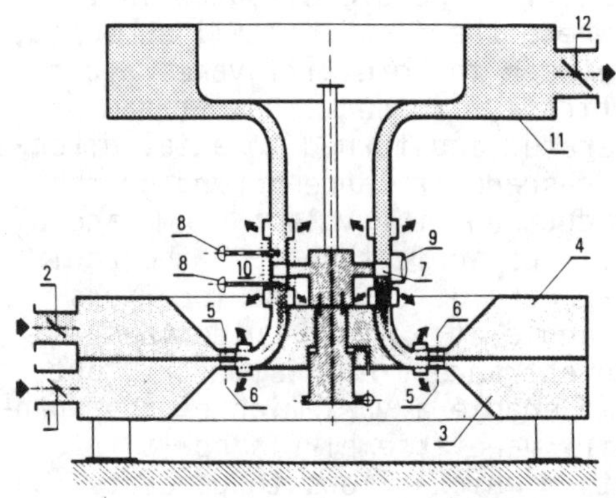

1,2 inlet valves
3,4 outer and inner settling chambers
5,6 outer and inner preswirl vanes
7 aerodynamic probes up- and downstream of the test cascade
9 cylindrical optic for Schlieren visualisation
10 static wall pressure tappings
11 outlet valve
suction

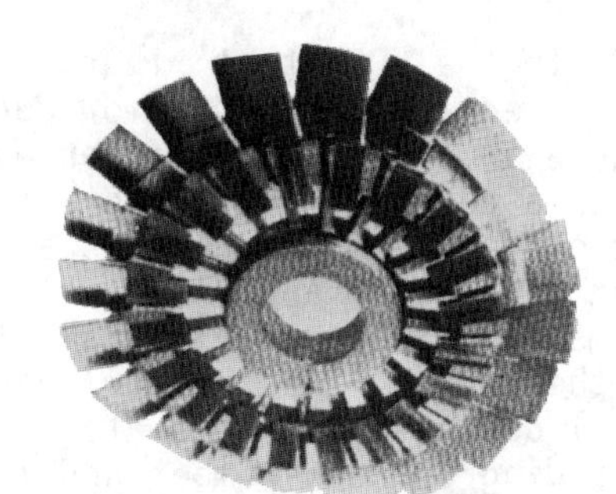

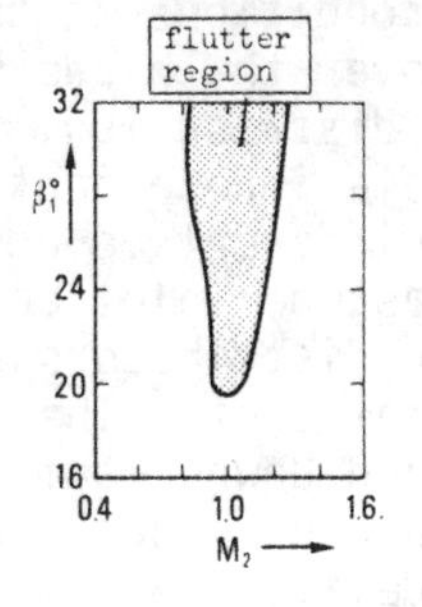

FIG. 13 - EPFL RADIAL-AXIAL ANNULAR CASCADE

the blade row axially and leaves it with a high swirl. The outlet Mach number varies from subsonic velocity at the tip to high supersonic velocity at the hub. The outlet swirl is removed by a stationary perforated plate placed downstream of the stator. The opening distribution of theplate was designed to reproduce the opening of a real rotor. Compared to measurements in full size turbines, the advantages of this purpose built cascade are :
(1) more detailed and more reliable measurements of the inlet and outlet flow field and
(2) measurements of the blade pressure distribution at several radial stations.
The flow field measurements are done with a disc probe for the static pressure, a single hole probe for the yaw angle and a cobra type three hole probe for the pitch angle and the total pressure.

2.1.2 EPFL radial-axial annular cascade [23]. Testing of annular fixed cascades for inlet flow directions other than axial is possible by generating the required inlet swirl with pre-swirl guide vanes. The major disadvantage of such arrangements is that the preswirl vanes generate non-axisymmetric inlet flow condisitons to the cascade. This is particularly the case when supersonic inlet flow conditions have to be simulated. Not only the cascade inlet flow will be non axisymmetric but the trailing edge shocks of the preswirl vanes will interfere with the blades of the following cascade [24].

In the EPFL test rig the preswirl vanes are situated in the radial entrance of the flow duct where the flow is still subsonic, figure 13. Shock interferences between the preswirl vanes and the test cascade are therewith automatically avoided. Downstream of the guide vanes the flow is accelerated and turned to axial direction. Velocity and flow angle at cascade entrance depend on the geometry of the radial-axial flow duct and the values of M_0 and β_0 at the exit of the preswirl vanes. The sonic throat in the radial-axial duct is self generated and appears at different locations depending on Mach number and flow angle at exit of guide vanes. The flow conditions at the cascade entrance are axisymmetric but variable over the blade height. To enable a variation of the spanwise flow distribution, the preswirl vane assembly is split into two separate independent distributors which allows to generate two flows of different velocities and angles. The spanwise distribution at cascade inlet depends on the initial conditions at the exit of the two distributors and the mixing between the two flows in the radial-axial duct. The flow can be further influenced by suction upstream and downstream of the cascade. The air supply system consists of a continuously running compressor with a maximum power of 2500 kW, delivering a maximum mass flow rate of 10 kg/s at a maximum pressure ratio 4.2. The facility serves both for steady and unsteady flow investigations. A typical example for the research

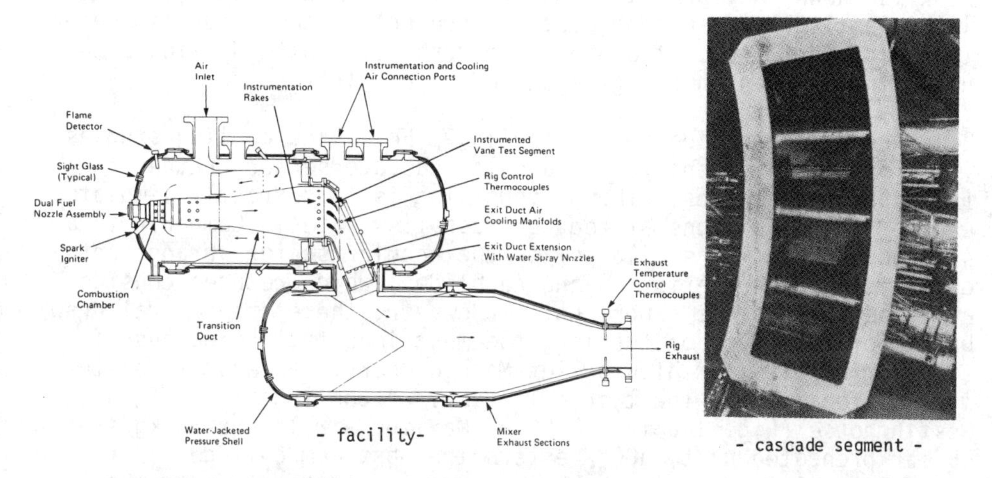

FIG. 14 - WESTINGHOUSE COOLING RESEARCH CASCADE FACILITY FOR INVESTIGATION OF FULL SCALE FIRST STAGE STATOR SEGMENTS

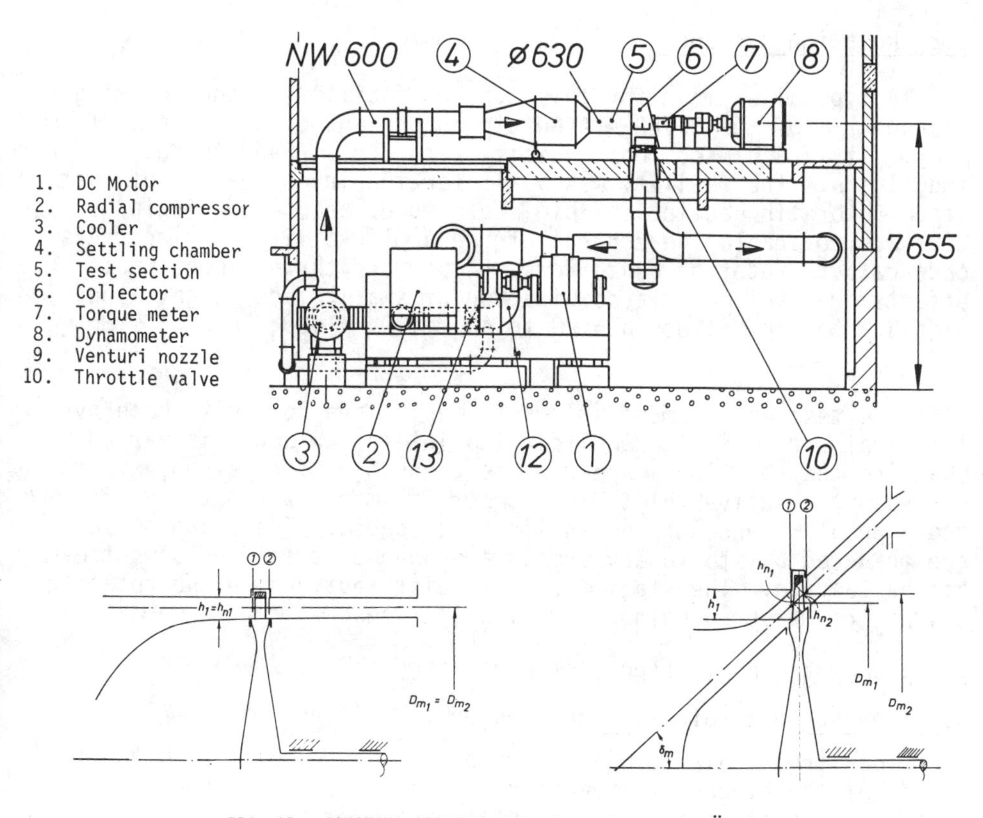

FIG. 15 - ROTATING CASCADE FACILITY - DFVLR AVA GÖTTINGEN

work is shown in figure 13. It presents an annular turbine cascade with elastically suspended blades representing the rotor tip section of a last stage of a big steam turbine. Unsteady flow behaviour in the transonic regime caused blade flutter.

2.1.3 Annular cascade sector testing. For blade cooling studies it is often convenient to use annular cascade sectors rather than full annular cascades. The obvious wish is to correctly simulate engine flow conditions at reduced mass flow rates. Compared to a straight cascade of similar size, the annular sector arrangement enables the use of existing engine blades, which reduces considerably the cost of the experiments. However, there are several drawbacks for this type of testing, the most important being one of flow periodicity for high outlet Mach numbers. An example of this type of facility is the twin-cylinder vane cooling test rig at Westinghouse (Fig. 14a from [25]). Maximum airflow of 38 kg/s at 15 bar preheated up to 480°K enters the upstream cylinder at the air inlet. A combustion chamber raises the temperature to the required inlet temperature. Figure 14b shows a view on the instrumented three-vane, four passage test segment assembly.

2.2 Rotating cascades

A typical example for this type of facility is the rotating cascade rig of the DFVLR-AVA Göttingen presented in figure 15 from [26]. It is a closed loop facility with dry air and freon as working fluids. The facility was built for the investigation of last stage rotor tip sections of big steam turbines. A particular objective was to enable the control of the applicability of linear cascade data to rotating blade rows. Important three dimensional effects are avoided by a combination of small blade heights and high hub to tip ratios ($h = 20$ to 40 mm, $D_M = 520$; $R_H/R_T = 0.85$ to 0.9).

The test arrangement allows to test blade rows with both cylindrical and conical endwalls. The inlet flow angle is varied by changing the rotational speed. The transition from a subsonic to supersonic relative inlet Mach number is smooth without any of those problems encountered in linear cascades. The blade losses are measured by stationary pressure probes upstream and downstream of the cascade. The static pressure distributions on the rotating blades are measured using mercury slip rings.

3. SINGLE AND MULTISTAGE TEST FACILITIES

3.1 Low speed turbine and compressor facilities

The large scale turbine stage of GE in figure 16 is without any doubt the biggest turbomachinery testing facility ever built. The main characteristics of the stage are summarized below :

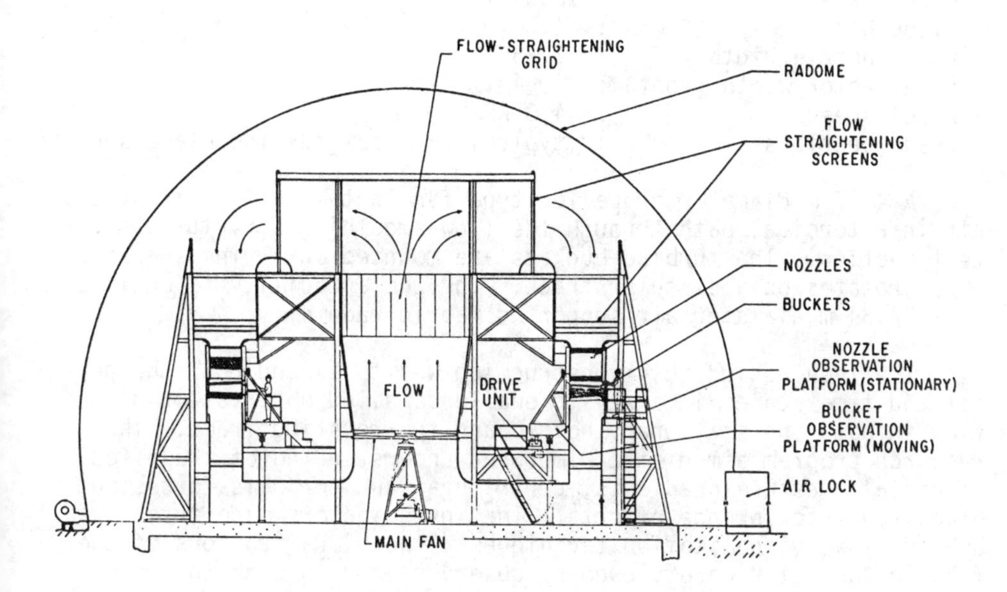

FIG. 16 - GE LARGE SCALE LOW SPEED TURBINE FLOW VISUALIZATION FACILITY

FIG. 17 - ONERA TRANSPARENT LOW SPEED AXIAL COMPRESSOR STAGE

- mean diameter : 15.04 m
- blade height : 1.36 m
- axial nozzle width : 1.25
- axial rotor width (root) : 0.64
- wheel speed : 4.3 RPM
- blade sections : typical IP steam turbine blade sections

A 4.27 m diameter propeller type fan in the center circulates air in a toroidal path through the flow-smoothing structures and test section. The turbine buckets are mounted on a ring shaped car which rotates on a circular track. The whole facility is enclosed in a 27.34 m diameter air supported fabric radome.

The idea behind this construction was to expand both the physical and time scales in a way to enable detailed observation of the flow in both the stationary nozzle and the rotating frame. The research program aimed at an improved understanding of the effects of radial pressure gradients, centrifugal and coriolis forces and blade row interference effects. The investigators made extensive use of smoke visualization techniques. The visualizations of the flow in the rotor were viewed by observers moving with the rotor.

Smoke visualization techniques have always greatly contributed to a better understanding of complex two- and three dimensional flow structures. Figure 17 shows an ONERA low speed flow visualization test rig (350 RPM; 500 m Ø) for the study of the unsteady flow field in axial compressors with emphasis on rotating stall. The smoke patterns at various regimes were recorded on a movie film [28].

Of much smaller dimensions than the GE large scale turbine facility but nevertheless of very large size is the large low speed single stage experimental compressor in the Whittle Laboratory of Cambridge University, figure 18 [29]. The tip and hub diameters of the free vortex stage are, respectively, 1520 mm and 610 mm. The chord length of the 24blade rotor blade row is 152 mm, while the 15 blade stator blade row has a chord length of 305 mm. The exceptionally long chord length allows detailed measurements of the development of both the time-mean and the instantaneous-mean blade boundary layer velocity profile along the chord (the term "instantaneous-mean" implies that all random velocity fluctuations - turbulence - are averaged out, while any ordered unsteadiness in the velocity is preserved). The purpose of these measurements is to control the applicability of straight cascade tests to an actual turbomachine blade.

A facility of similar size (tip diameter of 1520 mm as in the previous case) for testing both turbine and compressor stages is used at UTRC (United Technology Research Center) [30]. The turbine model has maximum three blade rows : inlet guide vane, rotor, second vane. Tip and hub diameters are constant for all blade rows.

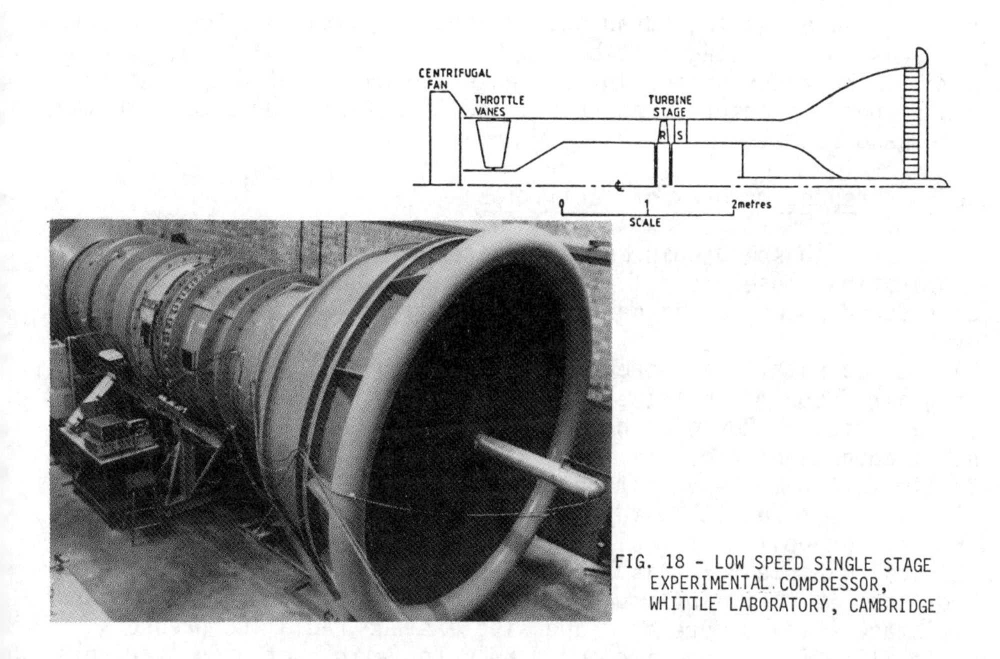

FIG. 18 - LOW SPEED SINGLE STAGE EXPERIMENTAL COMPRESSOR, WHITTLE LABORATORY, CAMBRIDGE

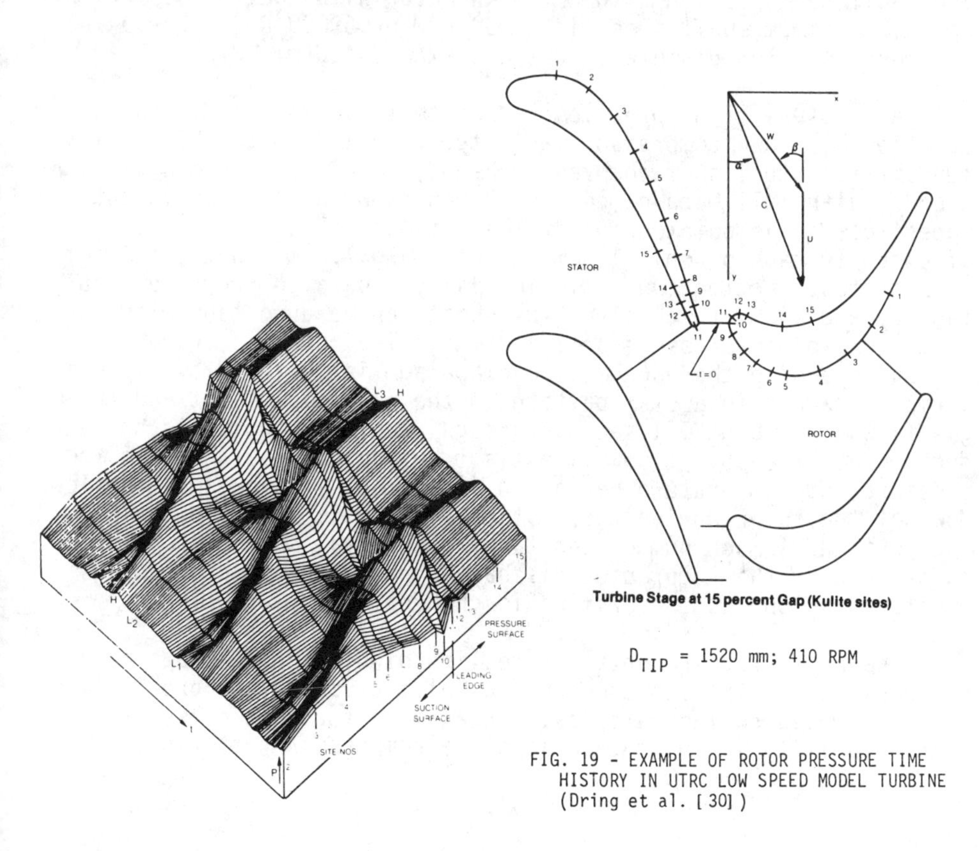

Turbine Stage at 15 percent Gap (Kulite sites)

D_{TIP} = 1520 mm; 410 RPM

FIG. 19 - EXAMPLE OF ROTOR PRESSURE TIME HISTORY IN UTRC LOW SPEED MODEL TURBINE (Dring et al. [30])

The large size has the advantage of simulating the Reynolds numbers in Hp-gas turbine stages at only 410 RPM and an axial velocity of 23 m/s. The rotor stator interference is investigated by means of high response pressure transducers and thin film gauges on both the stator and rotor.

3.2 High speed compressor facilities

Most high speed compressor research work is done in continuously operating closed loop facilities. The advantages are obvious :
(a) reduced power requirements by lowering the overall pressure level;
(b) reduced rotational speed for given peripheral Mach number by using air-freon mixtures;
(c) variation of Reynolds number.
The disadvantages are :
(a) limitation in Re-variation;
(b) model test results obtained in air-freon mixtures cannot directly be applied to real machines.
A typical example for this type of facility is the VKI high speed compressor test rig in figure 20 which allows testing of single and multistage axial compressors and single stage radial compressors. The facility is driven with a 700 kW motor with continuous variable speed. Maximum shaft speed is 25000 RPM or 69000 RPM. For axial compressors rotor diameters must not exceed 400 mm [31].

An unusual high speed compressor test rig is the MIT short duration blowdown compressor facility, figure 21 [32] which allows fundamental studies of compressor aerodynamics, both steady and unsteady, with full Mach number and Reynolds number simulation at modest costs, as pointed out by the authors. The facility consists of a supply tank of only 1.5 m diameter and 1.5 m length, the compressor test section, which is initially separated from the supply tank by a 0.64 mm thick aluminium sheet, and a dump tank with an initial absolute pressure of 1.4 N/mm^2. The rotor is brought to nominal speed in the vacuum, the diaphragm is ruptured by explosive strips attached in a star pattern to the aluminium sheet and the gas flows for about 0.1 second through the rotor, which is driven during this time by its own inertia. By a proper matching of the pressure and temperature variation in the supply tank and the slow down of the rotor during the test period, it is possible to keep the peripheral Mach number constant. The axial Mach number is controlled by the compressor dischage throttle plate and remains constant as long as the orifice is choked.

The power requirements for testing high mass flow rate axial compressors in air with atmospheric inlet pressure are prohibitive for most research laboratories. An example for this class of research facility is the NASA multistage compressor test rig in

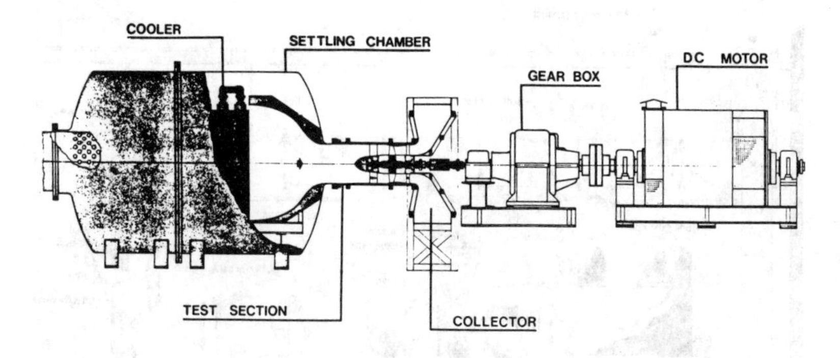

FIG. 20 - VKI HIGH SPEED CLOSED LOOP COMPRESSOR TEST RIG

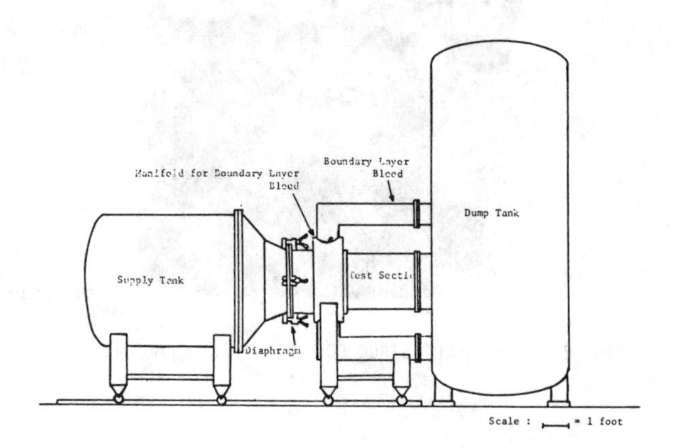

Scale drawing of blowdown compressor facility, sized for 23.25-in-dia rotor

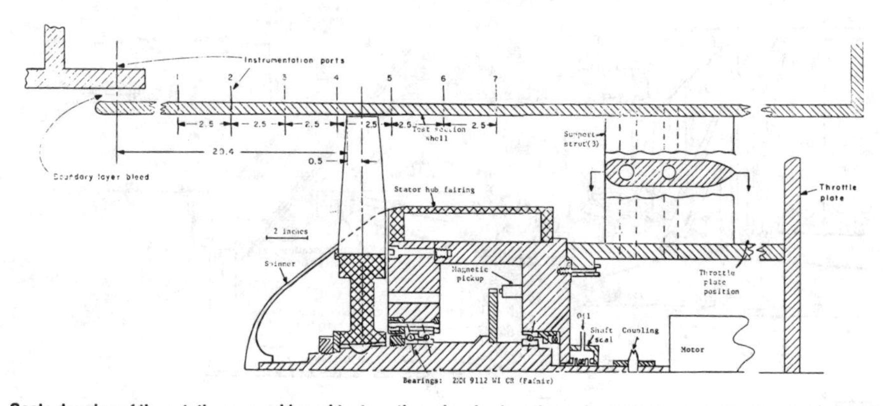

Scale drawing of the rotating assembly and test section, showing locations of rotor instrumentation ports, and throttle plate

FIG. 21 - MIT BLOWDOWN COMPRESSOR FACILITY

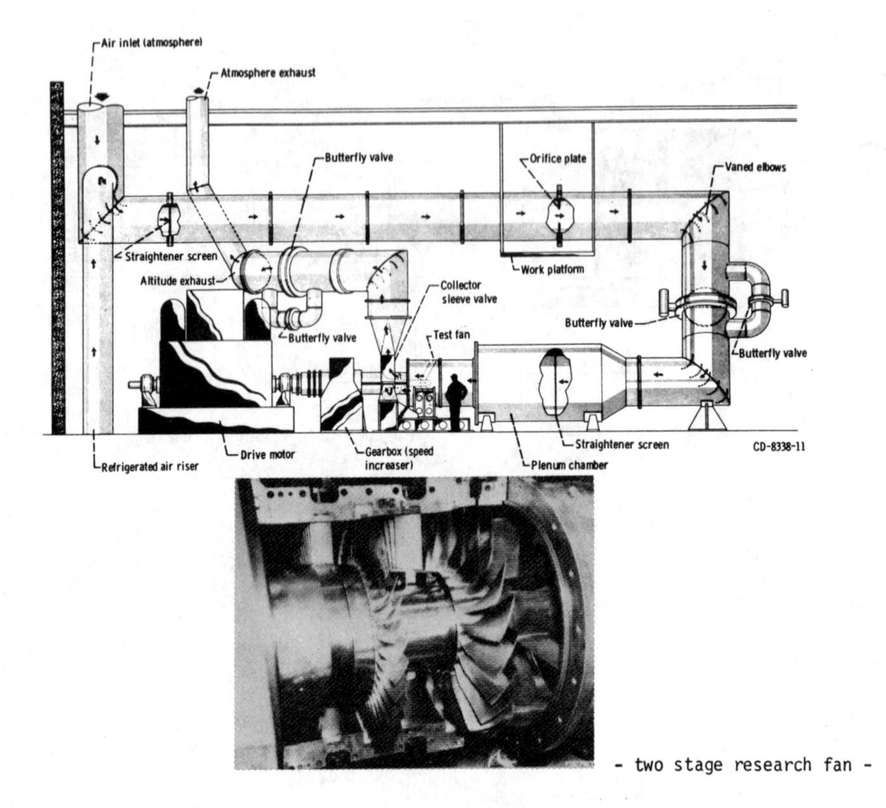

- two stage research fan -

FIG. 22 - NACA MULTISTAGE COMPRESSOR TEST FACILITY

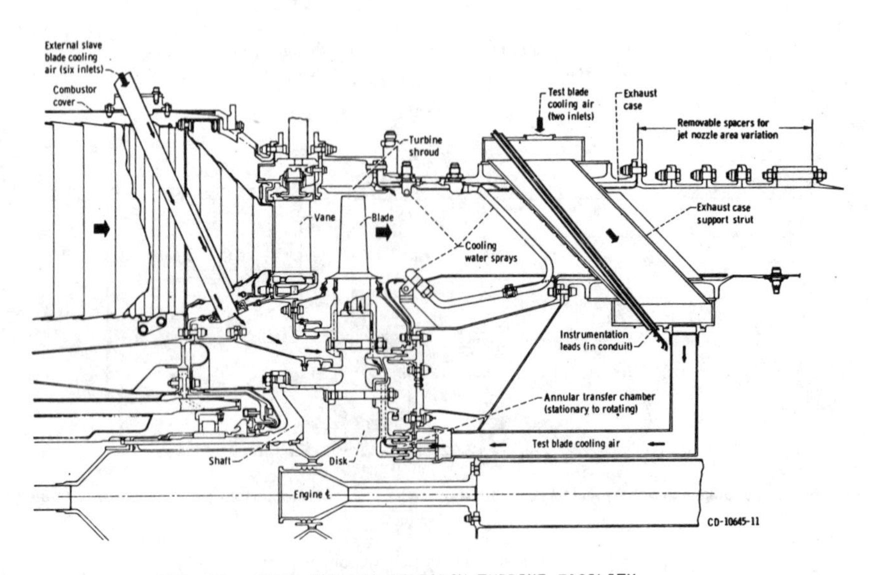

FIG. 23 - NACA COOLING RESEARCH TURBINE FACILITY

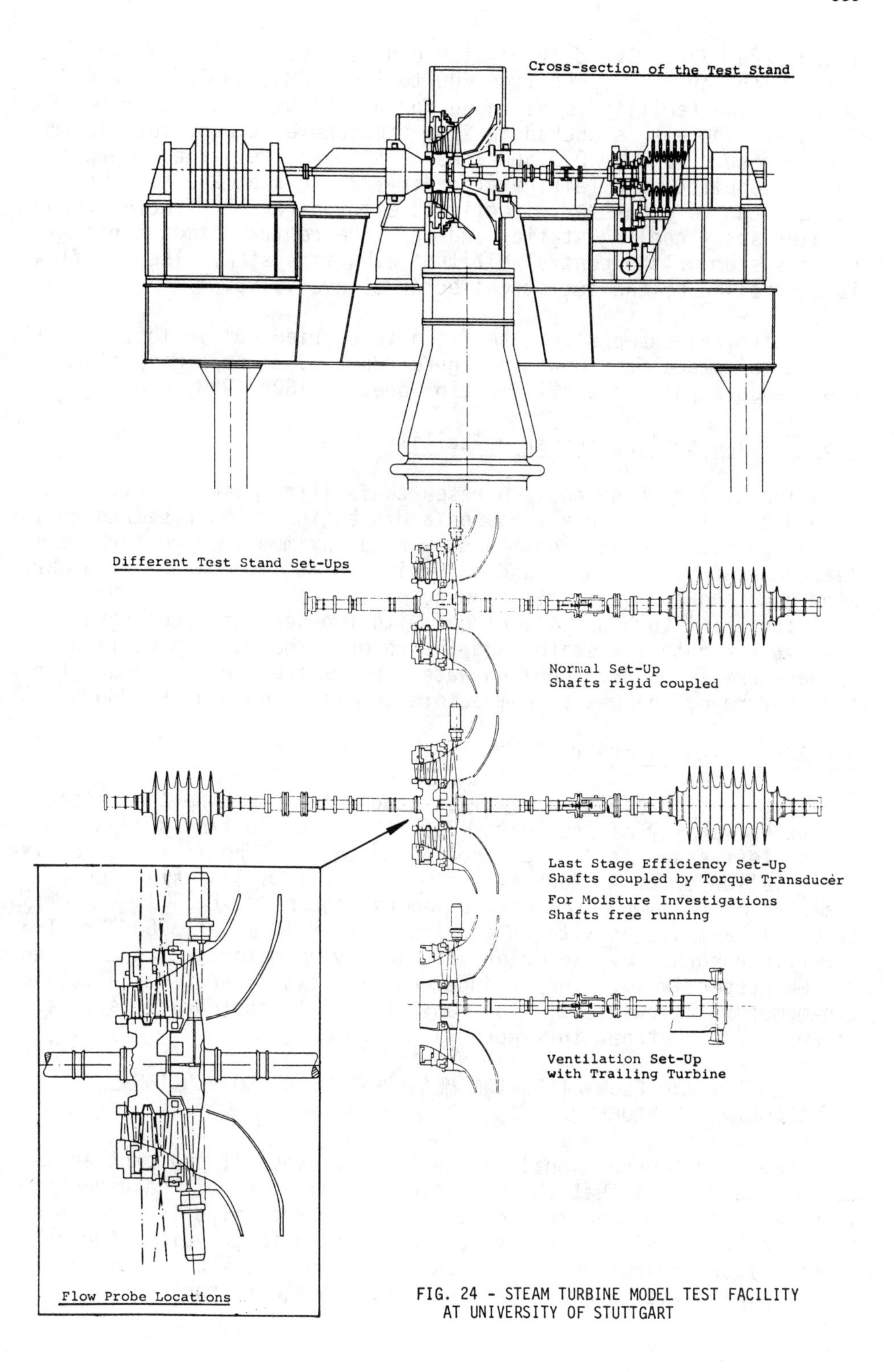

FIG. 24 - STEAM TURBINE MODEL TEST FACILITY AT UNIVERSITY OF STUTTGART

figure 22a [33]. The drive system consists of an 11.2 MW electric motor with variable speed from 400 to 3600 RPM, coupled to a 5.21:1 gearbox. The facility is designed for a maximum mass flow rate of 45 kg/s. The air is sucked in from atmosphere at a preset plenum tank pressure between 0.2 and 1 bar. Alternatively the compressor can be supplied with refrigerated air : 25 kg/s at 243°K. The outlet air can be exhausted directly to atmosphere, or directed through a water spray cooling station and into the central atmospheric exhaust system or the central altitude exhaust system. The air flow is controlled by the outlet collector sleeve valve.

A typical example for the research carried out in this facility is the two-stage fan shown in figure 22b : 33.25 kg/s mass flow, 2.4 pressure ratio and 427 m/s tip speed at 16042 RPM.

3.3 Turbine cooling research facility.

The NASA turbine cooling research facility [34] is built around the high pressure spool of a J75 engine. The research engine is designed to operate continuously at a maximum average turbine inlet temperature of 1644°K and a turbine inlet temperature of 3 bar. Figure 23 shows a schematic drawing of the hot section of this facility. The turbine is equipped with two separate cooling-air systems for both the stator and the rotor. The maximum cooling air temperature is 921°K. Cooling water sprays are used to reduce the temperature of the gas stream before entering the exhaust duct.

3.4 Model steam turbine.

Figure 24 presents a new last stage steam turbine test stand at the University of Stuttgart [35]. Steam of 70 t/h is supplied to the test stand at 2-6 bar and 130 to 250°C. The test rig permits installation of up to four stages with a maximum last stage outer diameter of 1000 mm. The maximum operating speed of 18000 RPM allows to model last stages with an annulus area of 18 m^2 at 3000 RPM. The power is absorbed by two water-friction dynamometers rated at 5 and 3.5 MW respectively. The arrangement with two separated shafts and dynamometers allows testing of many different configurations some of which are sketched in figure 24.

4. VELOCITY AND FLOW DIRECTION MEASUREMENTS BY CONVENTIONAL AERODYNAMIC PROBES

The word "conventional" has a light flavour of contempt and could make believe that the execution of this type of measuements is straightforward and free of any particular difficulty. However, this does certainly not apply to measurements in a typical turbomachine flow environment. The combination of small blade dimensions, small axial spacing between blade rows, transonic and supersonic velocities and strong gradient fields makes it very difficult indeed to obtain correct flow data. This is perhaps best illustrated by the fact that since 1969 every other year informal specialist meetings are organized in Europe to discuss about measuring techniques in transonic and supersonic flows in cascade and turbomachines.

The literature on velocity measurements is abundant and the reader is referred to surveys by Hill et al. [36], Erwin [37], Chue [38] and Wuest [39]. The present chapter deals in particular with the specific problems encountered in performance measurements of turbomachinery bladings. The material presented here is based to a great extent on AGARDograph 207 on "Modern Methods of Testing Rotating Components of Turbomachines" [40] and on the proceedings of the meetings on "Measuring Techniques in Transonic and Supersonic Flow in Cascades and Turbomachines"

4.1 General relations

The velocity of a flowing stream is most commonly determined by measuring the difference of total and static pressure. For incompressible flow the relation is given by the Bernoulli equation

$$V = \sqrt{\frac{2(P_0 - P_S)}{\rho}} \tag{4.1}$$

For compressible flow the magnitude of the velocity is expressed in general by the Mach numbers $M = V/a$ or $M^\star = v/a^\star$. The relation between the Mach number M and the total and static pressure is given by the equation

$$M^2 = \frac{2}{\kappa - 1}\left[\left(\frac{P_0}{P_S}\right)^{\frac{\kappa-1}{\kappa}} - 1\right] \tag{4.2}$$

In supersonic flow it is difficult (but not impossible) to measure directly the true total pressure. In most cases a bow shock stands in front of the total pressure probe and the probe head senses the total pressure P_{pit} behind a normal shock, the strength of which depends on the Mach number. The Mach number in the undisturbed flow field, i.e., upstream of the bow shock is then calculated from the static pressure P_S upstream of the bow wave and P_{pit} by the relation :

$$\frac{P_S}{P_{pit}} = \left[\frac{2\kappa M^2-(\kappa-1)}{(\kappa+1)\left(\frac{\kappa+1}{2}\cdot M^2\right)^{\kappa}}\right]^{\frac{1}{\kappa-1}} \tag{4.3}$$

To assure that the bow wave has indeed the characteristics of a normal shock across the mouth of the pitot tube, the outer to inner diameter ratio of the pitot tube should be bigger than 2 [36].

The shock strength is a function of Mach number and therefore the static pressure P_S has to be known to evaluate the true total pressure $P_0 = P_{pit}+\Delta P_{0,shock}$. The influence of an error in $P_{S,2}$, expressed as $\Delta P_S/(P_{02}-P_{S2})$, on the cascade efficiency η defined as

$$\eta = \frac{V_2^2}{V_{2,is}^2} = \frac{1-\left(\frac{P_{S2}}{P_{02}}\right)^{\frac{\kappa-1}{\kappa}}}{1-\left(\frac{P_{S2}}{P_{01}}\right)^{\frac{\kappa-1}{\kappa}}} \tag{4.4}$$

is shown in figure 15. The figure demonstrates clearly that :
(a) the accurate measurements of the static pressure in supersonic flow is as important as the correct measurement of P_{pit} and
(b) the error in η increases rapidly with Mach numbers.

4.2 Measurements in flows with strong pressure gradients

Figure 26 shows total, static and differential (ΔP_{LR} for flow angle) pressure traces behind a turbine cascade at subsonic, M_2 = 0.8, and supersonic, M_2 = 1.5, outlet Mach numbers. Apart from measuring errors which can directly be attributed to errors in the various pressure readings, the most frequent errors are due to the fact that the total, static and directional pressures are measured in different flow field points with non identical flow conditions.

In subsonic flow the only important pressure gradient exists in the wake where the total pressure undergoes large variations. The influence of the velocity gradient on the pitot probe reading may be taken into account by displacing the true location of measurement by δ = 0.15D, (D - probe stem diameter), McMillan [41]. However, a more severe error may occur for the measurement of the local wake flow angle. Probes of type A in figure 27 measure large sinusoidal variations of the flow angle because of the distance between the directional holes in the traverse plane. Probes of type B with all sensing holes in a plane normal to the traverse plane avoid this problem.

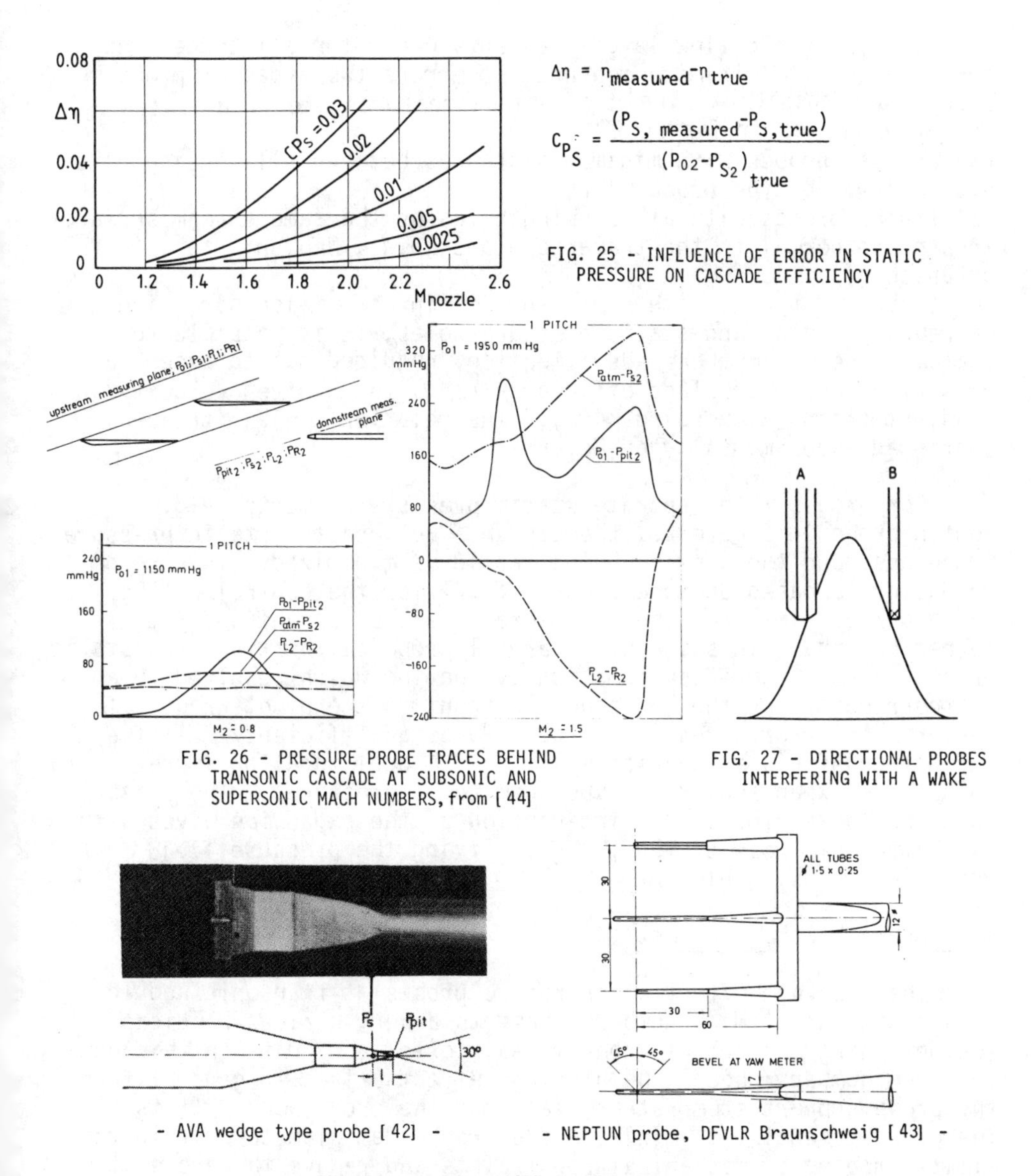

FIG. 25 - INFLUENCE OF ERROR IN STATIC PRESSURE ON CASCADE EFFICIENCY

FIG. 26 - PRESSURE PROBE TRACES BEHIND TRANSONIC CASCADE AT SUBSONIC AND SUPERSONIC MACH NUMBERS, from [44]

FIG. 27 - DIRECTIONAL PROBES INTERFERING WITH A WAKE

- AVA wedge type probe [42] -

- NEPTUN probe, DFVLR Braunschweig [43] -

FIG. 28 - COMPACT PROBE AND FINGER PROBE

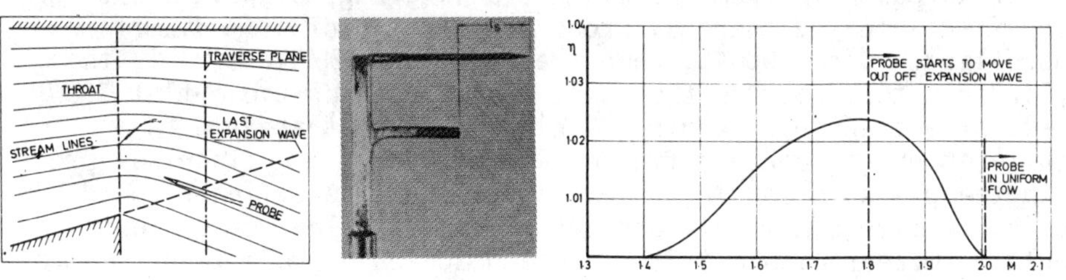

FIG. 29 - TRAVERSE OF A PRANDTL-MEYER EXPANSION WITH A NEEDLE STATIC PRESSURE PROBE (Sieverding [44])

In supersonic flow large gradients occur for all three pressure traces. To minimize measurement errors two types of probe designs are possible. Their principal features are illustrated by the two examples in figure 28 :
(a) compact probes with minimum distances between all sensing holes e.g. AVA wedge type probe [42] ;
(b) finger probes with all sensing holes on the same stream surface in a plane normal to the traverse plane, e.g., Neptun probe (DFVLR Braunschweig) [43].
For probes of type a with a distance ℓ (in streamwise direction) between the total and static pressure holes, it is possible to evaluate the minimum cascade dimensions required not to exceed a prescribed error in blade efficiency $(\Delta\eta)_\ell$ for a given velocity gradient field (Sieverding [44]). The relation $(\Delta\eta)_\ell = f(\ell)$ was confirmed experimentally by Meyer [45].

For probes with separate static pressure elements, e.g. Neptun probe in figure 28, the distance between the static pressure probe nose and the sensing holes should be minimized. This recommendation is based on model tests of VKI for the control of the behaviour of similar probes in a Prandtl-Meyer expansion. It turned out that, in spite of a careful probe calibration, the static pressure was measured systematically too low which resulted in an overcorrection for the bow shock in front of the pitot probe. The measurement error appears in figure 29 as an efficiency $\eta>1$ (for definition of η see equation 4.4); $P_{02} \simeq$ local total pressure in P. M. - expansion, P_{01} - upstream total pressure). The reason is to be looked for in the interaction of the expansion waves with the long needle nose of length ℓ_S modifying the pressure field around the sensing element.

4.3 Probe blockage effect

The blockage effect of intrusive probes at transonic and low supersonic outlet Mach numbers presents a very serious handicap for obtaining meaningful experimental performance data in this outlet Mach number range. The blockage effect arises in general from the probe support stem rather than from the probe head. It is therefore essential to minimize the probe stem diameter within the limits imposed by mechanical constraints and remove the probe stem as far as possible from the blade trailing edge plane. The blockage effect of a cylindrical probe stem (ratio of stem diameter to chord length d/c = 0.046, downstream distance X/C_{ax} = 0.52, the cylinder extends across the whole blade span) on the outlet flow field of a transonic turbine is clearly demonstrated in the Schlieren photographs in figure 30 from [46]. A cylindrical probe with the pressure sensing holes directly on the cylinder would obviously give completely erroneous results. The measurements would be more reliable if the probe head would project upstream of the probe stem, let us say to point A on the photographs.

FIG. 30 - BLOCKAGE EFFECT OF PROBE STEM ON TRANSONIC OUTLET FLOW OF TURBINE CASCADE (Sieverding [46])
- ratio of stem diameter to chord length $d/c = 0.046$
- axial distance of stem from blade $x/c_{ax} = 0.52$

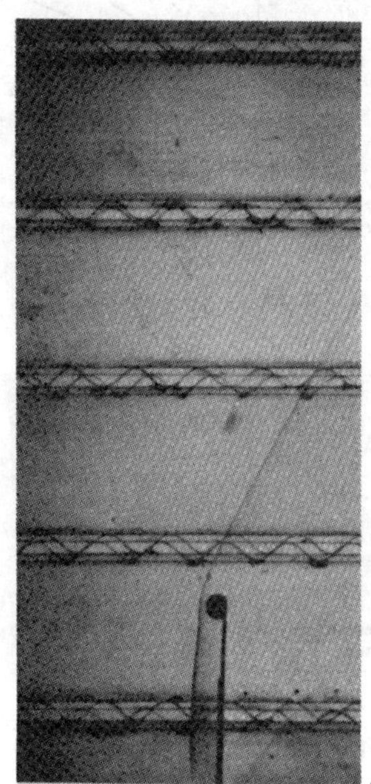

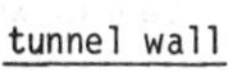

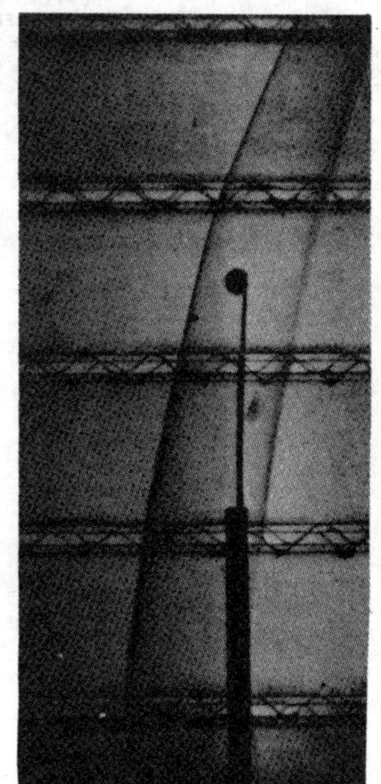

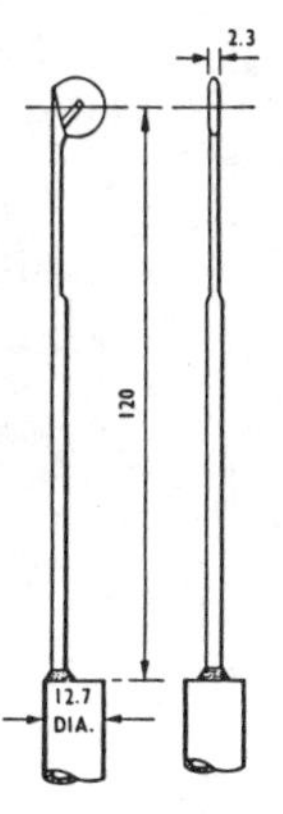

FIG. 31 - EFFECT OF PROBE STEM BLOCKAGE ON BOW SHOCK DISTANCE AT M = 1.2 (Langford, Keeley & Wood [47])

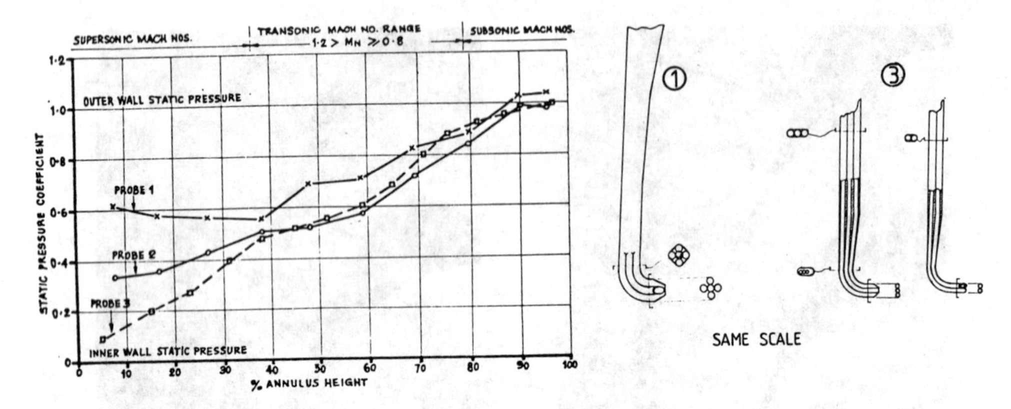

FIG. 32 - EFFECT OF PROBE BLOCKAGE AT STATOR EXIT OF LAST STAGE OF A BIG STEAM TURBINE (Cox [48])

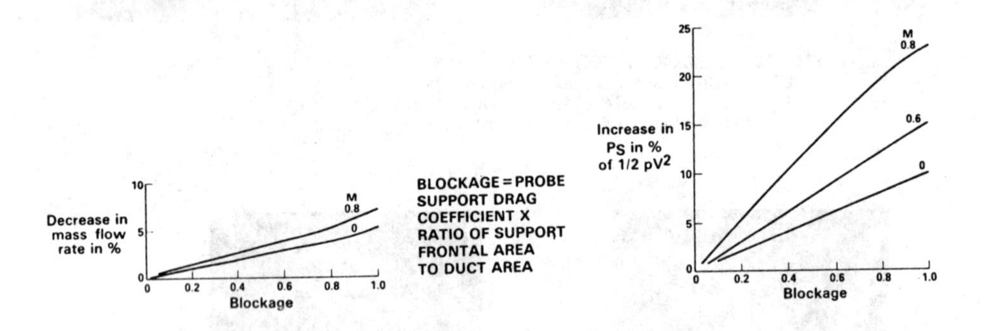

FIG. 33 - PROBE INSERTION EFFECTS IN CIRCULAR DUCT (Gettleman & Krause [51])

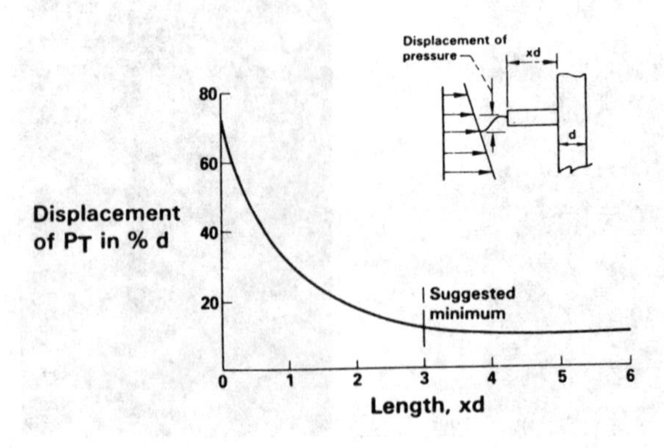

FIG. 34 - EFFECT OF PROBE LENGTH IN PRESSURE GRADIENT FIELD (Gettleman & Krause [51])

For in-stage space measurements in rotating machines the maximum downstream distance of the probe stem is necessarily much smaller and so is the upstream projection of the probe head. Langford, Keely and Wood [47] investigated several probe designs for traversing at the fixed blade exit of the final stage of big steam turbines. Figure 30 shows schlieren photographs of a probe comprising an isolated disc mounted on a slender support fixed in a one-half-inch probe stem at Mach number M = 1.2; right: the probe is completely immersed in the flow; left: the probe is flush-mounted with the tunnel wall. The most striking feature is the different bow shock stand-off distance for both cases. The authors concluded that the use of this probe at transonic Mach numbers was not to be recommended due to complex shock interaction effects.

The effect of probe blockage downstream of fixed blades in a model LP steam turbine was also studied by Cox [48]. The author points out that there is no absolute datum against which the detailed traverse results can be assessed. The accuracy of traverse data can only be gauged from comparisons with known values obtained from external measurements. Possible checks include (a) comparison of mass flow from traverse data with mass flow measurements obtained from accurately calibrated nozzle and venturi; (b) comparison of extrapolated static pressure traverse data with measured wall statics in the plane of traverse. The measurements were carried out with three five-hole probes (for design N° 3 the five-hole probe is split into two separate parts) with a probe blockage expressed in percentage of throat width of 73%, 41% and 13% respectively. The comparisons for the mass flow are given below. The comparison of the static pressure evolution over the blade height is presented in figure 32.

	Design 1	Design 2	Design 3
Traverse data/dall tube	1.045	1.072	1.002

Goutines [49] reported on probe immersion effects in an annular cascade of small blade height at subsonic flow conditions. The probe immersion induces :
(a) a radial or circumferential displacement of the streamlines due to the blockage of the probe;
(b) an abrupt change in the curvature of the streamlines at the extremity of the probe due to the sudden transition to zero obstruction causing a local static pressure gradient;
(c) secondary effects like corner vortices at the extremities of the probe.
These three phenomena depend on the proximity of the walls, on the wall shape and the dimension of the test section. The author points to the problem of using a calibration obtained under very different geometric conditions and flow conditions as those during the calibration.

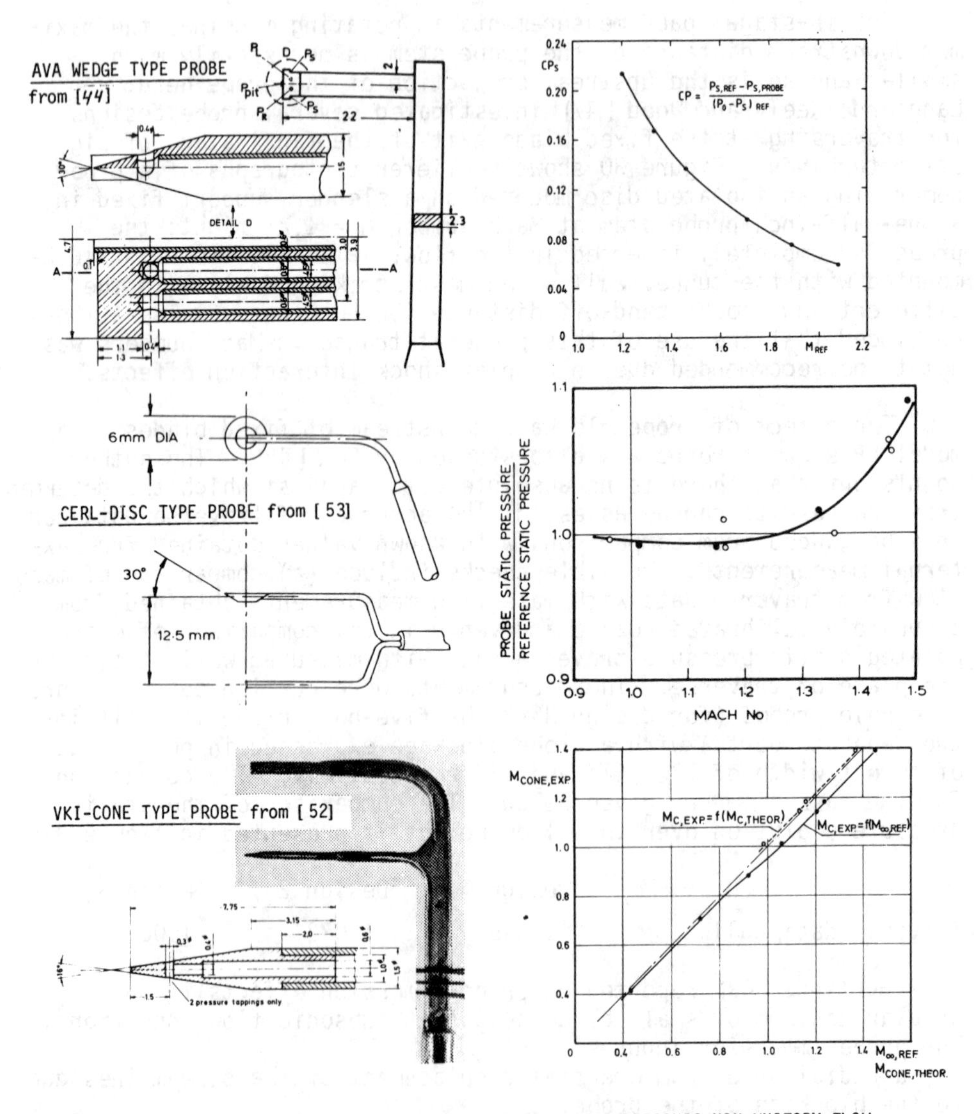

FIG. 35 - VARIOUS PRESSURE PROBES FOR MEASUREMENTS IN SUPERSONIC NON-UNIFORM FLOW

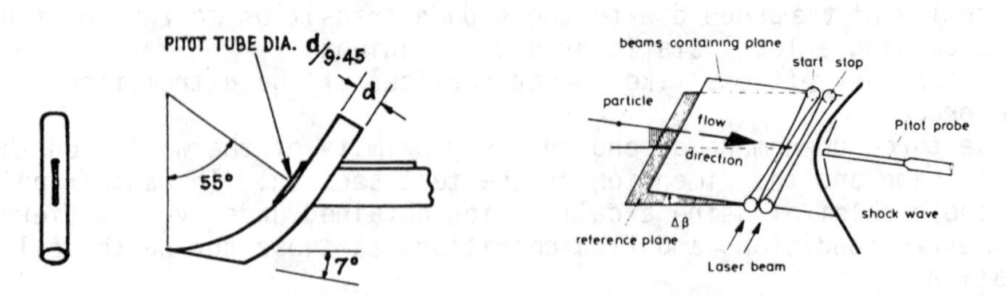

FIG. 36 - GOODYEAR PROBE FOR MEASURING TRUE TOTAL PRESSURE IN SUPERONIC FLOW [55]

FIG. 37 - COMBINATION OF LASER VELOCITY MEASUREMENTS WITH PITOT PROBE MEASUREMENTS, DFVLR [54]

Probe blockage effects near hub and tip endwalls in a transonic axial flow compressor were also experienced by Bois et al. [50]. However, a linear interpolation between the wall static pressures at hub and tip, as done by many investigators, can give rise to large errors.

Further examples of probe blockage effects are reported by Gettlemen and Krause [51]. The decrease of mass flow and increase of static pressure in a circular duct due to probe insertion is reproduced in figure 33, the effect of probe length in a pressure gradient field in figure 34.

4.4 Pressure probes for transonic cascade measurement

If the flow is two dimensional it is possible to combine total pressure and yaw angle measurements at mid span with static pressure readings from side wall pressure tappings. The use of wall tappings may prove to be the only solution if the blade height is so small that static pressure measurements in the flow become problematic. Wall static pressure taps require much more care in supersonic flow than in subsonic flow. The hole must be perpendicular to the surface. The diameter must be small with respect to the pressure gradients. The corners must be sharp and most important must have no burrs protruding into the airstream.

Figure 35 shows three examples of combined total, static and directional pressure probes, which seem to be at present the most interesting designs for use in supersonic non-uniform flow :
(a) AVA-Wedge-Type Probe, figure 35a (from [44]) : This probe incorporates all sensing holes for total and static pressures and yaw-angle in one single unit. The pitot tube is situated in the center of the wedge (wedge angle 30°), the directional holes lie on the wedge surfaces and the static pressure is measured on both sides of the wedge behind a step at 1.1 mm distance behind the wedge. This probe requires very detailed probe calibration due to the strong variation of the static pressure calibration coefficient with Mach number.
(b) Disc-Type Probe, figure 35b (from [53]) : The static pressure is measured with a flat sided disc probe. The disc is aligned with the cascade axis using the beam from a small He-Ne laser set to an orthogonal axis. The static pressure calibration coefficient remains very small up to Mach numbers of $M = 1.3$.
(c) Cone-Type Probe, figure 35c (from [52]) : The probe comprises a combined total-directional probe head and separated from it a cone probe for static pressure measurements. The total cone angle is 16°. The two static pressure holes positioned at 1.5 mm from the cone tip are aligned with the pitot probe mouth. The measured cone Mach number is very close to the reference tunnel Mach number for $M < 0.8$ and to the theoretical cone Mach number for $M > 1.0$.

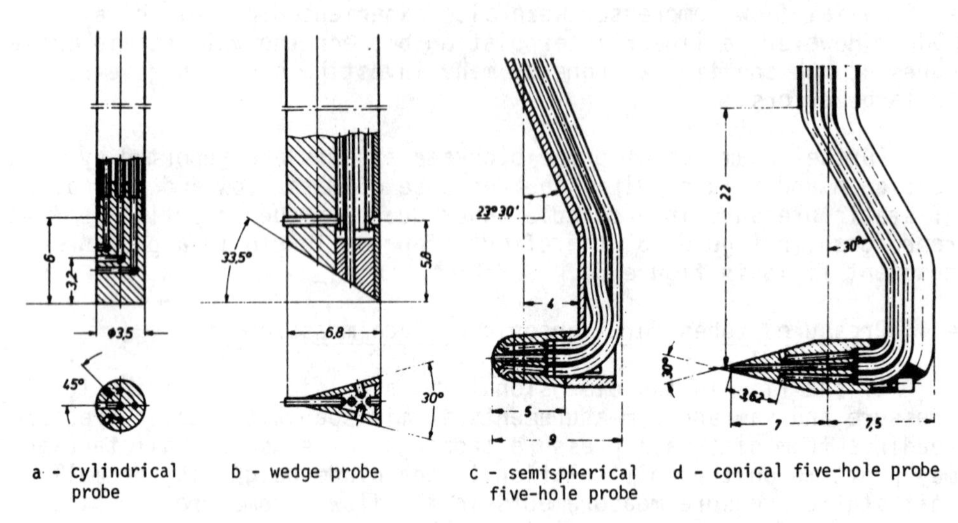

a - cylindrical probe

b - wedge probe

c - semispherical five-hole probe

d - conical five-hole probe

probe designs used at RWTH [56]

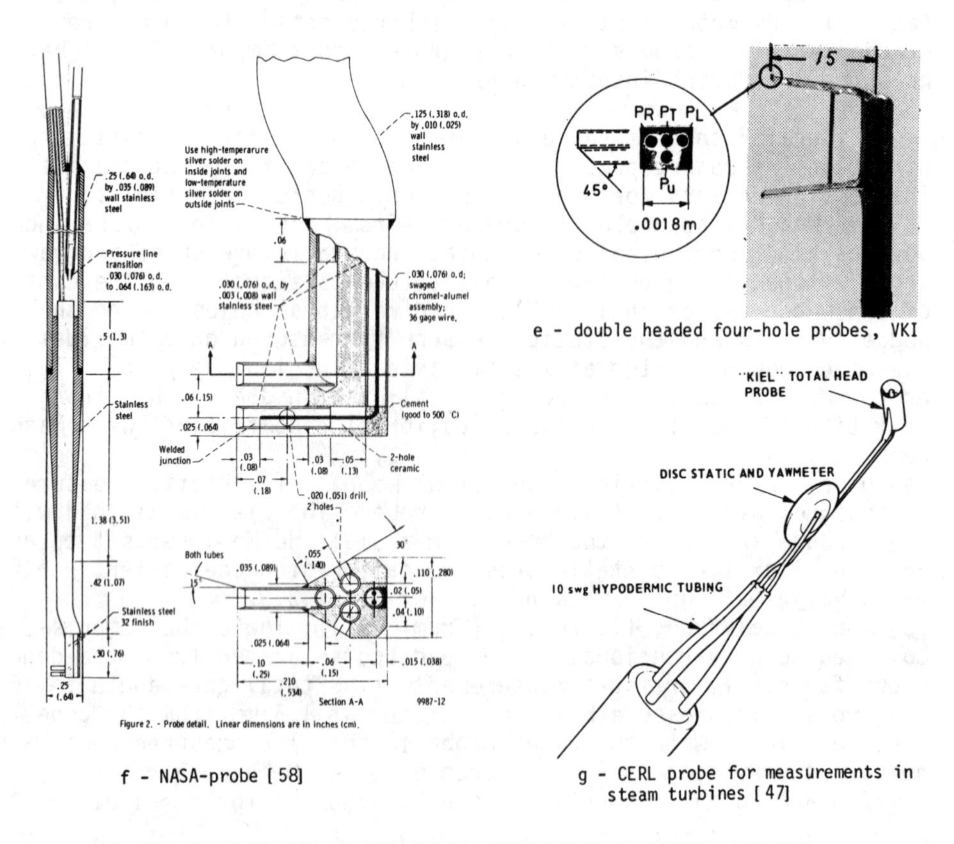

f - NASA-probe [58]

e - double headed four-hole probes, VKI

g - CERL probe for measurements in steam turbines [47]

FIG. 38 - PROBE DESIGNS FOR PERFORMANCE MEASUREMENTS IN ROTATING MACHINES

In most cases the static pressure probe calibrations are Re-number dependent. Some information on this effect is given in reference 44. It appears that cone probes are less sensitive to Re changes than the AVA wedge type probe and disc probes. The best way to overcome this problem is to calibrate the probes in the same tunnel (i.e. at the same test conditions) in which the cascade tests are carried out. Calibrating in the same tunnel has the further advangate that the probe blockage effects are similar for the calibraiton and the cascade tests.

Schreiber [54] proposed to overcome the problem in measuring the static pressure by combining pitot probe measurements with velocity measurements using the laser-2-focus method (Fig. 37).

Figure 36 shows an unusual total pressure probe design by Goodyer [55]. The probe allows to take measurements in supersonic flow without correction for a bow shock. The probe decelerates the supersonic flow isentropically by a curved compression surface. A pitot tube measures the total pressure in the locally subsonic region.

4.5 Pressure probes for measurements in rotating machines

The design of pressure probes for use in rotating machines is necessarily a compromise between what is aerodynamically desirable, in particular from the point of view of probe blockage effects (see section 4.3) and what is feasible in view of (a) the narrow axial clearances between the blade rows; (b) the save operation of the machines (probe destruction due to bending and/or vibration and (c) simplicity of access to reach the point of measurement and simplicity of execution of measurement. In view of (c) it is desirable to introduce the probe through a circular hole in the casing and to have the sensing holes near the axis of rotation to allow measuring in the zero balance mode.

Figure 38 shows a small collection of typical probe designs. Four of them are capable of measuring all three velocity components: probes b,c,d and e (probe e is particularly appropriate for measuring close to both endwalls). Probe g is capable of measuring in a three dimensional flow field, however, it does not provide the radial flow angle. Probe g is the only one to have a separate static pressure sensing element. For all other probes the static pressure has to be derived from the angle sensing holes. Probes c and d are equipped with temperature sensors beneath their probe heads (see T in figure 40). Probe f has a shielded thermocouple on top of the pitot probe.

Figure 39 from [56] shows for probe f (NASA small combination probe) the relation between the free stream static pressure and

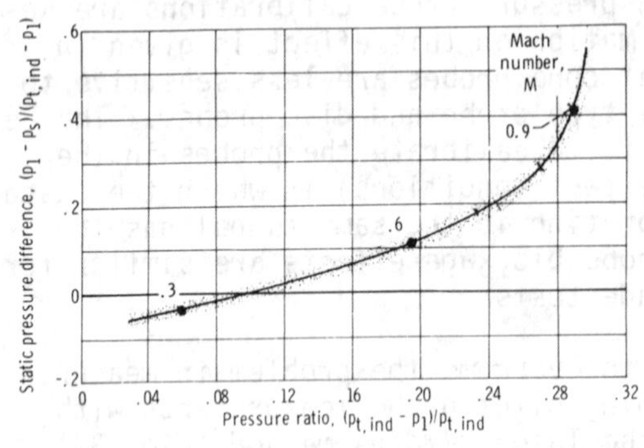

P_S - free stream static pressure

p_1 - pressure on wedge

$P_{t,ind}$ - pitot pressure

FIG. 39 - CALIBRATION CURVE FOR NASA PROBE IN FIG. 38b (Glawe et al. [57])

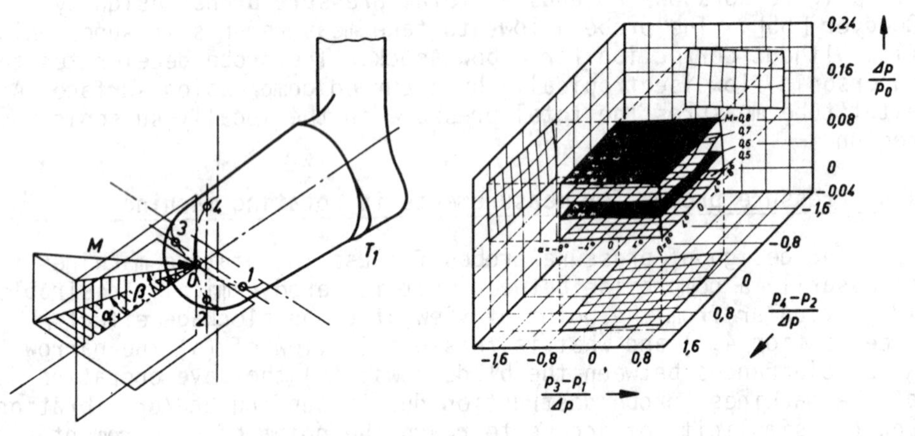

FIG. 40 - CALIBRATION SPACE FOR A FIVE-HOLE PROBE (Gallus & Bohn [58])

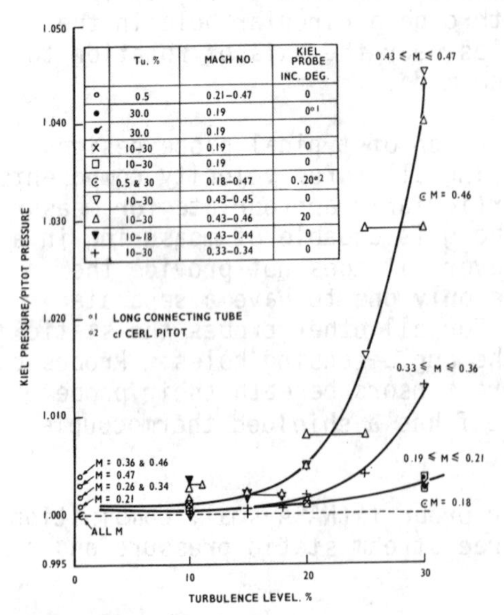

	Tu, %	MACH NO.	KIEL PROBE INC. DEG.
○	0.5	0.21–0.47	0
●	30.0	0.19	0*1
●	30.0	0.19	0
×	10–30	0.19	0
□	10–30	0.19	0
∈	0.5 & 30	0.18–0.47	0, 20*2
▽	10–30	0.43–0.45	0
△	10–30	0.43–0.46	20
▼	10–18	0.43–0.44	0
+	10–30	0.33–0.34	0

FIG. 41 - RATIO OF PRESSURES MEASURED BY KIEL PROBE AND 1 mm OD PITOT PROBE (preliminary results by Wood et al. [60])

measurable probe pressures when the probe is aligned with the flow. The graph indicates clearly the problem of using the probe for $M > 0.80$. The calibration curve was obtained in a 90 mm diameter free jet of constant static pressure. Traversing the jet with the probe resulted in a variation of the static pressure coefficient from -5 to +5 percent from the centerline value!

The calibration space of a five hole probe is shown in figure 40 (from [58]). The calibration parameters are :

$$k_\alpha = \frac{P_3-P_1}{\Delta P}; \quad k_\beta = \frac{P_4-P_2}{\Delta P}; \quad k_M = \frac{\Delta P}{P_0}; \quad \Delta P = P_0 - \frac{P_1+P_3}{2} \qquad (4.5)$$

The flow parameters total pressure, static pressure, Mach number, yaw angle and pitch angle are function of k_α, k_β and k_M. Gallus and Bohn propose in reference 58 a method for calculating the flow characteristics by multiparameter approximation functions :

$$\left.\begin{matrix}\alpha\\ \beta\\ M\\ P_0\\ P_s\end{matrix}\right\} F_i(k_\alpha, k_\beta, k_M); \quad i = 1,5 \qquad (4.7)$$

rather than by a interpolation-iteration procedure.

The use of three dimensional pressure probes over wide ranges of Re numbers requires additional calibration and the amount of calibration time becomes prohibitive. Broichhausen and Kauke [56] tested several conversion formulas to predict Reynolds number effects. The most promising results were obtained when treating the hole pressures in analogy to the total pressure

$$P_{ic} = P_i \frac{Re_c}{Re} \frac{(A/A^\star)_c}{A/A^\star} \sqrt{\frac{T_{0,c}}{T_c} \frac{\mu(T_c)}{\mu(T)}}; \qquad i = 1,5 \qquad (4.8)$$

with $A/A^\star$ - sonic area ratio

4.6 Measurement of mean fluctuating pressure in unsteady flows

Kiel head type probes like that in figure 38g are widely used for the measurement of total pressure in turbomachines because of their insensivity to incidence angles up to 40°. However, there is evidence in the literature that in unsteady flows the Kiel probe indicates higher pressures than the ordinary pitot probe [59,60]. Figure 41 from Wood and Langford [60] is a typical example. An error in measured pitot pressure will lead to an error in the mass flow via the velocity deduced from the pressure measurements. Based on figure 41 the authors conclude that dependent on Mach number the mass flow could be in error of 1% at 10% unsteadiness and up to 2% at 20% unsteadiness.

For flows which are at least approximately homogeneously turbulent the mean total pressure is given by :

$$\overline{P}_0 = \overline{P}_S + \frac{1}{2}\overline{\rho}\,(\overline{U}^2 + u'^2) \tag{4.9}$$

(U and u' in streamwise direction). When turbulent velocity levels are large then u'^2 has to be taken into account.

The time average which is the basis for the above relation is defined as

$$\overline{P} = \frac{1}{T}\int_0^T P_{(t)}\,dt \tag{4.10}$$

This simple time average of pressure is not necessarily the most significant quantity in the performance calculation of turbomachines. Instead, the performance analyst is more interested in the quantity

$$\overline{P.U}_{ax} = \frac{1}{T}\int_0^T P_{(t)}\cdot U_{ax(t)}\,dt \tag{4.11}$$

which presents the energy transport at any given moment. (Alwang: "Measurement of Steady and Unsteady Fluid Dynamic Quantities", VKI LS [61]). If $P_{(t)}$ and U_{ax} are each written as the sum of a time averaged value and a fluctuating component whose time averaged value is zero,

$$P_{(t)} = \overline{P} + p'_{(t)}$$
$$U_{ax(t)} = \overline{U}_{ax} + u'_{ax(t)} \tag{4.12}$$

and if these definitions are introduced in (4.11) then the result is :

$$\overline{P\cdot U}_{ax} = (\overline{P})\cdot(\overline{U}_{ax}) + \frac{1}{T}\int_0^T p'_{(t)}\cdot u'_{ax(t)}\,dt \tag{4.13}$$

The product $(\overline{P})\cdot(\overline{U}_{ax})$ is only equal to the desired quantity $\overline{P\cdot U}_{ax}$ if the fluctuations are not correlated with each other. However, the pressure and axial velocity changes resulting from passing blade blade wakes behind a rotating blade row may be strongly correlated and and the quantity $(\overline{P})\cdot(\overline{U}_{ax})$ is therefore generally lower than $\overline{PU}_{ax}$.

5. TOTAL TEMPERATURE MEASUREMENTS FOR STEADY STATE AERODYNAMIC PERFORMANCE MEASUREMENTS

Besides the measurement of pressure, velocity and flow direction, the gas temperature is the most important flow characteristic required for performance evaluation of turbomachines. The most commonly used device is the thermocouple which is based on the principle that an emf is generated when two dissimilar wires are connected with each other. The net emf of a thermocouple circuit is described by the equation (Moffat [62]) :

$$E = \int_{T_0}^{T_{\ell_1}} e_1 \cdot dt + \int_{T_{\ell_2}}^{T_0} e_2 \cdot dt \tag{5.1}$$

where E - output emf, mV
T_0 - temperature at junction of both wires
$T_{\ell_1}(\ell_2)$ - temperature at other end of wire 1 (2) of length $\ell_1(\ell_2)$
e - total thermoelectric coefficient (assumed to be constant through the wires)

Usually $T_{\ell_1} = T_{\ell_2}$ and (5.1) reduces to

$$E = \int_{T_0}^{T_\ell} (e_1 - e_2)\, dt \tag{5.2}$$

5.1 Thermocouple wires

Typical thermocouple wire materials in sequence of their thermoelectric characteristics are : chromel, copper, platinum, alumel, constantan. Figure 42 presents emf comparisons for several thermocouples. Their use in turbomachine applications is discussed by Fleeger and Seyb [63]. The main points are briefly summarized.

Copper-constantan thermocouples (type T) are extremely stable but their high thermal conductivity requires extremely large wire length to diameter ratios. They are sometimes used to measure the compressor inlet temperature.

Chromel-constantan thermocouples (type E) have highest emf output and good stability. They are used in compressor performance tests.

Chromel-alumel (type K) thermocouples are used by all engine manufacturers. They have a high emf and can be used up to 1250°C.

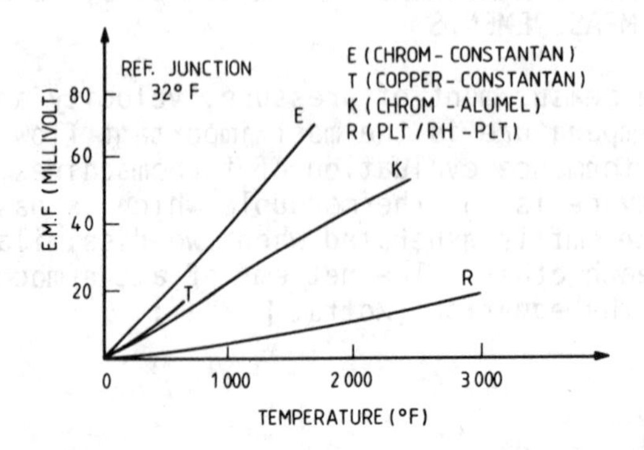

FIG. 42 - EMF COMPARISON FOR SEVERAL THERMOCOUPLES, from [63]

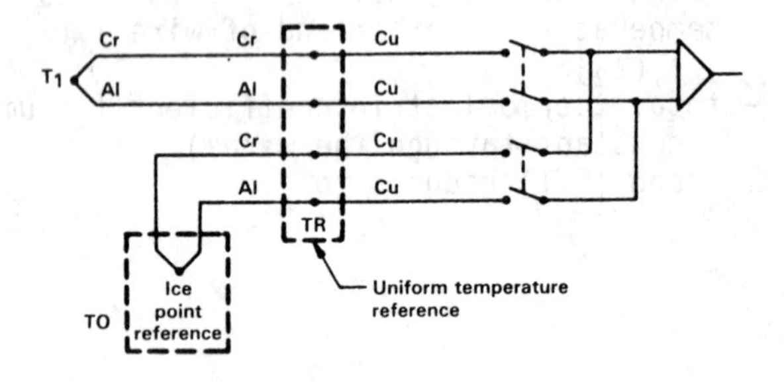

FIG. 43 - TYPICAL THERMOCOUPLE CHANNEL, from [61]

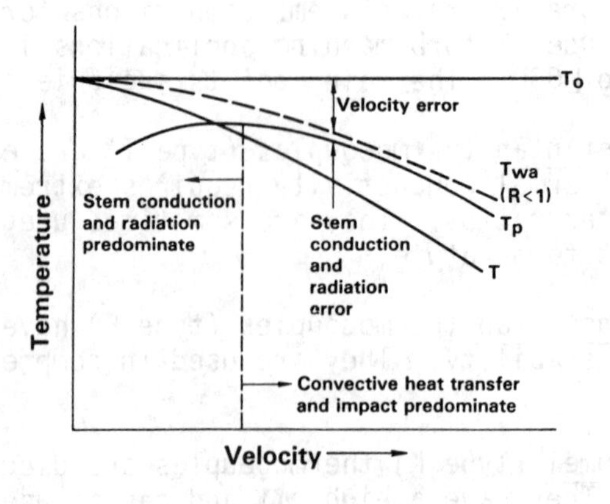

FIG. 44 - TEMPERATURE IN HIGH-VELOCITY, HIGH TEMPERATURE GAS STREAMS, from Alwang [61]

However, their use requires some precaution due to changes of their emf curve depending on the working conditions. Used for long time periods below 600°C the thermocouple output is affected by an aging process and it is advisable to pre-age type K wires to guarantee stable behaviour. If an aged thermocouple is used at temperatures above 600°C, it will revert to the standard calibration curve.

Platinum thermocouples are characterized by a rather low sensitivity but a very good stability over the whole temperature range (up to ∿1600°C). Their low emissivity reduces radiation errors, but the emissivity is affected by coating due to various types of impurities in the gas flow. The biggest problem in using platinum wires are their catalytic effects. If chemical reactions are still present when the gas flow passes over the probe, the probe will favour the completion of these reactions, causing therby a local increase of the gas temperature. This is in particular the case in engine transients.

In a thermocouple circuit one compares the unknown temperature with a known reference temperature. A typical thermocouple channel is shown in figure 43 (from Alwang [61]). All thermocouple leads are terminated in a UTR (Uniform Temperature Reference) box. The box is heavily thermally insulated to minimize thermal gradients in the assembly. The temperature in the UTR box is measured with respect to an ice point reference.

5.2 Gas temperature measurement

Due to the problem of measuring static temperatures in a gas stream it is common practice to measure the stagnation temperature and infer the static temperature from the equation

$$T_0 = T_S \left(1 + \frac{\kappa - 1}{2} M^2\right) \qquad (5.3)$$

To measure with a fixed probe the true stagnation temperature implies that all of the kinetic energy be recovered by the boundary layer as thermal energy. However, the effect of frictional heat on the thermocouple junction temperature is opposed by heat disssipation from the probe to the fluid. The fact that in the absence of any other errors the thermocouple junction temperature is lower than the stagnation temperature is expressed by the recovery ratio

$$R_1 = \frac{T_j}{T_0} \qquad (5.4)$$

or the recovery correction factor

$$\Delta = 1 - \frac{T_j}{T_0} \tag{5.5}$$

Another frequently used definition is the recovery factor

$$R_2 = \frac{T_j - T_S}{T_0 - T_S} \tag{5.6}$$

Equation (5.6) can be rewritten as

$$T_0 - T_j = (1 - R_2)\,\frac{V^2}{2 \cdot cp} \tag{5.7}$$

Assuming a perfect insulation of the junction from the probe support and no heat exchange with the surroundings, equation (5.7) shows clearly that the difference between the junction temperature T_j and the true stagnation temperature T_0 is a velocity error.

In addition to velocity errors, the temperature response of the probe is affected by conduction heat loss from the junction through the wires to the probe supporting structure. The temperature error due to this heat conduction loss is

$$T_0 - T_j = \frac{T_0 - T_M}{\cosh \ell \sqrt{\frac{4 \cdot h_c}{k \cdot d}}} \tag{5.8}$$

where T_M is the probe support temperature, h_C is the convective heat transfer coefficient and k is the thermal conductivity of the wire material. To keep this error small the wire length to diameter ratio ℓ/d must be large. The ratio is restricted by the effects of vibration, temperature and aerodynamic loading. The error can become important if the probe base is cooled (big temperature difference T_0-T_M). However, conduction effects for probes with a cooled probe support can be limited by improving the heat transfer from the gas to the probe by aspiration (increase of coefficient h_C).

If the gas temperatures are much higher than the local metal temperatures, radiation errors become important. Surface emissivity plays an important role for the amount of radient energy received and emitted by parts of the probe. Nobel metals have much lower emissivities than base metals. Errors due to thermal radiation are evaluated from

$$T_0 - T_j = \frac{K_R \cdot \sigma \cdot \varepsilon \cdot A_R}{h_c \cdot A_c} (T_j^4 - T_W^4) \qquad (5.9)$$

where σ is the Boltzman constant, ε the emissivity, K_R is a radiation correction coefficient, A_R is the radiation heat transfer area, A_C is the area for heat transfer by convection and T_W is the temperature of the surrounding walls [64].

Radiation errors are reduced using shielded thermocouples. The number of shields depends on the temperature difference but in most cases one does not use more than two shields. The magnitude of the error versus number of radiation shields for a case typical of a gas turbine combustor is shown below (from Alwang [61]):

gas temperature	1600°C;	Mach number	0.1
wall temperature	870°C;	Reynolds number	5.6×10^4
pressure	13.6×10^5 Pa	emissivity	0.6

No. of shields	Junction temeraure	% error
0	1400°C	14.0
1	1580°C	4.0
2	1625°C	1.3
3	1642°C	0.03

5.3 Probe designs

As for pressure probes, it is good practice to minimize all error sources when designing a new probe. This implies that attention has to be paid to the importance of each error for a given environment. This is qualitatively demonstrated in figure 44 from [61]. If the velocity error is predominant, then the velocity of the gas passing the junction has to be reduced, by designing the probe as a stagnation device. If radiation errors are predominant, the main emphasis is to be put on the shield design. Guidance for the design of radiation shieldings is given by Moffat [65].

Figure 45 shows a few examples of probe designs. The figure includes also combined temperature pressure probes.

Glawe et al. [66] carried out a comprehensive calibration program for different sizes of the unshielded and shielded probe designs (a) and (b) in figure 45, to determine their recovery and radiation corrections and time constants. The tests included variations of Mach number, pressure, temperature and pitch and yaw angle. The probe dimensions are given in table 1. The thermocouple was made of chromel-alumel wires. Some results related to the recovery correction factor Δ are shown in figure 46.

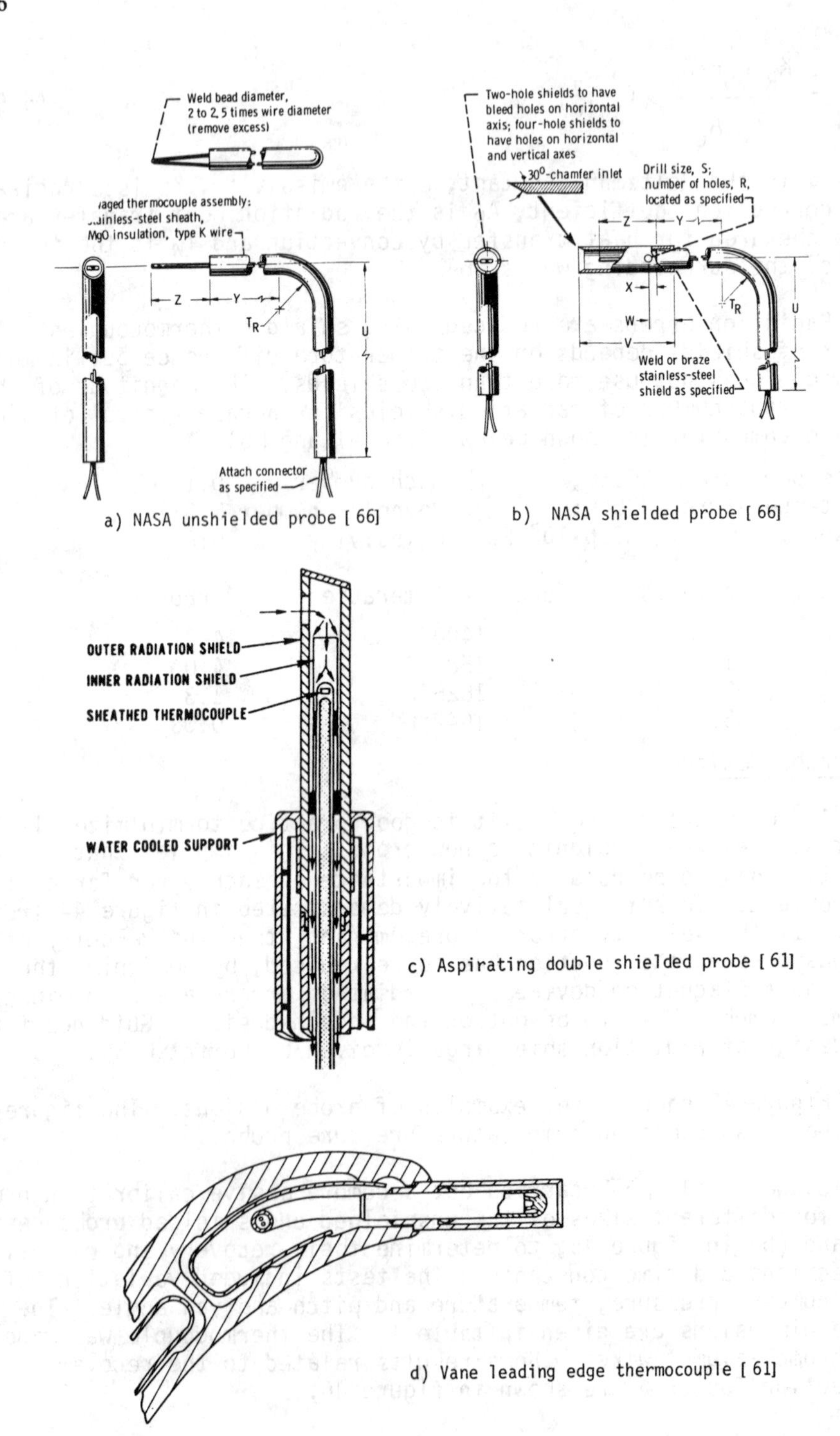

a) NASA unshielded probe [66]

b) NASA shielded probe [66]

c) Aspirating double shielded probe [61]

d) Vane leading edge thermocouple [61]

FIG. 45 - TEMPERATURE PROBE DESIGNS

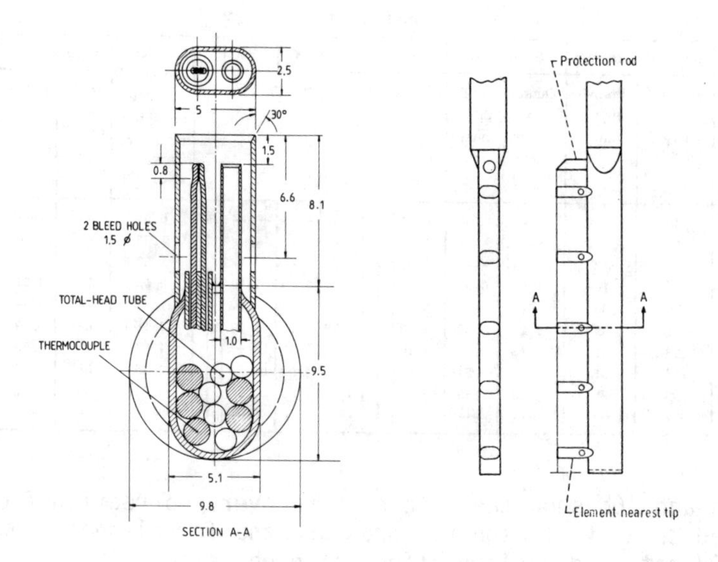

e) Combined total temperature and pressure rake [67]

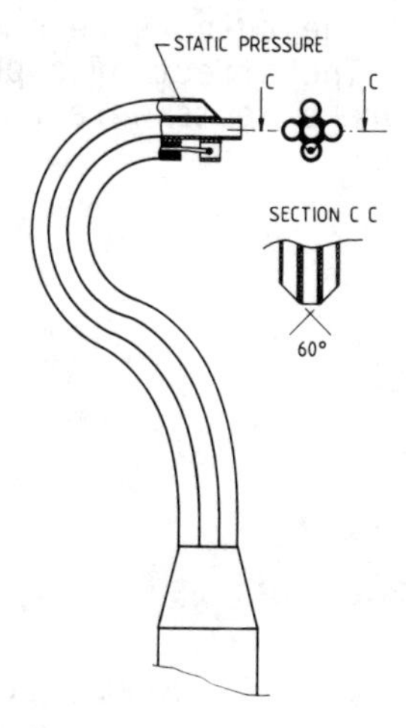

f) Combined temperature, pressure and flow direction probe [50]

FIG. 45 - TEMPERATURE PROBE DESIGNS (cont'd)

TABLE I. - NOMINAL PROBE DIMENSIONS

Probe	Swaged assembly outside diameter, mm	Nominal wire size		Shield size		Z	Y	X	W	V	U	Twice sheath outside diameter, minimum, T_R, mm	Drill size, S		Number of bleed holes, R
		Brown & Sharp gage	Diameter, mm	Outside diameter, mm	Wall thickness, mm	Dimension (table II), mm							Drill number	Diameter, mm	
1	0.25	48	0.032	----	----	0.51	(a)	----	----	-----	(a)	0.51	--	----	--
2	.50	38	.101	----	----	1.53		----	----	-----		1.02	--	----	--
3	.81	36	.13	----	----	1.91		----	----	-----		1.57	--	----	--
4	1.02	34	.16	1.63	0.30	2.41		0.51	2.03	5.59		2.03	76	0.51	2
5	1.27	30	.25	----	----	3.81		----	----	-----		2.54	--	----	--
6	1.57	28	.32	2.28	.32	4.83		.79	3.18	9.53		3.18	68	.79	2
7	2.29	26	.41	3.18	.46	6.10		.79	3.96	12.4		4.57	68	.79	4
8	3.18	24	.51	3.96	.51	7.62		1.07	5.03	15.5		6.35	58	1.07	4
9	4.78	20	.81	6.35	.89	12.2		1.57	7.92	24.9		9.53	52	1.57	4
10	6.35	18	1.02	----	----	15.2		----	----	-----		12.7	--	----	--

Figure 46a and 46b show the reference recovery correction factors Δ_0 (defined as Δ at atmospheric pressure and for aligned flow). The authors did not find a clear effect of probe size. The scatter for the shielded probe design is significant, but the 40% variation presents only 0.2% uncertainty of the absolute temperature. The effect of a yaw angle variation on the shielded probe design is presented in figure 46d. For $\beta > 10°$ the shield begins to cut the flow from the thermocouple junction and the angle effect takes very significant proportions. The authors recommend to align the probe with the flow direction. The effect of a pressure variation is seen in figure 46c. Temperature effects were not investigated but they are assumed to be small.

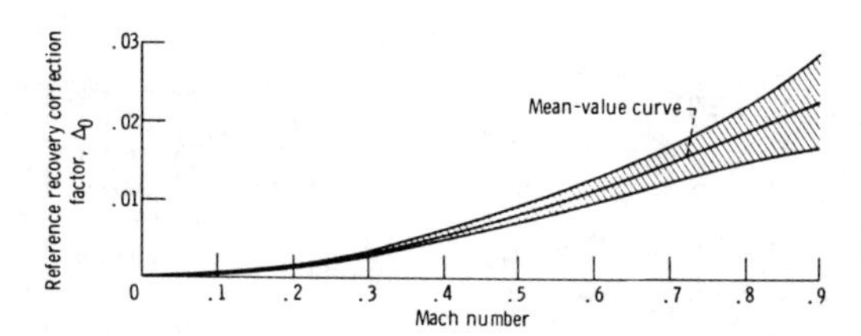

a) Effect of Mach number on recovery correction factor for unshielded probes

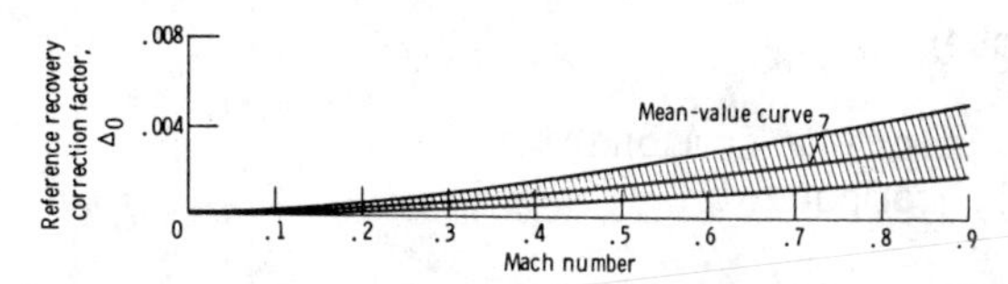

b) Effect of Mach number on recovery correction factor for shielded probes

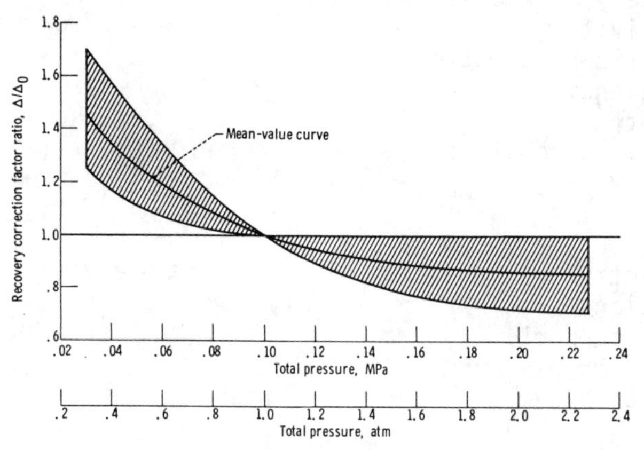

c) Effect of pressure on recovery correction factor for shielded probes

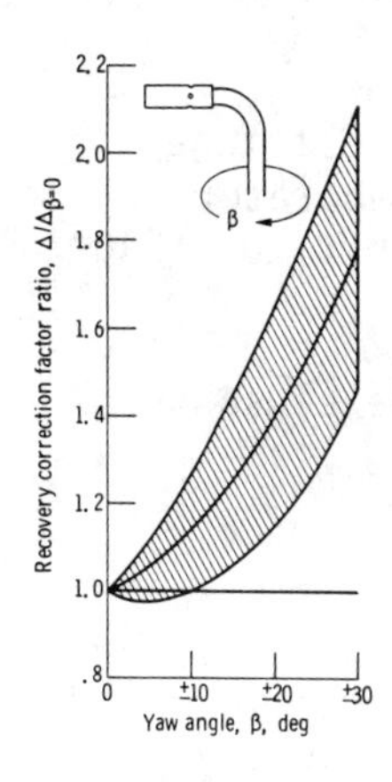

d) Effect of yaw angle on recovery correction factor for shielded probes

FIG. 46 - RECOVERY CORRECTION FACTOR AS FUNCTION OF MACH NUMBER, PRESSURE AND YAW ANGLE FOR NASA PROBES IN FIGURES 45a-b, from Glawe et al. [67]

LIST OF SYMBOLS

A	area
C_p	pressure coefficient
d,D	diameter
E	emf output
h	heat transfer coefficient
k	thermal conductivity
ℓ	length
M	Mach number
P	pressure
Re	Reynolds number
t,T	temperature
u'	fluctuating velocity component
U	mean velocity component
V	velocity
α	pitch angle
β	yaw angle
Δ	recovery correction factor
ε	emissivity
κ	specific heat ratio
η	efficiency
ρ	density
σ	Boltzmann constant

Subscripts

ax	axial
c	convective
j	junction
M	probe support
pit	pitot
S	static
W	wall
0	total

ABBREVIATIONS

AVA	Aerodynamische Versuchsanstalt Göttingen
CERL	Central Electricity Research Laboratory
DFVLR	Deutsche Versuchsanstalt für Luft- und Raumfahrt
EE	English Electric
ECL	Ecole Centrale de Lyon
EPFL	Ecole Polytechnique Federale de Lausanne
GE	General Electric
GEC	General Electric Company
MEL	Marshwood Engineering Laboratory
MIB	Mitsubishi
P&W	Pratt & Whitney
RWTH	Rheinisch Westfälische Technische Hochschule
UTRC	United Technology Research Center

REFERENCES

1. STARKEN, H.: Comparison between flows in cascades and rotors in the transonic range. Part 1 : Basic considerations. In: "Transonic Flows in Axial Turbomachinery", VKI LS 84, Feb. 1976.
2. STARKEN, H.: Untersuchung der Strömung in ebenen Überschallverzögerungsgittern. DLR FB 71-99, 1971.
3. FORSTER, V.T.: Turbine blade development using a transonic variable-density cascade wind tunnel. Proc. Inst. Mech. Engrs., P6/65, 1964.
4. KURAMATO, Y.; NAGAYAMA, T.; IMAIZUMI, M.: A study on the transonic turbine cascade by shock tunnel. MTB 101, July 1975.
5. OLDFIELD, M.L.G.; SCHULZ, D.L.; NICHOLSON, J.H.: Loss measurements using a fast traverse in an isentropic light piston tunnel cascade. In: Proceedings of the Symposium on "Measuring Techniques in Transonic and Supersonic Flows in Cascades and Turbomachines", at ECL Lyon, France, October 1981.
6. KIOCK, R.; LASKOWSKI, G.; HOHEISEL, H.: Die Erzeugung höherer Turbulenzgrade in der Messtrecke des Hochgeschwindigkeits-Gitter Windkanals Braunschweig zur Simulation turbomaschinenähnlicher Bedingungen. DFVLR FB 82-25, 1982.
7. KIRILLOV, I.I. & YABLONIK, R.M.: Fundamentals of the theory of turbines operating in wet steam. NASA TTF 611, Nov. 1970.
8. DECUYPERE, R.: The two dimensional steam tunnel of the Royal Military Academy of Belgium. In: Proceedings of the Symposium on "Measuring Techniques in Transonic and Supersonic Cascade Flow", at CERL, March 1979, CERL Report RD/LIN 166/79.
9. WOOD, N.B.: LP turbine tip section in wet steam flow. In: Proceedings of the Symposium on "Measuring Techniques in Transonic and Supersonic Cascades and Turbomachines", EPFL Lausanne, November 1976.
10. BREUGELMANS, F.A.E.: Prospectives and problems of transonic and supersonic compressors. In: "Advanced Components for Turbojet Engines, AGARD CP 34, Sept. 1968, paper 5.
11. STARKEN, H. & LICHTFUSS, H.J.: Cascade flow adjustment. In: "Modern Methods of Testing Rotating Components of Turbomachines-Instrumentaion", AGARDograph 207, 1975.
12. MIKOLAJZAK, A.A.; MORRIS, A.L.; JOHNSON, B.V.: Comparison of performance of supersonic blading in cascade and compressor rotors. ASME Trans., Series A - J. Engrg. Power, Vol. 93, No. 2, Jan. 1971, pp 238-248.
13. LEYNART, J.: Cascade test methods in wind tunnel at ONERA. In: "Measuring Techniques in Transonic and Supersonic Cascades and Turbomachines", at EPFL Lausanne, November 1976.
14. STARKEN, H.: A new technique for controlling the exit flow periodicity of supersonic cascades. In: Proceedings of the Symposium on "Measuring Techniques in Transonic and Supersonic Cascades and Turbomachines" at EPFL Lausanne, November 1976.

15. SIEVERDING, C.H.: Unsteady flow measurements in straight cascades. In: Proceedings of the Symposium on "Measuring Techniques in Transonic and Supersonic Cascades and Turbomachines" at EPFL Lausanne, November 1976.
16. MEAUZE, G. & THIBERT, J.J.: Méthode d'étude expérimentale de grilles d'aubes transsoniques à forte déviation. ONERA TP 1094, 1972.
17. MEAUZE, G.: A method for testing cascades with converging side walls. ASME Trans., Series A - J. Engrg. Power, Vol. 100, No. 2, July 1978.
18. SCHREIBER, H.A.: Comparison between flows in cascades and rotors in the transonic range. Part II : Investigation of two transonic compressor cascades and comparison with rotor data. In: "Transonic Flows in Axial Turbomachinery", VKI LS 84, February 1976.
19. FORSTER, V.T.: Measurements in transonic steam turbine cascades. In: Proceedings of the Symposium on "Measuring Techniques in Transonic and Supersonic Cascades and Turbomachines" at EPFL Lausanne, November 1976.
20. GOTTHARDT, H.: Theoretische und experimentelle Untersuchungen an ebenen Turbinengittern mit Pfeilung und V-Stellung. Dissertation Technische Universität Braunschweig, 1983.
21. BREUGELMANS, F.A.E.; CARELS, Y; DEMUTH, M.: Influence of dihedral on the secondary flow in a two dimensional compressor cascade. 1983 Tokyo Gas Turbine Congress, October 1983; also VKI Preprint 1983-08.
22. GRANT, R.J. & SPARR, A.: The MEL transonic annular cascade. In: Proceedings of the Symposium on "Measurement Techniques in Transonic and Supersonic Cascades Flows", CERL, March 1979. CERL Report RD/L/N 166/79.
23. BÖLCS, A.: A test facility for the investigation of steady and unsteady transonic flows in annular cascades. ASME P 83 GT 34.
24. DETTMERING, N. & BECKER, B.: Steps in the development of a supersonic compressor stage. In: "Advanced Components for Turbojet Engines", AGARD CP 34, 1968.
25. TOBERY, E.W. & BUNCE, R.H.: Cascade heat transfer tests of the air cooled W 501 D first stage vane. ASME P 84 GT 114.
26. MEYER, J.: Der Prüfstand für rotierende Schaufelgitter der Aerodynamischen Versuchsanstalt Göttingen. DFVLR-AVA 70A49.
27. FOWLER, J.E. & PERRY, J.J.: A facility for flow visualization in a large scale turbine stage. Paper presented at ASME Winter Annual Meeting, Chicago, November 1965.
28. Compresseur transparent FA24. Film ONERA No 693, 1971.
29. EVANS, R.L.: Boundary layer development of an axial flow compressor stator blade. ASME Trans., Series A - J. Engrg. Power, Vol. 100, No. 2, April 1978, pp 287-293.
30. DRING, R.P. et al.: Turbine rotor-stator interaction. ASME Trans., Series A - J. Engrg. Power, Vol. 104, No. 4, October 1982, pp 729-742.
31. Education and Research 1956-1976, von Karman Institute for Fluid Dynamics.

32. KERREBROCK, J.L. et al.: The MIT blowdown compressor facility. ASME Trans., Series A - J. Engrg. Power, Vol. 96, No. 4, October 1974, pp 394-406.
33. CUNNAN, W.S.; STEVANS, W.; URASEK, D.C.: Design and performance of a 427-meter-per-second-tip-speed two-stage fan having a 2.40 pressure ratio. NASA TP 1314, 1978.
34. CALVERT, H.F. et al.: Turbine cooling research facility. NASA TMX-1927.
35. WACHTER, J. & EYB, G.: The last stage test stand as a facility to investigate the phenomena in the LP-part of large steam turbines. ASME 82-GT-7, 1982.
36. HILL, J.A.F. et al.: Mach number measurements in high speed wind tunnels. AGARDograph 22, 1956.
37. ERWIN, J.R.: Experimental techniques. Section D of Aerodynamics of Turbines and Compressors. Editor, W.R. Hawthorne, Princeton U. Press, 1964.
38. CHUE, S.H.: Pressure probes for fluid measurement. Prog. Aerospace Sci., Vol. 16, No. 2, 1975, pp 147-223.
39. WUEST, W.: Measurement of flow speed and flow direction by aerodynamic probes and vanes. In: "Flight Test Instrumentation", AGARD CP 32, 1967.
40. Modern methods of testing rotating components of turbomachines (instrumentation). M. Pianko (Ed.), AGARDograph 207, 1975.
41. McMILLAN, F.A.: Viscous effects on pitot tubes at low speeds. J. Roy. Aeronaut. Soc., Vol. 58, 1954, pp 570-572.
42. AMECKE, J. & LAWACZECK, O.: Aufbau und Eichung einer neuentwickelten Keilsonde für ebene Nachlaufmessungen, insbesondere im transsonischen Geschwindigkeitsbereich. DLR Forschungsbericht 70-69, 1970.
43. KIOCK, R.: Description of a probe for measurements of two-dimensional wake quantities. DFVLR Braunschweig, IB 151-74/2.
44. SIEVERDING, C.H.; Pressure probe measurements in cascades. In: "Modern Methods of Testing Rotating Components of Turbomachines", AGARDograph 207, 1975.
45. MEYER, J.: Theoretical and experimental investigations of the flow downstream of two dimensional transonic turbine cascades. ASME P 72 GT 43, 1972.
46. SIEVERDING, C.H.: Probe blockage effects in transonic turbine cascade testing. Presented at Symposium on "Measuring Techniques in Transonic and Supersonic Flows in Cascades and Turbomachines", at ECL Lyon, October 1981.
47. LANGFORD, R.W.; KEELEY, K.R.; WOOD, N.B.: An investigation into the calibration characteristics of CERL disc-type static pressure probes in transonic flow. CERL, March 1979, CERL report RD/L/N 166/79.
48. COX, H.J.A.: Experimental development of bladings for large steam turbines. In "Steam Turbines for Large Power Outputs", VKI LS 1980-6, April 1980.

49. GAUTINES, M.: Comparison of two static pressure probes. Contribution to the meeting on "Measuring Techniques in Transonic and Supersonic Cascade Flow", ONERA, January 1974.
50. BOIS, G. et al.: Measurements of secondary flows in a transonic axial flow compressor. Proc. Sym. on "Measuring Techniques in Transonic and Supersonic Cascades and Turbomachines" at EPFL Lausanne, November 1976.
51. GETTLEMAN, C.G. & KRAUSE, L.N.: Effect of interaction among probes on pressure measurements in ducts and jets. ISA Proceedings, Vol. 7, paper 52-12-2.
52. SIEVERDING, C.H.: Calibration characteristics of a 16° cone probe. Proc. Symp. on "Measuring Techniques in Transonic and Supersonic Cascade Flow" at CERL March 1979, CERL report RD/L/N 166/79.
53. JACKSON, R. & WALTERS, P.T.: Design considerations for the CERL wet steam tip section cascade and first test results. Proc. Symp. on "Measuring Techniques in Transonic and Supersonic Cascade Flow" at CERL March 1979, CERL report RD/L/N 166/79
54. SCHREIBER, H.A.: Supersonic exit flow measurements downstream of a compressor cascade by a laser-2-focus method. Proceedings of Symposium on "Measuring Techniques in Transonic and Supersonic Cascade Flow" at CERL March 1979, CERL report RD/L/N RD/L/N 166/79.
55. GOODYER, M.J.: A stagnation pressure probe for supersonic and subsonic flows. Aeronautical Quart., Vol. XXV, Part 2, May 1974, pp 91-100.
56. BROICHHAUSEN, K.D. & KAUKE, G.: Investigation of the Reynolds number effects on probe measurements in supersonic flow. Proceedings of Symposium on "Measuring Techniques in Transonic and Supersonic Cascades and Turbomachines" at ECL Lyon, Oct. 1981.
57 GLAWE, G.E.; KRAUSE, L.N.; DUDZINSKI, T.J.; A small combination sensing probe for measurement of temperature, pressures and flow direction. NASA TN D-4816, 1968.
58. GALLUS, H.E. & BOHN, D.: Multi-parameter approximation of calibration values for multi-hole probes. Proceedings of Symposium on "Measuring Techniques in Transonic and Supersonic Cascades and Turbomachines" at EPFL Lausanne, November 1976.
59. SAMOILOVICH, G.S. & YABLOKOV, L.D.: Measurement of periodically fluctuating flows in turbomachines by ordinary pitot tubes. Thermal Engineering, Vol. 17, 1970, pp 105-110.
60. WOOD, N.B.; LANGFORD, R.W.; PRIDLY, W.J.: Effects of flow unsteadiness on the time-mean response of a Kiel-type total pressure probe : preliminary investigation. Proceedings of Symposium on "Measuring Techniques in Transonic and Supersonic Flows in Cascades and Turbomachines" at ECL Lyon, October 1981.
61. ALWANG, W.G.: -Measurements of steady state fluid dynamic quantities. -Measurements of non-steady fluid dynamic quantities. -Measurement of metal temperature, heat flux and strain. In: Measurement Techniques in Turbomachines", VKI LS 1981-7.

62. MOFFAT, R.J.: Understanding thermocouple behaviour : the key to precision. ISA Conference, October 1968, Paper 68-628.
63. FLEEGER, D.N. & SEYB, N.J.: Aerodynamic measurements in turbomachines. in "Modern Methods of Testing Rotating Components in Turbomachines (Instrumentation)", AGARDograph 207, 1975.
64. LIGRANI, P.: Measurement techniques. VKI CN 108, 1979.
65. MOFFAT, R.J.: Gas temperature measurement : Direct design of radiation shielding. ISA Paper N° 69-517.
66. GLAWE, G.E.; HOLANDA, R.; KRAUSE, L.N.: Recovery and radiation corrections and time constants of several sizes of shielded and unshielded thermocouple probes for measuring gas temperature. NASA TP 1099, 1978.
67. KRAUSE, L.N.; GLAWE, G.E.; DUDZINSKI, T.J.: A probe for measuring temperature and pressure at the same points in a gas stream. NASA TM X-2577, 1972.

OPTICAL TECHNIQUES FOR TURBOMACHINERY FLOW ANALYSIS

R. Schodl

Institut für Antriebstechnik, DFVLR Köln, Germany

1. INTRODUCTION

In the context of further research and development work on turbomachinery our knowledge of the aerodynamic performance of a blade row must be deepened. More accurate conceptual and mathematical models are required on which designs can be based.

Traditionally, measurements to improve the understanding have been available only upstream and downstream of each blade row and on the annulus wall. Although a considerable amount of knowledge about the performance of the blade row can be obtained from such measurements, detailed understanding can be obtained only by making measurements within the blade row. Windtunnel measurements in stationary cascades of blades are a source of data, but as the model improves, full three dimensional and rotational effects become important and measurements must be made in a fully representative situation. Great efforts were made in order to apply different kinds of measuring techniques to the real turbomachinery hardware. Optical measuring methods, as e.g. flow visualisation [1], Schlieren [2,3] and holographic interferometric techniques [4,5,6] have contributed greatly to our present understanding of the turbomachinery flow even though most of the methods give only qualitative information. Under particularly favourable laboratory conditions, quantitative measurements are also possible (e.g. velocity measurements by the spark traceing method in experimental compressors [7] and measurements of density at a point by using fluorescent light from an admixed gas [8]), but these methods are usually unsuccessful under actual test conditions in turbomachines.

As a result of developments in laser technology, some known physical principles have become useful in the development of new methods of measurement. Successful attempts have been made to register the extremely weak scattering of laser light on molecules, i.e. by measuring the density and concentration by Rayleigh scattering and by measuring the temperature by Raman scattering [9,10]. This method - in the meantime well stablished in basic combustion research - is not likely to be applied to turbomachinery components, especially because of the considerable problems with interfering radiation as e.g. background flare. More details about these and other measuring techniques, with respect to turbomachinery application are treated in ref. [11-18].

At present the laser Velocimetry (LV) is the most promising and most developed measuring technique and well established in the turbomachinery flow research. Two different forms of laser anemometers, the laser Doppler and the laser two focus (L2F) velocimeters, have achieved importance today and they are widely spread out in turbomachinery research laborationes. These techniques which allow to map a two or even threedimensional flow field by point by point quantitative measurements of the flow velocity and turbulence will be described and discussed in the following.

2. LASER VELOCIMETRY

Laser anemometers are optical systems with the capability to measure the speed of small micron sized particles. Such particles are naturally contained within all practical airflows or can be seeded if it's required. When they are carried along with the flow it is supposed that they follow the flow so exactly that by measuring the particles speed the flow velocity can be deduced.

Very sensitive optical photomultiplier detectors are available to registrate the weak light which is scattered by these micronsized particles in all directions when they are illuminated by a laser beam. On the basis of these light scattering properties different optical methods have been developed for measuring the particles speed.

A Laser anemometer consists mainly of three components: a transmitting device by which laser beams are directed to a certain point in the flow path where the probe volume is formed, a receiving device which is aligned with the probe volume in order to registrate the scattering light from the particles passing through and an electronic device which processes the photomultiplier signals and delivers velocity and turbulence information. Before going into details of the LD and the L2F-velocimetry method some remarks about the scattering behaviour of small particles should be made.

Particles of sizes comparable to the wavelength of light are of interest in laser velocimetry. Their scattering characteristic can be coordinated within the Lorenz-Mie scatter region.

The intensity of scattered light in this region depends fundamentally upon the size and shape of the particles, the intensity and wave length of the incident light, and at any rate upon the refractive index of the particle matter employed. Moreover, the intensity of scattered light varies seriously as the scattering angle changes with reference to the direction of the incident light (ref. [19], [20]).

In Fig. 1 a polar coordinate graph of the scattering behaviour of a micronsized particle is shown. From out of the direction indicated the light beam of the intensity I_o meets the particle. ε is the scattering angle with respect to the direction of the incident light and I_s the intensity of the scattered light.

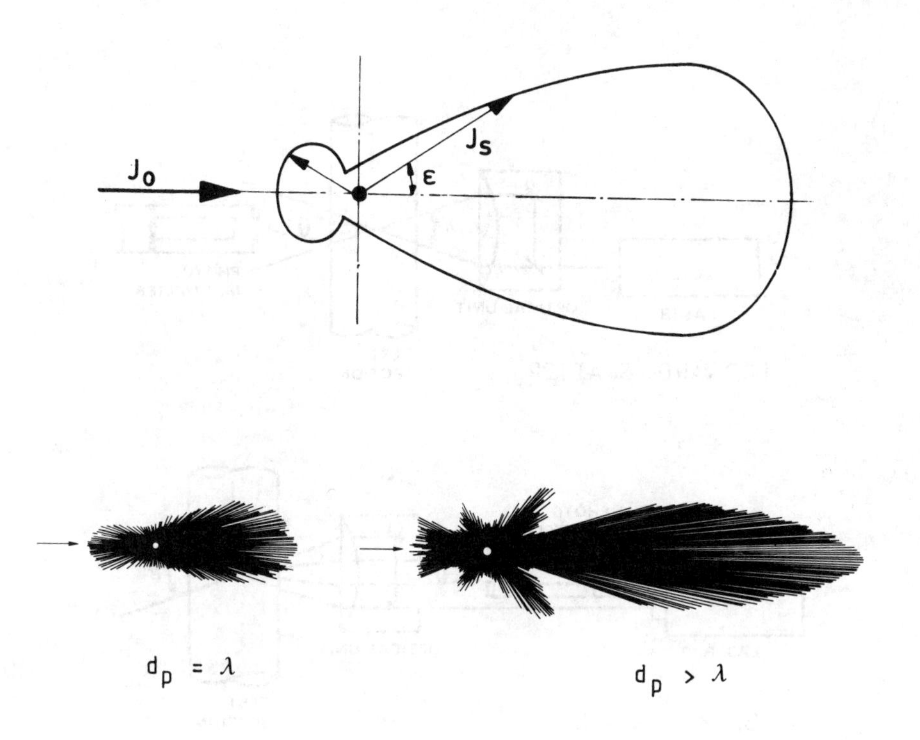

Fig.1 Typical distribution of light being scattered by small particles λ: wavelength of light, d_p: particle diameter.

Conspicuous is that the particle scatters far more light in the forward direction than in backward direction. The ratio of forward to backscattered light is in the order of 10^2 to 10^3.

This means that in general a forward scatter optical arrangement as it is shown in the upper part of Fig.2 provides a much higher signal to noise ratio and is therefore preferable to a backscatter arrangement (see lower part of Fig.2).

2.1 The Laser Doppler Velocimeter

The first publication [21] on the successful testing of a laser-Doppler method (LDV) in 1964 was followed by a large number of reports (see bibliography in [22]) on the further development of this method of measurement. During this period, two different forms of LDV crystallized out, i.e. the reference-beam method and the cross-beam or dual-scatter method. Laser Doppler anemometers are based on the Doppler shift of the frequency of light scattered from particles moving with the flow. Because of the very high frequencies it is very difficult to measure the shifted frequencies directly.

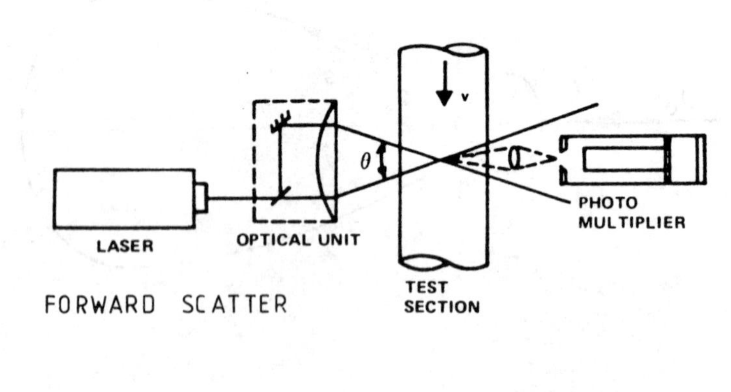

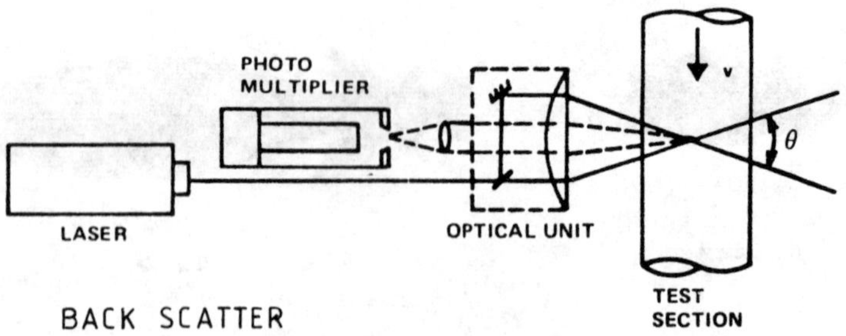

Fig.2 Optical Arrangements of Laser Anemometers.

By combining the frequency shifted scattered light with a reference beam from the same light source the resultant beat frequency lies then within an electronically detectable range. This original method is called the reference-beam method however it is not much used in turbomachinery work.

Another principle to get the frequency low enough to be detectable is to combine the scattered light from two incident beams. This so called dual-scatter method has achieved main importance today, especially in gaseous flow research. The dependency of the measured frequency from the particles velocity can be described by a detailed analysis of the doppler shift of laser light scattered from the two beams. However there is a much simplier way to explain this technique when the interference phenomenon of two coherent light beams is considered.

In the region of intersection between the two coherent light beams, an alternating system of bright and dark interference fringes is produced; if the light waves are plane, the fringes are parallel orientated in the probe volume (see Fig.3). If θ is the angle of intersection between the beams and λ the wavelength of light, the distance s between fringes is

$$s = \frac{\lambda}{2 \sin(\Theta/2)}$$

If a particle travels through this fringe system, it emits a scattered light signal which fluctuates periodically at the following frequency, owing to the difference in light intensity in the fringe system:

$$\nu = \frac{u_{\perp}}{s}$$

The measured component u which is perpendicular to the angular bisector of the two beams is given by:

$$u_{\perp} = \frac{\lambda}{2 \sin(\Theta/2)} \cdot \nu$$

The component $u_{\parallel}$ which is parallel to the angular bisector cannot be measured. In Fig.4 the optical set ups of a forward scatter and a backscatter Laser Doppler velocimeter are shown.

The photodetector delivers a sinusoidal voltage signal at the frequency ν, the number of periods in the signal corresponding to the number of fringes in the measuring space (see Fig.5).

The electronic processing of the signals is difficult, because of the high frequency, the limited length of the signal train and the broad-band noise of the photomultiplier. A further difficulty is the variable amplitude in each signal train, caused

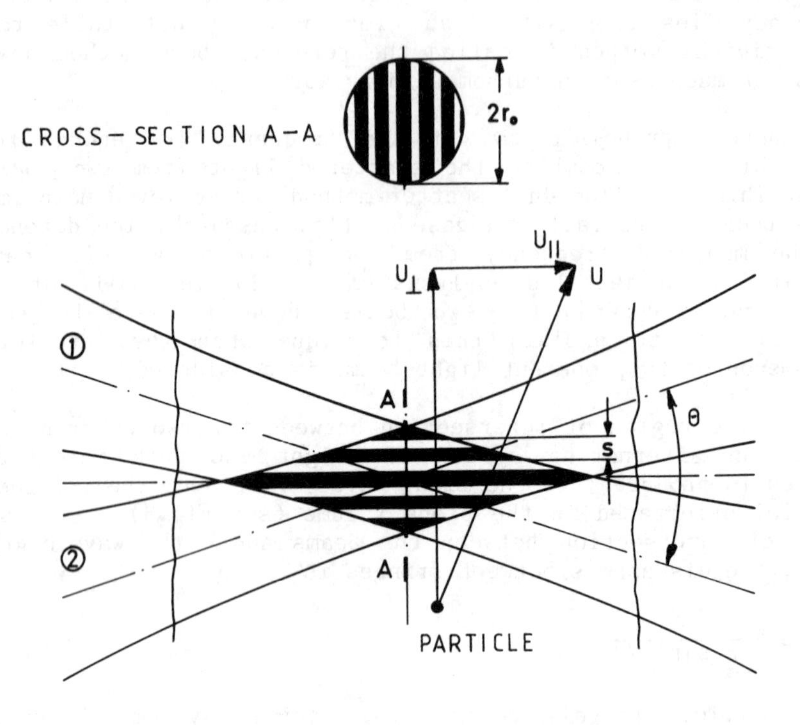

Fig.3 Intersection between two coherent light beams.

by the light intensity distribution (usually Gaussian) in the probe volume.

At present, use is made of four main signal-processing methods or devices - spectrum analysers and frequency-tracker systems, mainly for high proportions of particles in fluids, and rapid counter systems and photon correlators, mainly for low portions of particles in gas flows.

Nowadays counter processing systems are more often applied than any other data acquisition devices. Among various types of counter methods the dual counter system which was described in 1972 for the first time [23] has found the widest spread use so far in research and test institutions. In order to discuss the performance of this type of instruments its mode of operation will briefly be outlined by means of Figure 6: When the Doppler signals as they are show in Fig.5 are bandpass filtered the resulting signal ① will be processed as soon as a certain trigger leves is exceeded. The zero crossing detector ② delivers a pulse for each positive going zero crossing of the signal. Counter ③ then measures the time interval t_1 corresponding to

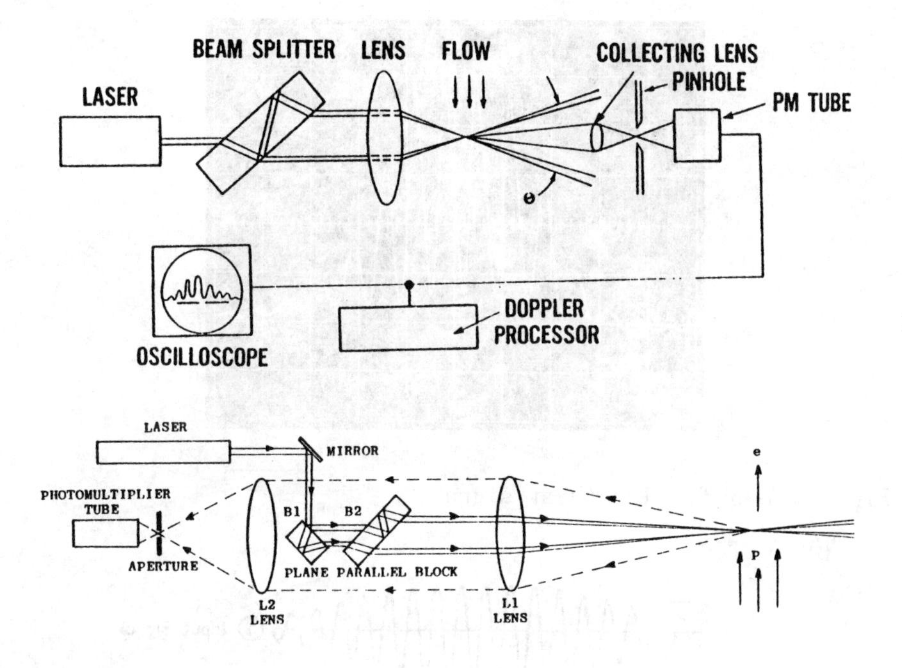

Fig.4 Dual scatter LDV-systems upper: forward scatter, lower: back scatter system.

a number n_1 of oscillations, whereas, for the same signal, counter ④ measures the time interval t_2 which corresponds to a number n_2 of oscillations. A signal is only processed if the following condition is met:

$$\left| \frac{n_1}{t_1} \cdot \frac{t_2}{n_2} - 1 \right| < \varepsilon$$

where ε is much less than unity.

From the time interval measurements the frequency, and on the basis of the aformentioned equation, the velocity can be calculated. Numerous individual measurements taken at a distinct point in the flow field will be gathered and arranged in a velocity histogram from which the mean velocity and the turbulence intensity can be deduced.

The L.D. technique,as described,does not allow for flow sense detection. If, however, an artificial frequency shift f_s is introduced then the fringe system will move. Thus the detected frequency will either increase or decrease depending on the sense of the particle velocity with respect to the moving fringes. This frequency shift is often achieved in practice with an acousto-optic

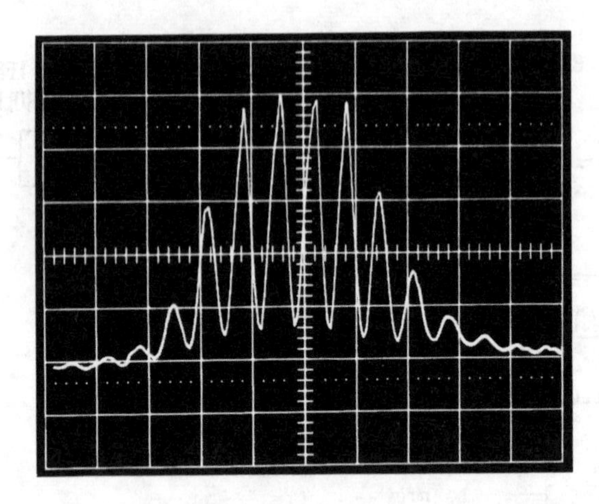

Fig.5 Laser Doppler burst signal.

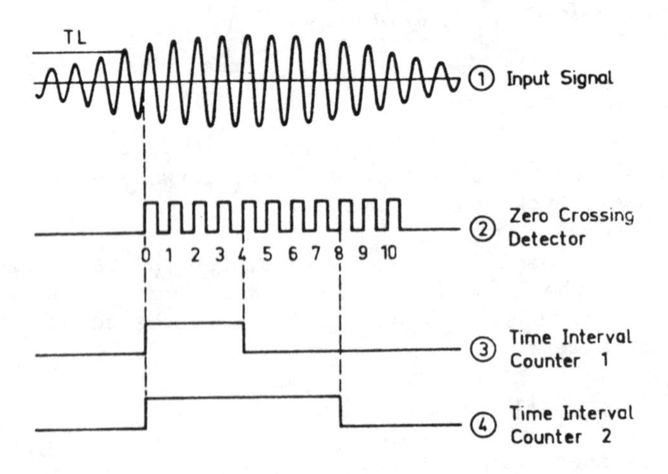

Fig.6 Signal processing of dual counter systems.

modulator (Bragg Cell). A train of acoustic waves, propagated by a series of bonded piezo-electric transducers, modulates the incoming laser light at the acoustic frequency (see Fig.7). The frequency shifting is especially advantageous when the L.D. is applied to high turbulent flows at low speeds.

The one component L.D., as described, has been further developed and two-component LD velocimeters are available today. They are basically constructed as two one-component systems, integrated within one device. In general one laser is used which is splitted either into light of different polarisation direction

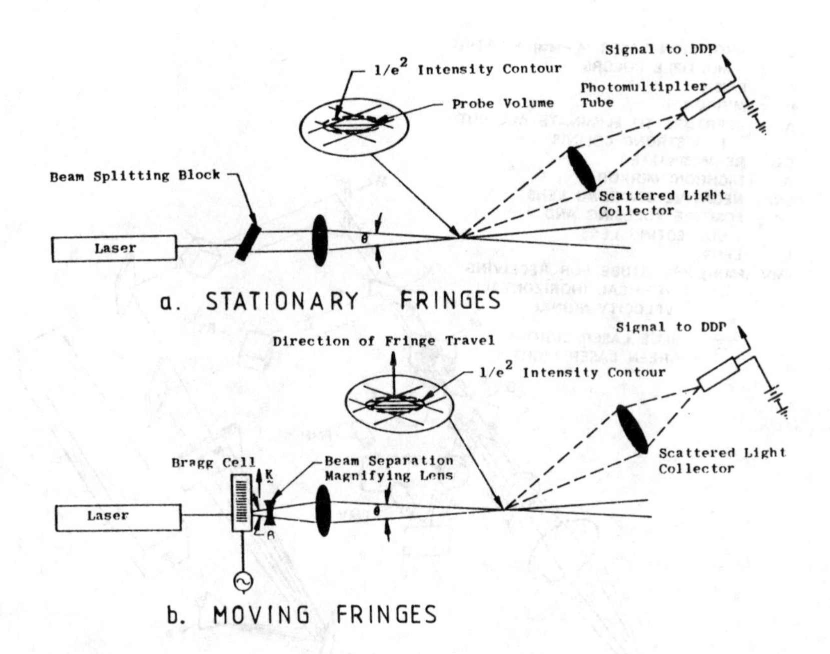

Fig.7 Comparison of moving and stationary fringe LDV systems.

or into different colors. An optical setup of a two color two component LDV is shown in Fig.8. Two sets of electronics are needed for the registration of the horizontal and vertical velocity components. These kind of two component backscatter LDV's are mainly used today in turbomachine applications. Some newly developed three component systems have not found practical importance up to now.

2.2 The Laser Dual Focus Method

In 1972 the DFVLR Institut of Airbreathing Engines started the development of a new laser velocimeter system for use in the study of high mach number turbomachinery flows which differs very much from usual LDV's. The fringe pattern in the probe volume is replaced by two discrete light beams, thus creating a light gate. It was supposed that the much higher intensity concentration in the probe volume could be used to overcome the problems of applicability in turbomachines. The basic idea of such a time of flight anemometer was introduced by Tompson [26], Mesch [27] and Tanner [28]. As Tanner's method was restricted to application in low turbulent flows it was necessary to improve it for use in turbomachinery. This development led to the Laser-Two-Focus (L2F) velocimeter (Ref. [29] and [30]), which can be used in any turbulent flow to measure the magnitude and the direction of the

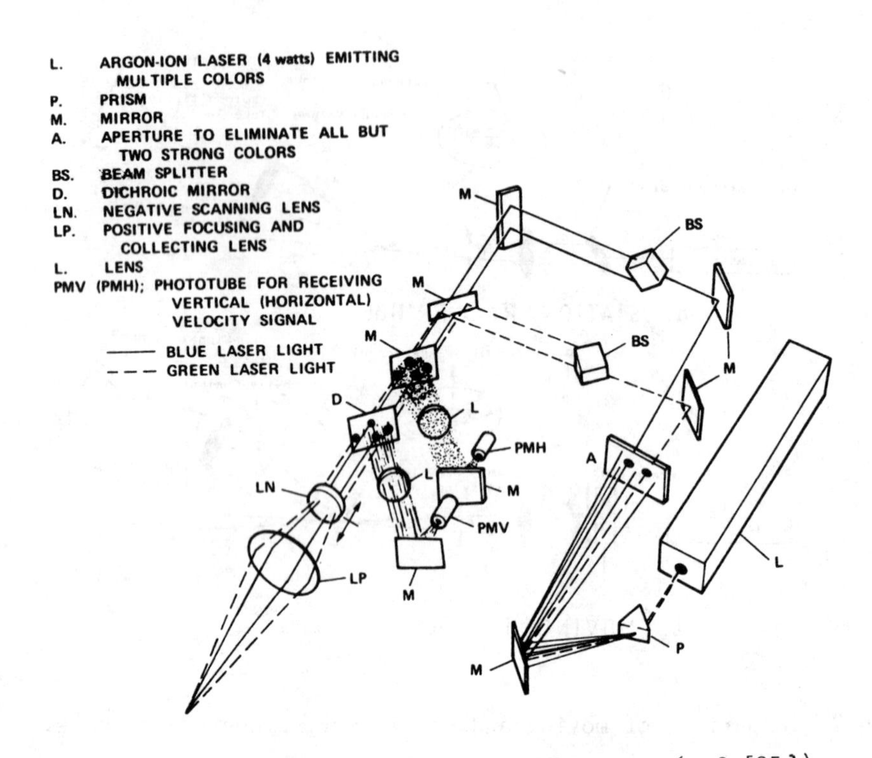

Fig.8 Schematic diagram of two-color LDV system (ref.[25]).

mean velocity as well as the turbulence intensities and the Reynold's shear stresses in the plane perpendicular to the beam axis.

The L2F-measuring device generates in its probe volume (Fig. 9) two parallel highly focussed light beams such that a "light gate" is created. The beam's narrowest diameter is about 10 μm the separation of the two beam axis is about 0.2 to 0.4 mm. A particle passing through both the light beams in the measuring section, along the focal (x_1 , x_2 -) plane, produces then two successive pulses of scattered light. The time elapsed between the pulses yields the component of the particle's velocity perpendicular to the optical (x_3 -) axis. However, this scattered light double pulse - which is necessary for time measurement - will only be generated once the plane of the beams has been established parallel to the flow direction. Consequently, the L2F-method responds to the flow in different directions.

Taking a single measurement of the time interval between two scattered-light pulses, one is never sure that this is a right measure. The reason is, that the two succeeding pulses can be generated in two different ways. First by one particle passing

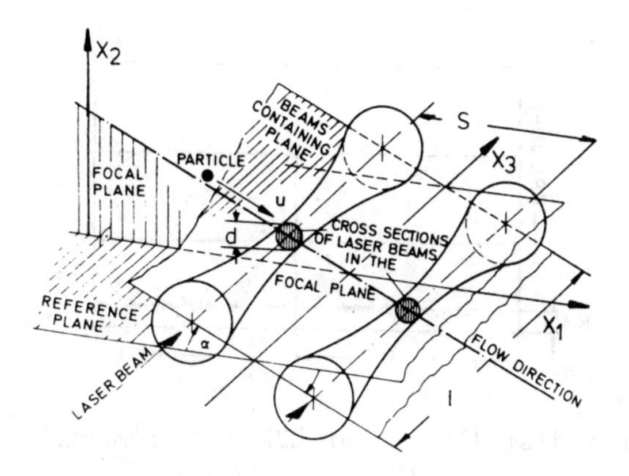

Fig.9 Laser-Two-Focus probe volume (principle sketch)

both the beams and causing a useful measurement and second by two different particles independently passing any beam causing wrong measurements. To distinguish between wrong and right measurements, many measured values taken at a certain probe position are necessary. In Fig.10, as an example, a probability histrogram of such data is plotted, the "noise band" (NB), which is approximately constant along the time axis, is due to non-useable measurements. The net distribution which is shaped like a gaussian curve - it has to be reduced by the averaged value of the noise-band - represents the right measurements.

In the statistical theory of turbulence, the properties of the flow are described by a probability density function. As it is shown in ref. [31] and [32] this function can be measured by a L2F-velocimeter if it is operated as follows: At a certain position in a turbulent flow the plane of the beams of the L2F-probe volume has to be approximately aligned with the mean flow direction. Due to the turbulent fluctuations of the flow vector only such particles which have this direction will be selected for measurements. Measuring the amount of velocity of a certain number of particles entering the probe volume leads to a probability histogram as shown in Fig.10. By setting the plane containing the beams to other slightly different angles and by accumulating the same number of particles entering the probe volume, more velocity histrograms can be measured as long as the chosen setting angles of the plane are within the range of the velocity angle fluctuations. This yields a two dimensional probability histogram (see Fig.11) which can be fitted by a two dimensional normalized probability density function $p(\alpha, t)$. As it is expounded by the theory of turbulence the function $p(\alpha, t)$ contains all information about the velocity components in the plane perpendicular to the optical axis (x_1, x_2-plane, see Fig.9). Any mean value of the fluctuating velocity components can be calculated by applying

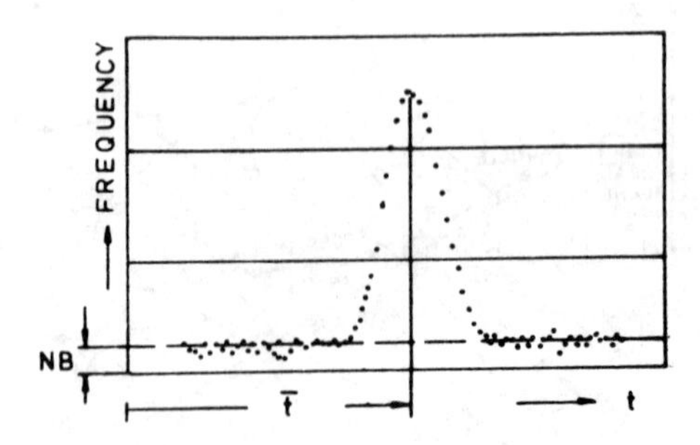

Fig.10 Frequency distribution of L2F-measurements.

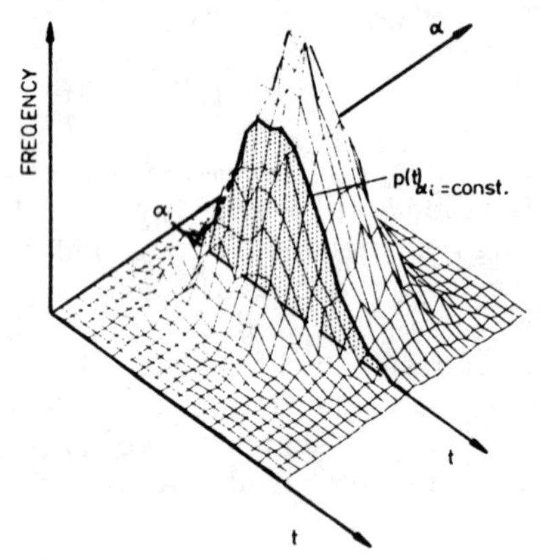

Fig.11 Two dimensional frequency distribution constructed by measured data.

$$\overline{F}_x = \int_0^{2\pi} \int_0^{\infty} F(\alpha,t) . p(\alpha,t) \; dt \; d\alpha$$

where in $\overline{F}_x$ is the mean value of the desired velocity function $F(\alpha,t)$. When inserting an appropriate velocity function $F(\alpha,t)$ into this equation the components of the mean flow vector, the turbulence intensities, the Reynolds' shear stresses as well as high order moments can be calculated (see ref. [31]). All this information is contained within the measured two dimensional probability distribution.

The arrangement of the L2F-set up is illustrated schematically in Fig.12. A polarization prism - Rochon prism - is used to split the initial laser beam (detail A) whose middle is located at the focal point of the immediately following lens. As a result, the beam axes leave the lens parallel to one another whereas the parallel beams are focused on the other focal plane. As shown in

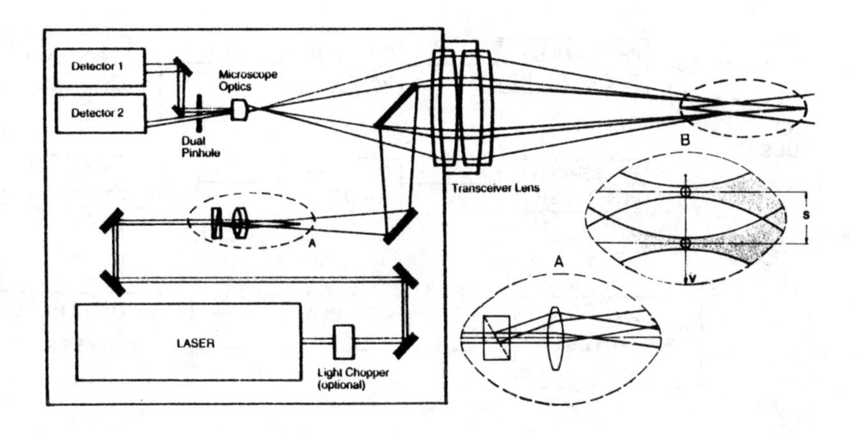

Fig.12 Schematic of the L2F optics.

Fig.12 - detail A - two parallel beams are formed, each of which is highly focused in the focal point so that the desired "light-gate" is created. After being deflected by two flat mirrors the two beams are figured by the internal part of the transceiver lens into the probe volume - detail B. The outer part of the transceiver lens picks up the back-scattered light and figures it into the observation plane of a microscope optics, where - corresponding to the laser beams in the measuring volume - two scattered light beams of very small diameter occur. By means of the microscope optics, these two beams may now be enlarged and projected on a double hole aperture. The following optics align each of the laser beams to a photomultiplier. The purpose of enlargement is to allow an exact adjustment of the aperture holes in order to minimize any background radiation.

To carry out measurements inside the rotating blade channels a light chopper is installed to switch the laser beam periodically.

As already mentioned the plane containing the two laser beams must be turned. This is accomplished with a rotation of the Rochon-prism and a synchronized turning of the dual hole aperture, while the respective position is indicated by use of a potentiometer.

In Fig.13 a block diagram of the data processing system is shown. The start signal for the time measurements is produced by photomultiplier 1 which receives the scattered light from the first laser beam. Photomultiplier 2 which is aligned with the second beam yields the stop signal. After the amplification of both signals discriminators enable the pulse type signals to be triggered at their maximum independent of the signal amplitudes. This is to avoid a broadening of the time measurement. The

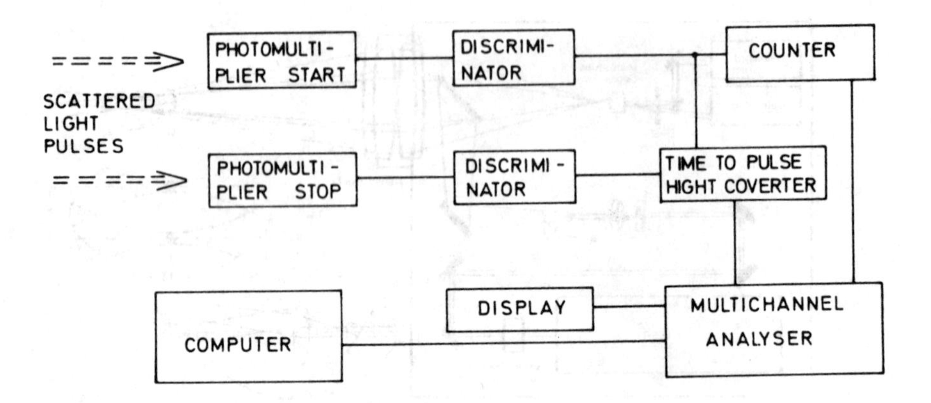

Fig.13 Processing of L2F signals.

discriminator output signals are well defined and have short rise times. They start and stop the time interval measurements which are carried out by a time-to-amplitude converter. A counter counts the started measuring events and after reaching the preset number the measuring cycle is stopped. The rectangular shaped output pulse of the time-to-amplitude converter containing the information on the measured time difference in its amplitude is digitized by the ADC of the following multichannel analyser. The data are stored and arranged along the time axis such that a probability histogram of time measurements is generated on the MCA display.

In Fig.14 an oscillogram is shown containing various probability density distributions corresponding to different angles α with respect to the mean flow direction. The t-axis represents the transit time of the particles between the two beams. The quantity of each time-measurement is arranged along the ordinate. Each distribution represents the same number of starting events at each measuring cycle. The probability that a particle traveling along the line of flow will be irradiated by both laser beams is maximum at the angle position $\alpha = 0^o$ corresponding to the mean flow direction. The peak of the distribution curve indicates the mean velocity, whereas the width of the curve near the baseline indicates the maximum velocity fluctuations. As α increases the probability of particle irradiation by both laser beams decreases rapidly and is - in this case - at $\alpha = 1.5^o$ practically non-existent. Negative values of α yield the same results. A combination of all results leads to a two dimensional probability density distribution depending on time difference (velocity^{-1}) and angle (setting angle α) as it is shown in Fig.11).

After the two dimensional probability distribution is measured and stored the data are transmitted to a computer for the calculation of the desired mean values.

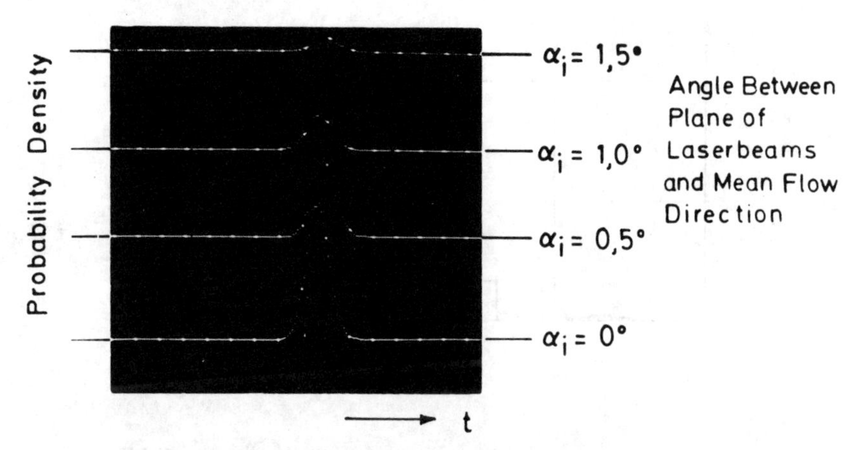

Fig.14 Histograms of time-measurements taken at different angle settings of the beam containing plane.

2.3 Measuring Errors and Seeding

During the development period of the laser velocimeters several systematical errors were pointed out and analysed. They are mainly assigned to the character of the laser beam, to the photodetectors, to the probe volume geometric and to the electronic processors. Most of these errors could be kept small enough that they do not practically affect the measurement accuracy, however, the statistical error must be taken into account.

One part of this error is due to the unavoidable limitation of the number of individual realizations or velocity measurements which can be taken at a fixed location in the flow. After classifying the data one gets a velocity histogram as it is shown in Fig.15. The mean velocity $\overline{u_\perp}$, the standard deviation σ_u and the turbulence intensity T_u result by use of the following equations

$$\overline{u_\perp} = \frac{1}{N} \sum_{j=1}^{n} u_{\perp j} N_j$$

$$\sigma_u = \left[\frac{1}{N} \sum_{j=1}^{n} (u_{\perp j} - \overline{u_\perp})^2 N_j \right]^{\frac{1}{2}}$$

$$T_u = \frac{\sigma_u}{\overline{u_\perp}}$$

$$N = \sum_{j=1}^{n} N_j$$

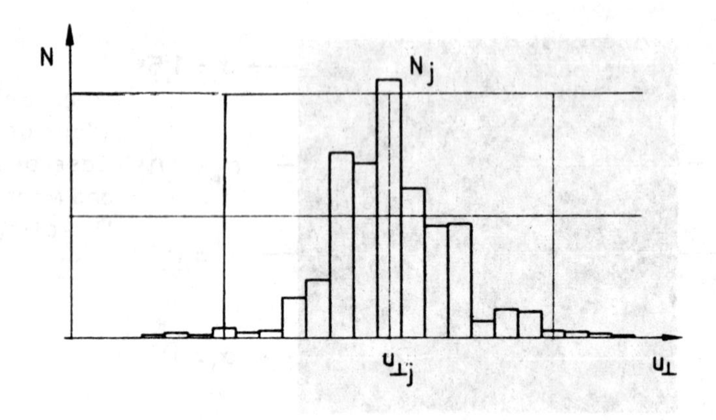

Fig.15 Histogram of the individual realisations of LV measurements.

The possible error in dependance of the number N of individual realizations is given by

$$\frac{\Delta \bar{u}_\perp}{\bar{u}_\perp} \leqq 2.58 \frac{T_u}{N}$$

and

$$\frac{\Delta T_u}{T_u} \leqq 2.58 \frac{1}{2N}$$

assuming a 99% confidence level.

The mean value error depends on the flow turbulence. If, say 500 individual measurements were taken and the turbulence is 10%, the resultant error of the mean value is about 1% and that of the turbulence about 8%.

The other important part of the statistical error is the so-called statistical bias. The samples of the velocity are not taken at equidistant time intervals but randomly according to the particles moving through the probe volume. Under the assumption that the particles are uniformly distributed in space it is evident that at higher velocities more particles per unit time are transported through the probe volume than at lower velocities. So, for an ensemble average the measured mean velocity is biased indicating a higher value than the true mean value. In contrast the measured turbulence intensity is lower than the true turbulence intensity. This phenomenon, which was described first by McLaughlin and Tiederman [33] , is still the subject of a long lasting discussion [34,35,36] . Various approaches have been published so far to find correction algorithms both for one and two component systems. It would go too much into the details to repeat all the highlights of this controversy. Nevertheless, the very first proposal for correction of statistical bias took into account that the number of realizations per unit time is proportional to the velocity vector. Therefore each individual realization must be weighted according to the reciprocal of the magnitude of this velocity vector. That means, according to the aforementioned equations, each N_j must be replaced by $N_j/u_{\perp j}$.

Many publications agree in the statement that this correction is valid for flows with turbulence intensities ranging up to 30%, but disagree for higher turbulences which usually do not occur in turbomachines.

Another possible error source is the following behaviour of particles within the flow. As it is shown theoretically the following of the particles depends on the ratio of particle density to flow density, onthe particle diameter and on the flow situation whether high accelerations occur or not. A good estimation for the selection of the right particles for a known application can be made with the results given in ref.[37]. There in it is concluded on the basis of theoretical calculations that in highly accelerated turbomachinery flows the particles should be smaller than 0.5 µm if a 1% measuring error is taken in account.

At lower accelerations somewhat bigger particles as they are often required for the LD velocimetry can be tolerated. Our experimental tests on transonic compressors have shown, that especially in shock regions particles bigger than 0.5 µm deliver wrong measuring results.

Therefore it is very necessary to control the particle size. A good way to do this is first to filter the test air at an upstream location and then to seed the air artificially by generators (mainly oil droplets were used) which are commercially available today.

2.4 Comparison of LD and L2F Velocimeter

For comparing both Velocimeters the probe volume dimensions should be considered (see Fig.16).

Along with doppler systems the diameter and the length of the probe volume depend on the maximum velocity u_{MAX} to be measured and on the electronics used. The reason is that there is a limiting maximum frequency ν_{MAX} for the different kind of electronics and a minimum number of periods N_{MIN} must be contained within the Doppler burst signal. The number of periods is comparable to the number of fringes in the probe volume.

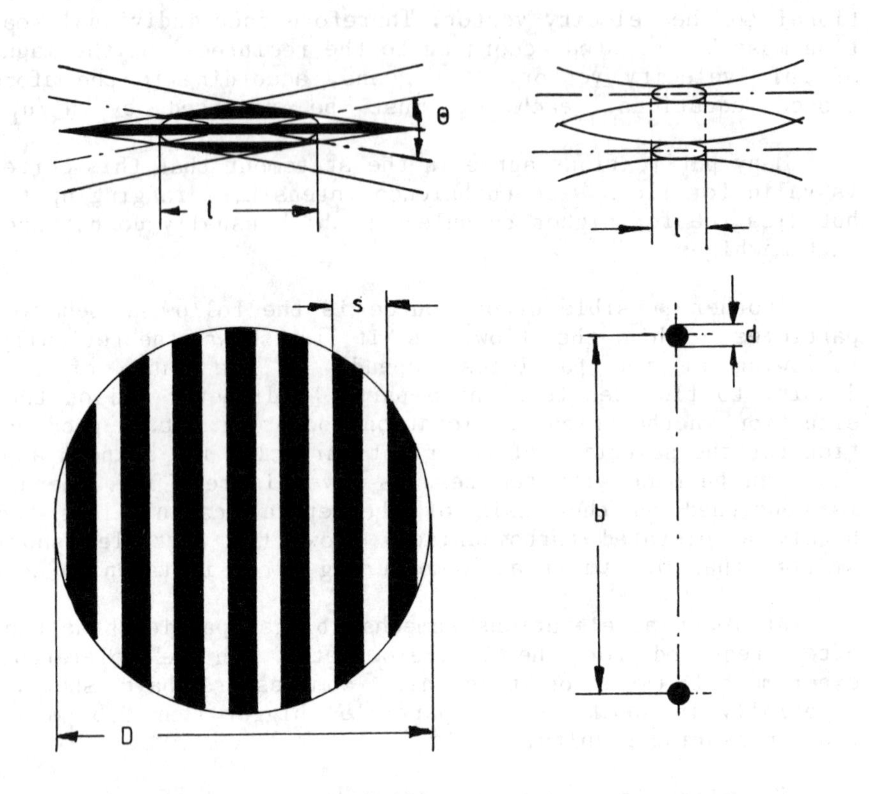

Fig.16 LD and L2F probe volume dimensions.

The minimum diameter d_{MIN} of the probe volume is

$$d_{MIN} = \frac{N_{MIN} \cdot u_{MAX}}{\nu_{MAX}} .$$

The length ℓ can be derived from the Doppler frequency equation and geometrical considerations

$$\ell = \frac{2 \cdot N_{MIN}}{\lambda} \left(\frac{u_{MAX}}{\nu_{MAX}}\right)^2 .$$

If as an example a counter is considered with an upper frequency limit of 50 MHz and a minimum number of fringes required of about 30. The probe volume dimensions then depend on the maximum velocity and on the wave length of light which is supposed to be 0.514 μm

u_{MAX} 100 m/s 500 m/s

d_{MIN}	0,06 mm	0,3 mm
ℓ	0,47 mm	11,7 mm

Considering L2F systems the probe volume geometry is not dependent on electronical limitations. With f the focal length of the first lens after the Rochon prism (see Fig.12) and d_L the diameter of the laser beam the length of the probe volume ℓ at the ℓ/e^2 intensity points and the narrowest diameter d are

$$d = \frac{u}{\pi} \lambda \frac{f}{d_L}$$

$$\ell = \frac{8}{\pi} \lambda \left(\frac{f}{d_L}\right)^2$$

The separation b of the two beams is usually chosen to be the same quantity as the length ℓ. The typical values of the probe volume geometry, as

$d = 10\ \mu m$

$\ell \approx b \approx 0{,}3$ mm,

are generally independent of the velocity range.

The comparison demonstrate that the light intensity concentration in the measuring region which is proportional to the square of the beam diameter is much higher at the L2F system, especially when the LD-system for high speed application is concerned. In that case the intensity ratio is about 1000 to 1. According to the higher light intensity the signal amplitude of the detected scattered light is increased.

Therefore the L2F-system is superior to the LD-system if small particles must be used to follow in high accelerated flows and when measurements close to surfaces or in narrow flow channel must be carried out, where the background flaregenerates a strong signal noise.

Using the same laser power of about 1 Watt and a confocal backscatter arrangement the L2F-systems are able to detect particles with a smallest diameter of 0.1 μm while the LD-system are restricted to particles bigger than 0.5 μm.

Due to the smaller dimensions of the probe volume the mass flow throught the L2F probe volume is lower. However this does not necessarily reduce the data rate in comparison to a LD-system. Because of the higher sensitivity, the L2F-system detects the small particles which are much more frequent in naturally seeded air than the bigger one.

The L2F-systems measures velocity in the plane in which both spots lie. They do not measure a component as LD-systems do but actual particles velocity in that direction.

The beam containing measuring plane must be adjusted with the flow direction and at several angle settings around the mean direction measurements have to be taken. This procedure requires especially in high turbulent flows much more time than is needed for LD measurements.

The decisions over which of the systems is the superior one depends very much on the applicational case. Hereby problems like the importance of the measuring time, the geometrical condition and flow acceleration have to be taken into account.

2.5 Turbomachinery Application Related Problems

Due to the very complicated geometrical conditions in turbomachines one is constrained in applying an optical backscatter arrangement. Thus, only the weak backscattered intensity of small particles can be employed by the photo-detectors leading to low signal amplitudes.

To measure the high velocities found within blade rows very small particles are required to ensure that they follow the real streamlines closely so that they provide an accurate measurement of the fluid velocity. This is especially important in flows which contain large gradients or shock waves. These submicron particles should have diameters below 0.5 μm. Unfortunately these small particles only scatter very weak intensities of light.

The most important restriction on the aplication of laser velocimeters in small turbomachines is the background radiation generated by laser light reflections at the hub and at the casing windows. Its order of magnitude is usually 10^6 times higher than the scattered-light of particles. If it is possible to chose a large angle of observation with respect to the axis of the incident laser beams one has a good chance to suppress the background radiation by spatial filtering. But especially in centrifugal machines such arrangements cannot be employed. It is necessary to have only one casing window for both the incident laser beams and the observed scattered light beam. This window should be flat so that it will not be restrictive in positioning the probe volume. Flat windows installed in the generally highly curved turbomachinery casing generate projecting edges which should be small to avoid flow disturbances. Fig.17 illustrate this conditions. The width of the protruding edge is ε, the radius of casing is ρ_A, D_F defines the window diameter and H the blade height from hub to tip.

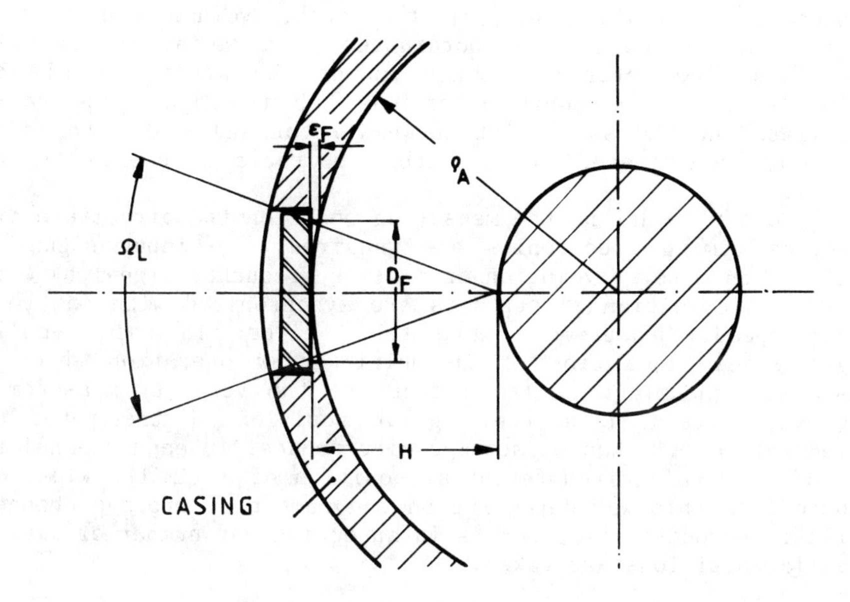

Fig.17 Insertion of plane-parallel windows in a curved housing.

Taking into account different types of turbomachines it was found that for the most critical measuring point close to the hub a cone angle of less than $\Omega_L = 10^o$ - when used simultaneously for the incident and scattered light - is required to avoid edges with ε greater than 0.2 mm. Due to this small spatial angle it is very difficult to suppress the mentioned background radiation.

As a result a confocal backscatter arrangement of an optical anemometer should be chosen as the best possible compromise under these limiting conditions.

Under usual applications of fixed flow channels the measuring position in the flow field is changed by moving the complete device on a x-y-z-coordinate platform. The probe volume is positioned at a certain distance in front of the measuring device. This distance remains constant as long as no refractive material is placed along the beam path. However this distance is altered if planparallel glass window have to be installed in the casings of flow channels or turbomachines. The result is a small change in the probe volume position which can easily be calculated from the glass thickness and its refractive index. But a better way is to adjust the coordinates with the help of a reference point inside the flow channel.

When applying the velocimeter to rotor flow measurements in turbomachines, displacements of the probe volume can only be carried out in one or two coordinates - in radial and in axial directions. The reference points inside the blade channels can be for example the position of the hub to adjust the radial coordinate and the small slot between rotor and stator to adjust the axial coordinate, or any other suitable reference points.

In order to define the measuring position the circumferential direction timing electronics are required. The input signal is usually a once per revolution or a blade frequency signal by which the data acquisition electronics are synchronized with the rotational speed. Measurement were taken either in the so-called single window operation or in multi-window operation which is less time consuming. In the latter one the velocity measurement event is correlated to the angular position of the rotor and stored into multi channel storage electronics. In each channel the data of a certain circumferential position of a limited width are gathered. In this way depending on the number of storage channels available velocity measurements in an equivalent number of circumferential positions are taken.

Before ending this consideration on the positioning conditions in turbomachines a problem must be amphasized which renders any measurements impossible in some regions of the blade channel. These regions - called blade shadow - result if the blade cross-section in the plane normal to the rotor axis is not orientated parallel to the optical axis of the velocimeter. The blade tips cut off the laser beams such, that depending on the respective orientation either close to the suction - or close to the pressure side measurements cannot be made. The way to overcome this restriction is to adjust the optical axis, at each axial position, parallel to the respective blade direction. Because these adjustments are often very difficult the small blade shadow regions are tolerated.

After a certain period of turbomachinery operation the glass windows in the casing became dirty, rendering the velocity measurements difficult or even impossible. Cleaning devices must be developed. The basic idea was to clean the windows during compressor operation just before each measurement by injecting a cleaning liquid upstream of the windows for a short time (2 seconds). Thus, a liquid sheet covers the glass and cleans it of any dirt. The measurements can then take place immediately after the liquid has evaporated.

The construction of a window mounting device including the window cleaning arrangement is shown in Fig.18. The cleaning liquid - usually Triclorathylen - is forced through the clearance between the window and the mounting device. Due to some slots in

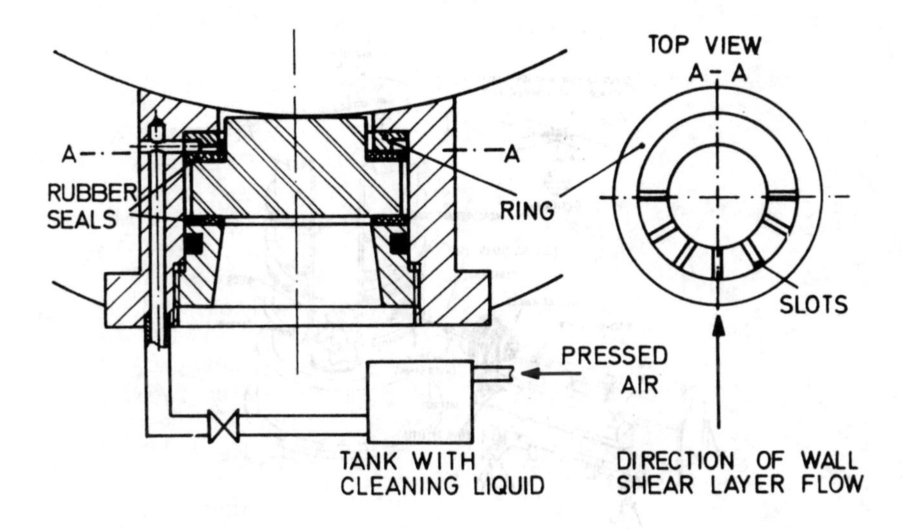

Fig.18 Mounting and cleaning device of a casing window.

a ring shown in the top view the liquid is directed to the upstream part of the window. The cleaning process is accomplished with both chemical and mechanical actions due to the turbulence within the casing shear layer which provides a scrubbing action.

3. REVIEW OF TURBOMACHINERY MEASUREMENTS

The first publication of laser Doppler velocimeter measurements in a turbomachinery application were those reported by Wisler and Mossey 1972 (see ref. [38]). The laser velocimeter was used to investigate the flow field within the rotating blade row of a low speed axial compressor. A sketch of the rig and laser velocimeter used in this work is shown in Fig.19. An example of the measurements made withing the rotating blade row are compared with velocity contours developed from analytic flow solutions in Fig.20. Based on the good agreement between the experimental results and the analytical prediction of the flow field, Wisler and Mossey concluded that the laser velocimeter showed promise in providing an accurate method for obtaining detailed flow field measuremaents withing the rotating blade rows of high speed fans, compressors, and turbines.

This work formed the basis for continued development of the laser velocimeter and its eventual use in a transonic turbomachine. A more recent paper by Wisler (see ref. [39]) decribes the application of the same basic technique to map the local flow velocity and to determine the shock wave locations within a high speed fan rotor.

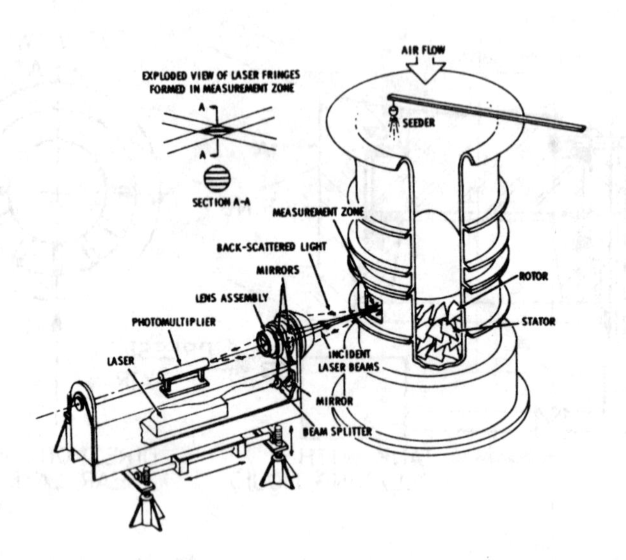

Fig.19 Wisler's low speed research compressor and the laser velocimeter.

Measurements of a similar nature have been carried out by Prat & Wnitney Aircraft. As reported by Walker, Williams and House [40] , a two-component LDV was used to measure flow velocities and determine shock locations in an experimental transdonic compressor blade row. Also in a research compressor Boutier, Guy Fertin and Larguier [41] performed measurements of the periodic flow downstream the rotor blades and Seasholtz [42] describes work going on at the NASA Lewis Research Center to make LDV measurements in axial turbine rotors.

These important papers have presented valuable results of turbomachinery flows however they also mention the difficulties encounteder in making measurements in regions which are located less than about 10 mm from solid surfaces. The relatively large dimensions of the LDV probe volume may introduce additional problems, when used in high speed turbomachinery applications.

Because of the rather large dimensions of the machines being examined these restricationes were not so severe. However in small machines or in the outlet part of centrifugal impellers operating at high speeds the already mentioned limitations make measurements very difficult.

Since the frequency limit of LDV electronics is mainly responsable for this drawback, Runstadler and Dolan [43] designed LDV-electronics which work up to about 150 MHz. In that way the probe volume diameter could be decreased and measurements closer

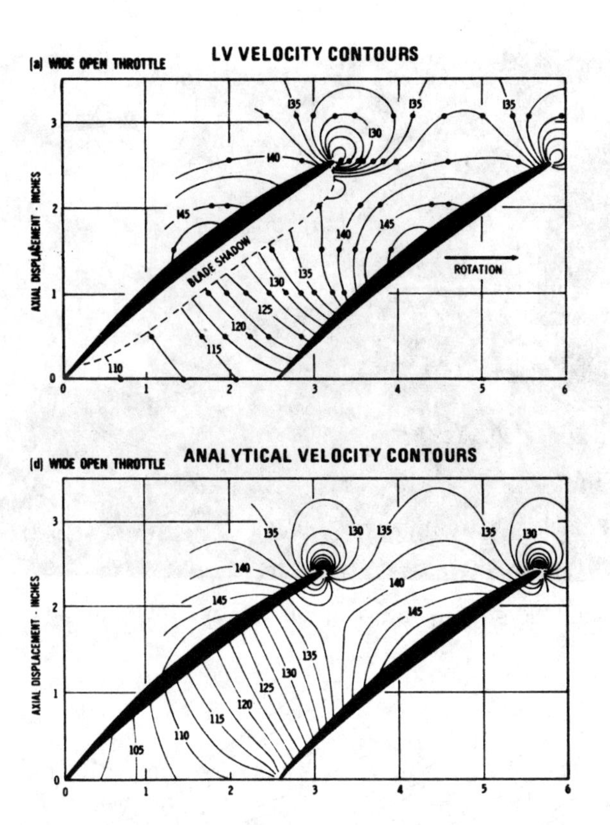

Fig.20 Comparison of experimental and analytical velocity contours (ref. [38]).

to a scattering wall became possible. Results of flow research at the inlet of a small high speed centrifugal impeller and behind the rotor are reported. An other way to overcome the unfavourable background flare is to seed the flow with fluorescent particles as it is reported in ref. [44], [45]. However the particles are not small enough to follow the flow sufficiently. This drawback could be widely eliminated be correction procedures which were used for defining the correct shock position.

The first L2F-measurements were made at the DFVLR in 1973 and the results were reported at the 1974 ASME conference in Zürich. In Fig.21 the L2F prototype is shown acting on low speed axial compressor. In the meantime far more measurements were taken in a high loaded centrifugal compressor (see ref. [46] and [47] and in a transonic axial compressor. An example of results taken in the latter one are shown in Fig.22. The flow field of the compressor rotor is maped by lines of constant Mach number. More details about the experimental research of this compressor can be

Fig.21 First L2F set up acting on a low speed axial compressor.

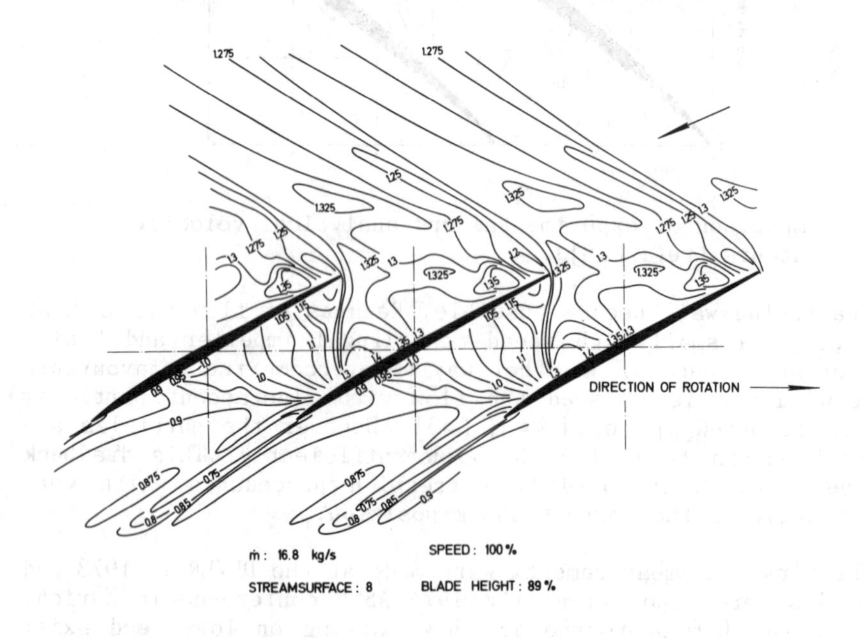

Fig.22 Measured rotor flow field maped by lines of constant Machnumber.

Fig.23 L2F-velocimeter-optical head.

found in ref. [48] and [49]. Recently the L2F system was further developed from the prototype system to a fully automated system of new design as it is shown in Fig.23. By the automation as well as by a further increased optical efficiency the long measuring time of the former device could be reduced by a factor of about 10.

In the Fig.24 and 25 some new measurements which are discussed in detail in ref. [50] are shown taken in a cold air turbine rotor at midspan. The relative velocity field as well as the turbulence field is maped. By looking at the contours of relative turbulence intensities the stator wake regions can be recognized and it can be observed how they were cutted by the rotor blades.

This only short review about the laser velocimeter in turbomachines is really not complete. Much more researchers have applied the LD and the L2F techniques and results of this measurements are published in ref.[11] to [18].

4. CONCLUSIONS

Since a few years laser velocimetry has become a most powerful tool for detailed internal flow studies in high-speed turbomachines. For the first time they make possible an accurate determination of the complex impeller flow pattern at realistic operation conditions. The presented L2F techniques has - due to its inherent superiors signal/noiser ratio certain advantages, especially where measurements in the immediate vicinity of light-reflecting surfaces are a must, e.g. in the shallow flow channels of high-pressure-ratio turbomachines.

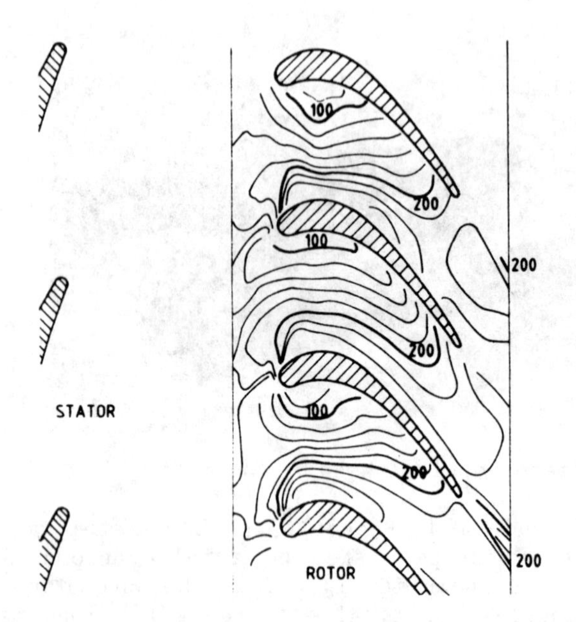

Fig.24 Instantaneous pattern of relative velocity $\tilde{w}$ (midspan, [m/s]).

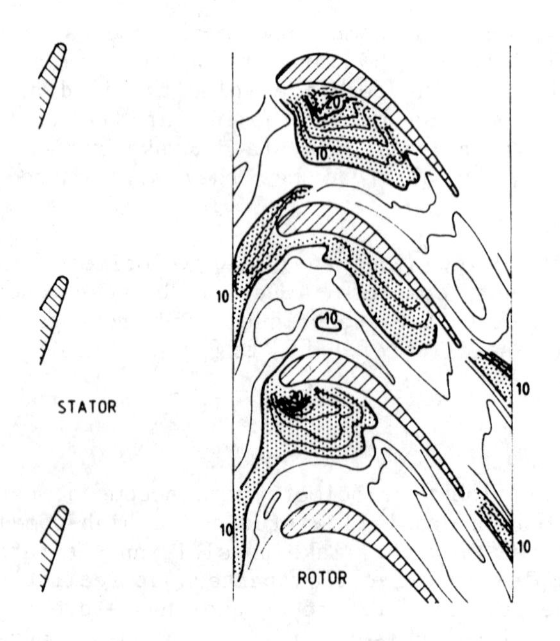

Fig.25 Instantaneous pattern of turbulence intensity Tu_{rel} (midspan, [percent]).

The laser velocimeters undergo further development and it is supposed that in the near future 3D-LD and L2F system can also be applied for a quantitatively accurate investigation of the complex flow pattern in high-performance turbomachinery and offer to the designer nearly all information, needed for a thorough theoretical flow analysis.

5. REFERENCES

1. Paulon, J. Optical Measurements in Turbomachinery, AGARD-AG-207, (1975), s.123-139.
2. Bohn, D., Simon, H. Anwendung des Schlierenverfahrens zur Beobachtung der Durchströmung ebener und rotationssymmetrischer Überschalverzögerungsgitter und der Umströmung von 5-Loch-Kegelsonden.
3. Philbert, M., Fertin, G. Schlieren Systems for Flow Visualisation in Axial and in Radial Flow Compressors, ASME No.74-GT-49, Zürich (1974).
4. Dodge, P. Laser Holography - Optical Design and Set Up, VKI Lecture Series 78 - Advanced Testing Techniques in Turbomachines, April 14-18 (1975).
5. Benser, W.A., Bailey, F.E., Gelder, T.F. Holographic Studies of Shock Waves within Transonic Fan Rotors, ASME No.74-GT-46, Zürich (1974).
6. Caussignac, J.M. Visualisation d'écoulements aérodynamiques dans les compresseurs par interférométrie holographique, ONERA Note Technique No.190 (1972).
7. Fister, W. Versuche zur Erfassung der Strömungsverhaltnisse an Radiallaufradern, AVA-Forschungsbericht Nr.63-01, (1963), s.106-128.
8. Epstein, A.H. Quantitative Density Visualisation in a Transonic Compressor Rotor, Massachusetts Institut of Technology, Internal Report (1975).
9. Schweiger, G. Laser in Aerodynamic Testing: Local Measuring Methods, DLR-Mitt. 74-09 (1974).
10. Schweiger, G., Fiebig, M. Möglichkeiten der Streulichtanalyse zur Dichte-und Temperaturmessung in nichtstrahlenden Gasen, BMBW-FB 72-20 (1972).
11. Moore, C.J. et al. Optical Methods of Flow Diagnostics in Turbomachinery, Proceedings of the Symposion Measuring Techniques in Transonic and Supersonic Flows in Cascades and Turbomachines held in Lyon, Oct. 15-16, 1981.
12. Adamson, T.C. jr., Platzer, M.F. Transonic Flow Problems in Turbomachinery Proceedings of the Projekt SQUID, Febr. 1976, Monterey Hemisphere Publishing Corporation, Washington 1977.
13. VKI LS-78, April 14-18, 1975, Advanced Testing Techniques in Turbomachines.
14. VKI-LS 1981-7, Mai 18-22, 1981, Measurements Techniques in Turbomachines.

15. AGARD-LS-90, Aug.-Sept. 1977, Laser Optical Measurement Methods for Aero Engine Research and Development.
16. Lakshminarayana, B., Runstadler, P. Jr. Measurement Methods in Rotating Components of Turbomachinery, ASME, New York, 1980.
17. AGARDograph No.207, on: Modern Methods of Testing Rotating Components of Turbomachines (Instrumentation), AGARD, London, 1975.
18. AGARD-CPP-281, Testing and Measurement Techniques in heat Transfer and Combustion, AGARD, London, 1980.
19. Van de Hulst, H.C. Light scattering be small particles, New York, John Wiley and Sons, Inc. 1957.
20. Giese, R.H. Tabellen von MIE-Streufunktionen, Part I: Gemische dielektrischer Teilchen, BMBW-FB W 71-23, 1971, Part II. Gemische absorbierender Teilchen, BMBW-FB W 72-19, 1972.
21. Yeh, Y., Cummins, H.Z. Located Fluid Flow Measurements with a He-Ne-Laser Spectrometer, Appl. Phys. Letters Vol.4 (1964), No.10, s.176-178.
22. Durst, F., Zare, M. Bibliography of Laser-Doppler Anemometry Literature, DISA-Information, Nor. 4208, Denmark, Okt. (1974).
23. Pfeifer, H.J. and Vom Stein, H.D. IEEE Transactions on Aerospace and Electronic Systems, Vol. AES-8, No.3, pp.345-349, 1972.
24. Lenner, A.E., Crosswy, F.L. and Kalb, H.T. Application of the Laser Velocimeter for Trailing Vortex Measurements, AEDC Report TR-74-26, Dec. 1974.
25. Grant, G.R. and Orloff, K.L. Two Color Dual-Beam Backscatter Laser Doppler Velocimeter, Applied Optics, Vol.12, No.12, pp.2913-2916, 1973.
26. Thompson, D.H. A Tracer Particle Fluid Velocity Meter Incorporating a Laser, J. Scientific Instruments (j. Phys. E.), Series 2, Vol.1, pp.929-932, 1968.
27. Mesch, F., Daucher, H.H., Fritsche, R. Geschwindigkeitsmessung mit Korrelationsverfahren, MeBtechnik 7, pp.152-157, 1971.
28. Tanner, L.H. A Particle Timing Laser Velocity Meter, Optics and Laser Technology, pp.108-110, June 1973.
29. Schodl, R. On the Development of a New Optical Method for Flow Measurements in Turbomachines, Paper presented at the ASME Gas Turbine Conference in Zürich, Switzerland, 1974.
30. Schodl, R. Advanced Testing Techniques in Turbomachines; A Laser Dual Focus Velocimeter for Turbomachine Applications, Von Karman Institute for Fluid Dynamics Lecture Series, April 14-18, 1975.
31. Schodl, R. Laser-Two-Focus Velocimetry for Use in Aero Engines, Laser Optical Measurement Methods for Aero Engine Research and Development, AGARD-LS-90, 1977, pp. 4.1-4.34.
32. Schodl, R. Development of the Laser Two-Focus Method for Nonintrusive Measurement of Flow Vecotrs, Particularly in Turbomachines, ESA-TT-528, 1979.
33. Mc Laughlin, D.K. and Tiederman, W.G. The Physics of Fluids, Vol.16, No.12, pp.2082-2088, 1973.

34. Buchhave, P. Proc. LDA-Symposium Copenhagen, pp.258-278, 1976.
35. Edwards, R.V. in Laser Velocimetry and Particle Sizing, Ed. Thompson & Stevenson, Hemisphere Publishing Corporation, pp.79-85, 1979.
36. Bogard, D.G. and Tiederman, W.G. in Laser Velocimetry and Particle Sizing., pp. 100-109, 1979.
37. Dring, R.P. Sizing Criteria for Laser Anemometry particles, Journal of Fluids Engineering, Vol.104, March 1982.
38. Wisler, D.C., Mossey, P.W. Gas Velocity Measurements Within a Compressor Rotor Passage Using the Laser Doppler Velocimeter, Presented at the Winter Annual Meeting of the American Society of Mechanical Engineers, New York, N.Y., Nov. 26-30, 1972. ASME Publication No. 72-WA/GT-2.
39. Wisler, D.C. Shock Wave and Flow Velocity Measurements in a High Speed Fan Rotor Using the Laser Velocimeter, Presented at the American Society of Mechanical Engineers Gas Turbine Conference and Products Show, New Orleans, La., March 20-25, 1976, ASME Publication No. 76-GT-49.
40. Walker, D.A., Williamms, M.C., House, R.D. Intra-Blade Velocity Measurements in a Transonic Fan Utilizing a Laser Doppler Velocimeter, Minnesota Symposium on Laser Anemometry, University of Minnesota, October 22-24, 1975.
41. Boutier, A., Fertin, G., Larguier, R. Laser Anemometry Applied to a Research Compressor, Proceedings of the ISL/AGARD Workshop on Laser Anemometry Applied to a Research Compressor, Proceedings of the ISL/AGARD Workshop on Laser Anemometry, May 5-7, (1976).
42. Seasholtz, R.G. Laser Doppler Velocimeter Measurements in a Turbine Stator Cascade Facility, Technical Paper presented at Laser Velocimetry Workshop, West La Fayette, Indiana, March 27-29, 1974, NASA TMS-71524.
43. Dolan, F.X., LeBlanc, F.R., Runstadler, P.W., Jr. Design, Development, and Test of a Laser Velocimeter for High Speed Turbomachinery, Presented at the LDA Symposium Technical University of Denmark, Lyngby, Denmark, August 25-28, 1975, Creare Technical Note TN-234.
44. Powell, J.A., Strazisar, A.J., and Seasholtz, R.G. Efficient Laser Anemometer for Intra-Rotor Flow Mapping in Turbomachinery, ASME Journal of Engineering for Power, Vol.103, No.2, April 1981, pp.424-429.
45. Powell, J.A., Strazisar, A.J. and Seasholtz, R.G. High-Speed Laser Anemometer for Intrarotor Flow Mapping in Turbomachinery, NASA TP-1663, February 1982.
46. Eckardt, D. Instantaneous Measurements in the Jet-Wake Discharge Flow of a Centrifugal Compressor Impeller, Trans. ASME, Journal of Engineering for Power, Vol.97, No.3, July 1975, pp.337-346.
47. Eckardt, D. Detailed Flow Investigations Within a High-Speed Centrifugal Compressor Impeller, ASME, Journal of Fluids Engineering, Vol.98, No.3, Sept. 1976, pp.390-402.

48. Strinning, P.E., Dunker, R.J. Aerodynamic- and Blade-Design of a Transonic Axial Compressor Stage (in German), DFVLR, Institute for Air Breathing Engines, Internal Report No. IB-352-75/7, 1975, p.87.
49. Dunker, R.J., Strinning, P.E., Weyer, H.B. Experimental Study of the Flow Field Within a Transonic Axial Compressor Rotor by Laser Velocimeter and Comparison With Through Flow Calculations, ASME-paper No. 77-GT-28.
50. Binder, A., Förster, W., Kruse, H., Rogge, H. An Experimental Investigation Into the Effects of Wakes on theUnsteady Turbine Rotor Flow. ASME-paper No. 84-GT-178.

INDEX